All Possible Worlds

The optimist proclaims that we live in the best of all possible worlds; and the pessimist fears this is true.

ALL POSSIBLE WORLDS
A History of
Geographical Ideas
(Second Edition)

Preston E. James
and
Geoffrey J. Martin

maps and illustrations by
Eileen W. James

John Wiley & Sons
New York • Chichester • Brisbane • Toronto

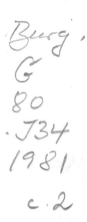
The quotation on page ii is from James Branch Cabell, *The Silver Stallion*
(New York: Robert M. McBride & Co., 1926), p. 129.

Library of Congress Cataloging in Publication Data:

James, Preston Everett, 1899–
   All possible worlds: A history of geographical ideas.

   Bibliography: p.
   Includes index.
   1. Geography—History. I. Martin, Geoffrey J.,
joint author. II. Title.
G80.J34 1981   910'.9   80-25021
ISBN 0-471-06121-2

Printed in the United States of America

10 9 8 7 6 5 4 3 2 1

Preface

THE HISTORY of geography is an account of the successive images that scholars have developed concerning the arrangement of things on the face of the earth. Long before the dawn of written history men who explored even just a short distance from home were aware of differences that distinguished one place from another. Some people felt the challenge to form mental images of what it was like beyond the horizon and then to communicate these images to others. The differentiation of the face of the earth is what John K. Wright called *geodiversity*. This is what geography is all about.

How can a mental image of geodiversity be formed with sufficient clarity so that it can be communicated to someone else? What things are combined in different places on the earth to produce the complex characteristics of the world's landscapes? In the first place, the world is much larger than man; and on its curved surface man's range of vision is narrowly restricted. Most of the fields of study in modern science seek to form mental images of things and events that are much too small to observe directly. But to form any kind of mental image of a world that is too big to see, it is necessary to generalize, to select certain features to build into the image, and to reject other features as not relevant. Furthermore, no coherent image of man's world could be put together unless the observed features are located in relation to a known point. One of the distinctive characteristics of a geographer is that he is always concerned about the relative location of things.

Along with the formation of a mental image of geodiversity, however, men have always felt the need to explain. Scholars have formulated many different kinds of explanations to make the mental images seem plausible or acceptable. And their explanations, in turn, often determined what features they chose to observe. Scholars developed ideas concerning sequences of events, or processes of change, that satisfied the need to understand; and they also sought and found mathematical regularities separate from the processes of change that, nevertheless, satisfied the urge to explain the image of geodiversity.

The mental images and explanations of one generation seldom were satisfactory to succeeding generations. There has been a continued search for new and more perfect images and for explanations that accord with contemporary beliefs. These changing images and the theoretical structures set up to give them plausibility make up the content of the history of geographical ideas.

The quotation on page xvii is from John K. Wright, "The History of Geography: A Point of View," *Annals AAG*, 15 (1925): 200–201.

There are certain built-in biases that should be recognized at the start. We are looking at the whole sweep of the history of ideas from the viewpoint of 1980; and we are looking at all the nations of the world from the United States of America. Inevitably the view is foreshortened. We devote much of our attention to the post-1960 period and more space to a consideration of geographical ideas in America than in any other one country. But we are not ignorant of the existence of scholars in other times, nor do we entirely overlook the geographical ideas and developments in the non-Western world.

Three chief periods are defined. The first period extends for thousands of years from the shadowy beginnings of geographical thought to the year 1859. This is the classical period during which relatively little attention was paid to the definition of separate fields of study. This was the period when the world's knowledge was not so great that a scholar could not become master of it. So it is that almost every Greek philosopher, often listed as a historian in the history books, may with equal justification be called a geographer. Even in the eighteenth century, when the separation of fields of study had started, there were men like Benjamin Franklin, or M. V. Lomonosov, or Montesquieu, who are usually not known as geographers but are nevertheless important links in the history of geographical ideas. The last person who could claim universal scholarship, however, was Alexander von Humboldt; when he died in 1859, no one could aspire to such preeminence in the world of scholarship as he had been able to reach. This classical period is discussed in Chapters 2 through 6.

The modern period began in the latter part of the nineteenth century. It is distinguished by the appearance of the professional field called geography—that is, a field of study in which trained students could earn a living by being geographers. To create a professional field three conditions had to be satisfied. There had to be a body of concepts or images that were accepted by the members of a profession, as well as an accepted way of asking questions and seeking answers. In other words, there had to be a model, or paradigm, of professional behavior. And no such paradigm could be formulated or passed on to later generations of scholars until there were departments of geography in universities offering advanced training in concepts and methods. And graduate departments of geography could not develop until there were paying jobs for students who had earned advanced degrees from these departments. But when all three conditions came into existence, the rise of a professional field could start. This is what we call the new geography. Chapter 7 asks what was new about it.

The new geography began in Germany in 1874, when departments of geography headed by scholars with the rank of professor were established in the German universities. Before that time students could attend lectures and then perhaps go on to give lectures themselves; but never before were there clusters of scholars qualified to offer graduate training in geography. When such positions became available in Germany in 1874 there were no trained geographers to take them; those who were appointed to the newly created professorships had to find their own answers to the question: What is geography? This German innovation spread rapidly to other countries—notably to France, Great Britain, and Russia. It was also transmitted by various routes to the United States. In each of these five countries [Germany, France, Great Britain, Russia (and later the Soviet Union), and the United States] distinctive national schools developed, depending on the answer given to the question concerning the nature of geography. From these five sources, the new professional geography spread all around the world. Chapter 8 reveals something of the development of the new geography in Germany; France is discussed in Chapter 9 and Great Britain is discussed in Chapter 10; the Soviet Union is discussed in Chapter 11. Then the spread of the new geography all around the world from these sources is outlined in Chapter 12. Chapters 13, 14, and 15 reveal something of the development of the new geography in the United States.

The third period in the history of geography is the contemporary one. Because of the importance to us of the last decade or so, this period is reviewed in Chapters 16 and 17. Chapter 18 attempts to unravel the twisted strands of innovation and tradition of the 1970s and the beginnings of the 1980s.

It must be clear from this brief resume that this is a study in breadth, not in depth. Yet at some point someone must put the whole sweep of things into perspective—if only for the purpose of locating desirable places for future detailed studies. We attempt to identify the major traditions as they appeared in the early classical period and as they reappear again and again in somewhat different form throughout the course of history. Ever since Pythagoras there has been a mathematical tradition in geography; ever since Anaximander there has been a cartographic tradition; and ever since Hecataeus there has been a literary tradition. It is our view that these traditions reinforce each other and serve to guard against a point of view that is too narrowly specialized.

For each chapter there are selected references to the subject of the chapter. These references are in no sense complete; nor do they do justice to the important geographical writings in languages other than English, French, and German. These references are heavily weighted toward books and articles published in the English language and easily available to American students. In the Index of Names, each entry is accompanied by dates and by a brief identification (some names listed do not appear in the narrative).

This book first appeared in 1972. Geoffrey Martin had aided significantly in the compilation of the first version. The two authors jointly founded the Archive and Association History Committee of the Association of American Geographers; co-authored *The Association of American Geographers: The First Seventy-Five Years, 1904–1979* and "On AAG History"; collaborated in the matter of biobibliographies for *Geographers: Biobibliographical Studies,* and otherwise repeatedly exchanged points of view regarding the history of geographical thought. Hence this co-authored volume.

Geographers too numerous to mention have helped along the way, but special mention must be made of Robert P. Beckinsale, Allen D. Bushong, Gary S. Dunbar, T. Walter Freeman, David J. Hooson, Ron J. Johnston, James R. McDonald, William D. Pattison, Theodore Shabad, William W. Speth, and Michael J. Wise. Especial thanks go to the American Geographical Society for permission to reproduce six of the photographs in their holdings.

Eileen James designed and drafted the maps and Norma Martin typed the manuscript and associated correspondence. To all, our gratitude.

Preston E. James
Geoffrey J. Martin

About the Authors

PRESTON E. JAMES, Maxwell Emeritus Professor of Geography at Syracuse University, was born in Brookline, Massachusetts. He received the B.A. and M.A. degrees from Harvard University and a Ph.D. from Clark University. His career as a teacher and research worker began in 1919 when he was still an undergraduate at Harvard. He has been a visiting professor at the Universities of Brazil, Edinburgh, and Puerto Rico. He has been councilor, secretary, president, and honorary president of the Association of American Geographers.

Professor James's travels have taken him all over the world—from field work in South America to Moscow, where he was a member of the U.S. Exchange Mission to the U.S.S.R. He has been a member of the National Research Council, Division of Geology and Geography; the Research and Development Board, Committee on Geophysics and Geology; and a U.S. National Member, Commission on Geography, Organization of American States, Pan-American Institute of Geography and History.

The National Council for Geographic Education presented P. E. James with its Distinguished Writing Award and Distinguished Service Award. He received the Pan-American Medal from the Pan-American Institute of Geography and History and the David Livingstone Centenary Medal from the American Geographical Society. He holds the Hon. Sc.D. from Eastern Michigan University and Hon. LL.D. from Clark University.

He is the author of several books, including *An Outline of Geography* (1935), *Latin America* (1942), *A Geography of Man* (1949), *One World Divided* (1964), co-editor of *American Geography: Inventory and Prospect,* and co-author (with G. J. Martin) of *The Association of American Geographers: The First Seventy-Five Years, 1904–1979.* He has additionally written geographies for elementary and secondary schools as well as more than 125 articles that have been published in professional geographical journals.

GEOFFREY J. MARTIN is Professor of Geography at Southern Connecticut State College. British by birth and upbringing, he read geography at the London School of Economics and Kings College (London). He has published *Mark Jefferson: Geographer* (1968), *Ellsworth Huntington: His Life and Thought* (1973), and *The Life and Thought of Isaiah Bowman* (1980). He designed this triad as an excursus into the ontography that characterized early twentieth-century American geographical thought.

With Preston James he has co-authored (and edited) *The Association of American Geographers: The First Seventy-Five Years, 1904–1979*. With P. E. James he helped to found the Archive and Association History Committee (Association of American Geographers), of which he has been chairman since 1977. He is a member of the History of Geographic Thought Working Party of the International Geographical Union, American representative for the U.S. biobibliographical contribution to *Geographers: Biobibliographical Studies* and councilor of the Association of American Geographers. He has amassed a very large privately maintained archival collection (history of geographic thought) and has otherwise contributed some 50 articles, which have been published in the geographical literature or delivered at professional meetings. Currently he is writing a history of geography in North America.

P. E. James and G. J. Martin

Contents

Maps and Figures

We may compare the mind of man to a mirror which has the ability not only to reflect but also to retain, record, and interpret more or less imperfectly the images that it reflects. It is not a clean, bright mirror which gives exact images, but too often is warped, clouded, spotted, cracked, and broken. The appearance of the image, no matter of what the reflection may be, is determined very largely by the nature of the mirror itself and by the spots, dust, and other foreign matter that may have accumulated upon it.

CHAPTER 1

A Field of Study Called Geography

. . . to understand the earth as the world of man . . .

In the beginning there was curiosity. We may assume that some of the earliest questions formulated by primitive man had to do with the character of the natural surroundings. Man, as well as many other animals, identifies a particular extent of territory on the earth's surface as his living space; and, like many other animals, he is irked by the possibility that the grass may be greener in someone else's living space. Curiosity compels him to find out what it is like beyond the range of hills that forms his horizon. But the world he discovers is a close reflection of his own mind, and in the long course of history men have discovered and described many different worlds. There are challenging limitations on man's capacity to observe and to generalize what he has observed. As man's capacity to observe and generalize is improved, the result is a new image of his world—yet all possible worlds are far from having been described.

Man's world includes what can be perceived on or from the surface of the earth. The earth is a medium-sized planet circulating around a medium-sized nuclear explosion that we call the sun. If the sun were the size of an orange, the earth, at the same scale, would be the size of a pinhead about one foot away. Yet this

The quotation above is from J. O. M. Broek, *Geography, Its Scope and Spirit* (Columbus: Charles E. Merrill, 1965), p. 79.

1

pinhead is just large enough to provide the gravitational pull to hold the thin skim of gasses called the atmosphere close to its surface. And the earth is just the right distance from the sun so that at the bottom of the atmosphere temperatures occur at which water can remain a liquid. Organic life originated in the seas, and when organisms came out of the water to live on the land they brought their water environment with them inside "space suits" made of skin. Similarly, when the astronauts traveled to the moon they took their gaseous environment with them inside artificial space suits. All the forms of life that we know about are dependent on water in the form of liquid and on energy received from the sun.

The face of the earth is a zone extending as far down below the surface as man has been able to penetrate and as far up above the surface as man normally goes. All science and all art are derived from observations made in this zone, which, until 1969, was man's whole universe. But it is a complex universe: there are things (phenomena) on it produced by physical and chemical processes; there are plants and animals produced by biotic processes; and there is man himself, influenced by his natural surroundings and also an agent of change in his surroundings acting through economic, social, and political events. All these things and the events of which they are the momentary signs exist in complex association and interconnection, forming what is called the man-environment system.

The Greek scholar Eratosthenes, who lived in the third century B.C., was the first to use the word *geography* (from *ge,* meaning the earth, and *graphe,* meaning description). But men were investigating geographic questions long before this. The Greek geographers credit the beginning of geographical writings to Homer, but the earliest known map was made by the Sumerians about 2700 B.C. The history of geographical ideas is the record of man's effort to gain more and more logical and useful knowledge of the human habitat and of man's spread over the earth: logical in that explanations of the things observed could be so tested and verified that scholars could have confidence in them; useful in that the knowledge so gained could be used to facilitate man's adjustment to the varied natural conditions of the earth, to make possible modifications of adverse conditions, or even to gain a measure of control over them.

WHERE IS IT?

Even a clear description of what man's world is like would not be of great value unless it were possible to answer the question: "Where is it?" Starting from known living space, in what direction does one travel and how far? How does one measure distance and direction? When you reach the place on the other side of the horizon, where are you?

This is no simple question to answer. To illustrate the difficulties involved in establishing location on the surface of a sphere, take a ping-pong ball with a black dot on it and then attempt to describe the location of the dot. Location must always be relative to something else, and we immediately think of the coordinates of latitude and longitude. It was the ancient Greeks who first formulated the theory of locating places with reference to a grid of imaginary lines based on the poles and the equator. But the Greeks lacked the instruments necessary to make exact measurements of latitude and longitude. Location still had to be described relative to some observable feature, such as a range of hills, a coastline, a river, or a town. In the course of time, as knowledge of the features of the earth's surface was increased, a greater variety of relative locations could be defined. Furthermore, in any specific question that is raised concerning the arrangement of things on the face of the earth, location relative to latitude and longitude may not be relevant. For example, to say that New York City is located at 40°42′N and 74°W (at City Hall) is meaningless if one is asking questions about the significance of location in the growth of the city. Asking questions about locations is one of the distinguishing characteristics of the field of study we call geography.

WHAT IS IT LIKE?

The little phrase—"What is it like?"—stands for a fundamental thought process. How does one go about observing and reporting on things and events that occupy segments of earth space? Of all the infinite variety of phenomena on the face of the earth, how does one decide what phenomena to observe? There is no such thing as a complete description of the earth or any part of it, for every microscopic point on the earth's surface differs from every other such point. Experience shows that the things observed are already familiar because they are like phenomena that occur at home or because they resemble the abstract images and models developed in the human mind.

How are abstract images formed? Man alone among the animals possesses language; his words symbolize not only specific things but also mental images of classes of things. Man can remember what he has seen or experienced because he attaches a word symbol to them. Man can look at a hill and, if he is familiar with hilly country, he can recognize the peculiar and unique character of each specific example, but he can also develop the mental image of "hill" in general.

During the long record of man's efforts to gain more and more knowledge about the face of the earth as the human habitat, there has been a continuing interplay between things and events observed through the senses and the mental images of things and events. The direct observation through the senses is described as a *percept;* the mental image is described as a *concept.* Percepts are what some people describe as reality, in contrast to mental images, which are theoretical, implying that they are not real.

The relation of percept to concept is not so simple as this. It is now quite clear that people of different cultures or even individuals in the same culture develop different mental images of reality; and what they perceive is a reflection of these preconceptions. There are some African cultures, for example, with color perception limited to red and blue because their language contains words only for these ends of the spectrum. As a result these people do not perceive intermediate colors, such as orange, yellow, or green (Krauss, 1968).[1] An artist trained in the perception of color and with a long list of word symbols can identify a great many colors. The direct observation of things and events on the face of the earth is so clearly a function of the mental images in the mind of the observer that the whole idea of reality must be reconsidered. The idea that one could approach a problem with a blank mind—that is, without preconceptions—is nonsense; for such an assertion means only that the person is not conscious of his preconceptions and is, therefore, a slave to them. Nor is it possible to dream a set of images not based on previous perception. Such a situation could only apply to very young infants.

We are faced then with an apparent paradox. Concepts determine what the observer perceives; yet concepts are derived from the generalization of previous percepts. What happens is that the educated observer is taught to accept a set of concepts and then sharpens or changes these concepts during a professional career. In any one field of scholarship, professional opinion at any one time determines what concepts and procedures are acceptable, and these form a kind of model of scholarly behavior. Such a body of doctrine determines what problems are considered worth investigating and what kinds of answers are professionally acceptable. But at all times the accepted concepts and the procedures based on them are subject to challenge. Progress is achieved when one working hypothesis is replaced by another.

WHAT DOES IT MEAN?

One concept that runs through the hisory of ideas since the earliest records is that of an ordered, coherent, and harmonious universe. Man has instinctively rejected the notion that he and his natural surroundings are the result of random events. Clarence J. Glacken writes as follows:

> What is most striking in concepts of nature, even mythological ones, is the yearning for purpose and order; perhaps these notions of order are, basically, analogies derived from the orderliness and purposiveness in many outward manifestations of human activity: order and purpose in the roads, in the grid of village streets and even winding lanes, in a garden or a pasture, in the plan of a dwelling and its relation to another [Glacken, 167:3].

There are many examples of man's effort to identify meaning as a reflection of the images in the human mind. With the concept of an orderly world firmly established, the perception of evidence to give it support was a natural consequence. The

[1]References are listed by chapter starting on p. 427.

Greeks distinguished betwees chaos (*kenos,* meaning void) and the cosmos, which refers to the universe conceived as a system of harmoniously related parts. Having accepted, almost without challenge, the concept of an ordered universe, the next step was to find some plausible way to account for it.

During the thousands of years since the earliest records of the history of ideas, learned people have accounted for the order they perceived in the universe in different ways. The accounts range along a continuum from arbitrary rule by man-like deities, through rule by a deity subject to law, through various kinds of cause and effect relations, to abstract mathematical law. These do not represent successive stages of increasing sophistication, for all of them can be found in the thinking of the ancient Greek philosophers as well as in the contemporary world.

Rule by a deity or deities is a very ancient concept. In Sumeria the religious leaders saw a world ruled by living beings like men but endowed with superhuman powers and with immortality (Glacken, 1967:4–5). Each of these beings was responsible for the control and maintenance of some feature of man's world, such as the flow of rivers, the rise and fall of the tides, the shift of the winds, the productivity of the harvest, the abundance of game animals. The deities competed with one another and reacted arbitrarily and often vindictively to the acts of man. Other cultures explained things in terms of a single deity whose acts were arbitrary and who needed to be frequently bribed if man was to be favored with his gifts.

A very different way of accounting for an ordered universe is the recognition of cause and effect sequences that take place in accordance with general law. In some cases the notion of a single deity is retained, but the acts of this deity are not arbitrary. Some would say that this god is the law. The idea of law, itself, is an anthropomorphism—that is, a reflection of human experience. Men who break divine laws are subject to punishment, but those who act in harmony with the law are rewarded. Of course there is a great difference between man-made law and scientific law: man-made law governs the behavior of things, and events are subject to law; but scientific law is a general description of events. Nevertheless, the word symbol, "law," carries unintended connotations that can impede clear thinking.

There are two different ways of interpreting cause and effect sequences, each represented by one of the two great philosophers of ancient Greece: Plato and Aristotle. Plato conceived of the world as having been created in perfection but now in the process of decline from perfection. Aristotle, on the other hand, conceived of the world as striving toward perfection. Aristotle was the father of the teleological concept, which sees the universe planned by its creator in the manner of a carpenter drawing up plans in advance for the house he is building. This doctrine of final cause differs from the idea that cause must be antecedent to effect. David Hume in the eighteenth century argued that cause and effect sequences could never be perceived or tested except in terms of human experience; yet the use of the concept continues (Brown, 1963; Ducasse, 1969).

At the other end of the spectrum is the use of mathematical regularities to explain the order in the universe. Order was conceived as mathematical even by Pythagoras, some six centuries B.C. Mathematical order can be illustrated by the regularity of the octave in sound, the principle that the circumference of any circle is the same as the length of the diameter times 3.1416, or that the force that attracts one body to another is proportional to the total masses of the two bodies divided by the square of the distance between them. Mathematical formulas describe the motions of planets, the flow of rivers, and the predictable flow of telephone calls between two cities. So all-pervasive are the principles of mathematics that one astronomer was led to remark that the universe was like a dream in the mind of a mathematician. It was Plato who insisted that to explain anything it was only necessary to identify the laws it obeys.

THE STUDY OF GEOGRAPHY

The study of geography for the purpose of gaining more and more logical and useful knowledge regarding the human habitat and man's interrelations with it has a record that goes back to the beginnings of man's scholarship. There are certain repetitive sequences in the interplay between concepts and percepts, hypotheses and observations. There are occasional periods marked by flashes of brilliant intuition when new and challenging concepts are set forth. New concepts presented as hypotheses result in a burst of new empirical observation, for the new concepts always broaden the range of man's perceptions. The new observations may demonstrate the inadequacy of the hypothesis, which is then withdrawn in favor of a new one or is substantially modified. These are periods of great progress. Then, when a conceptual structure becomes widely accepted and a paradigm of scholarly behavior is established, there is a period when observations increase so fast that they must be stored away for future use. These periods of intellectual stability are not periods of notable progress. Eventually, a new and sometimes radically different concept of the meaning of the observed data is set forth, and the sequence is repeated.

The first amazing period of intellectual ferment that is part of the written tradition of the Western world took place in ancient Greece, culminating in the fourth and third centuries B.C. The scholars of Babylonia had already collected a large number of data on the motions of the stars and planets, and they developed the concept that the position of the celestial bodies had a fundamental effect on human activities. In astrology if observed facts disagree with general principles, the facts are explained as exceptional and the principle remains unchanged. It was the Greeks who developed the procedures we describe as the scientific method. If observed facts differ with general principles, the principles must be revised. The Greeks developed the science of astronomy. This was a tremendous step forward in the whole history of ideas.

Many of the basic procedures were first set forth in the writings of Plato and Aristotle. Quotations from their works are as relevant today as they were thousands of years ago. Plato, who developed the deductive procedures, is the one most often quoted by those who prefer to give theory the position of chief importance. Aristotle, who developed the inductive procedures, preferred to formulate his concepts as generalizations of empirically observed "facts." Aristotle is the one who insisted on the importance of direct observation. Instead of making logical deductions from theory, he taught his students (including Alexander the Great) to "go and see."

Among the ancient Greek philosophers, almost all of whom made contributions to the study of geography, the two basic traditions of geographic study are to be found. There is the mathematical tradition, starting with Thales, including Hipparchus (who formulated the theory of locating things by latitude and longitude), and summarized by Ptolemy. And there is the literary tradition, starting with Homer, including Hecataeus (the first writer of prose), and summarized by Strabo.

There was a long period of decline during the Middle Ages, when geographic horizons closed in. The wide-ranging Greek and Phoenician explorers and the Greek geographical concepts were largely forgotten, except among Arabic scholars. Observations piled up in Christian monasteries, but the intellectual climate was not favorable for the formulation of new interpretations. Then the Age of Exploration began in the late fifteenth century, and the geographic horizons were again pushed back. The arrival of all these new observations in Europe was enormously stimulating and started the sequence of events that continues to the present.

Two fundamental innovations began to shake the world of scholarship in the sixteenth century, and this ferment reached its full proportions in the second half of the nineteenth century. In the first place the concepts derived from a literal reading of the Scriptures were challenged; and the battle to establish the principles of what we call academic freedom began. This is the right of professionally qualified scholars to seek answers to questions, to publish their findings, and to teach what they believe to be the truth, free from any controls except the standards of scholarly procedure established within their own professions. The battle started with Leonardo da Vinci and Copernicus in the sixteenth century and reached full fury as a result of the tradition-shattering concepts developed by Charles Darwin in the latter part of the nineteenth century. In 1809 the world's first university was founded at which neither faculty nor students were required by law to accept any particular religious creed or political doctrine—this was the University of Berlin. The idea of the university as a community of free scholars continued to gain acceptance throughout the world until the whole concept was challenged in the contemporary period.

The second fundamental innovation began in the seventeenth century and reached widespread development in the second half of the nineteenth century. This was the separation of the academic world into distinct fields, or disciplines, each devoted to the study of a specific group of related processes and each regulated by its own paradigm based on its own theoretical structure. From the time of ancient Greece, scholars were not concerned about restricting themselves too narrowly.

Herodotus made major contributions to both history and geography, and he is often called the father of ethnography. But the last person who could claim to be a universal scholar and who could speak with respected authority on the whole wide range of the world's scientific knowledge was Alexander von Humboldt. Humboldt and Carl Ritter, both of whom died in 1859, represent the culmination and the conclusion of ancient scholarship.

The rise of experimental science was a major cause of this separation of the disciplines. Such fields as physics and chemistry developed around the study of particular processes. Processes were isolated in laboratories, and as a result the knowledge of these processes was greatly expanded. This fragmentation of knowledge had a major impact on geography, which became known as "the mother of sciences." The movement started with the physical sciences, then with the biological sciences, and most recently with the various social sciences. Where processes could not actually be studied in laboratories, as in economics, they were isolated statistically; and the laboratory was represented symbolically by the phrase, "other things being equal."

Was there any place left for geography after Humboldt and Ritter? By 1870 it had become clear that there were aspects of knowledge not covered by the new sciences. Especially in the social sciences, processes never actually took place in isolation: they were modified in particular places by all the other aspects of the total environment that were not equal. Geography moved into the field of particular places, where a variety of things and events were examined in their natural but unsystematic groupings. Geography also undertook to bridge the widening gap between the physical and biological sciences on the one hand and the social sciences on the other. But an academic discipline cannot exist until there is a body of trained professional scholars formed around university faculties, where new generations of scholars can receive advanced instruction. Departments of geography headed by professors came into existence for the first time in Germany in 1874; but the professors selected to head these departments were people with training in geology, history, biology, or other fields. During the period from the 1870s until World War II, geographers all around the world were seeking to establish the status of geography as an independent discipline, distinct in its concepts and procedures from other disciplines.

World War II had a greater impact on the world of scholarship than is still generally recognized. Scholars were called upon to study the issues involved in the very complex problems of policy; and when they went to work on these pressing questions, they discovered that workers in any one discipline were usually ill-prepared to wrestle with essentially interdisciplinary matters. The workers in the separate fields were too narrowly trained and specialized—in some cases so specialized that they could not even appreciate the importance of work done in other disciplines. Geographers made important contributions in the form of maps and in analyses of the significance of location that other people were accustomed to over-look.

One outgrowth of the experience of the war was the development of General System Theory (Bertalanffy, 1968) in which the existence of complex associations of interconnected and interdependent elements was recognized; and methods were developed to solve problems of multiple causation, where predictions had to depend on probability theory. Fortunately, just at this time the electronic computer was developed. Without the computer the complex computations involving a great variety of factors could never have been carried out with speed and precision. Furthermore, new electronic devices for scanning the face of the earth from orbiting satellites revolutionized the methods of data gathering. These innovations, which came largely after midcentury, have introduced a third major period in the history of ideas.

What then does geography do? It is important to understand that since World War II this question does not call for a definition of geography that would establish its boundaries. The trend now is for all fields of study to come together around specific problems. The process of separation has now been replaced by a process of integration in which each professional field brings its own special skills and concepts to bear on such major difficulties as poverty, overpopulation, race relations, and environmental destruction. The special skills of geography are those related to the significance of location and the spatial relations of things and events. Geography has always had a holistic tradition so that it comes as no intellectual shock to study systems of interconnected and interdependent parts of diverse origin. Geography is closely involved with cartography in the development and use of maps, which are ideally suited to the study of complex location factors. A geographer is a person who asks questions about the significance of location, distance, direction, spread, and spatial succession. The geographer deals with problems of accessibility, innovation diffusion, density, and other derivatives of relative location.

MANY NEW WORLDS

There are many new worlds yet to be discovered, and the number of people engaged in discovering them has increased at an unprecedented rate since the end of World War II. As with many other fields of learning, it can be said of professional geography today that of all the scholars who ever lived and contributed to geographic knowledge, perhaps ninety percent are alive today. The volume of published material is at least a thousand times greater than the materials with which Humboldt and Ritter had to work. The electronic data bank has arrived just in time. But, with all these new scholars entering the field and with the variety of pressing problems to which geographic concepts and methods can be applied, there is no reason to narrow the field or to establish just one paradigm of scholarly behavior.

As a general assessment of the field of geography in the 1980s, we may say that there is still freedom to be curious, to inquire, to offer novel conceptual structures that reveal the outlines of yet another of all possible worlds. The mathematical and literary traditions are still being vigorously developed. The use of mathematical theory has been so clearly useful that graduate training in geography now usually involves preparation in mathematics through calculus and linear and matrix algebra. The literary tradition, threatened by the neglect of language training in the schools, continues to attract those who can command the skills of rhetoric and the manipulation of language. Cartographic skills are so interconnected with the analysis of location that, even when maps are made by computers, concepts regarding the use of maps as analytic devices have lost none of their importance.

There are at least five different kinds of questions of geographic character that can be investigated: (1) There are *generic* questions that have to do with the content of earth space but that cannot be effectively answered without a framework of concepts to guide the separation of the relevant from the vast complexity of the irrelevant. (2) There are *genetic* questions that have to do with the sequences of events leading from past situations through geographic changes to present conditions; these are studied by the methods of historical geography. (3) There are *theoretical* questions that deal with the formulation of empirical generalizations or of general laws, perhaps even with basic theory, and with the methods of drawing logical deductions. (4) There are *remedial* questions that have to do with the application of geographic concepts and skills to the study of practical economic, social, or political problems. (5) There are *methodological* questions that have to do with experiments in new methods of study, new techniques of observation and analysis, or new cartographic methods.

Although there are many in each generation of geographers who are inclined to establish their own methods of work as the only acceptable paradigm, these efforts have been resisted. The freedom to inquire wherever curiosity leads has been preserved. There are still new worlds to be discovered by the enthusiastic scholar.

PART ONE

CLASSICAL

The strands of classical geography are closely interwoven with those of all other fields of learning, for in those days very little was known for sure and a diligent worker could master the greater part of the world's knowledge. Any one person could contribute to a great variety of what we would now identify as separate fields of study. Most of the ancient Greek scholars are called philosophers or historians; yet a student such as Herodotus wrote both history and geography, and he is also claimed by the anthropologists as the "father of ethnography." These were the times of universal scholarship, and these times continued until the last quarter of the nineteenth century. They produced such notable figures as Edmund Halley and Francis Galton in Great Britain, the political philosopher Montesquieu in France, Alexander von Humboldt and Carl Ritter in Germany, Thomas Jefferson and Benjamin Franklin in America, and M. V. Lomonosov in Russia. The enormous versatility of Franklin is well known. These were the days when a curious scholar felt no need to cease inquiry at the border of his knowledge.

Thus it was during most of the long history of ideas.

2

The Beginnings of Classical Geography

The fundamental concepts and basic elements used by the Pythagoreans in their theory are odd in comparison to those used by the physicists, since they are not observable; they are arithmetical entities which (except in the case of astronomy) are not a part of our changing world. Despite this, the Pythagoreans are in fact interested only in natural phenomena. They talk about how the universe was created, and always check their theories against observation when discussing the behavior and interactions of objects in the universe, using their particular set of fundamentals solely for theoretical purposes to account for physical reality. Thus in essence they are just like the other natural philosophers, since they agree that the only real things are those that can be perceived with the senses and are part of this world of ours. Their modes of explanation and their fundamental concepts, however, give them direct access to realms which transcend the senses, and to which their way of thinking is really more adapted than to physics. For example, they give no hint how motion and change can possibly be deduced from the only underlying concepts they assume in their theory—from boundedness and unboundedness, or evenness and oddness. Nor do they indicate, since they do not provide for motion and change in their theories, how any physical process can take place— and in particular how the heavenly bodies can move in the way they do.

Geography as a field of learning in the Western world had its beginnings among the scholars of ancient Greece. Not that the study of the earth as the home of man did not excite curiosity outside of Greece, although this is an impression that

---

The quotation above is from Aristotle's *Metaphysica,* Book 1, trans. D. E. Gershenson and D. A. Greenberg (New York: Blaisdell, 1963), pp. 38–39.

can be gained by reading many of the histories of geography written by Europeans. It is clear that much attention was given to geographical study in ancient China, and Chinese explorers did as much to ''discover'' Europe as the Europeans did to reach the ''Far East.'' But Chinese scholarship did not form a major part of the stream of Western thought. The Greeks, like all innovative people, were great borrowers; and much of what they put together in logical and useful order came originally from the much older civilizations with which they were in contact, including Egypt, Sumeria, Babylonia, Assyria, and Phoenicia. The Greek scholars provided a framework of concepts and a model, or paradigm, of scholarly method that guided Western thinking for many centuries. Some of the Greek concepts had the effect of retarding Western scholarship so that it can be said that European science could not emerge until the influence of Aristotle had been overcome. But many of the basic procedures of scholarship still in use were first developed by the Greeks.

THE ROOTS OF GREEK SCHOLARSHIP

The Greeks were indebted to the world's earliest scholars in many ways. Egypt has been called the cradle of science because of the very early development of methods of observation, measurement, and generalization in that country. The Egyptian priests had to have a sound working knowledge of mathematics, astronomy, and geometry for the practical purposes of public administration. They developed ways to measure land areas and to identify field boundaries obliterated by the Nile floods so that they could collect taxes. They learned how to fix a north-south line so that their monuments and public buildings could be properly oriented. They invented the art of writing, and they found out how to manufacture something on which to write—papyrus, made from reeds that grew in the marshy Nile Delta.

The civilization of Mesopotamia also contributed to scholarship. The world's earliest mathematicians, who lived in Sumeria, had grasped the basic principles of algebra—although the algebraic symbols we use were not invented until the sixteenth century, some 3000 years later. Without symbols of the kind we use, the Sumerians understood and used such principles as:

$$(a + b)^2 = a^2 + 2ab + b^2$$

They also had enough knowledge of algebraic methods to be able to find the square root of any number.

The people of Egypt and Mesopotamia also developed a kind of mathematics based on multiples of six and sixty—a sexagesimal system. Both Egyptians and Sumerians at first believed that there were 360 days in a year. The Egyptians discovered their error and compensated for it by declaring a 5-day holiday period each year. They made additional adjustments every fourth year. The Sumerians divided the year into twelve months, each with 30 days. They also divided the circle of the zodiac into 360 parts. The idea that there are 360 degrees in a circle is a very ancient one.

The priests of these early civilizations also collected a large number of observations regarding the position and movement of celestial bodies. The Babylonians and Assyrians, seeking the meaning of all these observations, developed ideas regarding the influence of the moon and the stars on human affairs—a body of concepts that we call astrology.

The Phoenicians, whose homeland was in modern Lebanon, were among the earliest merchant explorers and navigators. Their voyages went far beyond the limits of the known world, but as merchants they were not anxious to report on what they had found. In a valley near modern Beirut there is an ore body in which copper and tin are naturally combined. The Phoenicians made and sold bronze. But although copper was plentiful in the Mediterranean region, tin was scarce. The Phoenicians made regular voyages to the Scilly Islands off Great Britain to find tin. They also sold cedar logs from the mountains of Lebanon. One of the oldest known pieces of writing is a bill of lading describing the cargo of cedar logs carried by forty ships that sailed from the Phoenician port of Byblos for Egypt some 3000 years before Christ (Casson, 1959:5). The Phoenicians established trading posts all around the shores of the Mediterranean, including the city of Carthage (near presentday Tunis) (Boyce, 1977).

The Phoenicians, too, developed the world's first phonetic alphabet. It was made up entirely of consonants, like the modern Semitic alphabet. The Greeks added the short vowels to the Phoenician alphabet.

GREEK GEOGRAPHY

Homer

The Greek geographers credited Homer with being the father of geography. This poet, whose existence is not known for certain. was the composer of the long epic poem, the *Iliad*, which describes episodes of the Trojan War sometimes between 1280 and 1180 B.C. This monumental poem, which is the earliest major literary work of Greek history, was probably put together during the ninth century B.C. A second great epic poem, the *Odyssey*, was written perhaps as much as a century later, but is also credited to a man named Homer (but perhaps not the same Homer). Whereas the *Iliad* is primarily historical, the *Odyssey* is a geographical account of the fringes of the known world. It records the efforts of Odysseus to return home to Ithaca after the fall of Troy. Blown off course by a storm, he spends twenty years wandering in distant places. Many historians of geography have attempted to identify the places described in the *Odyssey* and offer plausible evidence to suggest that the poet was indeed describing the Strait of Messina, or an island off the coast of Africa, or other well-known localities. There is one passage describing a land of almost continuous sunshine, where a shepherd setting out with his flock at daybreak hails another shepherd returning with his flock in the evening. Then, later, Odysseus comes to a land of continuous darkness, shrouded in mist. A Greek poet could not have imagined these scenes. Somehow word of the nature of the world in the far north during the long summer days and the continuous winter darkness had filtered back to Greece, to be woven with other geographical threads into the world's first adventure story.

The Greek sailors of the eighth century B.C. had no way of identifying directions at sea except by reference to the winds and associated weather types. In Homer's time they distinguished four directions: *Boreas* was the north wind—strong, cool, with clear skies; *Eurus* was the east wind—warm and gentle; *Notus* was the south wind on the front of an advancing storm—wet and sometimes violent; and *Zephyrus* was the west wind—balmy but with gale force (Bunbrey, 1883:1:36). Much later, in the second century B.C., the Athenians built a tower identifying eight wind directions (Schamp, 1955–56) with sculpture illustrating the weather types associated with each. The tower still stands in the midst of a Roman market at the base of the Acropolis (Fig. 1).

The names, Europe and Asia, do not appear in Homer as the names of land masses. But at some later time the name, Europe, was applied to the shore of the Aegean Sea toward the setting sun, and Asia was applied to the shore toward the rising sun. The origin of these names is not certain (Bunbury, 1883:1:38; Tozer, 1897/1964:69; Ninck, 1945:15–23).

Thales, Anaximander, Hecataeus

One of the earliest centers of Greek learning was the town of Miletus in Ionia on the eastern side of the Aegean Sea near the mouth of the Meander River (now the Menderes). Miletus became a major center of commerce and attracted Phoenician and Greek ships from all around the Mediterranean and the Black Sea. The sailors and merchants brought to Miletus a wealth of information concerning what things were like beyond the margins of Greek horizons: information about Europe north of the Black Sea, or about strange countries in Asia to the east, or about what could be found to the south of Egypt. Between 770 and 570 B.C. Miletus established some eighty Greek colonies around the shores of the Euxine (Black Sea) and along the Mediterranean shores to the west. In Miletus at this time there was not only a flow of geographical information, but there was also a group of thoughtful people to speculate about how all this miscellaneous information could be assembled in some kind of meaningful arrangement. To Miletus also came reports on Egyptian geometry, Sumerian algebra, and Assyrian astronomy.

The first of the Greek scholars to be concerned about the measurement and the location of things on the face of the earth was Thales, who lived in the seventh and

Figure 1. Tower of the winds.

sixth centuries B.C.[1] Thales was a practical businessman who at one time was able to corner the supply of olive oil to make a large profit for himself. But he was also a genius, who is credited with a great variety of innovations and is often likened to Benjamin Franklin in the breadth of his contributions and the fertility of his imagination. On a trip to Egypt Thales observed the priests at work measuring angles and base lines and computing areas. Thales returned to Miletus with his head full of mathematical and geometrical regularities that went far beyond the practical utility of trigonometry. There are six geometric propositions credited to him: (1) The circle is divided into two equal parts by its diameter. (2) The angles at either end of the base of an isosceles triangle are equal. (3) When two parallel lines are crossed diagonally by a straight line, the opposite angles are equal. (4) The angle in a

[1]Additional biographical data on persons mentioned in the text can be found in the Index of Names.

semicircle is a right angle. (5) The sides of similar triangles are proportional. (6) Two triangles are congruent if they have two angles and a side respectively equal (Sarton, 1952/1964:171). In the sixth century B.C. no one had ever stated these as general propositions before; but the most important contribution of Thales was his recognition that the solution of practical problems of measurement was less of an intellectual accomplishment than the rational generalization of the specific solutions.

Thales also made contributions in astronomy and reported on the magnetism of the lodestone. He speculated about the meaning of this fascinating universe and concluded that the material was made up of water in various forms. The earth he visualized as a disc floating in water. He was trying to offer an explanation of the universe in terms that could be checked by new observation, in sharp contrast to the traditional explanations in terms of manlike deities or astrological influences.

A younger contemporary of Thales in Miletus was Anaximander. He is credited with the introduction into the Greek world of a Babylonian instrument known as the *gnomon* (see also Heidel, 1937:57–58). This is simply a pole set vertically above a flat surface on which the varying position of the sun could be measured by the length and direction of the shadow cast by the vertical pole. This is what we call today a sundial. From the gnomon it was possible to make a variety of observations. Noon could be established by noting when the shadow was shortest; the noon shadow provided an exact north-south line, or meridian (from *merides,* meaning noon). The noon shadow varied from season to season, being shortest at the summer solstice and longest at the winter solstice. By observing the direction of the shadow at sunrise and sunset, it was possible to establish the time of the equinox, for at that time the sunrise and sunset shadows were colinear, but opposite.

Anaximander is reported by later Greek historians to have been the first ever to draw a map of the world to scale. To be sure, the Sumerians had drawn pictorial "maps" of some of their cities as early as 2700 B.C.; but a true map must show distance and directions to scale. Anaximander's map had Greece in the center and the other parts of Europe and Asia known to the Greeks were plotted around it. The map was circular and was bounded all the way around by the ocean. A copy of the map was supposed to have been cast in bronze and transported to Sparta in the effort to convince the Spartans that they should join in the war against the Persians. But the Spartans said the map proved that Persia was too far away to worry about.

The scholars who were seeking to explain their observations of the face of the earth and of the relative positions of the celestial bodies found difficulty in understanding how the sun could set in the west and yet get back to the east by the next morning. If the earth were a disc floating in water, how could the sun go under the water? Anaximander suggested that somewhere to the north there must be some very high mountains behind which the sun made the trip back again to the east. The shadow cast by these mountains would account for the night.

Anaximander was also one of the earliest philosophers to provide us with an example of how a word can be used to symbolize something that is not known and not observed. He did not actually reject the idea of Thales that water was the prime substance from which all observable features of the earth were made. But he used the word, *apeiron,* to symbolize this prime substance. Apeiron, which could not be experienced through the senses, nevertheless became a concept—a specific mental image that by the process of deduction could become a real substance. This thought process is possible for man because he uses words to symbolize abstractions; and it is still with him in the twentieth century, providing a semantic trap for the unwary who confuse observable reality with the reality of word symbols.

Thales and Anaximander can be recognized as the originators of the mathematical tradition in the study of geography. To Hecataeus goes the credit of originating the literary tradition. Hecataeus, who was born at about the time of the deaths of Thales and Anaximander and who died about 475 B.C., was the first to collect and classify the information brought to Miletus not only from the known world of the Greeks but also from the shadowy world beyond the Greek horizons. He was the first writer of Greek prose. One of the two prose works credited to him is the *Ges periodos,* or *Description of the Earth,* only fragments of which survive. But one fragment contains a kind of subtitle in which there is the first record of a "new geography." He says that he has written these things in his book because he believes they are true. "The narrations of the Greeks are many and in my opinion, foolish." He set the tone for writers of geography—that is, "new geography"— which has persisted for some 2500 years. Hecataeus divided his work into two parts, each dealing with one of the regional divisions of the earth. One book dealt with Europe, the other with the rest of the world—Asia and Libya. He followed what was apparently already tradition in separating Europe from Asia along the Hellespont, the Euxine, the Caucasus Mountains, and the Caspian Sea, which he thought was connected with the surrounding ocean (Fig. 2).

Hecataeus was not a theorist. His reaction to the speculations of his predecessors is similar to the reactions of countless generations to follow. He felt that discussions of whether water or apeiron should be accepted as the prime substance or whether there even was such a prime substance were futile. Before trying to solve the enigma of the universe, he insisted, we should take stock of what is around us and put the accumulated knowledge about the world together in usable form. The contrast in the approaches of these scholars of Miletus more than twenty-four centuries ago illustrates the apparent dichotomy between those who seek to formulate generalizations and those who seek to describe unique things. In modern times these two points of view are described as *nomothetic,* meaning law seeking, and *idiographic,* meaning descriptive of particular things. Down through the ages since Hecataeus there have been scholars who insist that geographical study must adopt one or the other approach; only rarely do we find anyone to proclaim that geographic study must—and, in fact, does—make use of both approaches and that the dichotomy exists only because of the word symbols we use.

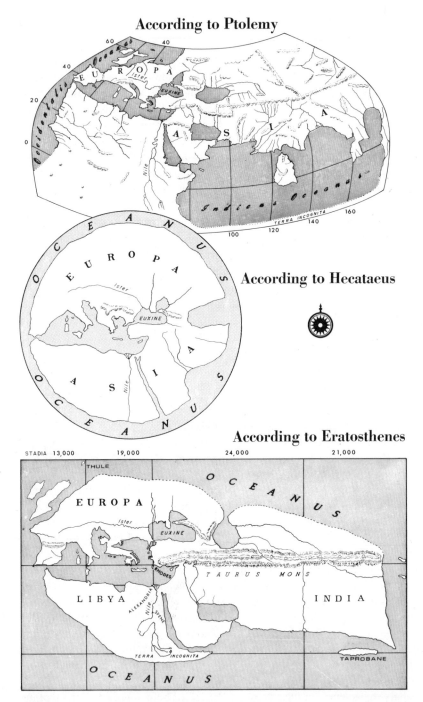

Figure 2. The world according to Ptolemy; according to Hecataeus; and according to Eratosthenes.

Herodotus

A century later the ideas of Hecataeus were ridiculed by another great scholar named Herodotus. His great work, which was written while he was residing in Italy, is a history of the Greek struggle with the barbarians and ends with the revolt of the Ionians against the Persians and with the Greek capture of the Hellespont (480–479 B.C.). But his history includes numerous digressions to describe the places he had visited and the people whose customs he had observed and recorded. In the fifth century B.C. no one was concerned to identify himself as a member of a separate profession. There were no historians, or geographers, or astronomers, and no professional societies to join. There were no academic departments. Herodotus is usually described as the first great historian, and his work was the first masterpiece of Greek prose. But Herodotus is identified as an historian chiefly because there are more historians than there are geographers, for a very large part of his work is easily identified as geography. In fact, Herodotus is credited with the very old idea that all history must be treated geographically and all geography must be treated historically. It is true that the notion of geography as "the handmaiden of history" came from Herodotus. Geography provides the physical background, the stage setting, in relation to which historical events take on meaning. Herodotus provides some excellent examples of what we would call today historical geography—that is, the re-creation of past geographies and the tracing of geographical change through time. But Herodotus is also identified as "the father of ethnography" because of his vivid portrayal of the culture traits of people strange to the Greeks.

The contributions of Herodotus to geography were based on his own personal observations during many years of travel. Westward he knew the Mediterranean shores as far as southern Italy, where he resided during the latter part of his life. He went through the straits into the Euxine (Black Sea), reaching the mouth of the Ister (Danube) and traveling for several days northward across the Russian steppes along the valley of the Don. He went eastward over much of the territory of the Persian Empire, visiting Susa and Babylon. Toward the south he visited Egypt many times and went up the Nile as far as the first cataract near Elephantine (Aswan).

In his discussion of Egypt he takes issue with the tradition of dividing Asia (the eastern side of the Mediterranean) from Libya (the southern side) along the Nile River, as Hecataeus had done. The Nile Valley, he insists, has been built by mud brought down from Ethiopia. This mud is dark colored and easily worked with the plow—quite unlike the light-colored clays of Syria or the red sands of Libya. He insists that Egypt is occupied by Egyptians and that they are not divided into Asians and Libyans along the river. Libya, he says, begins to the west of Egypt. This is one of the earliest discussions of regional boundaries and contains many of the arguments used over and over again by later generations.

Herodotus was well aware of some of the physical processes at work on the earth. He used the methods of historical geography to support the hypothesis that the Nile mud, deposited in the Mediterranean, had built the delta. He reconstructed the ancient shoreline and showed that many former seaports were now far inland. The process of delta building, said Herodotus, can be observed in many places—notably in the alluvial plain of the Meander River at Miletus. He also pointed out that the wind blows from cold places to places that are warmer. In the fifth century B.C. it was a significant accomplishment to explain how deltas are formed or to grasp the connection between temperatures and wind directions.

Not all the explanations suggested by Herodotus can be supported in the light of modern knowledge, but even where he was in error he supported his hypotheses with logic. Like all Greek geographers Herodotus was fascinated with the regularity of the summer floods of the Nile. In this river the water would rise suddenly in mid-May, reach its highest flood stage in September, and then decrease in volume, reaching its lowest stage in April or early May. Since all other rivers known to the Greeks, including the Tigris and the Euphrates, flooded from November to May and reached their lowest stages in summer, students of geography were faced with a challenging problem: What caused this distinctive characteristic of the Nile?

First Herodotus reviewed the explanations offered by other puzzled scholars and refuted them. For example, he rejected the idea that the strong north wind of winter (the Etesian Wind) blowing up the Nile caused the water to back up because floods and low water come whether the wind is blowing or not and no such effect can be observed in other rivers up which the wind is blowing. He rejected the suggestion that the Nile floods are caused by melting snow in Ethiopia, because Ethiopia is closer to the equator than Egypt. Egypt never has any snow, so how could there be snow in Ethiopia?

His own hypothesis was ingenious and illuminates the use of logic in Greek thinking. Like all the Greek scholars, Herodotus accepted as a fundamental principle that the world must be arranged symmetrically. The Ister (Danube), he believed, had its headwaters close to the western coast of Europe and flowed eastward before turning southward through the Euxine, the Hellespont, and the Aegean to reach the Mediterranean. The Nile, in accordance with the principle of symmetry, must follow a similar course, rising close to the western coast of what was then called Libya and flowing eastward before turning toward the Mediterranean to flow northward through Egypt. In winter, he continued, the cold north winds make the sun move along a more southerly course, passing directly along the valley of the upper Nile. The intense heat under the overhead sun draws up the river water, leaving the river with greatly decreased volume in winter. But during the summer, when the sun returns to its course "through the middle of the heavens," the volume of water rises again because the lower Nile crosses the sun's path at right angles and much less of its water is evaporated. Since this explanation was in agreement with both concepts (symmetry) and direct observations (time of flooding), it was generally accepted by scholars.

Herodotus also took exception with earlier writers who raised doubts about the existence of an ocean all around the margins of the world. Some had reported that

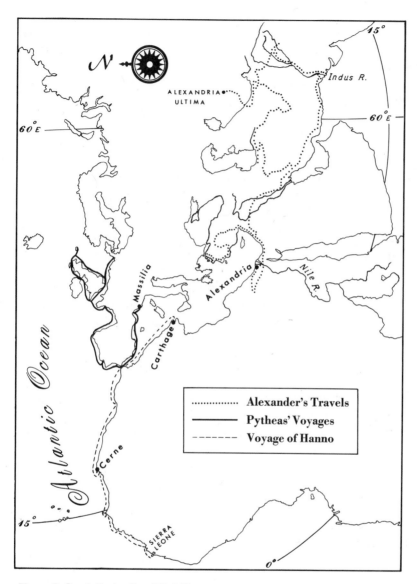

Figure 3. Greek Exploration 470–310 B.C.

there was no ocean to the south of Libya. But Herodotus, in talking with the Egyptian priests, had learned of a Phoenician expedition sent out by King Necho (who ruled Egypt from 610 to 594 B.C.) to sail around the southern end of Libya. The Phoenician ships, it was reported, sailed southward from the Red Sea along the east coast of the continent. They replenished their food supply by stopping from time to time to plant grain and remaining long enough to harvest the crop. It took three years to sail around the southern end of Libya; then northward along the western side; and, finally, to reenter the Mediterranean through the Pillars of Hercules (Gibraltar). This expedition proved that the land is entirely surrounded by water. Then he reported a circumstance that to him ''appears incredible, but others may believe,'' that while the expedition was near the southernmost part of Libya sailing toward the west, the sun was on their right hand. This observation led many scholars after Herodotus to discount the story of the circumnavigation of Libya. The reality of the Phoenician expedition is now generally accepted, and the circumstance that caused ancient scholars to doubt the story leads modern scholars to find the account plausible. There is also the possibility that some of the Phoenician ships, being caught in the west-flowing equatorial current south of the equator in the Atlantic Ocean, were carried across the relatively narrow ocean to the northeast of Brazil.

One difficulty with the interpretation of these ancient writings involves the things that are omitted. Considering the wealth of detail Herodotus provides concerning some places and some events, it is remarkable that he made no mention of another Phoenician voyage. This was the expedition led by Hanno about 470 B.C. The expedition was sent out by Carthage to establish trading posts and colonies along the Atlantic coast of Libya south of the Pillars of Hercules. Hanno's descriptions of the things he saw are detailed enough so that the voyage can be charted with confidence. After passing through the strait, he turned southward (Fig. 3). Near the present port of Safi in Morocco he passed a lagoon where elephants were feeding. Farther south on an island in the bay of Rio de Oro on which Villa Cisneros is now located, he established a base that he named Cerne. This remained a Phoenician trading post for many years. From Cerne, Hanno led two expeditions farther to the south. On the second of the two voyages he reached Sherbro Island, south of the present site of Freetown in Sierra Leone, almost seven degrees from the equator. Here the explorers came upon ''wild men and women with hairy bodies,'' who they were told, were called gorillas. They were unable to catch any of the men, but they did catch three women. These they killed and skinned and brought the skins back to Carthage. The record of the expedition was preserved on a bronze plaque in a temple at Carthage.

Plato and Aristotle

The two great Greek philosophers, Plato (428–348 B.C.) and Aristotle (384–322 B.C.), both made important contributions to the development of geographical ideas. Plato, who was a master of deductive reasoning, insisted that the observable things on the earth were only poor copies of *ideas,* or perfect predicates from which observable things had degenerated or were in process of degenerating (Popper,

1945/1962:18–34). At one time, he observed, Attica in Greece (the ancient territory of which Athens was the central city) possessed a very productive soil, capable of supporting the inhabitants in comfort. There were forests on the mountains that not only provided feed for animals but also held the rainwater from pouring down the slopes in floods during heavy rains. "The water was not lost, as it is today, by running off a barren ground to the sea. . . . What now remains, compared with what then existed, is like the skeleton of a sick man, all the fat and soft earth having been wasted away, and only the bare framework of the land being left" (Glacken, 1967:121). Arguing from the general theory to the particular situation in Attica, Plato used this as an example of the degeneration of things from their original perfect state. If Plato had argued from the particular to the general he might have realized that men make changes in the land they occupy and that soil erosion and land destruction are parts of culture history and are repeated in many places. The idea of man as an agent of change on the face of the earth was not formulated for thousands of years after Plato. As Glacken points out, Plato missed the chance to change the whole history of speculation concerning man-land relations by identifying man as destructive agent.

Plato related the story of Atlantis. The Greek world, he said, was about to be invaded in the year 9000 B.C. by a highly civilized people who lived somewhere to the west. The Greek armies, after a fierce struggle, were victorious; but just as the invaders were defeated their homeland suffered a disastrous earthquake and sank beneath the sea. It is possible, he reported, to sail over the sunken city of Atlantis if one is careful to avoid the shallow places. Explorers and popular writers have been searching for Atlantis ever since, even imagining it to have been a land bridge between Africa and America. Only in 1966 did skin divers identify the existence of a submerged city in the waters between Crete and the Greek mainland that may well have been the Atlantis to which Plato referred.

Was the earth round or flat? The great majority of the people living in those times did not question the evidence of their senses that the earth was flat; a few philosophers began to think of the earth as a ball on purely theoretical grounds. All the Greek thinkers accepted the idea that symmetry of form was one of the attributes of perfection, and the most completely symmetrical form was a sphere. Therefore, they argued, the earth, which was created in perfect form as the home of man, must be spherical. Pythagoras, who lived in the sixth century B.C., may have been the earliest philosopher to hold this view; at any rate he worked out some of the mathematical laws for the circular motions of celestial bodies; his pupil, Parmenides, applied these to observations made from the surface of a round earth. But Plato, who lived a century after Parmenides, seems to have been the first philosopher to announce the concept of a round earth located in the center of the universe with the celestial bodies in circular motions around it. Whether it was Plato's original concept or whether it was suggested to Plato by Socrates, whom he quotes, cannot be determined. Eudoxus of Cnidus, a contemporary of Plato, developed the theory of zones of climate based on increasing slope (*klima*) away from the sun on a spherical surface. All these formulations were deductions from pure theory—the theory that all observable things were created in perfect form and that the most perfect form was a sphere. But it was Aristotle who first looked for evidence to support the concept.

Aristotle was seventeen when he joined Plato's Academy near Athens. At this time (367 B.C.) Eudoxus was acting head during the temporary absence of the master. Aristotle remained at the Academy until Plato's death, at which time Aristotle was thirty-eight. During the next twelve years of his life he spent his time traveling widely throughout Greece and around the shores of the Aegean. In the year 335 B.C. when Aristotle was forty-nine, he returned to Athens and founded his own school, which he named the Lyceum (Sarton, 1952/1964:492). By this time he was convinced that the best way to build theory was to observe facts and the best way to test a theory was to confront it with observations. Whereas Plato built theory by intuition and reasoned from the general to the particular, Aristotle built theory by reasoning from the particular to the general. These two ways of thinking about things are known respectively as deduction and induction.

Aristotle recognized that observations made through the senses can never provide explanations. Our senses, he said, can tell us that fire is hot but cannot tell us why it is hot. Aristotle formulated four fundamental principles of scientific explanation—that is, of answering the question: "What makes this thing the way it is?" (Aristotle, trans. Gershenson and Greenberg, 1963:2:13). One way is to describe its nature, to tell its essential characteristics. A second way is to specify the kind of matter, the substance, of which it is composed. A third way is to tell what caused the process through which the thing became as it is. And a fourth way, which is complementary to the third, is to tell the purpose the thing fulfills. Unlike Plato, Aristotle assumed that things were in process of physical change leading to a final perfect state. This model for scientific explanation constituted the world's first paradigm for the guidance of scholars.

With regard to the matter, or basic substance, of which all material things are made, Aristotle followed Empedocles (490–430 B.C.), who a century earlier had improved on the single-substance idea of Thales (water) by postulating the existence of four basic substances: earth, water, fire, and air. All material objects on the earth are made up of these basic elements in varying proportions. Aristotle added a fifth substance, aether, which did not occur on the earth but was the material from which celestial bodies were made.

Aristotle pointed out that, to create the material objects on the earth and in the heavens, some kind of process of change had to take place. First there had to be empty space; the philosophers of that time recognized the existence of two kinds of space: celestial space and earth space—that is, space on the surface of the earth. There was some speculation concerning interior space within the earth, but there was little knowledge to guide these speculations. Aristotle, modifying the ideas of

Empedocles, developed the theory of natural places. Everything had its natural place in the universe, and, if removed from this place, it would seek to return. Earth space was the natural place of earth and water, and, if raised above the surface of the earth, these substances and things composed of them would fall back to the surface. Air and fire, on the other hand, had their natural places in celestial space, and, therefore, they tended to rise. Aether had its natural place in the celestial bodies far out from the earth.

Aristotle agreed in part with the teaching of Plato, derived from Pythagoras and Parmenides, that all things are patterned after numbers. The basic regularities of the universe are those of geometry and mathematics. He complains, however, that "nowadays everybody thinks that science is mathematics, and that it is only necessary to study mathematics in order to understand everything else" (Aristotle, trans. Gershenson and Greenberg, 1963:2:51). Aristotle insists that mathematics can be used to explain the process of change that makes things as they are, but it cannot answer the fourth question concerning the purposes or ideal states. Aristotle was the first teleologist in that he believed everything was changing in accordance with a preexisting pattern or plan, just as a carpenter building a house knows in advance what the house will be like when it is finished. All things, said Aristotle, are not deteriorating from an ideal state but rather are developing toward an ideal state.

Aristotle accepted Plato's concept of a spherical earth and began to seek an explanation of it and to test the concept with observations. His explanation was derived from the theory of natural places. When the solid matter of which the earth is made falls toward a central point, it must form a ball (Sarton, 1952/1964:510). Aristotle was the first scholar to recognize the significance of the observed fact that, when the shadow of the earth crosses the moon during an eclipse, the edge of the shadow is circular. He also recognized that the height of various stars above the horizon increases as one travels toward the north, which could only occur if the observer were traveling over the curved surface of a sphere. Strangely, he never realized that additional support for the concept of a round earth could be gained by noting the disappearance of a ship beyond the horizon—hull first. He must have had ample opportunity to observe this fact.

Aristotle's paradigm for scientific explanation did not include anything about controlled experiments or the verification of premises, but only the use of logic to formulate and give support for theory. And some of his logical explanations seemed so unassailable in the fourth century B.C. and have been so universally accepted for many generations since that time that his influence on the history of Western ideas has been enormous. It has been pointed out that modern science could not appear until Aristotle had been abandoned. Here we note a very common sequence of events in the history of ideas: the formulation of a new concept is enormously stimulating and results in an increase in the quality and quantity of observations, but the continued acceptance of the concept proves an obstacle to the progress of scholarship among succeeding generations.

An example of this in the field of geography was Aristotle's concept of the varying habitability of the earth with differences of latitude (Glacken, 1956). That habitability was a function of distance from the equator was a notion that seemed to accord with observed facts for people living around the shores of the Mediterranean. If the earth is a sphere and the sun is circulating about it, the parts of the earth where the sun is most directly overhead must be much hotter than places farther away from the sun. The Greeks were familiar with the excessively high temperatures to be experienced in Libya along the southern side of the Mediterranean. In modern times the world's record for high temperature observed in a standard thermometer shelter (136.4°F) is held by a place in present-day Libya about twenty-five miles south of the Mediterranean shore and more than 32°N of the equator. If the air gets that hot at this latitude, the Greeks reasoned, it must be very much hotter close to the equator. The people living in the northern part of Libya had black skins, and the Greeks assumed that they had burned black by exposure to the sun. At the equator then all life must be possible because any living thing would be burned in the intense heat. Aristotle reasoned that the parts of the earth close to the equator, the torrid zone, were uninhabitable; that the parts of the earth far away from the equator, the frigid zone, were constantly frozen and also were uninhabitable; and that the temperate zone in between constituted the habitable part of the earth. The *ekumene,* the inhabited part of the earth, was in the temperate zone; but much of it, said Aristotle, was not inhabited because of the ocean. Aristotle also postulated the existence of a south temperate zone, which could not be reached from Greece because of the intense heat of the torrid zone. Many scholars in antiquity accepted Aristotle's south temperate zone but doubted that it could be habitable because in the *antipodes* people would have to hang upside down. This notion of habitability as a function of latitude has had a long history and, in fact, is still widely accepted, especially by nongeographers.

Alexander the Great

Aristotle had many pupils, and in all of them he instilled a desire to test theory by direct observation. He taught them to "go and see" for themselves whether any one theory could or could not be accepted. And his greatest pupil was Alexander, who became the king of Macedonia at the age of twenty. Alexander studied with Aristotle for only three years (343–340 B.C.), between the ages of thirteen and sixteen, yet no one applied the master's teaching more effectively. When Alexander became king he started on a career of military conquest. But his treatment of conquered people was extraordinary: all people, said Alexander in the fourth century B.C., are brothers and should be treated as brothers. Furthermore, a primary objective in Alexander's conquests was to expand Greek geographic horizons—they were in the nature of armed explorations.

Alexander's conquests pushed Greek knowledge of the earth far to the east

(Fig. 3). After conquering the barbarian tribes living north of the Ister (Danube), in 334 B.C. he crossed the Hellespont into Asia. His first marches were close to the coast, where he could be supplied by ship. But then he grew bolder and invaded the central part of present-day Turkey, then part of the Persian Empire. Thence he continued southward along the eastern side of the Mediterranean to Egypt, where he established his rule. He founded the city of Alexandria in 332 B.C., which was destined to become one of the great commercial and intellectual centers of the ancient world. After some exploratory excursions to oases in the Libyan desert west of the Nile, he turned again toward the East, crossing into the heart of the Persian Empire (modern Iran) by way of Babylon and Persepolis. He pushed northward as far as the central Asian market town of Samarkand, then on eastward as far as—and across—the Indus River. Believing that he was only a short distance from the eastern limit of the ekumene, he wanted to march on farther, but his troops mutinied and insisted on returning to Greece. Alexander died in Babylon in 323 B.C., and the empire he had built and ruled with compassion fell to pieces in strife.

Few teachers have had the lessons they taught applied so well by their students. Alexander's staff included writers to describe the lands they crossed and astronomers to take observations of the height of the bright star Canopus to fix latitude, or distance north of the equator. There were trained pacers, whose duty was to measure distances on the march. As a result he sent back to the Greek world a wealth of new observations concerning what it was like beyond the Greek horizons and how far and in what direction it was necessary to travel to reach these strange places. At the time of his death he was planning to send out two additional expeditions to find answers to two geographical questions. One was to follow the shores of the Caspian Sea to settle the question whether or not this sea was connected to the open ocean, as some maps showed. The other expedition was to sail southward along the Red Sea from Egypt to find out whether Libya truly was surrounded by water on the south and whether human beings could survive in the intense heat of the equatorial regions. With his death both expeditions were abandoned.

Pytheas

While Alexander was extending Greek geographic horizons to the east, another Greek explorer was voyaging far to the northwest of the Greek world in western and northern Europe (Fig. 3). This was Pytheas, who is assumed to have made his remarkable voyage sometime between 330 and 300 B.C. Unfortunately, his original report did not survive and he is known to historians of geography only because of references made to his work by other writers (Bunbury, 1883:1: 589–601; Tozer, 1897/1964:152–164; Ninck, 1945:218–226; Sarton, 1952/1964:523–525).

The following account of his voyage is generally accepted today. Pytheas was a native of the Greek colony of Massilia (modern Marseilles), which was engaged at that time in a bitter rivalry with the Phoenicians of Carthage for control of the profitable trade in tin and amber. Whether Pytheas was sent out by Massilia to penetrate the screen of secrecy imposed by the Phoenicians or whether he financed his own voyage to satisfy his curiosity about what lay beyond the Greek horizons is not certain. At any rate he set out by ship from Massilia along the coast to the Pillars of Hercules and then managed to slip by the Phoenician naval base at Gades (Cadiz). He sailed along the coast of France to the English Channel and then around Great Britain.

Pytheas reported things that were so contrary to Greek experience that the geographic scholars of his day discredited him and treated his important information as pure fantasy. He told about the customs of the people of Britain with detail that could scarcely have been invented. He described the drinking of mead (fermented honey), the use of barns to thresh grain in wet weather, and the change in the character of agriculture from south to north in Great Britain. He also described a sea so full of ice that it could neither be traversed on foot nor in a boat—which exactly describes what the polar explorers call ice sludge. He also was the first Greek to tell about ocean tides (the tides on the Mediterranean are too small to be noticed), and he showed that the tides were related to the phases of the moon.

How much farther north he went is a puzzle, although he reported on the existence of a place called Thule, six days sailing north of Great Britain. It is probable that he sailed along the eastern shore of the North Sea, perhaps as far as modern Denmark. He is quoted as having been to a place where the length of the longest day was between seventeen and nineteen hours, which would place him 61°N, in the northernmost of the Shetland Islands. He is also quoted as reporting that at Thule the sun remained above the horizon during the whole of the longest day, which would put this place well to the north in Norway or possibly Iceland. It is certain, however, that before he left Massilia he observed the angle of the sun's shadow on a gnomon, and the latitude derived from this measurement was almost exactly correct (43°05'N, instead of 43°18'N).

The details that Pytheas is said to have recorded (and that led the scholars of antiquity to discredit him) lead modern scholars to believe that he was indeed reporting his observations correctly. Nowadays Pytheas is accorded his due place among the great explorers of all times.[2]

---

[2]There are many other important contributors to the development of Greek geographical ideas in classical antiquity. There was Hippocrates, the fifth-century B.C. physician, who, among many other writings attributed to him, produced the world's first medical geography. In his book *On Airs, Waters, and Places*, he was the first to present the concept of environmental influence on human character. There was Theophrastus, who succeeded Aristotle as head of the Lyceum and who was its director for thirty-five years. He wrote a prodigious number of books on a wide range of subjects, including meteorology, petrography, ethics, and religion. He is known as the "father of botany" because he described and classified more than 500 species of cultivated plants. There was Dicaearchus, who measured the heights of certain Greek mountains with a primitive theodolite and concluded that the highest mountains were only a slight roughness on the surface of the earth—this, in the late fourth century B.C. Aristarchus of Samos proposed the hypothesis that the sun was the center of the universe and

Eratosthenes, The Father of Geography

Eratosthenes is often identified as the "father of geography" because, among other contributions, he was the first to coin the word. But, as we have observed, many major contributors to geographical ideas are not identified as geographers. In many ways, however, Eratosthenes set a stamp on the study of the earth as the home of man that still persists.

Eratosthenes was born in the Greek colony of Cyrene in Libya. In Cyrene and later in Athens he received a broad education, including philology and rhetoric as well as mathematics and philosophy. He probably attended both the Academy and the Lyceum. In about 244 B.C. he accepted an invitation from the king of Egypt to become the royal tutor and was also named as "alpha fellow" at the museum in Alexandria. With the death of the chief librarian of the museum about 234 B.C., he was appointed to that coveted post, which was among the most prestigious in the Greek scholarly world. He remained as chief librarian until his death in about 192 B.C., at the age of eighty.

George Sarton gives some interesting sidelights on the attitude of the Greek scholars toward the chief librarian at Alexandria. Eratosthenes had two nicknames: he was called *beta*, which suggests that although he was the senior (alpha) fellow he was still only a second-rate scholar; and he was called *pentathlos*, a name given to athletes who performed well in five different games. Sarton points out that at this time there was a growing specialism in Greek scholarship of a kind that did not again appear until the seventeenth century after Christ. Specialists in a restricted field of study—then as now—are inclined to look with scorn on anyone whose scholarship is based on breadth rather than depth. Here is what Sarton has to say about this very human situation:

> The first nickname, *beta,* shows that the scientists and scholars of that age were already very jealous of one another and all too ready to deflate those whose superiority they misunderstood or resented. Now the professional mathematicians might consider him as not good enough in their field and be displeased with the abundance and variety of his nonmathematical interests. As to the men of letters and philologists they could not appreciate his geographic purposes. Eratosthenes might be second-rate in many endeavors, but he was absolutely first-rate in geodesy and geography and is to this day one of the greatest geographers of all ages. This his critics could not even guess, and therefore they pooh-poohed him. There was among them a man of genius but as he was working in a new field they were too stupid to recognize him. As usual in such cases, they proved not his second-rateness but only their own [Sarton, 1959/1965:101–102].

---

that the earth and planets were revolving around it. He explained day and night in terms of a rotating earth. The inquiring student who wishes to know more about the many other Greek and Roman geographers should consult the following: Bunbury, 1883; Tozer, 1897/1964; Heidel, 1937; Ninck, 1945; Sarton, 1952/1964, 1959; or Thomson, 1965.

Eratosthenes is perhaps most famous for his calculation of the circumference of the earth. He was able to do this, apparently, because he was the first scholar with the imagination to appreciate the significance of two separate observations of the position of the sun at the time of the summer solstice. One observation came from near Syene (Aswan). On an island in the Nile, just below the first cataract and opposite Syene, there was a deep well, and at the bottom of the well at the summer solstice the image of the sun was reflected in the water. The existence of this well had been known for a long time, and no doubt tourists in ancient times traveled up the Nile to witness this strange occurrence each year. This meant of course that on that date the sun was directly overhead. The second observation was made outside the museum in Alexandria, where there was a tall obelisk. Using the obelisk as a gnomon, Eratosthenes measured the length of the shadow at the solstice. He was thus able to measure the angle between the vertical obelisk and the rays of the sun. With these data in mind, Eratosthenes made use of the well-known theorem of Thales, which states that when a diagonal line crosses two parallel lines, the opposite angles are equal. The parallel lines were given by the parallel rays of the sun (Fig. 4). The rays of the sun at Syene, which were vertical, could be extended to the center of the earth (*SC*). Also the obelisk, which was vertical at Alexandria, could be extended to the center of the earth (*OC*). Then the angle between the sun's rays and the vertical obelisk at Alexandria (*BOC*) must be the same as the opposite angle at the center of the earth (*OCS*). The next question was this: How much of the whole circumference of a circle is subtended by the angle *OCS*? Eratosthenes measured this as one-fiftieth of the whole circumference. It was then only necessary to fill in

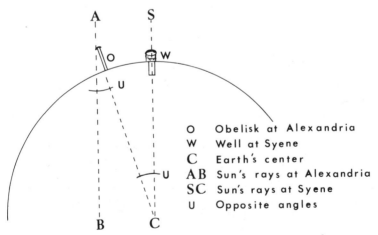

Distance OW subtended by angle OCS equals 1/50 of the circumference of a circle

Figure 4. Calculation of the earth's circumference by Eratosthenes.

the distance between Syene and Alexandria, which the Egyptians said was the equivalent of about 500 miles, and then multiply this distance by 50. Eratosthenes, therefore, concluded that the whole earth was about 25,000 miles in circumference (actually the circumference measured through the poles is 24,860 miles).[3]

The measurements in those times were far from exact. Eratosthenes assumed that Alexandria was due north of Syene, whereas, in fact, it is about longitude 3°W of Syene. The length of the road between Syene and Alexandria, which the Egyptians said was the equivalent of 500 miles, is actually 453 miles. And Syene is actually at latitude 24°5′N, a little to the north of the tropic. But all these errors canceled out so that the resulting calculation was amazingly close to the correct figure (Thomson, 1965:159–161).

Eratosthenes also wrote a book describing the ekumene, the inhabited earth, in which he accepted both the major divisions of Europe, Asia, Libya, and the five zones—a torrid zone, two temperate, and two frigid zones. He improved on Aristotle by giving the mathematical boundaries of these zones. The torrid zone he thought was forty-eight degrees of the whole circumference (twenty-four degrees north and south was calculated as the location of the tropics). The frigid zones extended twenty-four degrees from each pole. The temperate zone was between the tropic and the polar circles. In his book he was one of the few who accepted the reports from Pytheas: he extended the ekumene from Thule, near the Arctic Circle, to Taprobane (Ceylon) in the Indian Ocean. The ekumene, he reported, also extended from the Atlantic Ocean to the Bay of Bengal, which he assumed was the eastern limit of inhabitable land.

He also prepared a world map (Fig. 2). He made use of a frame of north-south and east-west lines; but these were not spaced regularly. Rather, he used the meridian of Alexandria, which he extended southward through Syene and northward through Rhodes and Byzantium, as the prime meridian; and he used the latitude of the Pillars of Hercules, which he thought also passed through Rhodes (Sarton 1959/1965:106–108). His map of the known world was plotted in relation to these lines.

Hipparchus

When Eratosthenes died he was succeeded as chief librarian by Hipparchus. The dates of his birth and death are unknown, but it is certain that he was working at the library in 140 B.C. Hipparchus was more of a mathematician and astronomer than he was a geographer; but he did show, in theory at least, the way to establish the exact position of every point on the earth's surface. He was the first to divide the circle into 360 degrees, based on Assyrian arithmetic; and for all nations the circle

[3]Eratosthenes gave his linear measurements in stades. The value of the stade is not known exactly, but it ranges around 10 stades to the modern mile. So his estimate of 5000 stades was approximately 500 miles.

continues to be so divided. Hipparchus defined a grid of latitude and longitude lines, as Eudoxos had done for celestial space. The equator, he pointed out, was a great circle (one that divides the earth into two equal parts) and the meridians that were drawn converging on the poles were also great circles. The parallels, on the other hand, became shorter and shorter as they approached the poles. Since the earth makes one complete revolution in twenty-four hours and there are 360 meridians drawn from equator to poles, each hour the earth turns through fifteen degrees of longitude.

Hipparchus had high hopes that geography could be made more exact through the plotting of locations in this theoretical grid. The Greeks did know how to make fairly good measurements of latitude by using the gnomon, but very few such observations had, in fact, been made.[4] Longitude, however, remained a matter of guesswork. There was no way to measure time, especially at sea. Hipparchus suggested that the local times of the start of an eclipse at different places could be compared. The time differences would provide a measure of longitude, but no such system of coordinated observations was even attempted for thousands of years after his time. Already in the second century after Christ, geographical studies had become too technical and too mathematical for the use of general readers or of others who wished to find information about particular countries. Polybius, the Greek historian, saw geography as an essential support for the study of history, as Herodotus had suggested.

Hipparchus also was the first to wrestle with the problem of showing the curved surface of the earth on a flat surface. It cannot be done because a spherical surface cannot be made to lie flat without cutting or stretching it. He devised two kinds of projections, however, so that the distortion of the spherical surface on a map could be carried out mathematically. He told how to make a stereographic projection by laying a flat parchment tangent to the earth and extending the latitude and longitude lines from a point opposite the point of tangency. The orthographic projection is similarly produced but by projecting the lines from a point in infinity. On the stereographic map the central portion is too small in relation to the periphery; on the orthographic projection the central portion is too large. These two projections, it should be noted, can only show a hemisphere, not the whole earth.

Posidonius

Another important Greek historian and geographer, who lived shortly before the time of Christ, was Posidonius. Two of his contributions to geographical ideas must be described: one, a wrong idea that persisted for centuries; the other, a correct idea that was overlooked.

---

[4]Hipparchus invented an instrument that was easier to use than the gnomon. This was the astrolabe. A circular dial was marked off into 360 parts and a rotating arm was fixed at its center. Hanging on the rigging of a ship, the astrolabe made possible the measurement of latitude at sea by observing the angle of the polestar.

It was he who recalculated the circumference of the earth and arrived at a much smaller figure than that of Eratosthenes. He felt no confidence in the work of Eratosthenes, so he undertook to make his own measurements. He observed the height above the horizon of Canopus (a star of the first magnitude) at Rhodes and Alexandria, which he assumed to be on the same meridian. He then estimated the distance between them based on average sailing time for ships. The figure he arrived at for the circumference of the earth was 18,000 miles. He also greatly overestimated the west-to-east distance from the westernmost part of Europe to the eastern end of the ekumene, then thought to be occupied by India. He declared, therefore, that a ship sailing westward across the Atlantic from western Europe would reach the east coast of India after a voyage of only 7000 miles. As we will see, when Columbus argued before the scholars at the Spanish court that he could sail west to India, he used the smaller circumference estimated by Posidonius, whereas the scholars favored the larger circumference calculated by Eratosthenes.

Posidonius, however, was right about another matter. He refused to follow Aristotle in believing that the equatorial part of the torrid zone was uninhabitable because of heat. The highest temperatures and the driest deserts, he insisted, were located in the temperate zone near the tropics and the temperatures near the equator were much less extreme. He arrived at this conclusion—amazing in the first century B.C.—on purely theoretical grounds, for he had no access to credible reports from anyone who had crossed the Sahara, including Hanno's voyage along the African west coast. The sun, he pointed out, pauses longest near the tropics and is overhead for a much shorter time at the equator. The interesting point is that Posidonius's incorrect estimate of the circumference of the earth was widely accepted by those who followed him, while his correct belief concerning the habitability of the equatorial regions was overlooked.

Strabo

A very large part of what scholars think they know about ancient geography came from Strabo. Most of the books written by earlier scholars have disappeared entirely or survive only in fragments. Much of the history of geographical ideas in ancient Greece and Rome must be pieced together from surviving cross-references. But Strabo's monumental work on geography was found almost intact, with only a very few minor parts missing; fortunately, the first part of Strabo's writing is a review of what other geographers since Homer had done. Strabo's work was another example of what has become almost standard practice—the proclamation by an author that what he has produced is, indeed, the "new geography." Here is the way Strabo outlined his task:

> Accordingly, just as the man who measures the earth gets his principles from the astronomer and the astronomer his from the physicist, so, too, the geographer must in the same way take his point of departure from the man who has measured the earth as a whole, having confidence in him and in those in whom he, in his turn, had confidence,

and then explain, in the first instance, our inhabited world—its size, shape, and character, and its relations to the earth as a whole; for this is the peculiar task of the geographer. Then, secondly, he must discuss in a fitting manner the several parts of the inhabited world, both land and sea, noting in passing wherein the subject has been treated inadequately by those of our predecessors whom we have believed to be the best authorities on these matters [Strabo, trans. Jones, 1917:429–431].

Strabo was born in Amasia (in what is today central Turkey, some 50 miles south of the Black Sea coast) about the year 64 B.C. He died in A.D. 20. His family was sufficiently well-to-do so that Strabo received a good education and was able to travel widely in the Greek world. He lived for several years in Rome and also worked in the library at Alexandria. His travels took him no farther west than Italy and no farther east than the borders of Armenia. He had sailed on the Black Sea; and, while he was at Alexandria, he made a trip up the Nile (in 24 B.C.) as far as Philae, a short distance above the first cataract. He wrote two major works after he returned to Amasia. One was a history from the fall of Carthage to the death of Caesar, of which only a few fragments have been found. It is possible that had this book survived, Strabo would be recognized as an historian rather than a geographer. But his *Geography*—almost all of the seventeen books—did survive.

Strabo's *Geography* is compiled from the writings of his predecessors. He defends Homer's knowledge of geography at great length but then discards Herodotus as a "fable-monger." He also discards Hanno's voyage along the western side of Africa and Pytheas's exploration of northwest Europe. He accepts Aristotle's zones of habitability, as defined by Eratosthenes, and then goes on to assert that the limit of possible human life toward the equator is a latitude 12°30′N—on what basis he does not say. He also places the northern limit of the habitable earth, where cold is the limiting factor, only 400 miles north of the Black Sea. No one can really be civilized if he lives north of the Alps in Europe because it is necessary to huddle around fires just to keep alive. He accepts the calculation of the earth's circumference made by Posidonius. On the other hand, Strabo gives a correct explanation of the floods of the Nile, attributing them to the heavy summer rains in Ethiopia.

Strabo wrote for a specific group of readers: for the educated statesman and the military commanders. His purpose was to provide a text for the information of Roman administrators and military commanders, and his work constitutes the world's first administrator's handbook. He is very critical of those geographers who try to copy Aristotle in the search for explanations; rather, he wants to provide an accurate description of the parts of the ekumene. The rest of the world does not interest him at all. He recognizes the geographer's need for a sound mathematical basis, and he derives this chiefly from Hipparchus and Posidonius. The major part of his work is devoted to detailed descriptions of the various parts of the known world. After two books of introductory material, including the discussion of his sources, Strabo devotes eight books to Europe, six books to Asia, and one book to what we would today call Africa. Most of this book deals with Egypt and Ethiopia; after completing this coverage he then says, "Now let me describe Libya, which is the only part left for the completion of my Geography as a whole" (Strabo, trans. Jones 1917:155). Clearly he is not much interested in any abstract argument about the placing of regional boundaries, but he accepts without discussion the idea of Herodotus that Libya begins west of the Nile Valley.

It is interesting that many centuries passed before Strabo's *Geography* was actually read. When Pliny the Elder wrote his encyclopedia of geography in A.D. 77, based on the reading of some 2000 volumes, he did not even mention Strabo. The administrators who might have benefited from the work never saw it. However, by the sixth century after Christ, Strabo's *Geography* had been "discovered" and had become a classic, as indeed it remained for many centuries thereafter.

ROMAN GEOGRAPHY

Unlike the Greeks, the Romans produced little that was new in the field of geography. Writing shortly before the time of Christ, one Marcus Terentius Varro wrote a compendium of geography that would scarcely have merited the adjective, new, had he not set forth a theory of culture stages that remained almost unchallenged until the nineteenth century. Varro describes man's culture as progressing through a regular sequence. Originally man derived his food from the things that the virgin earth produced spontaneously. From this original state man advanced through a stage of pastoral nomadism, then through an agricultural stage, and finally to the stage of contemporary (first century B.C.) culture (Glacken, 1956:72–73). Varro's stages were generally accepted until the nineteenth century when Alexander von Humboldt pointed out that there had been no pastoral stage in the Americas and that the theory of stages could not be applied everywhere.

There were other compendia of descriptive writings. Pomponius Mela, writing in A.D. 43, produced such a work; and he was widely quoted in the much larger encyclopedic collection of Pliny the Elder. There were also extensive sailing directions published for the guidance of ship captains that described the coastlines and ports with considerable detail and accuracy, such as the *Periplus of Scylax* for the shores of the Mediterranean and the *Periplus of Arrian* for the shores of the Euxine (Black Sea) (Bunbury, 1883:2: 384). The most complete work of this kind was an anonymous one that offered a guide for navigators and traders that covered the Red Sea, the east coast of Africa as far as Zanzibar (more than six degrees south of the equator), and the northern side of the Indian Ocean as far as the southern end of the Malabar Coast in India. This was the famous *Periplus of the Erythraean Sea,* which Bunbury dates as some ten years after the death of Pliny in A.D. 79. (Bunbury, 1883:2: 443–479). The merchants and sailors of the first century after Christ, who had not read Aristotle or Strabo, were happily not conscious of the horrible fate that would come to those who ventured within twelve degrees of the equator or of the impossibility of maintaining life in this central part of the torrid zone. At Zanzibar they carried on a flourishing trade with the inhabitants of the African mainland.

Ptolemy

Ancient geography really came to an end with the monumental work of Ptolemy (Claudius Ptolemaus), who lived in the second century after Christ. Nothing is known of the man's life except that he worked at the library in Alexandria between A.D. 127 and 150. He is the author of the great work on classical astronomy—the *Almagest*—which long remained the standard reference work on the movements of celestial bodies. His concept of the universe agreed with that of Aristotle: the earth was a sphere that remained stationary in the center while the celestial bodies moved around it in circular courses. This remained accepted doctrine until the time of Copernicus in the seventeenth century.

After completing the *Almagest,* Ptolemy undertook the preparation of a *Guide to Geography.* His teacher, Marinus of Tyre, had already started a collection of data regarding place locations on the basis of which the maps of the known world were to be revised. By this time, in the second century, much new information had been collected by the far-ranging Roman merchants and armies. Ptolemy went on with the work Marinus had started. He adopted the grid of latitude and longitude lines developed by Hipparchus, based on the division of the circle into 360 parts. Every place, therefore, could be given a precise location in mathematical terms. Ptolemy's guide contains some six volumes of tables and forms the world's first geographical gazetteer, on the basis of which he revised the world map. The difficulty is that in spite of the appearance of precision, the work really was a monumental collection of errors. In those days latitude could be determined only approximately, and most voyagers failed to make use of the few instruments available. There was no way to measure longitude. Therefore, each listing of latitude and longitude was, in fact, only a selection among estimates. Furthermore, Ptolemy followed Marinus in taking as his prime meridian a north-south line through the westernmost known islands in the Atlantic—either the Canaries or the Madeira Islands. Therefore, starting in the west with his estimates of longitude, the error accumulated toward the east. Ptolemy not only accepted the smaller estimate of the earth's circumference by Posidonius, but he also increased the error in the eastward extension of the land area. Using the authoritative work of Ptolemy, Columbus figured that Asia must lie very close to Europe on the west.

The *Guide to Geography* consisted of eight volumes. The first was a discussion of map projections together with a few corrections of the data from Marinus based on actual astronomical observations that he had carried out himself. Books 2 through 7 contained tables of latitude and longitude. The eighth book contained maps of different parts of the world based on the gazetteer (Fig. 2). Ptolemy repeated the commonly accepted idea that the parts of the earth near the equator were uninhabitable because of heat. He also indicated on his maps that the Indian Ocean was enclosed by land on the south, an idea he probably took from Hipparchus, but where Hipparchus found this information cannot be discovered. This *terra australis incognita* was not cleared from the maps until the voyages of Captain James Cook in the eighteenth century proved that such a southern land area did not exist.

   With the death of Ptolemy the geographic horizons that had been widened both physically and intellectually by the Greeks closed in again. It was many centuries before the effort to describe and explain the face of the earth as the home of man again attracted the attention of scholars.

CHAPTER 3

Geography in the Middle Ages

Afterwards they went to the country of Paris, to King Fransis (i.e. Philippe IV, le Bel). And the king sent out a large company of men to meet them, and they brought them into the city with great honor and ceremony. Now the territories of the French king were in extent more than a month's journey. And the king of France assigned to Rabban Sauma a place wherein to dwell, and three days later sent one of his amirs to him and summoned him to his presence. And when he had come the king stood up before him and paid him honor, and said unto him, "Why hast thou come? And who sent thee?" And Rabban Sauma said unto him, "King Arghun and the Catholicus of the East have sent me concerning the matter of Jerusalem."

During the fifth century after Christ the Roman world with its system of centralized administration fell apart. For the Greeks the geographic horizons (the limits of the area that was known at least to scholars and merchants) had been extended from the Indus River to the Atlantic and from the Russian steppes north of the Black Sea to Ethiopia. For the Romans the geographic horizons included the vast area brought under Roman jurisdiction. But now geographic horizons closed in again until many of those who lived in Christian Europe after the fifth century were really familiar only with their immediate surroundings. The worlds beyond were peopled with fantastic creatures conjured up by imaginations unfettered by facts.

The report above of the discovery of France is taken from a Chinese account of the visit of the Nestorian Christian Rabban Bar Sauma of Peking to the French king in 1287. *The Great Chinese Travelers,* ed. J. Mirsky (New York: Pantheon Books, 1964), p. 190.

Only in the shelter of monasteries were the flickering flames of the intellectual life preserved.

But this is not a complete picture of the medieval period, which extended from the fifth to the fifteenth centuries. In Christian Europe, although the word, geography, disappeared from the ordinary vocabulary, the study and writing of geography did not entirely cease (Tillman, 1971). Little by little curiosity concerning other possible worlds that might lie beyond the horizon again prompted some adventuresome people to travel and explore; and the crusades, organized to wrest the Holy Land from the control of the Muslims, took many people out of their localities and then brought them back again to tell of the strange people and landscapes that had been seen. From the thirteenth century on there were extended travels by missionaries and merchants that reached all the way to China.

And we must ask the question about this period: Who was discovering whom? Although the geographic horizons had closed in around the communities of Christian Europe, this was a period of greatly widened horizons for the Muslims. Not only did the Muslim conquests, which started with the conquest of Palestine and Syria in 632, carry the followers of Islam eastward to the islands of Southeast Asia, westward to the Atlantic and into Europe, and southward across the Sahara but also Muslim missionaries and merchants traveled far beyond the limits of Muslim control. Furthermore, Muslim scholars in the great centers of learning were busily engaged in translating the works of the Greek writers into Arabic. It was through the Arabic that Greek learning eventually became known to the Latin world of the Christians.

Meanwhile, in the far north of Europe the intrepid Norsemen were sailing across the stormy North Atlantic to Iceland, Greenland, and the continent of North America. Since the Norsemen did not write books, news of these discoveries was a long time getting back to the rest of the world.

Of the greatest importance were the accomplishments of the Chinese. Actually Europe and India were "discovered" by Chinese missionaries long before the Christian travelers reached the Orient. According to Joseph Needham (Needham, 1963:117), in the period between the second century before Christ and the fifteenth century after Christ, the Chinese culture "was the most efficient in the world in applying knowledge of nature to useful purposes." The study of geography in China, as a part of a wider scholarly tradition, was well advanced beyond anything known in Christian Europe at this time.

GEOGRAPHY IN THE CHRISTIAN WORLD

The scholars who gathered together in monasteries in Christian Europe were not studying the earth as observers or experimenters. Rather, they were compilers of information from documentary sources and commentators, whose primary effort was to reconcile the geographic ideas recorded in documents with the authority of the Scriptures, especially the Book of Genesis. In the early medieval period the European scholars could work only with Latin documents and only in the latter part of this period did a few of them master the Arabic language. The Greek materials remained entirely unknown, except in translation.

John K. Wright, in his masterly study of the geographical ideas available to the Christian scholars of this period, points out the kinds of information that could be found in Latin (Wright, 1925:88–126). The Roman geographers, such as Pomponius Mela and Pliny the Elder, were widely used sources. Both of these writers as we have seen compiled their books from Greek sources, and through them the medieval scholars had a kind of secondhand and quite incomplete access to Greek concepts (Kimble, 1938). Two medieval scholars—Martianus Capella and Ambrosius Theodosius Macrobius—provided translations of Plato as early as the fifth century. Through the writings of Capella and Macrobius the medieval Christian scholars had access to the concept of a spherical earth. Although there were many, like Cosmas, who conceived of the world as a round disc rather than as a sphere, there were always a few scholars who accepted the idea of a spherical earth as demonstrated beyond dispute.

Ptolemy became the major authority in the medieval Christian world for matters pertaining to astrology and astronomy. His work dealing with the effect of the positions of the celestial bodies on human affairs—the *Quadripartitum*—was translated from the Arabic into Latin by Plato of Tivoli in 1138; and his *Almagest*—the great work on astronomy—was made available in Latin by Gerard of Cremona in 1175 (Kimble, 1938:75–76). As a result, Ptolemy's geocentric model of the celestial universe remained the accepted model for many centuries, and most of the ideas still used by astrologers can be traced back to him.

The geographical ideas of Aristotle were first made available in Christian Europe by translation from the Arabic in the twelfth century. The first medieval writer to make use of Aristotle was Albertus Magnus (Tillman, 1971), whose book on the nature of places combined astrology with environmental determinism. The Greek theory of equating habitability with latitude became strongly implanted in medieval writings. Albertus even went beyond the Greeks: from them he accepted the idea that people who lived too close to the limits of the habitable earth turned black, but then he insisted that if black-skinned people should move into the temperate latitudes they would gradually turn white (Glacken, 1967:265–271).

There was no really good way to evaluate the conflicting ideas that these translations from the Arabic made available. Furthermore, it was almost impossible to trace the sources of the ideas since in those times it was standard practice to include in one's own writings whole passages taken verbatim from earlier writers without any kind of credit. Isidore of Seville, who compiled a sort of geographic encyclopedia during the seventh century, took long passages from Solinus,[1] who, in

---

[1] Solinus was the first to describe these seas as "mediterranean" (in the midst of the land), and Isidore was the first to use the descriptive term as a proper name (Wright, 1925:307).

turn, had taken them from Pliny. When the medieval scholars did seek explanations for natural events, the kinds of events with which they were concerned were spectacular ones, such as earthquakes, volcanic eruptions, or floods. No hypotheses from the Greeks had been presented in Latin concerning the slower and less obvious natural processes, such as the erosion of mountains or the building of deltas, and, in the absence of a background of theory, these slower processes were not perceived.

Another characteristic of this period in Christian Europe was the deterioration of mapping. The once fairly accurate delineations of the better known coastlines were lost, and instead maps became pure fancy. This was the period of the so-called T-O maps. The inhabited world was represented by a circular figure surrounded by the ocean. The figure was "oriented" toward the east (Wright, 1925:66–68). In the midst of the land area was a T-shaped arrangement of water bodies. The stem of the

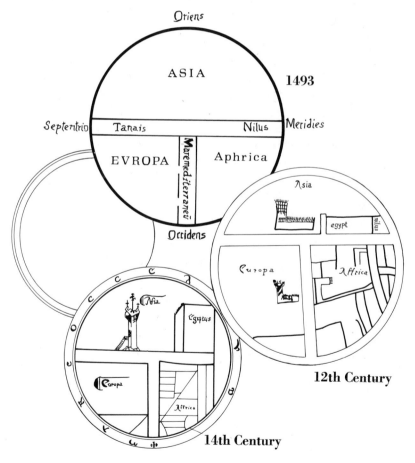

Figure 5. T-O maps.

T represented the Mediterranean. The top of the T represented the Aegean and Black Seas on the one hand, and the Nile River and Red Sea on the other. The three divisions—Europe, Asia, and Africa—were accepted as standard. The center of the inhabited world, just above the center of the T, was Jerusalem. At the far east, beyond the limit of the inhabited world, was paradise (Fig. 5).

Medieval Christian Travelers

Meanwhile, outside the monasteries there were Christians who did travel and make observations but who had no knowledge of the existence of theoretical concepts regarding the nature of the earth as the home of man. In A.D. 326, Helena, the mother of the Emperor Constantine, made one of the earliest pilgrimages from Rome to the Holy Land. Silvia of Aquitaine, a Roman lady, was one of the earliest woman geographers. She traveled overland to Jerusalem and then on to Egypt, Arabia, and Mesopotamia; eventually, she wrote an account of her travels. As the number of pilgrims increased, itineraries were compiled to guide them on the routes to Jerusalem (Beazley, 1897–1906/1949).

By the eleventh century the passage of pilgrims overland through what is today Turkey and Syria had become more and more difficult and dangerous. As a result the Christians of Europe organized a series of military invasions of the Holy Land. Between 1096 and 1270 there were eight separate crusades, each with the objective of recapturing the Holy Sepulcher at Jerusalem from the Muslims. Some went by sea, some by land; one crusade was even successful in occupying Jerusalem for a short period before the Muslims drove out the invaders. After the eighth formal crusade, there were other military invasions of Muslim-held territory, one of which in 1365 sacked Alexandria and burned the famous library, where Eratosthenes and other Greek geographers had worked.[2]

The crusades had a major impact not only on Christian Europe but also on the Muslims. From almost all parts of Europe men had been recruited for the war against the infidels and had made the trip to the Holy Land. When they returned to Europe, they not only brought with them many new kinds of machines—such as the windmills, later adopted by the Dutch for pumping water—but also exciting stories about strange people and strange landscapes beyond the geographic horizons. The result was a great stimulation of interest in the description of unfamiliar places, which is still preserved in our library catalogues as "popular description and travel." For people who knew nothing about geographic theory, popular description and travel became, in essence, geography. Meanwhile, the Muslims, who at first were notably tolerant of people of other faiths, reacted to the violence of the

---

[2]It is believed, however, that the collection of manuscripts that constituted the major record of Greek geography had long since been ruined due to lack of care even before the destruction of the library by the Christian invaders.

crusaders by becoming aggressively intolerant of unbelievers. One result was the closing of the routes across North Africa—Southwest Asia by which the merchants of Venice and Genoa could make contact with the traders of the East.

Marco Polo

Yet in spite of the blocking of the eastern sea routes, Christian Europe did, in fact, make contact by land with the centers of Chinese culture by following a route to the north of the main Muslim strongholds. The route was followed both by missionaries sent out from Rome and by merchants. The most celebrated of these travelers were the Polo brothers and the son of one of them, Marco Polo. In 1271,

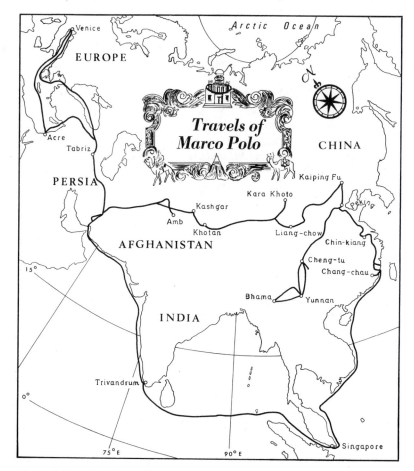

Figure 6. Travels of Marco Polo.

when Marco was seventeen, he started out from Venice with his father and uncle to make the long journey to China (Fig. 6). The Polo brothers had already visited China on a trip that lasted from 1260 to 1269; and the Great Khan, the Mongol emperor of China, had invited them to return. The return journey to China took four years, and the Polos remained there for seventeen years. Marco served the Khan as ambassador to various parts of China and in various other official capacities, as a result of which he was able to gain intimate knowledge of Chinese culture. In fact, the Polos were so useful to the Khan that he was reluctant to permit their departure. Finally, in 1292 the Khan provided the Polos with a fleet of fourteen large ships, some so large that they required a crew of more than a hundred sailors. Along with the Polos there were some 600 other passengers. The fleet set sail from a port in southern China, probably the modern Lungch'i, and took three months to reach Java and Sumatra, where it was held up for five months. The expedition then continued to Ceylon and southern India and thence along the west coast to the ancient port of Hormuz on the Persian Gulf. Of the 600 passengers, only 18 survived the voyage. Most of the ships were lost. But the Polos finally returned safely to Venice in 1295 after an absence of twenty-five years.

Marco Polo, while being held a prisoner in Genoa some years later, dictated his book of travels to a fellow prisoner (Polo, 1930). His descriptions of life in China and of the perils encountered on the route to and from China were vivid—so vivid, in fact, that they were commonly regarded as the products of a heated imagination. In addition to descriptions of the places he actually visited, he included reports on Cipangu, or Japan, and on the island of Madagascar, which, he said, was near the southern limit of the habitable earth. Since Madagascar is well south of the equator, here was abundant evidence of the fact that the torrid zone was not torrid and was in fact inhabited.

It is important, however, that Marco Polo was not a geographer and had no knowledge of the existence of such a field of learning. Nor was he aware of the major disputes then going on (1) among those who believed in a torrid zone that was not habitable and those who disagreed with this notion or (2) among those who accepted the smaller estimate of the earth's circumference derived from Posidonius, Marinus, and Ptolemy and those who preferred the larger figure of Eratosthenes. Nor was Marco Polo aware that the Greek geographers thought that the eastern end of the ekumene was near the mouth of the Ganges. Nor was he aware that Ptolemy had said that the Indian Ocean was enclosed by land to the south. It is doubtful if Marco ever thought of measuring the latitude and certainly not the longitude of the places he visited, but he did report that to reach a place required a journey of a certain number of days in such and such a direction. He makes no comments concerning previous geographic ideas. Today we can see that his book must stand among the great records of geographic exploration; yet in medieval Europe it seemed much like many other books of the time, filled with wild but interesting stories. Columbus had a copy of it, marked with notations in his own writing.

Brighter Spots in Medieval Scholarship

Toward the end of the medieval period of Christian Europe, there were a few scholars who began to insist on the need to confront authority with reason. If God gave man the gift of reason, they insisted, there could be no excuse for refusing the use of it. William of Conches, who died about 1150, was one of the earliest to portray a universe governed by law rather than by the unpredictable acts of a divine authority (Kimble, 1938:79). He presented some remarkably modern ideas concerning the heating of the atmosphere from below and the formation of clouds by the cooling of air. Robert Grosseteste, the Bishop of Lincoln, was one of the earliest Christian scholars to master the Arabic language and who, therefore, had access to a much wider range of geographic materials than those written in Latin. As the teacher of Roger Bacon, he was at least in part responsible for confronting the notion of a torrid zone that was uninhabitable with Arabic reports on an inhabited east coast of Africa extending at least as far as 20°S (Wright, 1925:163–165).

Cardinal Pierre d'Ailly, writing in the early fifteenth century, is one of the later medieval scholars who had a major influence on the age that followed. Although he derived his material chiefly from Latin sources, his book *Tractatus de imago mundi* did represent a kind of summary of the work of the period. In a second edition of his book in 1414, he is one of the first to make use of the Latin translation of Ptolemy's *Geography* (published in 1406). He repeats the different opinions concerning the habitability of the torrid zone but without taking any stand on the matter. But he does dispute Ptolemy's idea of an enclosed Indian Ocean. He quotes numerous reports that indicate the existence of an open ocean around southern Africa, and this had great influence on the Portuguese geographers and navigators, who soon thereafter began to seek a way to India that could avoid the Arabic territory. D'Ailly also accepts the smaller estimate of the earth's circumference, and he was among the first to insist that India could be reached by sailing west, which was influential in building up Columbus's determination to do just this (Kimble, 1938:208–211). It is important, also, that the invention of movable type about the middle of the fifteenth century made possible the publication of books in large editions. The works of Pierre d'Ailly were popular, as were all the geographic writings then available in manuscript form.

Another Christian geographer, whose writings belong in this period, was Pope Pius II (Aeneas Silvius). While he was Pope between 1458 and 1464, he wrote a book on Europe and Asia in which he suggests the possibility that the torrid zone is inhabited and in which he agrees with Pierre d'Ailly that there is abundant evidence that the Indian Ocean is not enclosed on the south, as reported by Ptolemy.

Navigation and Cartography

During this period there were also several important advances in the arts of navigation at sea. Some of these new skills were first developed, as we shall see, at

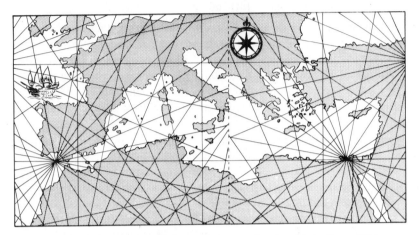

Figure 7. A Portolano chart of the Mediterranean, after Juan de la Cosa, 1500.

the University in Palermo in Sicily, where the Norman King Roger II and the Muslim geographer Edrisi were beginning to learn how to navigate away from the land. The first mention of the magnetic compass occurs in Christian Europe in the writings of Alexander Neckam about 1187. The earliest Arabic reference to a compass was in 1230. Yet there is some evidence that this instrument was in use much earlier and was perhaps independently invented by the Vikings (Kimble, 1938:223). It was certainly in wide use by the fifteenth century and had become indispensable for long voyages away from the land. In this period, also, the astrolabe had been improved, and it came into common use as an aid to navigation by making possible a more accurate fix on the altitude of the polestar (Kimble, 1938:223–225).

The late fourteenth century also witnessed a notable improvement in the art of mapping. The Portolano charts, of which the earliest is about 1300, became standard equipment for sea captains. The term Portolano means handy or easily available. Instead of a grid of latitude and longitude, the Portolano charts were covered by a network of overlapping lines radiating from several centers in different parts of the chart (Fig. 7). The radiating lines conform to the eight or sixteen principal directions of the compass, each corresponding to a wind direction (Bagrow and Skelton, 1964:62–66). Sailors laid out compass courses along these lines (Taylor, 1957:112–114). With the lines to indicate directions from key points, the coastlines, especially around the Mediterranean, were drawn with considerable accuracy.[3]

[3] With the use of the magnetic compass, it became the usual practice to draw maps with north at the top. When a map is laid on the ground so that directions on the map correspond with compass directions, the map is still said to be "oriented." In the year 800 Charlemagne, hoping to end the confusion about directions, decreed that henceforth in France there would be only four cardinal directions: north, east, south, and west. North was indicated with the *fleur de lis,* as it still is.

The famous Catalan map of the world, made in 1375, incorporates the material from numerous Portolano charts. It also includes the west coast of Africa to the south of Cape Bojador, which had not then been reached by European sailors. It also shows east and southeast Asia based on reports by Marco Polo. This was the first map ever to give a proper outline to Ceylon and the Indian peninsula (Kimble, 1938:193). But the scholars in the monasteries, who were still gathering their information from written documents, had little connection with the map-makers, whose purposes were strictly practical rather than theoretical. The map-makers were working for the merchants, and perhaps they were not even aware of the differences of opinion regarding such questions as the habitability of the torrid zone or even the existence of a torrid zone.

GEOGRAPHY IN THE MUSLIM WORLD

One of the events of far-reaching importance in the medieval period was the spread of the Muslims. Muhammad, the prophet who died in 632, was the founder of the religion of Islam, whose followers are known as Muslims. These Arabic-speaking people from Arabia were previously grouped in small, isolated tribes and had no feeling of unity. They were given a common purpose, if not complete unity, by the teaching of the prophet and by the holy book—the Koran. This was the first book written in the Arabic language. It provided not only a religious orientation but also gave detailed prescriptions concerning all aspects of life—how to govern, how to carry on commercial transactions (including a prohibition against the payment of interest on loans), how to organize the family life, and many other matters. The Koran describes the world in detail, providing explanations for natural phenomena that all "true believers" accept without question.

The followers of Islam embarked on a conquest of the world outside of Arabia. In 641 they conquered Persia and in 642 took control of Egypt. The Muslims swept westward across the Sahara and by 732 all of the Great Desert was under Muslim control. They crossed through the Iberian Peninsula into France and were defeated and turned back only in the Battle of Tours (732). For some nine centuries the Muslims ruled most of Spain and Portugal. Muslim rule was also extended eastward into India and eventually to some of the islands of Southeast Asia.

Baghdad

In 762 the Muslims founded the new city of Baghdad near the ruins of Babylon, and for more than a century Baghdad was the center of the intellectual world. With the patronage of the Caliph Harun al-Rashid a project was started for the translation of the works of the Greek philosophers and scholars into Arabic. The project was continued under Caliph al-Mamun (813–833), who employed learned men of all faiths to make the translations. Books were collected from all available sources, and the translators were paid the weight of their books in gold (Ahmad, 1947:5). From Baghdad, therefore, a flood of new ideas from varied sources began to spread throughout the Muslim world. Eventually, the innovations were brought into Christian Europe as a result of Latin translations from the Arabic. Among other innovations was the use of the decimal system in arithmetic, which was brought into Baghdad from the Hindus, who had adopted it from the Chinese.

Al-Mamun directed his scholars to recalculate the circumference of the earth. They made use of the same method devised by Eratosthenes some ten centuries before. On the level plain of the Euphrates they established a north-south line and fixed the latitude at either end by observations of the stars. They then measured the distance between the fixed points and decided that the length of a degree was 56⅔ Arabic miles. The scholars made several other measurements, one near Palmyra in Syria, and arrived at almost the same results. These values were much too small, owing to errors in the linear measurements (Wright, 1925:395).

Muslim Contributions to Climatology and Geomorphology

The Arabic geographical writings in the period between 800 and 1400 were based on a much greater variety of sources than were those of Christian scholars in the same period. The Muslims had access not only to their translations from the Greek but also to the reports of their own travelers. As a result they had a much more accurate knowledge about the world than the Christian scholars had. One of the earliest of the great Arab travelers was ibn-Haukal, who spent the last thirty years of his life between 943 and 973 visiting some of the most remote parts of Africa and Asia. On his voyage along the African east coast to a point some twenty degrees south of the equator, he observed the fact that people in considerable numbers were living in those latitudes that the Greeks thought to be uninhabitable Yet the Greek theory persisted and keeps appearing in different form again and again, even in modern times.

The Arabic scholars made some important observations regarding climate. In 921 al-Balkhi gathered the observations of climatic features made by Arab travelers in the world's first climatic atlas—the *Kitab al-Ashkal*. Al-Masudi, who died about 956, had gone south as far as modern-day Mozambique and wrote a very good description of the monsoons. He described the evaporation of moisture from water surfaces and the condensation of the moisture in the form of clouds—this, in the tenth century. In 985 al-Maqdisi offered a new division of the world into fourteen climatic regions. He recognized that climate varied not only by latitude but also by position east and west. He also presented the idea that the southern hemisphere was mostly open ocean and that most of the world's land area was in the northern hemisphere.

Two Arab geographers offered important observations regarding the processes shaping the world's landforms. Al-Biruni wrote his great geography of India (*Kitab al-Hind*) in 1030. In this book he recognized the significance of the rounded stones he found in the alluvial deposits south of the Himalayas. The stones became rounded, he pointed out, as they were rolled along in the torrential mountain streams. Furthermore, he recognized that the alluvial material dropped close to the mountains was relatively coarse in texture and that alluvium became finer in texture farther away from the mountains. He quotes the Hindus as believing that the tides are caused by the moon. He also includes the interesting observation that toward the south pole night ceases to exist—which suggests that some explorers had voyaged to the far south before the eleventh century.

The other contributor to a knowledge of landforms was Avicenna, or ibn-Sina, who took advantage of his opportunities to observe mountain streams in the act of cutting down their valleys in the mountains of central Asia. He formulated the idea that mountains were being constantly worn down by streams and that the highest peaks occurred where the rocks were especially resistant to erosion. Mountains are raised up, he pointed out, and are immediately exposed to this process of wearing down, a process that goes on slowly but steadily. It was eight centuries later before James Hutton presented similar ideas concerning the process of erosion; he had never heard of Avicenna and could not read Arabic. Avicenna also noted the presence of fossils in the rocks in high mountains, which he interpreted as examples of nature's effort to create living plants or animals that had ended in failure.

Edrisi and Palermo

The most extensive corrections of the erroneous ideas handed down from Ptolemy were made by the Muslim geographer Edrisi, or al-Idrisi. Educated at the University of Cordoba in Spain, Edrisi was one of the scholars that Roger II of Sicily brought to Palermo. King Roger dispatched observers to many parts of the world where Edrisi said there were uncertainties concerning the actual arrangement of mountains, rivers, or coastlines. The observers brought back much new information to Palermo. As a result Edrisi was able to write a "new geography" that was really new. In 1154 he completed a book with the title *Amusement for Him Who Desires to Travel Around the World.* He corrects the idea of an enclosed Indian Ocean and the idea of the Caspian Sea as a gulf of the world ocean. He also corrects the courses of numerous rivers, including the Danube and the Niger, and the position of several major mountain ranges. As Kimble points out, it is strange that such an important book was not translated into Latin until 1619, at which time the translator did not even know the name of the author (Kimble, 1938:59; Ahmad, 1947:39).

Other important innovations were made at Palermo. Improvements were made in the methods of navigation, including the wide use of coast charts, which were the forerunners of the Portolano charts of the fourteenth century. It is said that the sailors of Genoa learned the arts of navigation from the Sicilians and that the Genoese passed on this knowledge to the Portuguese at Sagres in the fifteenth century. The first steps leading to the Age of Exploration were taken in Sicily in the eleventh and twelfth centuries (Wright, 1925:81).

Ibn-Batuta

One of the great travelers of all time was the Muslim, ibn-Batuta. He was born at Tangier in 1304 to a family whose members had traditionally served as judges. In 1325, at the age of twenty-one, he set out to make the usual pilgrimage to Mecca, where he proposed to complete his studies of the law. But on the way across North Africa and through Egypt he found himself fascinated more by the people and lands he passed through than by the law. After reaching Mecca he decided to devote himself to travel, and, in his comings and goings through Muslim territory, he carefully avoided following the same route twice. His travels took him to many parts of Arabia never before visited by one person. He sailed along the Red Sea, visited Ethiopia, and then continued southward along the coast of East Africa as far as Kilwa, nearly 10°S of the equator. At Kilwa he learned of the Arab trading post at Sofala in Mozambique, south of the modern port of Beira and more than twenty degrees south of the equator. Ibn-Batuta confirmed what ibn-Haukal had implied— that the torrid zone in East Africa was not torrid and that it was occupied by a numerous native population that justified the establishment of Arab trading posts.

After returning to Mecca, he set out again to visit Baghdad and Persia and the land around the Black Sea. He traveled in the Russian steppes and thence eventually to Bukhara and Samarkand. Then he crossed the mountains through Afghanistan into India. He served the Mongol emperor in Delhi for several years, traveling widely in India during this time. The emperor appointed him as ambassador to China. But delays kept him from reaching China for several more years, during which time he visited the Maldive Islands, Ceylon, Sumatra, and, eventually, China. In 1350 he returned to Fez, the capital of Morocco. But his travels did not end. He made a trip into Spain and then crossed the Sahara to Timbuktu on the Niger River, gathering important information about the Muslim Negro tribes living in that part of the world. In 1353 he settled in Fez, where at the Sultan's command he dictated a lengthy account of his travels (ibn-Batuta, 1958). During some thirty years he covered a linear distance of about 75,000 miles, which in the fourteenth century was a world record. Unfortunately, his book, written in Arabic, made little impact on the Christian world. Even today, when some of our schools teach children about the intense heat of the torrid zone, reference could be made to ibn-Batuta, who, six centuries ago, pointed out that the climate along the equator was less extreme than the climate in the so-called temperate zone in North Africa.

Ibn-Khaldun

The last of the Muslim scholars to make a major contribution to geography was
the great Arabic historian, ibn-Khaldun. Like ibn-Batuta he was born on the
Mediterranean coast of northwest Africa. He lived most of his life in the cities of
what is today Algeria and Tunisia and also for a time in the Muslim part of Spain.
He spent his later years in Egypt. In 1377 when he was forty-five he completed a
voluminous introduction to his world history—known as the *Muqaddimah*. This
work begins with a discussion of man's physical environment and its influence and
with man's characteristics that are related to his culture or way of living rather than
to the environment. He discusses various stages of social organization, identifying
the desert nomad as the most primitive and purest, and he suggests that the seden-
tary city dweller is dependent on luxuries and becomes morally soft. He discusses
the forms of government, describing a sequence of stages that mark the rise of a
dynasty to power, followed by its decline through corruption to its fall. Ibn-
Khaldun discusses cities and their proper location. Finally, he discusses the various
ways of making a living—commerce, the crafts, the sciences—all of which are
shown as both conditions and consequences of urban life. Many of the ideas he
develops in his effort to provide a theoretical model of national growth and national
decay were ideas that appeared later in nineteenth-century Europe. Yet only recently
have the writings of ibn-Khaldun appeared in English (ibn-Khaldun, 1958).

Although ibn-Khaldun, according to Kimble, "may be considered to have
discovered—as he himself claimed—the true scope and nature of geographic inquiry," the fact remains that his knowledge of the physical earth is based largely on
Greek theory; and his ideas about environmental influence are not highly sophisticated (Kimble, 1938:180). He accepts the traditional seven zones of climate running
parallel to the equator. Strangely, for an Arabic scholar, he repeats the idea concerning an uninhabitable zone along the equator and an uninhabitable polar zone. He
repeats the old idea that people turn black when they live too close to the sun and
that when black people move to the temperate zone they gradually turn white or
produce white children. The physical environment impressed its characteristics on
people in many subtle ways. Such naïve environmentalism, however, is somewhat modified by the recognition of different cultural traditions.

It can be said that ibn-Khaldun was the first scholar to turn his attention
specifically to man-environment relations.

GEOGRAPHY IN THE SCANDINAVIAN WORLD

The Scandinavians had never heard of Aristotle, or Strabo, or Ptolemy, or Isidore, or ibn-Khaldun. They had no idea that the lands they inhabited were
uninhabitable. Among the Scandinavians, the Swedes sent exploring parties far to
the east into what is now the central part of Russia. In the ninth century Othar of

Helgoland sailed in a Viking ship around the northern tip of Norway and far eastward into the White Sea.

But the greatest accomplishment of the Scandinavians from Norway—the Vikings—was the crossing of the North Atlantic Ocean to the American mainland. In 874 the Vikings reached Iceland and established a settlement there that grew and prospered. In 930 the world's first parliament was organized in Iceland.

Among the people of the Iceland colony was one especially violent and disturbing character named Eric the Red. In 982 he and his family and retainers were banished. Having learned of the existence of land farther west, Eric set sail across the stormy North Atlantic and came upon the southern part of Greenland. In fact the name, Greenland, which he gave to this new land, was perhaps one of the world's earliest examples of real estate promotion—for there was nothing very green about it. Nevertheless Eric's colony attracted additional settlers from Iceland. Regular voyages were made back and forth between Greenland, Iceland, and the mainland of Norway.

About the year 1000 Eric's son, Leif Ericson, returning to Greenland from a trip to Norway, encountered a severe storm that blew him far off his course. When the skies cleared he found himself off a strange coast that extended as far as he could see to the north and south. He landed and found fine stands of timber as well as vines on which wild grapes grew. Returning to Greenland he described this new land farther to the west.

In the year 1003 a man named Karlsefni organized an expedition to take another look at this new land. He set sail with a crew of 160 people—men and women—together with cattle and food supplies. There is no doubt that he reached the coast of North America. The large bay with a strong current of water coming out of it was probably the St. Lawrence estuary, and somewhere along its shore the party made a landing and spent the winter. Here the first European child was born in the Americas. The next summer the party sailed southward, certainly as far as Nova Scotia, probably as far as Cape Cod, and possibly even as far south as Chesapeake Bay. They liked the land they found, but the Indians proved warlike. Their attacks were such a nuisance that eventually the Vikings gave up the effort to settle on the strange shore and sailed back to Greenland. The story was passed on by word of mouth as *The Saga of Eric the Red*. To this day efforts are still being made to identify the places where Karlsefni and his people landed. It is quite possible that there were other expeditions even before the eleventh century, but geographical scholars in the European world heard only rumors of such voyages for several centuries (Sykes, 1961; Morison, 1963, 1971; Cassidy, 1968; Sauer, 1968).

GEOGRAPHY IN THE CHINESE WORLD

During all the time that the study of the earth as the home of man was being pursued in ancient Greece and Rome and later in Christian Europe, among the Muslims, and by the remote Scandinavians there was another major center of geographic study in the world. This was China. Essentially the European and Chinese worlds remained isolated, each discovering the other step by step. Yet there are certain fascinating parallels in the concepts and methods of study that seem to require the existence of contacts, however indirect and remote.

It is important for students steeped in western history to keep in mind that, from about the second century before Christ until at least the fifteenth century after Christ, the people of China enjoyed the highest standard of living of any people on earth (Needham, 1963:117). The Chinese mathematicians had discovered the use of zero and had developed the decimal system, which was vastly superior to the sexigesimal system of Mesopotamia and Egypt. The decimal system was introduced into Baghdad about 800 from the Hindus; but it is generally believed that the Hindus derived their decimal system from the Chinese.

The Chinese philosophers had a basically different attitude toward the natural world than that held by the Greeks. To the Chinese the individual is not separate from nature—he is a part of nature. There is no law-giving deity who created the universe for man's use in accordance with a preconceived plan. Death in China is not followed by life in a new paradise or punishment in a hell; rather, the individual hopes to be absorbed in the all-pervading universe of which he is an inseparable part. Confucianism developed a way of life that was highly effective in minimizing the frictions among individuals, but it remained relatively indifferent to the development of scientific knowledge.

Joseph Needham repeats the following story to illustrate this attitude:

> When Confucius was travelling in the east, he came upon two boys who were disputing, and he asked them why. One said "I believe that the rising sun is nearer to us and that the midday sun is further away." The other said "On the contrary, I believe that the rising sun and setting sun is further away from us, and that at midday it is nearest." The first replied "The rising sun is as big as a chariot-roof, while at midday the sun is no bigger than a plate. That which is large must be near us, while that which is small must be further away." But the second said "At dawn the sun is cool but at midday it burns, and the hotter it gets the nearer it must be to us." Confucius was unable to solve their problem. So the two boys laughed him to scorn saying "Why do people pretend that you are so learned?" [Needham and Ling, 1959:225–226.]

Consider what Socrates might have done in this situation, and a very fundamental difference in cultural attitude becomes clear. But this does not mean that there was no interest in finding out what it was like beyond the horizon or in developing methods of recording what was discovered. In fact, the record of geographical work in China is impressive, but it is concerned more with observable things and processes and less with the formulation of theory.

Geographical Work

Chinese geographical work was based on the development of methods for making accurate observations and for using these in the construction of useful inventories. For example, there are weather records that date back to thirteen centuries before Christ. The oldest piece of geographical writing is a survey of the resources and products of the nine provinces into which ancient China was divided in the fifth century B.C. For each province the nature of the soil, the kinds of products, and the waterways that provide routes of transportation are described (Needham and Ling, 1959:500). In the second century B.C. the Chinese engineers were making accurate measurements of the silt carried by the rivers. In A.D. 2 the Chinese carried out the world's first census of population. Other technical inventions included the making of paper, the printing of books, the use of rain and snow gauges to measure precipitation, and the use of the magnetic compass for navigation.

There was also a record of progress in the understanding of processes. By the fourth century B.C. the nature of the hydrological cycle was understood. At about the same time that Plato was observing the effects of forest clearing in Attica, the Chinese philosopher Mencius (Meng-tzu), who lived two centuries after Confucius, was pointing out that forests once cleared from mountain slopes could not reseed themselves as long as the slopes were grazed by cattle or goats (Glacken, 1956:70).

The Chinese had also learned much about the work of running water in wearing down mountains and forming alluvial plains. At about the same time that Avicenna was writing down his ideas about the erosion of mountains, the Chinese scholar Shen Kua was presenting the same idea (in 1070). Referring to a rugged mountain range with jagged peaks and steep-sided valleys he wrote:

> Considering the reasons for these shapes, I think that (for centuries) the mountain torrents have rushed down, carrying away all sand and earth, thus leaving the hard rocks standing alone.
> ... Standing at the bottom of the ravines and looking upwards, the cliff face seems perpendicular, but when you are on the top, the other tops seem on a level with where you are standing. Similar formations are found right up to the highest summits.
> ... Now the Great River (i.e., the Yellow River) ... (and certain others) are all muddy, silt-bearing rivers. In the west of Shensi and Shansi the waters run through gorges as deep as a hundred feet. Naturally mud and silt will be carried eastwards by the streams year after year, and in this way the substance of the whole continent must have been laid down. These principles must certainly be true [Needham and Ling, 1959:603-604].

Chinese geographical writings, according to Needham, were of eight major kinds: (1) studies of people, which we might classify as human geography; (2) descriptions of the regions of China; (3) descriptions of foreign countries; (4) accounts of travels; (5) books about the Chinese rivers; (6) descriptions of the Chinese coast, of special value to ship captains; (7) local topographies, including special descriptions of areas tributary to and controlled by walled cities, or famous mountains, or certain cities and palaces; and (8) geographical encyclopedias. There was a very considerable attention to the origin of and changes in Chinese place-names (Needham and Ling, 1959:508).

Chinese Exploration

The discovery of the rest of the world by Chinese travelers is an aspect of the history of geography that is often overlooked in western writings. Travels beyond the far Chinese horizons were undertaken by orders of an emperor or by missionaries and traders (Fig. 8, p. 58–59).

The earliest record of Chinese travels is a book of uncertain age that was probably composed sometime between the fifth century and the third century B.C. It was found in the tomb of a man who ruled a part of the Wei Ho Valley about 245 B.C. The books found in this tomb were written on strips of white silk pasted on bamboo slips and because of their bad condition were recopied in the late third century B.C. The travel books are known as *The Travels of Emperor Mu*, who ruled from 1001 to 945 B.C. Emperor Mu, it is recorded, had the ambition to travel all around the world and to leave the marks of his chariot wheels on every land. Like the *Odyssey* of Homer, this is a story of high adventure, surely embroidered by the imagination of the writer but with details that could scarcely have been invented by fancy. The Emperor traveled in forested mountains, encountered snow, and engaged in hunting expeditions. On his return he crossed a wide desert where water was so scarce that he had to drink the blood of a horse. There can be no doubt that at a very early date the Chinese travelers had gone far beyond the original culture hearth in the Wei Ho Valley (Mirsky, 1964:3–10).

The discovery of the Mediterranean civilizations is credited to the geographer Chang Ch'ien in 128 B.C. (Sykes, 1961:21; Needham, 1963; Mirsky, 1964:13–25; Thomson, 1965:177–178). The book he wrote describes the land route across inner Asia to Bukhara and thence to Persia and the Mediterranean shore. Over this route traders traveled regularly and had probably made contacts with the west long before the west was discovered officially. The Chinese products that were carried westward included peaches, almonds, plums, apricots, silk, and, eventually, silkworms. The Mediterranean products that were carried back to China included alfalfa, wheat, and the grapevine.

There were a number of other Chinese travelers whose records are complete enough to insure their place in history. One of the most distinguished was the Buddhist monk, Hsüan-Tsang (Sykes, 1961:24–30). In the seventh century after Christ he was able to cross the high, windswept plateaus of Tibet and the world's highest mountains on the way to India. After studying in the centers of the Buddhist faith for several years, he returned to China, carrying on the backs of animals a large collection of Buddhist relics and manuscripts. He was the Chinese discoverer of India (Mirsky, 1964). In that same century another Buddhist monk, I-Ching,

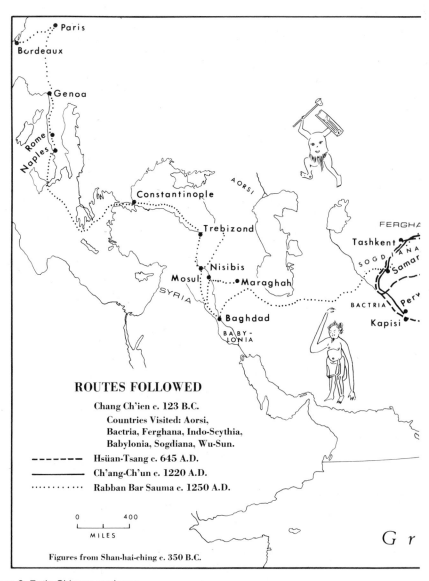

Figure 8. Early Chinese explorers.

reached India by sea, stopping first for eight months in Sumatra in 671. When he returned to China he carried with him more than 10,000 rolls of Sanskrit Buddhist texts, which he undertook to translate into Chinese (Sykes, 1961:30). Several centuries later another Chinese traveler crossed the deserts of inner Asia, encountered endless difficulties, and finally made contact with the Mongol leader, Genghis

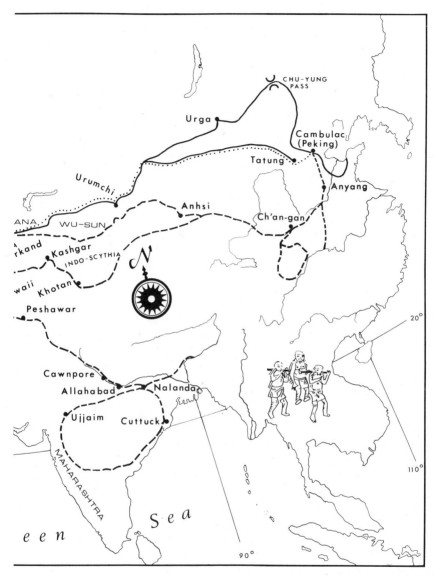

Khan, in Samarkand. This was in 1220. And there was the Nestorian Christian monk, Rabban Bar Sauma, who in 1287–88 made a pilgrimage to Rome. Finding the Pope dead and the new Pope not yet selected, he on through Genoa to Paris and Bordeaux, meeting the kings of France and England. Imagine the amazement of people in thirteenth-century France when they found themselves being discovered by a Christian from China. In 1288 he returned to Rome and received the blessing of the new Pope and thereafter journeyed back to Peking. This was several decades before the similar journeys of the Polo brothers to China. And in 1296 the Chinese traveler, Chou Ta-kuan, visited Cambodia and wrote a detailed account of the strange customs of the Cambodians.

Other Chinese explorers also went by sea. There was never any recorded effort to go far out into the Pacific, although Chinese expeditions did reach Japan and Taiwan. By the thirteenth century Chinese merchants were sailing their junks to Java and Malaya and even as far as India. Marco Polo found them in the Persian Gulf port of Hormuz. But the major work of exploration in this direction was carried out by Cheng Ho, a Chinese admiral, between 1405 and 1433. He led seven separate expeditions, each including a fleet of vessels. His voyages opened regular trade routes to Java, Sumatra, Malaya, Ceylon, and the west coast of India. He also reached the Persian Gulf, the Red Sea, and the east coast of Africa south of the equator. In the east he went as far as Taiwan. On his last expedition, from 1431 to 1433, he returned with ambassadors to China from more than ten countries. It is even possible that he sent one of his ships to the north coast of Australia (Hsieh, 1968).

Cartography

The Chinese were also expert in the making of maps. The engineer, Chang Heng, in the second century after Christ, was probably the first to introduce the grid system into China, although the maps he made have not survived. There is a reference to him, however, in which he is identified as one who "cast a network [of coordinates] about heaven and earth, and reckoned on the basis of it" (Needham and Ling, 1959:538).

The "father of Chinese cartography" was Phei Hsiu, who was appointed Minister of Public Works by the Chinese emperor in A.D. 267. He produced a map of the politically organized part of China on eighteen rolls of silk. To make the surveys on which the map was based, he measured several base lines and then located places distant from the base lines by the intersection of lines of sight, just as the Egyptians had done long before. He made use of a grid of east-west and north-south lines crossing at right angles to provide the frame on which to plot the rivers, the coastline, the mountain ranges, the cities, and other features. Did he, or Chang Heng, or someone much earlier get the idea of using triangulation to locate places and of using a grid of lines from contacts with the Greeks and, perhaps, through them with the Egyptians? It is entirely possible, although no such transmission has ever been established. And it is by no means impossible that many of these methods were devised much earlier in China and diffused westward, as the decimal form of arithmetic certainly was.

Two beautiful examples of Chinese maps were carved in stone in A.D. 1137 based on data that probably had been surveyed before 1100 (Needham and Ling, 1959:547–549). One, "A Map of China and the Barbarian Countries," extended from the Great Wall of China north of Peking southward to the island of Hainan and westward to the mountains of inner Asia. The other, entitled, ''The Map of the Tracks of Yü the Great,'' covers essentially the same area but is even more accurate in showing the courses of the great rivers and the coastline from the Gulf of Chihli, north of the Shantung Peninsula, to the island of Hainan. Neither of these maps shows Taiwan. Both maps—like other Chinese maps—were drawn with north at the top. The geographers who made them are not known.

THE MEDIEVAL WORLD IN RETROSPECT

From the fifth to the fifteenth centuries was a remarkable period. In various parts of the world important new data were being collected by direct observation, but the lack of close contacts among different peoples, in large part resulting from the barrier of language, meant that geographic knowledge gained by one group was diffused only slowly to others. No rumor of the Viking voyages reached Christian Europe. But Christian Europe did have a variety of documents that contradicted each other. Those who could read Arabic were aware that people were living along the equator and beyond it and that people with white skins had visited these regions without turning black. But there were also reputable authorities, ranging from Aristotle to ibn-Khaldun, who insisted that one would turn black at these very low latitudes. The basic Greek theory, which equated habitability with latitude, seemed firmly established, in spite of the evidence of direct observation. But who could be sure which documents were sober and reliable descriptions of actual conditions and which ones were purely imaginative?

In 1410 two important books appeared. One was the *Imago mundi* of Pierre d'Ailly, which summarized much of the writing of Christian Europe in this period. The other was the Latin translation from the Arabic of Ptolemy's *Geography,* which included a large number of errors. Ptolemy did present one important idea: that maps should be based on the precise location by latitude and longitude of specific points. The trouble was that none of the places he located were correctly placed; and he gave the weight of authority to several major errors, including the enclosed Indian Ocean.

Starting with Edrisi in Sicily there had been definite improvement in the arts of navigation, including the wide adoption of the magnetic compass. The Genoese made use of the compass when they rediscovered some of the groups of islands in the Atlantic off Africa, which had been known to the Phoenicians and then lost.

The stage was set for the next step in the discovery of the world. Scholars knew that the earth was a sphere and some of them insisted that no part of it was beyond the reach of man. The estimates of the circumference differed considerably, and those who wanted to believe that India was only a short distance west of Spain could find ample authority (Goldstein, 1965). If one could believe Marco Polo, China and Japan were well east of India, which would place them even closer to Spain. Pierre d'Ailly said you could sail westward to reach the East. It was the Florentine physician and scholar, Paolo del Pozzo Toscanelli, who sent the following remarks to Columbus in 1474:

> On another occasion I spoke with you about a shorter sea route to the lands of spices than that which you take for Guinea. And now the Most Serene King requests of me some statement, whereby that route might become understandable and comprehensible, even to men of slight education.
>
> Although I know that this can be done in a spherical form like that of the earth, I have nevertheless decided, in order to gain clarity and to save trouble, to represent (that route) in the manner that charts of navigation do.
>
> Accordingly I am sending His Majesty a chart done with my own hands in which are designated your shores and islands from which you should begin to sail ever westwards, and the lands you should touch at and how much you should deviate from the pole or from the equator and after what distance, that is, after how many miles, you should reach the most fertile lands of all spices and gems, and you must not be surprised that I call the regions in which the spices are found "western" although they are usually called "eastern," for those who sail in the other hemisphere always find these regions in the west. But if you should go overland and by the higher routes we should come upon these places in the east.
>
> The straight lines, therefore, drawn vertically in the chart, indicate distance from east to west; but those drawn horizontally, indicate spaces from south to north.
>
> From the city of Lisbon westward in a straight line to the very noble and splendid city of Quinsay 26 spaces are indicated on the chart, each of which covers 250 miles. (The city) is 100 miles in circumference and has 10 bridges. Its name means City of Heaven; and many marvelous tales are told of it and of the multitude of its handicrafts and treasures. It (China) has an area of approximately one third of the entire globe. The city is in the province of Katay, in which is the royal residence of the country.
>
> But from the island of Antilia, known to you, to the far-famed island of Cippangu (Japan), there are 10 spaces. The island is very rich in gold, pearls, and gems; they roof the temples and royal houses with solid gold. So there is not a great space to be traversed over unknown waters. More details should, perhaps, be set forth with greater clarity but the diligent reader will be able from this to infer the rest by himself [Morison, 1963:12–14].

CHAPTER 4

The Age of Exploration

Terra Incognita: these words stir the imagination. Through the ages men have been drawn to unknown regions by Siren voices, echoes of which ring in our ears today when on modern maps we see spaces labelled "unexplored," rivers shown by broken lines, islands marked "existence doubtful. . . ."

The Sirens, of course, sing of different things to different folk. Some they tempt with material rewards: gold, furs, ivory, petroleum, land to settle and exploit. Some they allure with the prospect of scientific discovery. Others they call to adventure or escape. Geographers they invite more especially to map the configuration of their domain and the distribution of the various phenomena that it contains, and set the perplexing riddle of putting together the parts to form a coherent conception of the whole. But upon all alike who hear their call they lay a poetic spell.

The sudden increase in exploring activity in Europe in the fifteenth century was a major turning point in world history. In China exploration supported by the emperor ceased after the seventh expedition of Cheng Ho (1431–33); and there was no great Muslim traveler after ibn-Batuta in the fourteenth century. But in Europe exploration for the first time was planned and supported by governments or by merchant companies and, for the first time, was directed to the open oceans.

The quotation above is from John K. Wright's presidential address to the Association of American Geographers, 1946, "Terrae Incognitae: The Place of the Imagination in Geography," *Annals AAG*, 37 (1947): 1-15; ref. pp. 1-2.

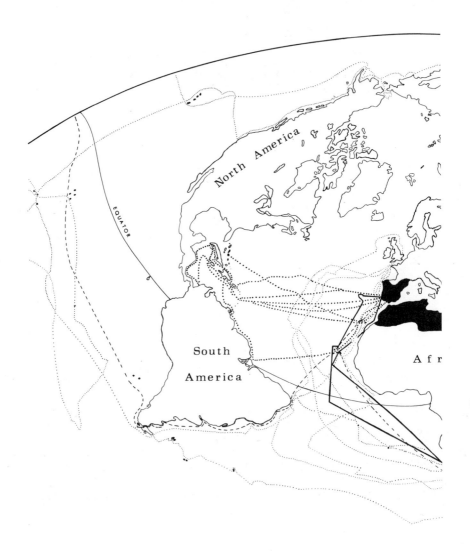

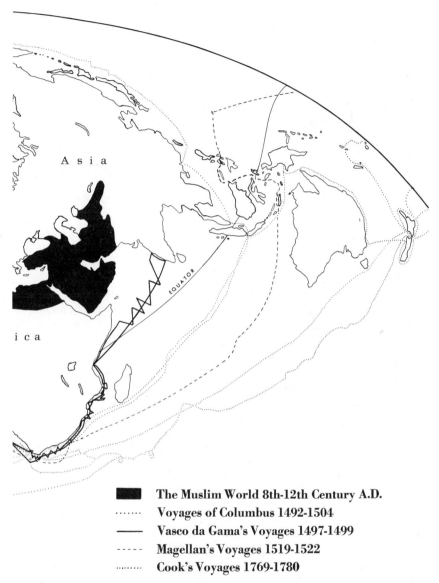

The Muslim World 8th-12th Century A.D.
······· **Voyages of Columbus 1492-1504**
——— **Vasco da Gama's Voyages 1497-1499**
- - - - **Magellan's Voyages 1519-1522**
········ **Cook's Voyages 1769-1780**

Figure 9. The Age of Discovery, Bartholomew's polar projection: The Muslim World 8th-12th centuries; Columbus; Vasco da Gama; Magellan; Cook.

What were the motives? In this case the challenge of the unknown—the urge to find out what it is like somewhere else—was subordinate to two other compelling objectives. One was the zeal to spread the Christian faith; the other was the critical need to replenish the European supplies of precious metals and spices. The two motives were conveniently blended and were never permitted to interfere with each other.

The zeal to spread the Christian faith was in part a response to the long series of conflicts with the spreading Muslims. The greater part of Spain and Portugal had been under Muslim rule since the eighth century. But the Portuguese had freed themselves of Muslim rule by the middle of the thirteenth century thus forming the world's first example of a nation-state—that is, a politically organized territory occupied by people with a unified national consciousness. From 1391 to 1492 the Spaniards waged continuous warfare with the Muslims, and the last Muslim stronghold (Granada) was taken by the Christians just before Columbus set sail for America. Young people, for that whole century, were brought up in an atmosphere of religious warfare, and, when the last Muslim foothold was pried loose, there was urgent need to find other infidels to kill or convert. The native peoples of America provided what was needed.

The amazing legend of Prester John should not be overlooked. In 1170 a letter was sent to the Pope in Rome and to the Emperor in Byzantium asking for help in warfare against the infidels. The letter purported to come from a powerful and wealthy Christian ruler located somewhere in the east. To find and bring aid to Prester John became a very real motive for sending out exploring expeditions. The difficulty was to be sure where to look for him. In Marco Polo's time Prester John was associated with the Nestorian Christians in China, perhaps a reflection of the visit of Bar Sauma to Rome and Paris in 1287–88. By 1340 his location had been shifted from Asia to Ethiopia, and one of the objectives of the Portuguese voyages of the fifteenth century was to make contact with Prester John. An ambassador was actually dispatched to represent the Portuguese king.

The need for precious metals and spices in Europe and the desire for personal wealth by the explorers was never far from the surface. Europeans were increasing their trading activities, and, to provide financial support for trade, there was urgent need for a supply of gold and silver or valuable gems. The Europeans also needed spices. In those days there was no adequate supply of sugar for sweetening, spices were used instead. Without refrigeration meat spoiled quickly unless it was dried or salted. But drier or salted meat is almost unpalatable unless spices can be used in cooking. The spices included cloves and nutmegs, both of which came from the Spice Islands (the Moluccas) in what is today Indonesia. The trade in precious metals, gems, and spices had been under the control of the merchants of Genoa and Venice, but the Arabs blocked direct contact between the Europeans and the Asian sources. The Arabs, in fact, had thrown a screen between Europe and Asia and between Europe and Africa by their control of the great dry land area of North Africa–Southwest Asia (Fig. 9). The Arabs were in a position to control the supply of these products to Europe and to collect the larger share of this profitable trade.

PRINCE HENRY AND THE PORTUGUESE VOYAGES

The first initiative toward wider exploration came from Portugal. The individual leader was Prince Henry (called the Navigator), the third son of the Portuguese king. In 1415 Prince Henry commanded a Portuguese force that attacked and captured the Muslim stronghold on the southern side of the Strait of Gibraltar at Ceuta. This was the first time a European power took possession of territory outside of Europe—in a sense the occupation of this portion of Africa began the period of European overseas colonization. At Ceuta Prince Henry learned from Muslim prisoners that the gold, ivory, ostrich feathers, and slaves sold in the Arab markets of North Africa had been brought across the great desert by caravan from Africa south of the Sahara. Why would it not be possible to reach Guinea by sea and thus capture some of this profitable trade for Portugal?

The Institute at Sagres

In 1418 Prince Henry established the world's first geographic research institute. It was probably, but not certainly, built on the rocky promontory known as Cape St. Vincent, not far from the port of Lagos in Portugal (Oliveira Martins, 1914:79–81). At Sagres (Fig. 10), Prince Henry built a palace, a chapel, an astronomical observatory, buildings to store collections of maps and manuscripts, and houses for the institute staff. He brought scholars of all faiths to Sagres from all around the Mediterranean: geographers, cartographers, mathematicians, astronomers, and experts in the reading of manuscripts in many languages. His workers included Christians, Jews, Muslims. Master Jacome of Majorca was appointed chief geographer. The purpose was to improve and teach the methods of navigation to Portugal's sea captains, to teach the new decimal mathematics, and to shift the evidence from documents and maps concerning the possibility of sailing southward along the African coast and thence to the Spice Islands. Were the equatorial regions uninhabitable or not? Would people turn black or was that nonsense? And how big was the earth—Was it as large as Eratosthenes thought? Or as small as Marinus of Tyre thought and as the Muslim geographers had found it to be from new measurements in the vicinity of Baghdad? These were very practical questions.

At Lagos, Prince Henry's naval architects were busy designing new and better ships. The width, or beam, of the older ships had been about half the length from bow to stern; but the new designs made the beam only a third to a quarter of the length. The new Portuguese caravels were lateen rigged with two or even three

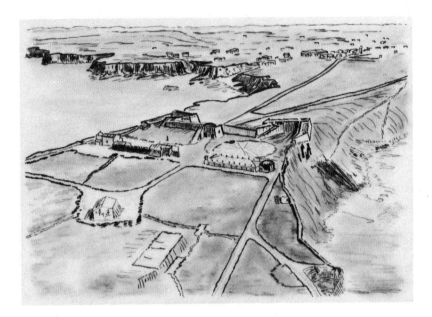

Figure 10. Sagres.

masts. The ships, with high castles fore and aft and of about 200 tons burden, were slow but seaworthy.

Navigation aids included the magnetic needle, which was mounted on cardboard—as taught by the sailors of Genoa and Venice, who had probably been previously instructed by the geographers at Palermo. Improved astrolabes were made of brass, permitting the more accurate measurement of star altitudes at sea; and tables were prepared to give the altitudes of various stars in different seasons and different latitudes. Portolano charts were drawn on sheepskin.

Prince Henry's captains were able to gain experience and confidence by sailing to the three groups of islands in the Atlantic off North Africa and Europe. The easiest voyages were to the Canary Islands (Ptolemy's Fortunate Islands, which he used for the 0° meridian on his maps). These were less than 100 miles off the African coast. The Madeira Islands offered more of a challenge, requiring a voyage of some 400 miles into the open ocean. After some unsuccessful efforts the Madeira Islands were reached in 1420 (they had been reached by some English ships in 1370). The Azores were still more of a challenge, requiring 1000 miles of sailing. To be sure the Azores were shown on the Portolano chart of 1351, but the bearings shown on these charts were not accurate, and at first the ships missed them. Then Prince Henry made the necessary corrections and in 1432 the islands were "discovered."

Cape Bojador

Meanwhile Prince Henry sent his more experienced captains southward along the African coast. The first probe in that direction was sent out in 1418 but soon turned back because the crews were afraid of what might lie closer to the equator. In spite of repeated efforts it took sixteen years for the Portuguese ships to get south of latitude 26°7′N. At this latitude, a little south of the Canary Islands, there is a low, sandy promontory on the African coast known as Cape Bojador. The strong current of water flowing south along this coast formed eddies with much foam off the end of the cape. Each time the ships approached the cape, the crews demanded that the voyage be stopped: here, surely, was the boiling water that the Greek geographers had described. Beyond this point people would turn black. And, moreover, on the Arab map of this coast just south of Bojador there was a drawing of the hand of Satan rising out of the water. To be sure the Portolano chart of 1351 showed nothing unusual about Bojador, which was, after all, just a minor cape. And at Sagres there was the report about the Phoenicians under Hanno who had sailed far to the south of Bojador thousands of years before. But who could be sure?

One of Prince Henry's most capable captains was Gil Eannes. He had tried to sail around Bojador in 1433 but his crew mutinied and he had been forced to return to Sagres. In 1434 he tried a new method, perhaps suggested by Prince Henry. From the Canary Islands he sailed boldly out to sea, far out of sight of land. When he had sailed south of the latitude of Bojador, he turned eastward. When he reached the shore the water was not boiling and no one turned black. At last the barrier of Bojador had been passed. The very next year the Portuguese ships sailed 390 miles south of Bojador (Beazley, 1895; Oliveira Martins, 1914:222).

By 1441 Prince Henry's ships had sailed far enough south to reach the southern zone of transition between the desert and the wetter country beyond. South of Cabo Blanco, in the area today included in Mauritania, the Portuguese explorers captured a man and a woman and then a group of 10 men and women. They also found some gold. When they returned to Portugal they created a sensation, and all at once there were hundreds of volunteers who wanted to go south. Between 1444 and 1448 some forty ships sailed along the African coast, and 900 Africans were brought back to be sold as slaves. The business of exploring was forgotten in the rush to profit from the capture and sale of slaves. In those years the African slave trade began in earnest.

Around Africa to India and Beyond

After a decade or so, however, Prince Henry was able to get his sailors to return to the business of exploration. Now he realized that there was a bigger prize beyond if ships could sail all the way around Africa to India. The Guinea coast was explored in 1455 and 1456, and the Cape Verde Islands were visited. Although

Prince Henry died in 1460, the work he had started was carried on and new voyages were sent southward. In 1473 a Portuguese ship crossed the equator without burning up. A few years later the Portuguese went ashore and set up stone monuments to indicate their claim to the African coast. Monuments set up at the mouth of the Congo were reported still standing in the present century. In 1486–87 Bartholomew Dias sailed southward from the equator, meeting headwinds and a north-flowing current. To avoid stormy weather he sailed far out westward away from the land and only when the weather improved did he again turn eastward. But after sailing for a longer time than he thought he should, he turned toward the north hoping to find land. He came upon the coast of southern Africa at Algoa Bay (Port Elizabeth). Returning, he passed Cape Agulhas, the southernmost point of Africa, and then the Cape of Good Hope.

The great voyage of Vasco da Gama took place between 1497 and 1499 (Fig. 9). Like Dias he avoided the strong north-flowing Benguela current and the head-winds along the coast south of the equator. He made a wide circle out into the Atlantic before turning eastward along the latitude of Cape Agulhas. He then fol-lowed the east coast northward to Mozambique, where, for the first time in these waters, the Portuguese came into contact with the Arabs. With the aid of an Indian pilot, Vasco da Gama sailed all the way across the Indian Ocean to Calicut in twenty-three days, thereby completing the "end run" around the Arab-held lands. But for some reason Vasco da Gama was not informed about the monsoons, which had been known and used for navigation for thousands of years. On the way to Calicut, in late April and May, the northeast monsoon was breaking up and he had favorable southwest winds. But when he tried to return in August, he encountered steady headwinds against which he had to tack. It took him three months to beat back to the African coast, during which time his crews developed scurvy. So many died that he had to destroy one of his three remaining ships. When he finally returned to Lisbon he had sailed 24,000 miles in more than two years; and of the 170 men who started with him, only 44 returned.

In the sixteenth century the Portuguese voyages continued and were extended farther and farther to the east. In 1510 the Portuguese took Goa and made this port into a major trading center on the west coast of India. In 1511 they established a base at Malacca on the strait between the Malay Peninsula and Sumatra. By 1542 the Portuguese even reached Japan. In 1557 they leased Macao from the Chinese as a base from which to control the profitable trade in a wide variety of new products. In 1590 they reached Taiwan and gave it the Portuguese name, Formosa.

CHRISTOPHER COLUMBUS

Christopher Columbus was born of Spanish parents in Genoa in 1451. He was fascinated by ships and by the problems of navigation; he himself said he started navigating at the age of fourteen. As a young man he served on ships in the eastern

Mediterranean, and in 1476 he arrived in Portugal to study at what was left of the institute at Sagres. In that same year he sailed on an English ship, which probably went as far as Iceland. In 1478 he was married and lived for a few years on the island of Madeira; but he never ceased to increase his knowledge of the ocean and was training himself to be one of the most highly skilled sailors of his time. After the death of his wife he went to sea again for the Portuguese, sailing several times after 1482 to the Guinea coast (Morison, 1942, 1963).

The Geographical Ideas of Columbus

At an early age Columbus had started to think about the possibility of sailing westward to Asia. There were at least five books that he read with enough care to write marginal notes in them. These were the *Imago mundi* of Cardinal Pierre d'Ailly, in which the conclusion was reached that China was little more than 3000 miles west of the Canary Islands. He read books by Aeneas Silvius (Pope Pius II), who reported on the Greek and Roman concepts regarding the earth. He also read Latin translations of Ptolemy's *Geography*. He was thoroughly familiar with the first Latin edition of Marco Polo and also with the popular favorite supposed to have been written by the fictitious Sir John Mandeville (p. 95).

Columbus had no doubts about the round earth. In those days no educated person, and no person with experience in navigation, had any illusions about the earth being flat. Columbus faced two problems, both of them related to the distance he would have to sail westward to reach Asia. If the whole circumference of the earth were divided into 360 parts, as Hipparchus had done and as had been common practice since his time, then what was the length of one part, or 1°? And the second problem was what authority to believe about the eastward extension of the known lands of the earth.

Regarding the first of these problems, Columbus had a wide range of calculations from which to choose (Nunn, 1924). Eratosthenes had shown how the circumference of the earth could be calculated. Just three measurements were needed: the altitude of the sun on any given day at two places on the same meridian and the distance between the two places. Eratosthenes had come very close to reaching the correct figure; Posidonius, using a different set of observations, reached a much smaller figure. Marinus of Tyre favored the smaller figure. Ptolemy reported both, but himself followed Marinus. During the ninth century the Muslim geographers had made several new sets of observations in Mesopotamia, and from these data they estimated the length of a degree was about 1480 meters. The actual length of a degree along a great circle is about 1829 meters.

Of course Columbus accepted the smallest estimate of the distance around the world. Furthermore, his belief was strengthened by a calculation carried out by Prince Henry's scholars at Sagres. The north-south line they used extended from Lisbon to a place in Guinea near the modern port of Conakry. In those days when the precise measurement of longitude was impossible, this was accepted as a north-south line (Conakry is actually 4°33' west of the longitude of Lisbon). The distance along this line was estimated on the basis of many voyages, and the estimate was actually quite close to the correct distance. But there was real trouble with the latitudes at either end. Lisbon was placed by Ptolemy at 40°15'N and by the Muslim geographers at 42°40'N (actually it is 38°42'N). The southern end of the line was placed at 1°5'N (actually it is 9°30'N). Based on the data accepted at Sagres, the length of the degree was almost exactly the same as that calculated by the Muslims of Baghdad—about 56⅔ Italian nautical miles. Columbus used this confirmation of the size of the earth to convince himself that reaching Asia would not be problematic. Would he have started if he had known how far away Asia was located and had not known of any intervening land? But Columbus reckoned that the east coast of Asia was located just about where the east coast of Mexico is actually located. Columbus had received additional support for his estimates in the correspondence with Toscanelli.

From Ideas to Action

But when it came to convincing the scholars in the Spanish and Portuguese courts that Asia was not very far west of Europe—that was another matter. King João II of Portugal, already committed to finding a way around Africa, rejected Columbus's proposal in 1484. In Spain Columbus did not fare much better. When he finally secured an audience with Ferdinand and Isabella, they appointed a royal commission to study the matter, which is a time-honored way of postponing a difficult decision. The commission did not report until 1490, at which time they rejected Columbus's plan because they had no confidence in the estimates of the earth's circumference nor of the east-west extension of Europe and Asia. They accepted a figure of 70 Italian nautical miles for the length of a degree, and, therefore, they figured that any ships attempting the long westward voyage would surely be lost. Isabella, however, was impressed with Columbus's unshakable faith that he could make the voyage. In spite of the adverse reports she agreed to make ships and supplies available to him (Davies, 1967:341).

The story of the four voyages of Columbus is too well known to be repeated here (Fig. 9). He came upon land just about where he expected to find it; although he did not find any advanced civilization, such as that reported by Marco Polo, he did remain convinced that he had found Asia. His belief was strengthened when he found the southern coast of Cuba and the coast of Central America trending toward the southwest, just as the coast of Asia was shown on Ptolemy's map. When he heard from the Indians of Central America that there were sources of gold only a short distance west and that there was another great ocean beyond, he was sure it must be the Indian Ocean. Furthermore, noting the great flow of water along the northern coast of South America, he figured that all this water had to go somewhere.

Figure 11. The Treaty of Tordesillas.

There must be a strait connecting the Caribbean with the Indian Ocean. Of course he did not know of the Gulf Stream. Instead of turning northward to reach the latitudes where Marco Polo had reported the existence of a Chinese civilization, he continued his exploration toward the southwest. One reason was that he expected to reach the opening into the Indian Ocean. But there was also another reason, as Sauer points out. It was commonly believed at this time that gold was produced by the heat of the tropical sun and thus more gold would be found closer to the equator (Sauer, 1966:23–24).

Columbus was the first explorer to discover and make use of the wind systems of the Atlantic. From his previous experience he knew of the presence of easterly winds in the low latitudes and of westerlies in the higher latitudes. He sailed westward on his first voyage along the latitude of the Canaries with the wind at his back. But, when it was time to return to Spain, he sailed northward to the latitude of the Azores, and then with the westerlies behind him, he returned to Europe.

The Treaty of Tordesillas

According to the analysis of Columbus by Arthur Davies (Davies, 1967), he was not only a skilled sailor but also a skilled diplomat. To gain possession of the lands he discovered, he arranged to have them confirmed by the Pope; and he negotiated a treaty with King João II of Portugal. This was the Treaty of Tordesillas signed by Spain and Portugal in 1494. The world was divided between these two countries, and the dividing line was to be drawn 270 leagues west of the Azores, or 370 leagues west of the Cape Verde Islands (Fig. 11). Portugal was to have the exclusive right to the lands east of that line, Spain to the west of it. This gave Portugal a free hand in the Indian Ocean, and it gave Columbus a free hand in the lands he had found west of the Atlantic.

Several problems resulted from this treaty. Since longitude could only be estimated, no one really knew where the dividing lines should be drawn. Soon it was clear that part of eastern South America was in the Portuguese hemisphere. The line passed through the Guianas if leagues were estimated on the basis of the longer measurement of a degree or it passed east of the mouth of the Amazon if the shorter measure were used. Actually the agreed-upon line is approximately along 48°W of Greenwich. On the other side of the world the boundary would be approximately along 132° E of Greenwich. The Philippines, which became Spanish, lie well within the Portuguese hemisphere; the Spice Islands (the Moluccas) are just about on the border.

LOOKING FOR ASIA

Columbus died in 1506 still believing that the lands he had discovered were a part of eastern Asia. By that time, however, almost all the geographers and explorers were convinced that the world was really much larger than Columbus had figured and that a new continent lay between them and Asia. This was confirmed when Pedro Alvares Cabral, sailing on a Portuguese voyage to India and setting a course southwestward from the Cape Verde Islands, came upon land within the Portuguese hemisphere, which he promptly claimed for Portugal. In 1501–1502 Amerigo Vespucci, sailing for Portugal, explored the coast of Brazil probably as far as La Plata River. He spent a month exploring the disputed area on the border between the Portuguese and Spanish hemispheres, which is now called Uruguay. Perhaps because he was the first to announce that the land he was seeing was in fact a new continent and not a part of Asia, a German cartographer wrote the name, America, on a map published in 1507.

Within two decades explorers had sketched in the whole eastern coast of the newly discovered continent. Greenland and Labrador had been rediscovered by John Cabot, and Cabot's son, Sebastian, sailed into Hudson Bay. The Cabots were born in Genoa and Venice, but they sailed for England. Other voyages sent out by France included one by Jacques Cartier, which proved that the St. Lawrence was a river, and one by Giovanni da Verrazano, who was the first European to sail into New York harbor (in 1524).

Magellan

It was the Portuguese explorer, Fernão de Magalhães, or Magellan, who was the first to reach eastern Asia by sailing west. Like Columbus he tried first to interest the Portuguese king in supporting such a voyage, but the king refused to have anything to do with the project. In Spain, however, he found greater interest, and in 1518 the Spanish king agreed to support the voyage. The next year Magellan set sail with a fleet of five ships, all in poor condition (Fig. 9). Looking for an opening through the Americas, he examined the Brazilian coast carefully, entering every bay to be sure that it was not the strait they were seeking. The expedition spent the winter (from March to August, 1520) in southern Patagonia, during which time some of his men mutinied. When the blustery winter winds subsided, the expedition started southward again; on October 21, 1520 they found the entrance to the strait that bears Magellan's name. It took thirty-eight days for the ships to pass through the 360-mile strait, and it is believed that only an exceptional period of light winds made the passage possible at all because normally the westerly winds are too strong, even in summer. Leaving the strait, Magellan sailed in a northwesterly course for ninety-eight days before he reached the island of Guam. It is amazing that in that long voyage through an ocean dotted with islands, he came upon only two small and desolate rocks (St. Paul's Island and Shark Island). His men were sick with scurvy, and his supply of food and water was exhausted to the point that the explorers had to eat rats, oxhides, and sawdust and drink putrid water before they reached Guam. After replenishing supplies and regaining health they started again westward, reaching the Philippines on April 7, 1521. Magellan, on a previous voyage for Portugal, had reached the Spice Islands (the Moluccas), which are east of the Philippines. Therefore, at this point he had become the first man to sail all the way around the earth.[1]

Magellan was killed in a fight between Philippine natives on April 27, 1521. Thereafter one of his remaining ships, the *Vittoria* commanded by Juan Sebastian del Cano, continued the voyage across the Indian Ocean after taking on a load of cloves at the Moluccas. They rounded the southern end of Africa and then sailed northward, reaching Seville on July 30, 1522. The *Vittoria* was the only one of the original five ships that returned to Spain. But the whole cost of the expedition was more than covered by the sale of the cloves.

[1]The second voyage around the world was led by Sir Francis Drake, 1577-80.

Outlining the Continents

Between 1521 when Magellan was killed by the natives in the Philippines and 1779 when James Cook was killed by the Hawaiians, the exploration of the world was directed primarily at fixing the outlines of the land and water on the world map. At first there was no single answer to the question about what to observe. Everything was novel and the sober descriptions of new lands could scarcely be distinguished from the imaginative writings of those who followed Sir John Mandeville. But the outlines of the map could be filled in. Gradually, as the Age of Exploration continued, scholars began to try to digest the flood of new information and to formulate scientific generalizations. In the next chapter we will consider the impact of exploration on the shape of geographic ideas.

FIVE PROBLEMS

Meanwhile, the explorers themselves faced five major problems, all of which were solved in the period between Magellan and Cook. These were: (1) the problem of maintaining health on long sea voyages; (2) the problem of perfecting the accuracy of navigation; (3) the problem of making precise determinations of longitude; (4) the problem of showing the whole spherical surface of the earth, or large parts of it, on flat pieces of paper; and (5) the problem of eliminating from the world map the many erroneous concepts of Ptolemy.

Scurvy

Scurvy was the disease that posed the greatest health problem on long voyages away from land. The problem did not appear as long as navigation was close to the shore where fresh food could be taken on board frequently. But when the Portuguese learned how to sail far away from the land, scurvy began to appear among the crews. The first expedition to suffer seriously from scurvy was that of Vasco da Gama when he was returning across the Indian Ocean from Calicut to Malindi in Africa in 1498. No one knew why the men died.

Scurvy is a disease of dietary deficiency and is easily avoided or cured when adequate amounts of vitamin C are supplied by fresh vegetables, such as potatoes, cabbages, onions, carrots, and turnips or by such fruits as lemons or limes. When these fresh foods are lacking, the skin becomes sallow, gums are tender, and the victim is miserable with muscular aches. Eventually, the gums become like putty, teeth fall out, and massive hemorrhages occur. When Magellan sailed for ninety-eight days from the Strait of Magellan to Guam, most of the members of his crew had scurvy and either died or were too weak to do any work. There are many stories of ships drifting helplessly at sea with no one able to handle the sails (Penrose, 1952:203–207).

Apparently the British were the first to discover a remedy. In 1601 a British captain on a voyage to the Indian Ocean had to stop in southern Africa because so many of his men had scurvy. Perhaps by chance he served them some fresh lemon juice and miraculously the symptoms of scurvy promptly disappeared. Yet not until 1607 did another British captain try the same remedy, with encouraging results. Reports of this way of controlling scurvy were not widely read or believed because even in the late nineteenth century explorers in the Arctic were still suffering from the ravages of scurvy. But the first voyage on which fresh fruit and vegetables were regularly supplied for the purpose of controlling the disease was the second voyage of Captain James Cook to the Pacific, 1772-75. On this voyage the method of controlling scurvy was demonstrated beyond doubt, although it was still many decades before it was widely adopted and still longer before the function of vitamin C was understood.

Navigation

There remained the ancient, unsolved problem of measuring distance and direction and fixing position on the surface of the earth. The Greeks figured out how to do this in theory, but they never possessed instruments accurate enough to carry out the observations they knew were needed. When long sea voyages were undertaken the need for precise determinations of distance and direction became critical.

The magnetic compass proved to be one of the best ways for determining direction, especially when skies were cloudy. The Chinese are credited with the invention of this device, long before the time of Marco Polo; but the first mention of the compass in European writings was in 1180. A magnetized needle was first attached to a card mounted on a pivot during the thirteenth century, but not until the eighteenth century was an improved compass for use at sea made available. Columbus was the first explorer to mention the fact that the needle does not always point north and that it varies from north according to longitude. In fact there was a period when it was believed longitude could be identified by noting the variation of the compass needle. In 1699-1700 Edmund Halley, the British astronomer, made a long ocean voyage for the purpose of plotting the magnetic variation on a map. His use of lines to show equal magnetic variation resulted in the first isarithmic map ever made. He proved clearly that the variation was not the same for each longitude.

The determination of latitude depends on observations of the height of certain stars or of the sun above the horizon. The astrolabe that Hipparchus had invented had been improved by expert workmanship, and for centuries it was the only way to determine latitude at sea. In the sixteenth century the cross-staff was invented. This is a pole about 3 feet in length that the observer points toward the horizon. A rod mounted vertically on the staff can slide back and forth. The observer places his eye at the end of the staff and moves the vertical piece until the sun or star is exactly at the tip of the rod. There were two important difficulties with this device. The

observer had to look at two things at once—the horizon and the sun or star. And if
he were using the sun's altitude, he had to look directly at it. In 1594 the polar
explorer, John Davis, invented the back-staff, in which the sun's altitude is measured by its shadow. This was later improved by using a mirror instead of a shadow
and was known as a quadrant. John Hadley's octant was invented in 1731. The
sextant, still in use, was a still later improvement based on the principle first
adopted by Davis.

Another problem was the measurement of speed at sea. Sailors voyaging
within the narrow confines of the Mediterranean developed skills at estimating a
ship's speed, but on the open ocean the errors of such estimates became intolerable.
At one time a navigator would throw some object overboard near the bow of the ship
and then measure the time required for the object to reach the stern. The time
interval could be estimated by the speed of pacing or by observing the sand in an
hourglass. The British sailors are credited with the invention of the log. A thin but
strong cord is attached to a heavy log and the latter is thrown into the water. The log
remains stationary, causing the line to play out at a speed that can be measured by
the time interval between knots on the line. The log was thrown out whenever there
was a change in wind direction or of the ship's course (Taylor, 1957:201).

Longitude

None of these methods of observing distance and direction were accurate enough
to permit a precise determination of longitude; and until longitude could be determined, navigation remained dangerously inaccurate. The measurement of longitude required some way of keeping accurate time at sea. The Greeks knew that the
observation of the time of an eclipse at two places could be used to check estimates
of longitude. But eclipses did not always occur when they were needed. As early as
1522 the principle was clearly understood that a dependable timepiece would solve
the longitude problem. The difficulty was that the clocks of the sixteenth century
required frequent winding and would gain or lose up to fifteen minutes a day.
Furthermore, when clocks were carried into climates with different temperatures,
the metal parts would expand or contract, thus making measurements of time
unreliable. In 1657 Christian Huygens invented the pendulum clock, which could
keep time with great accuracy; but the pendulum clock was no good at sea.

The need for an accurate clock was underlined in 1707 when an English fleet
was wrecked on the Scilly Islands because the navigator was far off in his estimate
of longitude. The great loss of life on this occasion brought the need to the attention
of the public. In 1714 the British Parliament offered a reward of £20,000 to any
person or persons who could devise a way to measure time at sea with sufficient
accuracy. To receive the award a clock had to be built that, on a voyage from
England to the West Indies and back, would not gain or lose more than two minutes.
For many years no clocks were offered.

At this time in England there was an expect clockmaker named, John Harrison, whose excellent pendulum clocks were well known. In 1729 he decided to try for the reward, and he went to work on a clock suitable for use at sea (Quill, 1966). One of his earliest models did well enough on a voyage to Lisbon so that the Board of Longitudes of the British Government gave him a grant to help continue the work. Not until 1761, however, did Harrison complete clock number four, which he felt could meet the requirements. The clock was tested on a voyage to Jamaica and back and lost less than two minutes on the trip. Another trial was ordered, and this time the clock was mounted on a movable bracket. On a trip to Barbados and back it lost only fifteen seconds in 156 days. The Board of Longitudes still held back the prize because Harrison, they insisted, had not told anyone else how to make such a clock. Another clockmaker was commissioned to make a duplicate to Harrison's specifications. It was this clock that was carried to the Pacific on the second voyage of Captain Cook (1772-75), and it proved to be so accurate that for the first time in man's long effort to gain useful knowledge about the earth, an explorer could tell exactly where he was. Harrison, then eighty-two years old, was finally given the reward in 1775, a year before his death. By this time French and Swiss clockmakers were almost ready with their models. After this date navigators could measure longitude with the same precision with which they could measure latitude (Taylor, 1957:260-263).

The Need for New Kinds of Maps

Most of the cartographers of the fifteenth century lived either in Venice or Genoa because it was from these two places that Europeans departed on voyages to the eastern Mediterranean to pick up cargoes of valuable items from the east. The fifteenth-century geographers generally started with Ptolemy, and the new maps they produced began the job of correcting the work of the ancient cartographers.

One of the earliest corrections was made in 1459 by a monk named Fra Mauro who lived near Venice. His map showed the Indian Ocean open to the south, thereby breaking with the Ptolemaic tradition of enclosing that ocean. But how did Fra Mauro find this out in 1459? Bartholomew Dias did not sail around southern Africa until 1486-87. The Venetians must have learned about the Indian Ocean from the Arab traders, and Fra Mauro believed the reports he received. Presumably his map gave new confidence to the Portuguese voyagers just at the time of the death of Prince Henry. Incidentally, Fra Mauro's map, like most of the maps of this time, was "oriented" toward the south. But since the maps were usually read by laying them flat on tables rather than hanging them on walls, it apparently did not become customary to speak of "going up south."

The first cartographer to produce a globe showing the whole round earth was Martin Behaim of Nuremburg (Bagrow and Skelton, 1964:103-109). He was one of the scholars who joined the seminar at Sagres or at Lisbon after the death of Prince

Henry. He was among those who advised King João II of Portugal regarding the arts of navigation and the most likely directions in which exploration might bring important results. It is thought that Behaim was the first to make a brass astrolabe to replace the wooden ones in use earlier. He may have sailed southward to Guinea on one of the Portuguese expeditions. In 1490, however, he went back to Nuremberg. There, with the help of a painter named Jörge Glockendon, he produced the world's first globe. He did show the coastline of South Africa in accordance with the reports from Bartholomew Dias, but the other Portuguese discoveries—along the Guinea coast, for example—were inaccurately portrayed. The important thing about Behaim's globe, however, is that he showed the east coast of Asia just about where the east coast of the Americas is located. He scattered the Atlantic with mythical islands, some not far to the west of Europe. His globe was based on the small estimate of the earth's circumference, which also had been adopted by Columbus. Yet Columbus could never have known of Behaim's globe. Presumably Behaim gained his concept of the earth from those who had been influenced by Toscanelli, just as Columbus did.

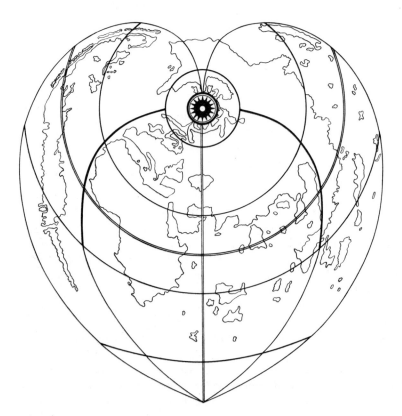

Figure 12. Apian's heart-shaped world, 1530.

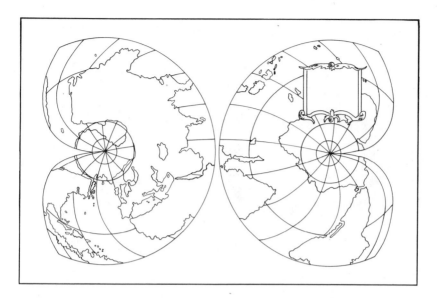

Figure 13. Outline of Mercator's world map, 1538.

There is a long list of cartographers who produced maps during the early period of exploration, making use of new data. Juan de la Cosa in 1500 drew a map using the observations from the first three voyages of Columbus and also from John Cabot's voyage to North America. But the first world map that shows America as a separate continent and not the eastern part of Asia was drawn by Martin Waldseemüller in 1507. He also made use of the name, America, for the first time, either because he thought that Amerigo Vespucci had reached the new continent before Columbus or because Amerigo was the first explorer definitely to identify the newly discovered lands as a separate continent. As a result of this decision by Waldseemüller, the new continent was not named after the European who first reported seeing it. Although Waldseemüller's map was labeled *Carta Marina,* it was no more useful for navigation than other maps of this period that made use of the Portolano principle of design. Explorers had already found that when they followed any of the straight lines on these maps for long voyages, they did not arrive at expected destinations.

A search began for new kinds of projections that would make it possible to show the curved surface of the earth on flat paper or parchment (Bagrow and Skelton, 1964). Peter Apian in 1530 produced a heart-shaped map of the earth (Fig. 12) in which both lines of latitude and longitude are curved. On this projection, however, both distance and direction are distorted. Apian was the teacher of Gerhard Kremer, who adopted the Latin name, Gerardus Mercator. Mercator in 1538 made a world map by joining two heart-shaped projections, one for each hemisphere (Fig. 13).

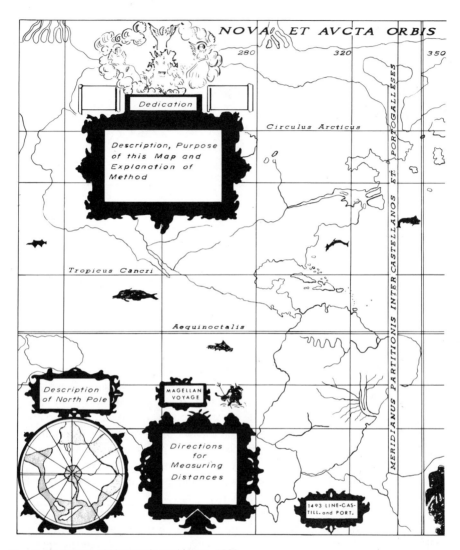

Figure 14. Outline of Mercator's world map, 1569.

Mercator is famous for the world map he produced in 1569, which became the only map used for navigation in the low and middle latitudes. Mercator's distorted continental shapes became the most widely known of all portrayals of the world as a whole. What was needed was a projection on which a compass bearing could be drawn as a straight line so that navigators could plot their courses without having to draw curves. Of course a straight line on a flat map is not the shortest distance

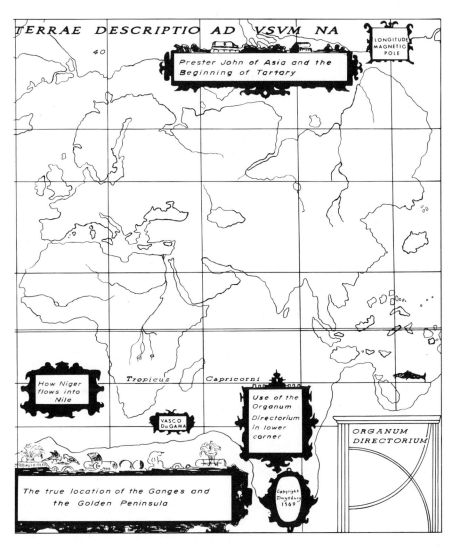

The following labels appear on the map:

TERRAE DESCRIPTIO AD VSVM NA

40

LONGITUDE MAGNETIC POLE

Prester John of Asia and the Beginning of Tartary

How Niger flows into Nile

Tropicus Capricorni

VASCO Da GAMA

Use of the Organum Directorium in lower corner

ORGANUM DIRECTORIUM

The true location of the Ganges and the Golden Peninsula

Copyright Duysburg 1569

between two points unless it follows the equator or a meridian. To follow other great circles, navigators draw a series of short, straight rhumb lines, approximating the curve of the great circle. The nature of a rhumb line on a flat map had been demonstrated by Pedro Nuñes (Curator of Maps and Charts for the King of Portugal) in 1534. But sailors not trained in mathematics found difficulty in believing that a straight line was not the shortest distance between two points on the surface of a sphere. On Mercator's projection the lines of latitude and longitude are straight lines and cross each other at right angles. This means that the east-west distance between meridians increases with increasing latitude until the poles are stretched out into lines. But if the parallels of latitude are drawn with an even spacing, as they are on a globe, straight lines plotted on them are not true compass bearings. Mercator's solution was to spread the parallels of latitude in the same proportion as the spread of the meridians. The scale of the resulting map varies from latitude to latitude, increasing progressively away from the equator. The actual amount of the spread was determined by trigonometric tables.

The map that Mercator published in 1569 was, as E. G. R. Taylor puts it, essentially a scholar's map (Taylor, 1957:222). The coastlines were not drawn with care but were generalized from the written reports of the explorers (Fig. 14, pp. 82–83). Furthermore, Mercator did not explain the method used in making his projection. One gets the impression that here was a man with a theoretical solution to a problem, who was not at all concerned with practical applications. The preparation of an accurate world map based on his projection he left to others.[2]

As a result Mercator's map was not widely adopted as navigators continued to make frustrating use of the Portolanos, always finding that at the end of a long voyage they had reached the wrong place. It remained for an English geographer to explain Mercator in words understandable to people who might make use of his projection. Edward Wright in 1599 produced the trigonometric tables that made it possible for other people to reproduce Mercator's projection. But how to explain this projection to ordinary sailors not trained in mathematics? Wright devised a simple, if incorrect, explanation. Think of a spherical balloon, he·said, tangent at the equator with the inside surface of a hollow cylinder. The balloon is marked off with lines of latitude and longitude. Now blow up the balloon. As it stretches it comes in contact with the inside of the cylinder, the lines of longitude lie as straight lines against the cylinder, and the lines of latitude are stretched apart in proportion. Of course the high latitudes will pop out of the ends of the cylinder, and the poles can never be shown at all no matter how far the balloon is stretched. This was a mechanical, nonmathematical demonstration of how such a projection could be made that satisfied the mathematically illiterate—which is most people (Brown 1960:95). By 1630 Mercator's projection had supplanted all others for use in navigation in low and middle latitudes, and it is still the only projection useful for this purpose.

Mercator and his friend Abraham Ortelius of Antwerp both conceived the possibility of preparing the world map in sections and binding them together in atlas form. The Ortelius atlas—*Theatrum orbis terrarum*—was the first to be published, in 1570. Thereafter a large number of atlases were prepared. Mercator started publishing sheets of his atlas in 1585, but the whole work did not appear until after

[2]Mercator was not the first to make use of this principle of projection. In 1511 Erhard Etzlaub of Nuremburg prepared a map of Europe and Africa as far as the equator on which he spread the lines of latitude and longitude proportionally. It is not clear whether Mercator knew of this or not (Bagrow and Skelton, 1964:148–150).

his death (Bagrow and Skelton, 1964:179–189). Amsterdam became a major center for the publication of atlases and wall maps, both of which became very popular, especially in the seventeenth century. King Charles II of England in 1660 commissioned a set of some forty wall maps, each 6 feet in height, on which the new world revealed by the explorers could be viewed. The Amsterdam publishers also began to make globes for popular use. In France the first producer of world atlases was Nicholas Sanson d'Abbeville, who in the seventeenth century founded a "dynasty of cartographers" which produced maps and atlases for over a century (Bagrow and Skelton, 1964:185).

During the seventeenth and eighteenth centuries the mapmakers only slowly revised their maps to include the newest information and to acknowledge the existence of gaps in their information (Babcock, 1922; Tooley, 1949; Crone, 1950). The mapmaker who is credited with being the first to eliminate the drawings of weird creatures that traditionally occupied the blank spaces on the maps was Jean Baptiste Bourguignon d'Anville. Jonathan Swift, the author of *Gulliver's Travels* (1726), had pointed out this practice in a well-known jingle:

So geographers, in Afric maps
With savage pictures fill their gaps,
And o'er uninhabitable downs
Place elephants for want of towns.

It was d'Anville in 1761 who first left out the elephants and showed blank spaces where blank places belonged. It is interesting that even today some map publishers cover their maps with a uniform spread of pictorial symbols, believing, no doubt, that balance and symmetry are more important than the careful portrayal of an asymmetrical nature.

Eliminating Erroneous Ideas from the World Map

The fifth problem that was solved between 1521 and 1779 was the removal from maps of the many erroneous ideas inherited from Ptolemy. The latter had become the almost undisputed authority in Christian Europe during the Middle Ages after his *Geography* had been translated into Latin. Each of the numerous errors in his work had to be eliminated despite the strong resistance of those who treasured the stability of tradition in an insecure age. To overcome the idea that the torrid zone is torrid and that people near the equator turn black took Price Henry's sailors many decades of effort.

One of Ptolemy's most influential errors had to do with the great southland, which he called the *terra australis incognita*, the unknown southern land. On his world map (Fig. 2) this unknown land encloses the Indian Ocean on the south. The Arabs knew that the Indian Ocean was open to the south, and Fra Mauro believed them. But there were many others who either did not know about the Arabic writings or discredited them. The Portuguese navigators had to find out for themselves things that certainly had been found out by others—even by the Phoenicians of the seventh century B.C. Still, the Portuguese voyages around southern Africa did not eliminate the idea of a great unknown land mass in the Southern Hemisphere. When Magellan passed through the strait near the southern end of South America, he was certain that Tierra del Fuego was a part of Ptolemy's southland. Later mapmakers extended the southland all the way across to Tasmania. Step by step the explorers sailed into higher and higher latitudes of the Southern Hemisphere without encountering the unknown land.

Captain Cook was the explorer who finally drew the outlines of the Pacific Ocean and eliminated Ptolemy's southland (Fig. 9). On his first voyage (1768–71) he sailed to Tahiti to carry out observations on the transit of Venus across the sun. The observations he made proved not very useful, but he did sail southward from Tahiti to about latitude 40°S, and, finding no land, he turned westward to New Zealand. He proved that New Zealand was not a part of any southland and then went on to make a careful chart of the east coast of Australia. He passed through the Torres Strait between Australia and New Guinea, rediscovering a passage that a Spanish explorer had reported in 1606.

Cook's second voyage is remarkable for several reasons. He was able to keep his crew from contracting scurvy through the regular consumption of fresh fruits and vegetables. He had a chronometer that permitted him to fix his position accurately at all times. He was also accompanied by such trained observers as the Forsters, father and son, of whom we shall hear more later. This was the first of the exploring expeditions to start gathering information about matters other than just coastlines. But it was also the voyage on which the southland was finally eliminated. Cook left England in 1772, and from December, 1772, to March, 1773, he sailed in the far south of the Atlantic and Indian Oceans. From November, 1773, to February, 1774, he sailed in the far south of the Pacific Ocean. His farthest point south was 71°10'S. Still there was no land; although Cook expressed the belief that there was an ice-covered land farther south, he was unable to reach it. From February to October in 1774 he sailed through the tropical South Pacific, discovering and locating numerous islands that Magellan had missed. When he returned to England in 1775, he brought the first authoritative report to reach Europe regarding the vast expanse of water in the Southern Hemisphere.

Cook's third voyage ended in disaster, but not before he had completed the outlines of the Pacific Ocean. In 1776 he sailed into the North Pacific, reaching the Hawaiian Islands in 1778. He followed the coast of North America northward through the Aleutian Islands and through Bering Strait into the Arctic Ocean, where he was stopped by ice floes at latitude 70°44'N. He then returned to Hawaii for the winter. After sailing away from the islands in early 1779, he was forced to return to repair a mast on one of his ships. He was killed in a fight with some of the native islanders. Thereafter his second in command again took the ships to the far north to prove that there was no Northwest Passage. The ships returned to England in 1780. The world map had been essentially completed, at least in its major outlines.[3]

THE SHAPE OF THE EARTH

With the general dimensions of the earth established, the problem of the shape of the earth became increasingly important. Columbus had thought that it was pear-shaped and that it was necessary to climb toward the equator. In 1687 Isaac Newton and Christian Huygens arrived at the conclusion mathematically that the earth must be flattened at the poles and that it must bulge at the equator. But this concept was challenged by Jacques Cassini in 1720. We need to become acquainted with the Cassini family.

There were four generations of Cassinis who were in charge of the astronomical observatory in Paris and who carried out the first major topographic survey of a large country—France. Giovanni Domenico Cassini, an Italian, became the director of the observatory in 1667 and began to carry out a series of measurements to fix certain locations within France. His son, Jacques Cassini, who became director in 1712, in 1713 carried out a measurement of the arc of the meridian extending from Dunkerque through Paris to Perpignan. His son, César-François Cassini de Thury, used the meridian surveyed by his father to start a network of triangulation on which to fix the position of the sheets of the topographic survey. He started the survey in 1744 but died before it was completed. The work was finally carried through to completion by his son Jacques Dominique Cassini (Fig. 15) (Crone, 1950).

In 1720, the first Jacques Cassini, who made the measurement of the Paris meridian, published a paper entitled *De la grandeur et de la figure de la terre*. His measurement did not show any indication that the earth was not a regular sphere and was, therefore, a direct challenge to the mathematical concepts of Newton and Huygens. The French Academy decided to settle the controversy by carrying out measurements of the arc of the meridian at different latitudes. From 1735 to 1748 a French expedition was working in the high Andes of Ecuador, which was joined by the French scientist Charles Marie de La Condamine together with Louis Godin and Pierre Bouguer. Sometimes the surveyors who had set up their instruments on the tops of the high peaks were forced to wait for weeks for the rare clear days when observations could be made. But, eventually, the job was done, and de La Condamine returned by way of the Amazon, carrying out the first scientific exploration of that river. Meanwhile another French expedition, headed by Pierre de Maupertuis, went to Lapland. In spite of the hardships imposed by the very low temperatures they encountered, the survey of a meridian was completed in 1736–37. These

---

[3]For accounts of the explorers and the routes they followed see Parks, 1928; Penrose, 1952; Taylor, 1957; Sykes, 1961; Rogers, 1962; Hakluyt, 1965; Parker, 1965; Hanson, 1967; Friis, 1967.

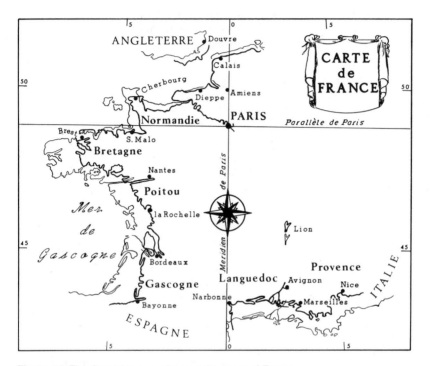

Figure 15. The Cassini survey changes the coast of France.

two surveys proved that the earth was flattened at the poles and did bulge at the equator and that Cassini was wrong. These measurements were accepted as correct until the modern period of satellite technology, which shows the earth to be dented and bulging like an old baseball, with a major bulge south of the equator. The deviations from the perfect sphere revealed by remote sensing could not have been detected by the eighteenth-century instruments.

RETROSPECT

During all this time between Magellan and Cook, explorers were finding out more and more about the earth. Expeditions had penetrated the Arctic, seeking either a Northwest Passage or a Northeast Passage between Europe and the Orient. Edmund Halley had plotted the trade wind zones of both hemispheres. He had also mapped the monsoons of Asia and had set forth the hypothesis that these seasonally shifting winds were the result of the differential heating of land and water. In winter the land was cold and the ocean warmer, so the winds moved from land to water; in summer the reverse of the winds was produced when the land became warmer than the water. A vast amount of knowledge flowed into Europe, and the impact of all this on the intellectual life was enormous.

At first the explorers tried to describe everything they found in the strange lands. But, having no concepts regarding the new phenomena they found, an attempt was made to describe things by analogy. Unfamiliar animals were described by reference to familiar ones. For example, when a fifteenth-century Florentine traveler first saw a giraffe he described it as "almost like an ostrich save that its chest has no feathers but has very fine white wool. . . . It has a horse's feet and bird's legs. . . . It has horns like a ram" (Hale, 1966:164). Only later did the explorers begin to describe things in terms of dimensions, texture, or color, and, finally, in terms of abstract categories. Fantastic stories of strange creatures gradually gave way to sober accounts that could be, and were, verified. After the first contacts with unfamiliar places when observers reported on anything— especially on strange and unusual things—geographers began to ask questions about what should be observed; how they should be observed; and, especially, how the things observed could be related to some generalization of the empirical perceptions. This, in turn, led to new and sharper perceptions.

In the next chapter we will examine the impact of all this on the geographic ideas inherited from centuries past.

CHAPTER 5

The Impact of Discoveries

Furthermore, men avidly made use of the accumulating voyages and travels which had been and were being published. What is the state of nature, the primitive stage in mankind's development? What are primitive peoples like? What influences, according to the travelers, determine the character of far-off peoples? These questions were asked in the seventeenth century also, but its thinkers still leaned heavily on the classical writers. From the middle of the eighteenth to the early part of the nineteenth century, in the writings of Buffon, Montesquieu, Herder, and Malthus ... one sees what a refreshing and inexhaustible well these voyages and travels had become....

... They built on the past, but the world became a richer place for their departures. Even today their questions suggest our questions, but it would be a miracle if they did more than that. For they lived in a world which resembled the past more than what was to come, at least as far as problems of human culture and the natural environment are concerned.

What happened to geographical ideas in Europe as a result of the unprecedented flood of new information coming from the voyages of discovery? What kind of meaning could be discerned in this rapidly expanding new picture of an unfamiliar world? The cosmographers, pursuing their ancient goal of increasing their knowledge of the ordered universe, the cosmos, assumed that the coherence and harmony that they observed was the creation of a divine being (Tuan, 1968).

The quotation above is from C. J. Glacken, *Traces on the Rhodian Shore* ... (Berkeley and Los Angeles: University of California Press, 1967), pp. 502–503.

Traditionally the divine being was conceived in terms of human experience: like a man he must have a conscious purpose and a plan of action. Answers to questions about the universe were given as logical deductions from the basic and generally accepted theory that the world was created specifically as the human habitat. But from this assumption two quite different interpretations of man's place in the universe are possible: man may be seen as a creature of his habitat—his activities and even his physical character controlled by his natural (nonhuman) environment, including the astrological influences that were seen as guiding human affairs; or man may be seen as a creature commanded by the divine plan to conquer and control his natural environment, and in a sense to complete the work of creation out of the raw materials of his natural surroundings. These are two quite different interpretations, and each is logically derived from the concept of a planned creation.

From the early fifteenth century on, as new information poured over the European intellectual world, the cosmographers struggled with these questions. At first they followed the traditional patterns, seeking evidences of the divine plan. As in the medieval period preceding the fifteenth century, most of the scholars accepted almost literally the account of creation presented in Genesis; and they went to great trouble to make the new findings about the earth fit the Scriptures. But these methods and traditions became more and more difficult to sustain. The period we are examining was one of groping, when men struggled to shake loose from the weight of old paradigms and to find new answers to old questions about order, harmony, and meaning.

There was still very little specialization in the kinds of problems scholars investigated or wrote about. The development of special branches of knowledge came later. But of course there were some whose approach was primarily mathematical, and there were others whose approach was literary. There were some who asked questions and looked for answers, and others who were fully occupied—and satisfied—with the task of establishing the validity of the facts that were reported by the explorers and of putting together a general picture of what things were like.

NEW CONCEPT OF ORDER IN CELESTIAL SPACE

There were a few brilliant flashes of genius during these centuries that illuminated the road ahead. One of these had to do with the question of whether the earth was fixed in the center of the universe with the sun and stars revolving around it or whether the sun was the center of the universe, with the earth revolving about it. The idea of a heliocentric universe was not new: it had been offered as a hypothesis in the third century B.C. by Aristarchus of Samos. But Ptolemy had accepted the geocentric universe, and he was the great authority who could not be easily questioned. Nevertheless, by the fifteenth and sixteenth centuries there were some who doubted Ptolemy. One of these was a Polish scholar, Nicolaus Copernicus, who, between 1497 and 1529, carried out numerous observations of the movements of the planets, the moon, and the stars. The Ptolemaic system was too complicated; but when he calculated the movements of these bodies with the sun as a fixed center, the results were more satisfactory. In 1543 he published his great work, *De revolutionibus orbium coelestium,* in which he presented the picture of a heliocentric universe. He still followed Ptolemy in describing the movements of the planets around the sun as circular.

Three other scholars were of primary importance in breaking the Ptolemaic tradition concerning the planets and the laws of motion. In 1618 Johannes Kepler, a German astronomer, recognizing that the planetary motions were elliptical rather than circular, presented his work on the laws of motion. In 1623 Galileo presented the proof that Copernicus was right about the heliocentric universe and thereby brought a storm of abuse from those who thought that the findings of science might undermine the authority of the Church. In 1632 his book comparing the universe conceived by Ptolemy with that conceived by Copernicus was acclaimed by scholars throughout Europe not only for the brilliance of its ideas but also for the literary quality of the presentation. Galileo for the first time formulated the concept of a universal mathematical order, a universe that could be described in terms of mathematical law rather than in the verbal and logical terms of Aristotle. Finally, in 1686 Isaac Newton presented his laws of gravitation. With Copernicus, Kepler, Galileo, and Newton—in about a century and a half—the seeds of the scientific revolution had been planted; and with this revolution came the beginnings of specialization, which would end the work of the cosmographers.

THE ACCOUNTS OF THE VOYAGES OF DISCOVERY

There was still the real possibility, in sixteenth-century Europe, that a considerable proportion of the written accounts of the voyages of discovery might be lost—as so many similar documents in ancient times had been lost. To be sure the printing of books in Europe began in the middle of the fifteenth century, but there were not enough printing establishments to meet the demand. There were three geographers and publishers in the sixteenth and seventeenth centuries who started what became important collections of such reports. The earliest was a Venetian named Giovanni Ramusio, whose three-volume work, *Navigationi et viaggi,* was published between 1550 and 1559. It contains an edition of Marco Polo and a little-known Portugese report, *All the Kingdoms, Cities, and Nations from the Red Sea to China,* written in 1535. The collection was completed by the publisher after Ramusio's death in 1557.

Most widely known of these collections of voyages is the one started by Richard Hakluyt in England. The first small edition of Hakluyt's voyages appeared in 1598; the full three-volume collection came out between 1598 and 1600 (Parks, 1928). Hakluyt was a lecturer at Oxford after 1574 and introduced the "new geography" of the late sixteenth century in that university. He was also advisor to many navigators of his time. His recommendation to the polar explorers was to stop trying to find a way from Europe to eastern Asia by the Northeast Passage and to concentrate on the search for a Northwest Passage, north of North America. He also urged the Britons to establish colonies along the eastern side of North America.

Hakluyt was influential in getting the third collection of reports on voyages started. This was done by the German publisher, Théodore de Bry. The de Bry collection included some twenty-five volumes published between 1590 and 1634.

THE CHANGING IMAGE OF THE EARTH

The traditional way of writing about the earth as the home of man, which was followed by the early cosmographers of the sixteenth and early seventeenth centuries, was inherited from Strabo. The idea was to bring together all that was known about the different parts of the world. Strabo was very critical of his sources and indicated clearly that he had confidence in some and little or no confidence in others. His judgments were made on the basis of then current preconceptions and established dogma, and in the long run they did not always prove to be right—as, for example, when he rejected Pytheas as a storyteller. As new information came in during the Age of Exploration and as earlier writers were shown to be in error, there was ample opportunity for writers to offer their versions of the "new geography." The only new items were the reports on what the explorers had found; the methods of observation, the ideas or concepts that guided the observations, and the kinds of questions asked remained the same.

Not all the writers of this period were as careful to be critical of their sources as Strabo had been. In 1490, almost twenty years after the first Portuguese ship crossed the equator into the Southern Hemisphere without burning up, an Italian writer published a compendium in which he described the torrid zone as sterile and uninhabitable, almost in the same terms used by Aristotle and Ptolemy. He gave a demonstration of the method of thinking deductively from astrological principles when he insisted that there could be no land in the Southern Hemisphere because the heads of the animals in the zodiac were all pointed toward the north (Kimble, 1938:219). And there were whole books purporting to describe new lands discovered by travelers that were, in fact, wholly fiction, like the famous *Travels of Sir John Mandeville*. This book, which was pieced together by an unknown author from earlier fourteenth-century travel accounts (with the trivial and dull passages removed and with the addition of descriptions of the regions traversed and of the strange customs and physical appearance of the inhabitants) was a widely popular but wholly false literary production of the early Renaissance (Fig. 16, p. 94). Even in the seventeenth and eighteenth centuries Mandeville made good reading and was not always recognized as fiction.[1]

[1]The Dover edition, published in 1964 and edited by A. W. Pollard, gives a brief summary of the history of this famous work (Pollard, 1964).

Figure 16. Drawings adapted from Sir John Mandeville.

Sebastian Münster

The first important compendium of geography published after the early voyages of discovery was the one by the German cosmographer, Sebastian Münster. Münster, who had previously established his reputation as a classical scholar by translating several geographical works from the Latin into German, spent eighteen years (and made use of the services of some 120 writers and artists) in preparing his *Cosmographia universalis*, published in 1544. The work is strictly in the Strabo tradition, and, in fact, Münster was known to his contemporaries as "the German Strabo" (Glacken, 1967:363–366).

The *Cosmography* is in six books. The first presents a general picture of the universe in Ptolemaic terms, to which the author adds a discussion of the dispersal of mankind over the earth after the flood. The other five books deal with the major divisions of the earth. Of these the best ones have to do with southern and western Europe and with Germany. He divides the old world into the traditional three parts—Europe, Asia, and Africa—a practice whose origin was previously discussed in Chapter 2. Europe he separates from Asia along the Don River; and the line between Asia and Africa is placed along the Nile—presumably he never read Herodotus on this subject. His treatment of Asia and of the Americas is amazingly uninformed, considering that the German map-makers had access to the new information from the explorers and this information was surely available in Germany while Münster was working. He enlivens the weird stories of America and Africa with woodcuts of men with heads in their chests or with only one leg—creatures that come right out of Sir John Mandeville.

Münster's *Cosmography* was regarded as the authoritative work on world geography for more than a century. New editions with few changes were published in 1545, 1546, and 1548. In 1550 there was a new edition that included many additions and corrections. After his death (from the plague in Basel) in 1552 there were numerous later editions that included much supplementary material.

Cluverius and Carpenter

Writers of books on "universal geography" in the seventeenth century had better access to the new materials than Sebastian Münster had. In 1616 Cluverius (Philipp Clüver) published a book on the historical geography of Germany, of which Carl Sauer remarked that he skillfully united "knowledge of the classics with knowledge of the land" (Sauer, 1941:11). A similar work on Italy was published posthumously in 1624. Also in 1624 his six-volume compendium of geography appeared.[2] In this work he started with the traditional picture of the universe as portrayed by Ptolemy, showing that he had no knowledge of the work of Coper-

---

[2]*Introductionis in universam geographiam* (Leiden); translated into French in 1639 and into German in 1678.

nicus, which had been published seventy-nine years before the death of Cluverius. The other five volumes of the compendium, however, are much more up to date. His work on Europe was especially well done (Partsch, 1891).

The first geographer to attempt a compendium of geography in English was the British geographer, Nathanael Carpenter. He was at Oxford during the period after 1609, when Cluverius was a frequent visitor, and must have received many ideas from this source. He also reports numerous examples of how human character is determined by climate, ideas which he derived from the writings of the French scholar, Jean Bodin (p. 103). Both Carpenter and Bodin accepted the old Greek theory that the habitability of a place was a function of its latitude.

Varenius

The European scholar who profoundly influenced the content and scope of geography for more than a century was Varenius (Bernhard Varen). Varenius set forth, more clearly than anyone had done before his time, the relationship between geographical writings that describe the characteristics of particular places and those that describe the general and universal laws or principles that apply to all places. The first he called *special geography,* the second *general geography.* This intellectual problem of the relation between the specific and the general became of major importance in the early part of the seventeenth century as a result of the flood of new information about specific places and the effort to generalize this information. Cluverius made no attempt to connect the general ideas in his first book with the specific observations about places in the other books; Carpenter did very little with descriptions of particular places. Varenius saw that there was a need to demonstrate the close interconnection between these two viewpoints: special geography is of great practical importance for government and for commerce but it leaves out the fundamentals of this field of study; general geography provides these fundamentals, but to be of maximum utility they must be applied. Therefore, special geography and general geography did not suggest a dichotomy, but rather two mutually interdependent parts of a whole. It is indeed unfortunate that Varenius lived only twenty-eight years and was not able to amplify these ideas.[3]

Bernhard Varen was born near Hamburg in Germany in 1622. In 1640 he entered the university at Hamburg where he studied philosophy, mathematics, and physics (Dickinson and Howarth, 1933:100). After three years he went to Königsberg to study medicine and shortly thereafter he went to Leiden. In 1647,

[3]J. N. L. Baker points out that Bartholomew Keckermann, a German geographer, used the terms, *general geography* and *special geography,* in lectures at Danzig in 1603 and in a book published in 1617. Since Varenius made use of Keckermann's work, it is likely that he adopted this distinction from the older writer and then provided a clear demonstration of the relation between these two points of view (Baker, 1955b:56; 1963:113).

however, he took a position as private tutor in a family living in Amsterdam. In this busy commercial city he came into close contact with merchants who were sending their ships to trade in the most distant parts of the earth, including Japan, where the Dutch had a trading post on an island in the harbor at Nagasaki. Responding to the practical need of the Dutch traders to know much more about the people with whom they were doing business, Varenius in 1649 published a book entitled *Descriptio regni Iaponiae et Siam*. This work contains five parts: (1) a description of Japan, compiled from information available in Amsterdam; (2) a translation into Latin of a description of Siam by J. Schouten; (3) an essay on religions of Japan; (4) some excerpts from the writings by Leo Africanus on religion in Africa; and (5) a short essay on government. This was a book aimed at providing the merchants of Amsterdam with useful knowledge about places and people.

At this point Varenius saw that descriptions of particular places could have no standing as contributions to science if they were not related to a coherent structure of general concepts. He began work on his *Geographia generalis* in the fall of 1649 and completed the work in the spring of 1650. In it are references to a series of books on geography that he proposed to write, but in that same year he died in Leiden.

The *General Geography* of Varenius was the standard text in this field for more than a century. There were four Latin editions published in Amsterdam (1650, 1664, 1671, and 1672). Isaac Newton was so impressed with the work that he edited two more Latin editions published at Cambridge University in 1672 and 1681. In 1693 there was an English translation by Blome. Another Latin edition was published at Cambridge in 1712, with comments and corrections by J. Jurin. The edition of 1712 was also translated into English by Dugdale and Shaw, and this translation went through four editions between 1736 and 1765. One of the Latin editions, edited by Isaac Newton, was used as a text at Harvard in the early years of the eighteenth century (Warntz, 1964:117).

Unlike the writers of the earlier cosmographies, Varenius was careful to include the most recent ideas. He accepted the heliocentric universe of Copernicus, Kepler, and Galileo. On the basis of this he was the first to note the difference in the amount of heat received from the sun in the equatorial regions as compared with the higher latitudes. The sun's heat, he suggested, thins the air close to the equator and, therefore, the cold, heavy air of the polar regions must flow toward the equator. This was the first step toward the explanation of the world's wind systems (Peschel, 1865:396).

Varenius had a conception of the field of geography that was far in advance of his time. He insisted on the practical importance of the kind of knowledge included in special geography, but he also insisted that special geography only becomes intelligible when the specific features are explained in terms of abstract concepts or universal laws. Geography, he wrote, focuses attention on the surface of the earth, where it examines such things as climate, surface features, water, forests and deserts, minerals, animals, and the human inhabitants. The human properties of a place include "a description of the inhabitants, their appearances, arts, commerce, culture, language, government, religion, cities, famous places, and famous men" (Dickinson and Howarth, 1933:101). In general geography, he continued, most things can be proved by mathematical or astronomical laws; but in special geography, with the exception of celestial features (climate), things must be proved by experience—in other words, by direct observation through the senses.

QUESTIONS AND HYPOTHESES

The mounting mass of new observations and the startling new theories concerning the motions and behavior of physical bodies led to new speculation concerning the origin of the earth. But the methods of thinking had not yet broken free from the traditions inherited from ancient Greece and from the accounts of the creation in the Bible. Even brilliant innovators, such as Galileo and Newton, were as much concerned to demonstrate that their theories did not really depart from the Scriptures as they were to demonstrate that they were supported by observations.

During the second half of the seventeenth century there were groups of scholars taking important first steps in the use of the scientific method. In England there were men like Robert Boyle, discoverer of the law concerning the behavior of a given volume of gas with changes in temperature and pressure; Robert Hooke, developer of laws concerning the elasticity of solid bodies; John Flamsteed, first director of the Royal Observatory at Greenwich (established in 1675); and Edmund Halley, astronomer and author of the first scientific explanation of the world's wind systems. These men were making use of the methods of experimental science in arriving at hypotheses concerning the operation of natural processes. In 1663 some of these innovators formed the Royal Society of London "for the improvement of natural knowledge." Amsterdam was a center of intellectual freedom, where observations brought back by the Dutch merchants and explorers were stimulating new kinds of inquiry.

One of the problems that bothered people during this period had to do with the impact of the new science on the traditional concepts of the Greeks. Did fire and air rise in seeking their natural places in the universe, as Aristotle had believed and as everyone had believed since his time, or did all these elements obey the universal law of gravitation as proposed by Newton and Boyle? Controlled experiments had to compete against the weight of long-established tradition. Yet there were some who realized that the emphasis on the classics in education would have to be modified by greater attention to mathematics.

The Origin of the Continents and Oceans

Much speculation was aroused concerning the origin of the earth and of its surface features as a result of the publication of a two-volume work by Thomas Burnet in 1681 entitled *Telluris theoria sacra* (Taylor, 1948). Burnet was a clergyman whose hypothesis regarding the origin of the earth and its present condition came entirely from his own unfettered imagination. When God created the earth, said Burnet, and set it spinning on its axis the earth became egg-shaped. Since its axis was then perpendicular to the plane of the ecliptic (its orbit around the sun), there were no seasons but only a perpetual spring at the latitude of England. The surface of the earth was smooth. But people like Methuselah lived so long and with so much leisure time that they were able to develop much evil among men. In anger, God decreed the destruction of the earth. The surface began to crack open and break up into ugly mountains and valleys. Then a flood of water covered the whole earth, released from deep inside where the water had been confined. The shock to the earth knocked its axis away from the perpendicular so that thereafter there were seasons. The surface was broken into continents, mountains, and deep depressions into which the water drained off to form the oceans.

The appearance of Burnet's *Sacred Theory of the Earth* set off a controversy that lasted for decades. Several new theories regarding the origin of the earth were offered. In 1695 John Woodward presented the idea that the flood that God sent in anger dissolved the rock material of the earth, and this material was later deposited in the form of layers, or strata, some of which contain the fossil remains of vegetable and animal life. William Whiston, who was much impressed by the observations made by Edmund Halley in 1682 regarding the comet that bears his name, developed the theory that the earth itself was made from the debris of a comet. Furthermore, the near approach of a second comet was the cause of the flood, of the elliptical (rather than circular) orbit of the earth around the sun, and of the continents and ocean basins on the earth's surface. The comet raised tides in the rock crust of the earth similar to the tides raised by the moon on opposite sides of the earth. The tidal crests were represented by the continents, the troughs by the basins of the Atlantic and Pacific oceans. Whiston supported his theory by impressive mathematical equations showing how a comet could produce such tides in the rock crust of the earth. However, there were so many things left out in his calculations that he also was immediately attacked. The theologians based their attack on the Bible: How could the sun already be in existence before the earth started moving around it, when Genesis clearly states that God created light on the fourth day?

In Germany Abraham Gottlob Werner taught that the material dissolved in the waters of the flood was deposited over the earth to form a series of layers like the skin of an onion. All the rocks of the earth's surface, he said, were formed in this way. Critics asked how he explained the disappearance of all that water; but the failure to find a plausible explanation for the disappearance of the flood did not make Werner abandon his theory, which was almost universally accepted well into the nineteenth century. In fact, the Swiss scholar, Horace Bénédict de Saussure (who was one of the first men to climb Mont Blanc—August 8, 1786), supported Werner by suggesting that the Alpine valleys had been cut by the violent torrents when the floods drained off into the present ocean basins.

On the Origin of Landforms

A great difference of opinion persisted throughout the period from the fifteenth to the eighteenth centuries and indeed on into the nineteenth century concerning the origin of the landforms of the earth's surface. Were the landforms created by divine purpose? Were the hills really everlasting as the Bible insists? Or were the mountains and valleys and ocean basins formed by cataclysms of nature, as Burnet believed? Or were the landforms carved out by the slow processes of erosion? The various theories some inherited from the Greek and Roman geographers, that explain the surface features of the earth as resulting from violent catastrophes, such as earthquakes or volcanic eruptions, are included under the general term, *catastrophism*. Theories based on the idea that all the processes of change observable today also operated in the past and can account for all the world's surface features are included under the general term, *uniformitarianism*.

In spite of the widespread belief in the principle of catastrophism, there were a number of students of the earth's surface features who rejected the whole idea of convulsions of nature. Of course neither the Arabic student, Avicenna, nor the Chinese student, Shen Kua, were known in Europe. One of the First Europeans to ridicule the idea of a universal flood was Leonardo da Vinci, who argued that running water could level off the heights of land until the earth was a perfect sphere. Bernard Palissy in France used a soil augur to observe the nature of the soil and suggested that rivers could easily wash the soil away if it were not held in place by forests. During the seventeenth century John Ray who is famous for his pioneer work in classifying plants and animals, also argued that water running down the slopes could slowly wash away the mountains.

During the seventeenth century there were increasing numbers of students who examined the rocks of the earth's crust and who developed new ideas about the way rock structure is reflected in the forms of surface features. In 1719 John Strachey showed how landforms reflect the underlying rock structure, and his work was given support by such field observers as Johann Gottlob Lehman, who published a study of the rocks and landforms of the Harz Mountains and the Erzgebirge in central Europe in 1756. Georg Christian Füchsel made a similar study of the Thüringerwald in Germany in 1762. In 1777 Simon Pallas published geological maps to show that many mountain ranges had granite cores. His expedition to

Siberia from 1768 to 1774 brought back many observations of the relation between rock structure and landforms. Meanwhile, the Italian scholar, Giovanni Arduino, in 1760 offered a classification of the rocks of the earth in which he used four major categories: primitive, secondary, tertiary, and volcanic. He explained that unconsolidated alluvium might cover all four.

Furthermore, there were some who developed basic ideas concerning the mechanics of river flow and valley development. Domenico Guglielmini, who died in 1710, studied the laws of river flow; and in 1786 the French scholar, Louis Gabriel Comte de Buat, worked out the mathematical equation to describe how the flowing water of a river can establish an equilibrium between velocity and the load of alluvium being transported. Thus he established what later came to be known as a "graded valley"—one that slopes just enough to maintain the flow of water. All this was presented in 1786.

Uniformitarianism had its first great supporter during the eighteenth century in James Hutton, the Scottish geologist. It was he who provided the first comprehensive treatment of the origin of landforms by processes that can be observed today. The processes that shape the surface of the earth, he wrote, indicated a world of continuing change with no vestige of a beginning and no prospect of an end. James Hutton, brilliant as were his ideas, wrote obscure prose. His ideas were communicated to the scholarly world through the writings of John Playfair, who, in 1802, published his book *Illustrations of the Huttonian Theory of the Earth* (Playfair, 1802/1956). This set off a major dispute between the supporters of Werner and those who began to accept the ideas of uniformitarianism (Chorley, Dunn, and Beckinsale, 1964:3-94).

It is important to note in these discussions that nothing promotes progress more rapidly than presenting a hypothesis that meets with a highly critical reception. This leads scholars to make new observations and to formulate new hypotheses. In 1704 Edmund Halley wrote that "the detection of error is the first and surest step toward the discovery of truth" (Taylor, 1948:112).

Methods of Classifying Plants and Animals

No less productive of scientific progress were the efforts to find more useful methods of classifying plants and animals. By the seventeenth century the voyages of discovery had flooded Europe with descriptions of the lands and organisms of previously unknown parts of the earth and with a steadily mounting volume of collections of plants and animals. These had to be organized and classified. If the suggested order was a manifestation of a divine plan, so much the better. One of the earliest scholars to offer a new way to classify organic forms was the English student, John Ray, who graduated from Cambridge University in the middle of the seventeenth century. In a work published in 1682 he outlined a way of classifying plants, and this was followed by a suggested classification of fishes and other animals. His great work on the classification of organic life was published in 1691, *The Wisdom of God Manifested in the Works of Creation.*

John Ray was a major influence on the Swedish botanist Carolus Linnaeus (Carl von Linné), who first suggested the systems of classification based on classes, orders, genera, and species, with increasing attention to detail. His categories of plants, published between 1735 and 1753, provided a simple and useful scheme of putting the new plant specimens in some kind of order.

Lamarck, however, was the first to show the need for a system of classification for plants and animals based on their natural characteristics. In a paper on invertebrate animals in 1801 he first set forth the idea that animals could develop new organs and new characteristics in response to their needs and that organs so developed would be inherited. In his *Histoire naturelle des animaux sans vertèbres,* published in Paris between 1815 and 1822, he further developed the concept of the evolution of animals and of their adaptation to environment. In his lectures on zoology he used the example of the giraffe, an animal that developed a long neck and long front legs as a result—so he believed—of stretching to browse high on the trees. Lamarck challenged the widely held dogma that all plants and animals were created in their present forms and had not changed since the creation. Lamarck formulated the first concept of evolutionary change—to which Darwin later added the mechanism of natural selection as the cause of evolution, rather than need or use.

Can Man Be Studied Scientifically?

In this period also the first steps were taken toward a scientific study of population. As early as 1662 William Petty and John Graunt suggested the kinds of statistical studies that could be made if adequate data were available. Graunt showed that certain statistical regularities made prediction of births and deaths possible on the basis of probability. Petty is described as "the father of political economy" because he blazed new paths in the study of population and economy, even where statistical data were lacking (Glacken, 1967:398-399). The first mortality rates were calculated for 1687-91 for the city of Breslau by the astronomer Edmund Halley (Peschel, 1865:685). Halley showed how life insurance rates could be calculated on the basis of probability.

Stimulated by these pioneer statistical studies, it was the German scholar, J. P. Süssmilch, who first demonstrated the existence of certain statistical regularities in population data. The sexes, he found, tended to remain more or less balanced, and birth and death rates could be forecast. Süssmilch, a Prussian clergyman, pointed to the evidence of God's planning in his book *The Divine Ordinance Manifested in the Human Race Through Birth, Death, and Propagation,* 1741. Thereafter governments undertook the collection of population data, and statistical studies based on these data were rapidly improved. However, it was a century later before Lambert Quetelet published his work *On the Social System and the Laws Which Govern It*, 1848, in which he showed that numerical information concerning individuals tends to group around averages in accordance with the theory of probability.

The Influence of Environment on History and Government

The enlargement of geographic horizons during the Age of Exploration also produced much speculation regarding the influence of the natural environment on human behavior. Although there was not yet any clear separation of the world of scholarship into disciplines, nevertheless the scholars who speculated about the origin of the earth and its surface features were not the same as those who speculated about the effect of the physical earth on man. The latter were mostly historians or students of government. And at the start these observers of environmental influence were only repeating the concepts set forth centuries earlier by the Greeks. They included a large amount of astrology inherited from the Greeks through Ptolemy and also from the writers of the Middle Ages.

One of the earliest was Jean Bodin, a French political philosopher who lived in the sixteenth century. Bodin's major work, published in 1566, lay in the search for universal principles of law. Anarchy, he wrote ten years later, was the supreme catastrophe, and he investigated various ways in which political order could be established and maintained. Accepting the Greek concept of climatic zones, he sought for the influence of the planets on the behavior of the earth's inhabitants. The people of the southern parts of the world, influenced by the planet Saturn, were given to lives of religious contemplation. The people of the northern regions, influenced by Mars, became warlike and excellent in the use of mechanical devices. The people of the middle regions, influenced by Jupiter, were able to achieve a civilized way of living under a rule of law. Nathanael Carpenter, who published the first geographical work in English in 1625, included much material from Bodin. He contributed to the persistence of the concept of the three zones of climate and of the effect of living in these zones on the character of the people. By this time there were numerous reports on the equatorial climate that supported the idea first developed by Posidonius that temperatures were not so high near the equator as they were beyond the tropics; but these reports were not read or, if they were read, they were not believed.

During the centuries after Bodin and Carpenter, scholars continued to seek examples of the influence of climate on human character and behavior. The Abbé de Bos, writing in 1719, found that weather had a definite effect on suicide and crime rates in Paris and Rome (Glacken, 1967:556–558). Suicides are most common either just before the beginning of winter or just after winter, when the wind comes from the northeast. Most of the crimes in Rome are committed in the two hottest months of summer. He also observed that works of art are produced only in the zone between latitudes 25° and 52°N.

One of the most influential of the eighteenth-century writers was the French political philosopher, Montesquieu.[4] One of the major themes in his work on laws had to do with the influence of climate on politics. The sterility of the ground in Attica, he wrote, resulted in the establishment of a popular form of government in Athens, whereas the fertility of the soil around Sparta was reflected in the establishment there of an aristocratic government. People develop different characteristics in cold climates than in hot ones, but by making use of proper laws the effects of climate can be minimized. Kriesel points out that a careful reading of Montesquieu's work demonstrates that he recognized the importance of other factors than climate alone—such factors as religion, the maxims of government, precedents, and customs. In any one country as some of these factors act with stronger force, the others are weakened. Kriesel, therefore, describes him as a possibilist rather than an environmental determinist (Kriesel, 1968).

Nevertheless, Montesquieu was very persuasive in his discussion of the effect of differences of climate on behavior. His famous experiment with a sheep's tongue and the conclusions he drew from it illustrate his method of thought:

> I have observed the outermost part of a sheep's tongue, where, to the naked eye, it seems covered with papillae. On these papillae I have discerned through a microscope small hairs, or a kind of down; between the papillae were pyramids shaped towards the ends like pincers. Very likely these pyramids are the principal organ of taste.
>
> It caused the half of this tongue to be frozen, and observing it with the naked eye I found the papillae considerably diminished: even some rows of them were sunk into their sheath. The outermost part I examined with the microscope, and perceived no pyramids. In proportion as the frost went off, the papillae seemed to the naked eye to rise, and with the microscope the miliary glands began to appear.
>
> This observation confirms what I have been saying, that in cold countries the nervous glands are less expanded: they sink deeper into their sheaths, or they are sheltered from the action of external objects; consequently they have not such lively sensations.
>
> In cold countries they have very little sensibility for pleasure; in temperate countries, they have more; in warm countries, their sensibility is exquisite. As climates are distinguished by degrees of latitude, we might distinguish them also in some measure by those of sensibility. I have been at the opera in England and in Italy, where I have seen the same pieces and the same performers; and yet the same music produces such different effects on the two nations: one is so cold and phlegmatic, and the other so lively and enraptured, that it seems almost inconceivable (Quoted from *De l'esprit des lois* in Glacken, 1967:569).

[4]Charles Louis de Secondat, baron de la Brède et de Montesquieu. His great work was *De l'esprit des lois* . . . (Paris, 1748).

Montesquieu made a number of mistakes that can be recognized today. He had no understanding of the method of confronting theory with observations or with controlled experiments. And in spite of the existence in France during the eighteenth century of up-to-date geographic concepts and information, he followed the ancient Greeks in dividing the world into Europe, Asia, and Africa (but adding the Americas, which the ancients did not know about). Europe, he said, has a variety of climates but no extremes; Asia, on the other hand, is either very hot or very cold and has no temperate climates. In other words he drew up generalizations about climatic conditions by continents rather than by zones of latitude. Yet he made the theory of climatic influence so plausible that these ideas persisted long after his time, and indeed are still to be found firmly embedded in some school curricula. The literary quality of his writing and the importance of his ideas about government gave him enormous prestige in the scholary world.

The Beginnings of Natural History

All these efforts were "new" in the sense that they offered new hypotheses, or new methods of classification, or new ways to make use of mathematics. But one of the newest of the "new" ways of viewing the earth came with the ground-breaking work of Count Buffon.[5] Buffon was director of the Jardin du Roi, the botanical garden in Paris, between 1739 and 1788; here he had access to a vast number of specimens of plants and animals and descriptions written by travelers and explorers from all over the world. His *Histoire naturelle* was written with the aid of many collaborators. It represents one of the first works resulting from the reports of voyages of discovery in which attention was turned from the oddities and marvels to a search for regularities and for the laws governing processes of change. His approach was nonmathematical and was not based on deductive reasoning: this was a strictly inductive study aimed at finding some kind of order in the flood of new information.

Buffon accepted the idea of a divinely created earth, but he rejected the notion that the complete and final plan of the creation was in the mind of the creator from the beginning. Man is one of the animals; but man differs from the other animals because with his mind he can remember experiences and learn from them. Man is commanded to conquer the earth and transform it. Wild nature, said Buffon, is ugly: man changes the face of the earth in the process of developing a civilization. Buffon was the first to focus attention on man as an agent of change—a focus that Plato failed to develop in spite of his observations on man-made changes in Attica and that had escaped the attention of other writers since Plato, except for the Chinese philosopher Mencius.

[5]Georges Louis Leclerc, comte de Buffon. His great work was the *Histoire naturelle, générale et particulière*, 44 vols. (Paris, 1749–1804). (Glacken, 1960; Glacken, 1967:655–685, 720–721.)

Buffon developed the idea of a cooling earth. The warmth at the earth's surface, he believed, came from the interior; for the heat received from the sun was not enough to counteract the loss of heat from the cooling earth. On the other hand, Buffon gathered evidence to support the notion that when forests are cleared the sun's heat can establish an equilibrium at the earth's surface between incoming and outgoing heat. As evidence of the beneficial results of forest clearing he pointed out that although Quebec and Paris are both at about the same latitude, Paris, in a cleared area, is much warmer than forested Quebec.

Climate, Buffon believed, affects the people who are exposed to it. But the examples he gave showed how vague was his concept of actual distributions in the world and how completely he missed discovering the regularities of geographic patterns over the earth. He made use of the traditional regional generalizations inherited, as we have seen, from the earliest Greek writers: white people occupy Europe, black people are in Africa, yellow people are in Asia, and red people in America. In spite of the writings of Marco Polo and the reports of the Dutch voyages of the sixteenth and seventeenth centuries, Buffon did not conceive of a similarity of climate in similar continental positions. In America, he continued, there are no black people because the tropical regions are not so extremely hot as those of Africa. This he explained by pointing to the easterly winds that cross the Atlantic and sweep unobstructed across the American tropics, cooling the air and bringing heavy rainfall.

On the other hand, Buffon insisted that man could adjust to any climate on earth. He was not compelled to react to any climate the way uncivilized native people would react. With proper clothing he could protect his skin color. With a typical European viewpoint he insisted that man could adjust more easily to life in cold climates than in hot ones.

Buffon developed the curious notion that nature in America was relatively weak compared with the natural conditions of other continents. The forests, he insisted, were less dense, the animals smaller, the potentialities for human settlement poorer. This brought forth strong reactions from Americans. Thomas Jefferson, who made numerous contributions to the study of geography, visited Buffon in Paris to protest at such an interpretation; but he was only able to get Buffon to acknowledge his error when he had a friend in Maine ship the skeleton and hide of a bull moose to the French scholar. Jefferson's geographical study, *Notes on the State of Virginia*, was one of the results of this dispute (Jefferson, 1787).

Scientific Travelers

When trained scientists were included in voyages of discovery and could make their own observations, knowledge of the earth increased rapidly. The earliest of these scientific travelers was the British astronomer, Edmund Halley. But Halley was interested in many things besides astronomy: his great genius consisted in making some kind of order out of complex data. It was he who made the first mortality tables (for the city of Breslau in 1693). He was the originator of many graphic methods for showing the geographical distribution of physical features of the earth (Tooley, 1949:54–55). His maps and discussion of the trade winds of the Atlantic, published in 1686, provided the first illustration of wind directions and wind shifts.[6] On this same chart he was the first to map the equatorial westerlies of the Gulf of Guinea. In 1698–1700 he was on the first voyage undertaken for purely scientific purposes. On the basis of observations from all around the world, he prepared the first map of magnetic variations, using lines of equal variation to show the pattern on a world scale (isogonic lines). These maps were published in 1701–1702 (Thrower, 1969).

Two pioneer field observers were Johann Reinhold Forster and his son Georg F. Forster. They accompanied Captain James Cook on his second voyage during which he sailed far to the south in the Indian and Pacific Oceans. On many islands of the South Pacific, where the expedition stopped, the Forsters carried out botanical observations and made collections. Georg Forster was the first to identify the pattern of temperatures on the eastern and western sides of continents at the same latitude and to point out the climatic similarity of western Europe and western North America. The Forsters, who derived their ideas of physical geography from Buffon and their ideas about the classification of plants from Linneaus, had the great advantage over their predecessors in that they could make their own observations and were no longer dependent on descriptions of what other observers had seen. They were very critical of ideas about the influence of climate on man, supporting their criticism with careful observations of the Dutch in South Africa and of the Polynesians. Both had important influence on the scholars who followed them. Of special significance was the influence of Georg Forster in the early life of Alexander von Humboldt.

Among the scientific travelers of this period mention should be made of Major James Rennell, onetime surveyor-general of India (1767–77), student of ancient geography, and one of the founders of the study of oceanography. At the age of twenty, in 1762, he embarked on a voyage that took him through the Strait of Malacca and on to the islands today included in Indonesia. In his journal he gives the results of his observations of the people, the climate, and the coasts of the islands. In 1764 he was appointed to carry out surveys of the East India Company's lands in Bengal and in 1767 succeeded to the post of surveyor-general of India. His *Atlas of Bengal* went through numerous editions between 1779 and 1788 and remained the standard work on Bengal until 1850. His professional papers include descriptions of the Ganges and Brahmaputra rivers and an account of how the alluvium brought by these rivers has built the huge delta at the head of the Bay of Bengal. He also collected a wealth of information regarding the currents of the Indian Ocean and the Atlantic Ocean. He carried out the world's first systematic

---

[6]Halley explained the deflection of these winds toward the west by the apparent westward movement of the sun. The first explanation in terms of the earth's rotation was given by John Hadley in 1735.

108 CLASSICAL

observation of ocean water; and two years after his death, in 1832, his daughter brought out the book he had almost finished, *An Investigation of the Currents of the Atlantic Ocean and of Those Which Prevail Between the Indian Ocean and the Atlantic Ocean.* This provided the first comprehensive view of the movements of ocean water in the Atlantic (Baker, 1963).

Problems of Population and Food Supply

Toward the end of the eighteenth century an increasing number of scholars (like the Forsters) began to seek new answers to old questions concerning man and his universe. Thomas Robert Malthus was one of these. His father followed the ideas common in earlier generations concerning the design of the earth by divine will and concerning the perfectability of human society as seen by Jean Jacques Rousseau, Condorcet, and others. The son took issue with the father, insisting that the construction of a truly happy society would always be hindered by the tendency of population to increase faster than the food supply. Population tends to increase geometrically, he said, whereas the food supply can only be increased arithmetically; and population always increases until it reaches the limits of subsistence, after which it is checked by war, famine, and pestilence.[7] A population close to the limits of subsistence suffers widespread misery. Malthus published his first essay on population in 1798; and in 1803 he issued a second edition, much enlarged and amplified with examples from various parts of the world. A sixth edition of his essay appeared in 1826.

The work of Malthus on population and food supply is recognized as one of the brilliant achievements of this period, when new concepts were being formulated to accommodate the new knowledge about the earth and man. One phrase in his essay on population referred to the "struggle for existence." Decades later Charles Darwin and Alfred Russel Wallace both realized independently that this was the keynote of the process of natural selection among organisms. Malthus, in seeking an explanation of the impossibility of increasing agricultural production to keep pace with population growth, became the first to formulate the economic law of diminishing returns from inputs of capital and labor.

NEW PERSPECTIVES

By the middle of the eighteenth century the vastly increased volume of new information about the world required a search for new perspectives and new ways of organizing what had been called cosmographies. This was the time when "universal geographies" began to appear, replacing the older type of descriptive geography. In some cases the "new geographies" were really innovative, as when Halley made

[7]Glacken points out that the Italian scholar Giovanni Botero in 1588 wrote that "the population of a city, or of the whole earth, will increase to the number permitted by the food supply" (Glacken, 1967:373).

use of isogonic lines to reveal the pattern of magnetic variation in the world, or when Philippe Buache, in 1737, used lines of equal depth (isobaths) to bring out the shape of the English Channel (Dainville, 1970). But in many cases only the facts were new; the method they followed was in the tradition of Sebastian Münster.

Buache and Büsching

Among the long list of authors of world geographies in the eighteenth century two were of special importance and had considerable impact on later writers. These were the French geographer, Philippe Buache, and the German philosopher and writer of geography, Anton Friedrich Büsching. Buache is best known for his concept of an earth marked off into major basins bordered by continuous ranges of mountains. On the land these are drainage basins, and the mountains form the drainage divides between different river systems. The basins continue under the oceans, and here the mountains form strings of islands or submerged sand banks.[8]

The persistence of his idea that drainage basins are bordered by mountains is amazing in view of the easily accessible examples in Europe of rivers that rise in one basin and flow through mountains to other basins. Perhaps Buache's concept would not have gained such wide support had it not been for the persuasive writing of the German geographer, Johann Christoph Gatterer, who identified the drainage basins as natural regions and used them as the frame of organization for geographical texts. From Gatterer the idea was picked up by several authors in Great Britain. The river basin was widely used as the framework for the identification of what we would now called systems of interrelated elements.

It is not so well known that Buache was the first geographer to identify the existence of a land hemisphere. He pointed out that a hemisphere with Paris at its center contains the greater part of the world's land. He presented this idea in 1746, presumably guessing that Ptolemy's *terra australis incognita* did not really exist and that Captain Cook (who would not enter the Royal Navy until 1755) would eventually prove that the land hemisphere concept was correct (Beythien, 1898; Wagner, 1922:268)—see p. 86.

Another "new geography" was offered by the German philosopher, Anton Friedrich Büsching, whose *Neue Erdbeschreibung* was published in six volumes in 1792. This was a discussion of Europe, organized in the traditional manner by political divisions. In many ways it was strictly in the Münster tradition, although the information it contained was up to date and reliable. Büsching did present two new concepts. He was the first to make use of population density as a geographic element (Peschel, 1865:xv). He was also ahead of his time when he pointed out that the transportation of goods by water could free man from dependence on local resources. This was the first suggestion of the principle of economic interdependence among countries, and it was written before the invention of the steam engine.

[8] *Essai de géographie physique . . . (1752); Considérations géographiques et physiques sur les nouvelles découvertes au nord de la grande mer* (1753). (Baker, 1955a.)

Büsching died in 1793 before he could complete the other parts of his proposed world geography. He was one of the best-known geographers of his time. His influence on Russian geography will be discussed later (pp. 224–225).

Malte-Brun

Perhaps the most important of the "universal geographies" of this period was the one written by Conrad Malte-Brun.[9] Malte-Brun was a Dane whose name was Malthe Conrad Bruun; but, when he was banished from Denmark in 1800 for liberal activities, he went to Paris and changed the form of his name. His universal geography was published in eight volumes between 1810 and 1829. He starts with a discussion of the history of geography, then proceeds in Volume 2 to an outline of geographic concepts, including the shape of the earth, the types of projections, and the astronomical relations. He reviews the various theories regarding the origin of the earth and the controversy between the catastrophists and the uniformitarianists, and then observes that the only way to treat physical geography in a useful way is to remain purely descriptive.[10] He believes that it is absurd to suggest that human characteristics are determined by climate. His treatment of the wind systems includes the latest observations from Cook's voyages.

In the second volume of the *Précis,* Malte-Brun devotes much space to a discussion of the land hemisphere concept. The center of the land hemisphere he places in France, to the west of Paris. In a footnote he refers to Père Chrysologue de Gy,[11] whose celestial planispheres were exhibited in Paris in 1778. One of his planispheres was centered on Paris and showed the land hemisphere. Malte-Brun makes no reference to Buache.

Kant

Another scholar who contributed to the development of geographical ideas in the eighteenth century was the German philosopher, Immanuel Kant. He lectured on a wide range of subjects at the University of Königsberg (in East Prussia, now Kaliningrad) from 1755 until a few years before his death in 1804. In 1770 he was appointed Professor of Logic and Metaphysics at that university. In his famous *Critique of Pure Reason* (1781) he rejected the teleological idea of final causes and insisted that explanations must be sought in what is chronologically antecedent. In this view Kant was opposed to such thinkers as Linnaeus, Leibniz, Süssmilch, Büsching, and Herder, but supported and amplified the ideas of Buffon, Maupertuis, Hume, and Goethe.

[9]*Précis de la géographie universelle* . . . (Paris, 1810–29); English translation by J. G. Percival, 3 vols. (Boston: Samuel Walker, 1847).
[10]"Il vaut mieux revenir à la marche purement descriptive de la géographie physique, la seule méthode vraiment scientifique et instructive." *Ibid.,* 1:495.
[11]Chrysologue de Gy (Noël André), *Théorie de la surface actuelle de la terre* (Paris, 1806).

Kant is also important in the history of geographical ideas because of the lectures he gave in physical geography at Königsberg between 1756 and 1796.[12] Physical geography, as the term was commonly used in Kant's time, included not only the features of the earth produced by natural processes but also the races of man and the changes on the face of the earth resulting from human action. Kant found that knowledge about the earth as the home of man was a necessary support for his philosophical studies, but he also found the subject very inadequately covered (May, 1970). "He devoted a great deal of attention to the assembly and organization of materials from a wide variety of sources and also to the consideration of a number of specific problems—for example, the deflection of wind direction resulting from the earth's rotation" (Hartshorne, 1939:38).

It was Kant's custom in the first lecture of the series each year to comment on the place of geography among the fields of learning. He pointed out that there are two different ways of grouping or classifying things for the purpose of studying them. Things that are similar because of similar origin, regardless of where or when they occur, can be grouped together in what Kant called a *logical classification*. Grouping things of diverse character and origin together because they occur at the same time or in the same place is called a *physical classification* of knowledge. A description or classification of things in terms of time is history; a description or classification of things in terms of area is geography (Adickes, 1925:2:394).

Hartshorne points out that Kant did not offer these ideas as something new. Rather this was a generally accepted classification of knowledge that seemed obvious and beyond dispute. Even by the end of the eighteenth century there was still no great focus of attention on the definition of fields of study or the separation of learning into disciplines. Like Strabo many centuries before, Kant was simply stating his purpose, not presenting an argument. In his lectures Kant followed Büsching in organizing his materials by political units. Kant saw man and his works in intimate association with the physical surroundings, and he also recognized human action as one of the principle agencies of change on the face of the earth. But he made no distinction between human and natural processes.[13]

---

[12] Kant never published these lectures but several versions were published on the basis of students' notes. Hartshorne points out that for over a century there was uncertainty about the authenticity of the several versions but that the question was largely answered by the careful studies of Erich Adickes, who first published on Kant's ideas about physical geography in 1911 (Hartshorne, 1939:38–39). Adickes is the most reliable authority on what Kant really said (Adickes, 1924–25).

[13] J. A. May's study of Kant's concept of geography makes it clear that the distinction between geography as the study of things arranged in earth-space and history as the study of things arranged in succession or sequence is a secondary division of his whole classification of the sciences. Physics, said Kant, is a theoretical science whereas geography "provides systematic knowledge of nature [the world that is external to man]. This implies that geography studies the relations among particular and concrete things rather than among abstract and general characteristics of things, and that it concentrates upon the differentiations rather than upon the similarities of nature." But Kant did not call such differentiations unique. Geography is an empirical science, seeking to present a "system of nature," and is a law-finding discipline. These ideas that Kant presented are analyzed in May, 1970, especially pp. 147–151.

CHAPTER 6

An End and a Beginning: Alexander von Humboldt and Carl Ritter

The fear of sacrificing the free enjoyment of nature, under the influence of scientific reasoning, is often associated with an apprehension that every mind may not be capable of grasping the truths of the philosophy of nature. It is certainly true that in the midst of the universal fluctuation of phenomena and vital forces—in that inextricable network of organisms by turns developed and destroyed—each step that we make in the more intimate knowledge of nature, leads us to the entrance of new labyrinths; but the excitement produced by a presentiment of discovery, the vague intuition of the mysteries to be unfolded, and the multiplicity of paths before us, all tend to stimulate the exercise of thought in every stage of knowledge. The discovery of each separate law of nature leads to the establishment of some other more general law, or at least indicates to the intelligent observer its existence.

The two great masters of German geography—Alexander von Humboldt (1769-1859) and Carl Ritter (1779-1859)—loom large across the pages of the history of science. Both lived and worked in Berlin for more than thirty years and both died in the same year. They were acquainted but not intimate. Never before or since have geographers enjoyed positions of such prestige, not only among scholars but also among educated people all around the world.

Many writers refer to Humboldt and Ritter as the founders of modern geography. But there are also good reasons for thinking of them as bringing the period of

The quotation above is from the translation by E. C. Otté of Alexander von Humboldt's *Cosmos* (London: H. G. Bohn, 1849-58) vol. 1, p. 20.

classical geography to an end. Using the large volume of new information resulting from the voyages of exploration, Humboldt and Ritter, each in his own way, produced massive syntheses. Although these syntheses made use of the new concepts and methods of study developed during the preceding two centuries, they, nevertheless, sought to present universal knowledge, just as Strabo had done and as had been attempted during the Age of Exploration by Münster, Varenius, Büsching, and others. But since 1859 the volume of recorded observations about the world and man's place in it has increased many thousands of times. In the nineteenth century the Age of Specialization came into being. No longer could any one scholar hope to embrace universal knowledge. The classical period had come to an end (Hartshorne, 1939:48–84).

ALEXANDER VON HUMBOLDT

Alexander von Humboldt was born into the Prussian landowning aristocracy. His father was an officer in the Prussian army and died when Alexander was ten years of age. He and his older brother Wilhelm were brought up by their mother. The mother has been described as "a very aloof and self-contained woman who provided for the education of her sons but gave them no intimacy or warmth. The sons were expected to show her respect and follow her directions" (Kelner, 1963:6). Alexander came to dislike the cold, constrained atmosphere of his home and lavished his affection on his brother and later on his brother's children. Alexander never married.

The brothers were educated at first by tutors, from whom they received an excellent grounding in classical languages and mathematics. Alexander had little interest in science but rather decided to undertake a career in the army. This was opposed by his mother, who insisted that he study economics as a preparation for a position in the civil service. However, events outside of his formal schooling combined with an almost insatiable curiosity about a great variety of matters led him toward a career in science. In Berlin his mathematics tutor introduced him to a group of liberals and intellectuals who gathered at the home of the Jewish philosopher, Moses Mendelssohn (the grandfather of Felix Mendelssohn, the composer). Jews and gentiles joined in discussions of the social inequities of an aristocratic society and drew up plans to do something about these things. Alexander also met the physician, Marcus Herz, a disciple of Immanuel Kant, who organized a series of lectures on scientific subjects, including demonstrations of scientific experiments.

When Alexander was ready to attend a university, he was already excited about the various aspects of the physical world. After a short time at a small university at Frankfurt on the Oder, he returned to Berlin to take a course in factory management at his mother's insistence. He also used the time to increase his knowledge of Greek and even began to study botany. In 1789 he went to the University of Göttingen, where he studied physics, philology, and archeology. It is well to remember that in the late eighteenth century studying a subject meant taking a course of lectures in which a scholar would impart to his students everything that was up to date in the field.

At Göttingen he met Georg Forster, lately returned from his voyage around the world with Captain Cook. From Forster Humboldt became excited about the study of botany. In 1790 these two started on a hiking trip down the Rhine to the Netherlands and thence by ship to England. The notes that Humboldt took on this trip show his interest and his ability to observe carefully such diverse matters as the varying price of wool or the effect of crop rotation on soil. He had a "success experience" in asking questions about the physical earth and man's use of it and in finding answers to his questions. Humboldt said later that his interest in geography started with his acquaintance with Georg Forster.

Humboldt then decided to attend the School of Mines at Freiberg in Saxony, where the celebrated scholar A. G. Werner was lecturing. Werner was the originator of a widely supported hypothesis that all the rocks of the earth had been formed by precipitation under water and had been deposited in layers. Humboldt attended lectures in physics, chemistry, geology, and mining. In 1792 he was appointed to an administrative post: first as inspector and later as director of mines in the Prussian state of Franconia. But his active mind was always formulating new questions about almost everything that caught his attention. He studied the effect of different rocks on magnetic declination. And underground in the mines he carried out experiments with subterranean plants he found growing there. His first scientific paper was on the results of these studies, published in 1793 (Humboldt, 1793). He also established a school for the miners and in various ways attempted to improve their living conditions. When he heard of experiments carried out by the Italian scientist, Luigi Galvani, regarding the electrical and chemical stimulation of muscles, he undertook some experiments of his own, as a result of which he came very close to finding out how to make an electric battery. There seemed almost no limit to the range of his curiosity. He also wanted to travel and see for himself what different parts of the world were like. He visited Bavaria, Austria, Switzerland, and Italy, on which trip he observed the rock structure of the Alps and tested some of the ideas of the Swiss scholar, Horace Bénédict de Saussure, who thought the deep Alpine valleys had been cut by the rush of water in the receding flood.

In 1796 Humboldt's mother died, and he came into possession of a small fortune. His part of the family inheritance was an estate on the east bank of the Oder River, known as Ringenwald. Income from this estate freed Humboldt from the necessity of earning a living. It paid for his travels in America and for the expensive publication of his many reports on these travels. In 1797 he resigned his government position and began to plan for traveling.

Humboldt's preparations for studies in the field were unprecedented. In Paris he gathered together an amazing variety of instruments and learned how to make use of them:

> He had been provided with an eight-inch Hadley sextant by Ramsden with a silver circle graduated at twenty-second intervals and a two-inch Troughton sextant, a kind of pocket edition which he called his snuffbox sextant. It was extremely accurate and very convenient to carry in difficult terrain. His barometers and thermometers had been standardized before his departure with those of the Paris observatory. The longitude determinations were made with a Dollond telescope and a chronometer by Berthoud whose rate of variation had been carefully checked. Three different kinds of electrometer, provided with pith spheres, straws, and gold-leaf, allowed him to observe atmospheric electricity. He also possessed a Dollond balance for the measurements of the specific gravity of sea water, an eudometer for the analysis of atmospheric gases, a Leyden jar and the necessary chemicals and glass bottles as well as a cyanometer designed by Saussure. This was an instrument by which the blueness of the sky could be determined through comparison with prepared gradations of blue colours and correlated with the hygrometrically determined humidity. The magnetic measurements were carried out with a Borda magnetometer, a rather cumbersome instrument [Kellner, 1963:62].

Before leaving Paris he learned from Pierre Simon Laplace how to use the aneroid barometer to determine elevations above sea level.

Several opportunities to join expeditions going overseas did not work out. One was an expedition to Egypt that was called off when Napoleon occupied that coun-

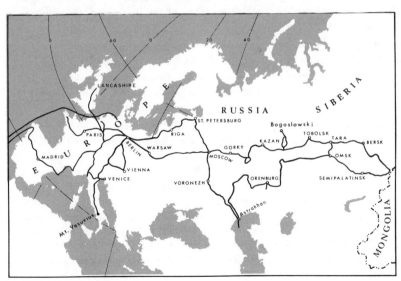

Figure 17. Humboldt's Travels in Europe and Russia.

try. Another was a trip to the Pacific, following the steps of Captain Cook. In 1798 Humboldt and a French botanist named Aimé Bonpland decided to go to Marseilles and there to take passage on a ship for Algiers, from which place they intended to travel overland to Egypt. Unfortunately—or furtunately we might say—these plans also fell through when the ship was wrecked off the coast of Portugal before it ever reached Marseilles. Humboldt and Bonpland then figured they might be more successful in getting passage on a ship from a Spanish port, so they set out overland for the city of Madrid, where all such passages would have to be arranged (Fig. 17). On the way to Madrid, Humboldt made daily observations of temperature and altitude, and he was the first to make an accurate measurement of the elevation of the Spanish Meseta.

In Madrid Humboldt's position in the Prussian aristocracy gave him access to the ruling aristocracy of Spain. He made a good enough impression on the Spanish prime minister that the latter granted permission for the travelers to visit the Spanish colonies in America—the first such permission granted to any non-Spanish Europeans since the expedition of C. M. de La Condamine to measure the arc of the meridian along the equator in 1735. Humboldt and Bonpland sailed in June, 1799.

Humboldt's American Travels

Humboldt's travels in the "equinoctial regions of the new continent" began at Cumaná in Venezuela (Fig. 18). First the two men went to Caracas and began exploring this long-settled part of the country. One of the first places they investigated was the Basin of Valencia in the midst of which is the Lake of Valencia, some 50 miles southwest of the capital. Humboldt noted that at one time the lake was much deeper and had an outlet to a tributary of the Orinoco but that in 1799 the lake had no outlet. Crops were being grown on the flat lakebed soils from which the lake waters had receded. Why should this event have taken place? The connection between the removal of forests and the drying up of rivers had been presented by Buffon and others, but Humboldt was the first to test this theory by confronting it with observed facts in a particular place. Here is what he had to say about the Lake of Valencia:

> Felling the trees which cover the sides of mountains, provokes in every climate two disasters for future generations: a want of fuel and a scarcity of water. Trees are surrounded by a permanently cool and moist atmosphere due to the evaporation of water vapor from the leaves and their radiation in a cloudless sky. They have an effect on the incidence of springs, not as was long believed by a peculiar attraction for the atmospheric vapor but because they shelter the soil from the direct action of the sun and thereby lessen the evaporation of the rainwater. When forests are destroyed, as they are everywhere in America by the hands of European planters, the springs are reduced in volume or dry up entirely. The river beds, now dry during part of the year, are transformed into torrents whenever there is heavy rainfall in the mountains. Turf and

moss disappear with the brushwood from the sides of the hills; the rainwater rushing down no longer meets with any obstructions. Instead of slowly raising the level of the rivers by progressive infiltration, it cuts furrows in the ground, carries down the loosened soil, and produces those sudden inundations which devastate the country. It follows that the destruction of the forests, the lack of springs, and the existence of torrents are closely connected phenomena [Humboldt, 1814–25, Williams translation, 1825:4, 143].

Around the Basin of Valencia, Humboldt observed the once continuous cover of tropical forest had been entirely removed and the lands were used for agriculture. The Lake of Valencia became a famous example of the application of a concept formulated by earlier writers but without carefully recorded direct observations to support it. Curiously, the idea that forests cause an increase of rainfall still persists.

During the year 1800 Humboldt and Bonpland carried out what must be one of

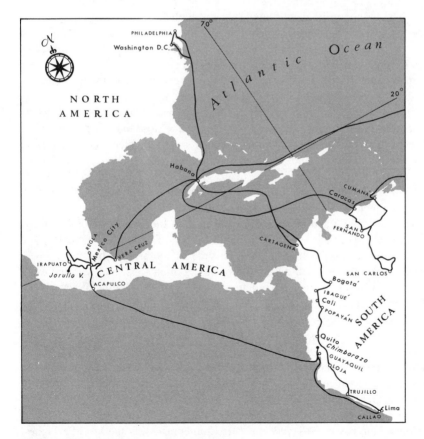

Figure 18. Humboldt's Travels in the Americas.

the great exploratory efforts in man's continuing urge to see beyond the horizon. They mapped some 1725 miles of the Orinoco River, mostly through uninhabited forests. In small boats and canoes they paddled upstream from the junction of the Orinoco and the Apure. Many years earlier de La Condamine had reported a story told by a Jesuit missionary—Father Manuel Ramón—that the water of the upper Orinoco split into two channels, one of which spilled over to reach a headwater of the Rio Negro and the Amazon. This was the Rio Casiquiare. But Philippe Buache, in compiling maps and reports on various parts of the world in accordance with his theory of continuous mountain chains, rejected de La Condamine's report. He showed a range of mountains along the divide between the Orinoco and the Amazon drainage basins. In 1800 Humboldt surveyed the Casiquiare and confirmed Father Ramón's observation of the bifurcation of the Orinoco. Modern geomorphologists recognize this as an example of river capture currently going on, in which, over long periods of time, the upper Orinoco will be cut off to become a part of the Amazon drainage. The Orinoco, we would now say, will be beheaded.

The trip up the river and along the Casiquiare imposed severe hardships. The travelers had to subsist largely on bananas and fish and were constantly exposed to the bites of clouds of mosquitoes, ants, and other insects as well as to poisonous snakes, man-eating fishes, and crocodiles. Almost everyone came down with fever, but Humboldt himself seemed immune and retained the necessary vigor to undertake whatever travel was necessary for his observations. With his instruments he established the exact latitudes of places and came very close to correct longitudes. He collected thousands of plant and rock specimens, all of which were transported back to Caracas and then to Cuba. Among his specimens were plants from which the poison, curare, is extracted. This kind of poison was first reported by Sir Walter Raleigh, but Humboldt brought back to Europe the first specimen. In November, 1800, the two men returned to Cumaná and sailed for Cuba.

In 1801 Humboldt and Bonpland arrived in the Colombian port of Cartagena and from there began their exploration of the Andes of Colombia, Ecuador, and Peru. With altitudes established for the first time by the aneroid barometer, with temperatures actually recorded by the use of the thermometers, and with the exact location of every observation fixed by latitude and longitude, Humboldt was able to give the first scientific description of the relations of altitude, air temperature, vegetation, and agriculture in tropical mountains. His description of the vertical zones of the northern Andes is a classic. He also examined the numerous volcanoes of Ecuador, descending again and again into active craters for the purpose of collecting the gasses emanating from within the earth. Looking closely at the rocks of the Andes, he decided that A. G. Werner was quite wrong about the origin of rocks and that granite and gneiss and other crystalline rocks were of volcanic origin.

Humboldt climbed most of the volcanoes of Ecuador. Since the expedition of de La Condamine it was believed that Mt. Chimborazo was the world's highest mountain. Attempting to reach its summit, Humboldt and Bonpland, on June 9,

1802, reached an altitude of 19,286 feet, which was the highest altitude that had been reached by man up to that time. This remained the record for twenty-nine years until in 1831 Humboldt's protégé, Joseph Boussingault, reached an altitude of 19,698 feet on the same mountain (Chimborazo, 20,561 feet high, was finally conquered by the British mountaineer, Edward Whymper, in 1880). Among the high peaks Humboldt was able to observe and report on the effect of altitude on human beings and to describe the symptoms of mountain sickness, or *soroche*. He explained the feeling of dizziness as resulting from low air pressure (it is now known to be due to lack of oxygen).

Humboldt and Bonpland finally reached Lima. Here Humboldt was able to observe the transit of the planet Mercury across the sun. This gave him an exact measurement of the longitude of Lima and made it possible to check his chronometer, which proved to be quite accurate. On the Peruvian coast he investigated the chemical properties of guano, or bird droppings. He sent some samples back to Europe and as a result started the export of guano as a fertilizer. On a sea voyage from Callao in Peru to Guayaquil in Ecuador he measured the temperature of the ocean water and for the first time described the movement of ocean water, including the upwelling of cold water from below. He named this the Peruvian Current and always objected to its being called the Humboldt Current because, he said, he had not discovered the current but only measured its temperature and velocity. Modern oceanographers have agreed to name all currents by geographical names—therefore, it is officially known today as the Peru Current.

In March of 1803 Humboldt and Bonpland sailed from Guayaquil to the Mexican port of Acapulco. The Viceroyalty of New Spain, as Mexico was then called, was at the peak of its prosperity due to relaxation of the restrictions on trade, to the investment of new capital in mining, and to the presence of an unusually capable group of administrators and ecclesiastical leaders. In 1794 New Spain was the first Latin American country to take a census of population. Humboldt updated the population figures to 1803 by consulting with the parish priests. He also found a rich collection of statistical data on production and trade. Traveling throughout the country he continued to climb the mountains, measure altitudes, fix locations by latitude and longitude, and investigate the many questions about man-land relations that occurred to his imaginative mind.

In 1804 the travelers sailed to Havana, Cuba. Humboldt now faced a problem that travelers have always faced—how to avoid the loss of notes and specimens painfully collected in the field. He and Bonpland had accumulated a large number of boxes containing notes written on their travels and specimens of plants and rocks of inestimable value. He dispatched the whole lot to Europe by different ships, some destined for Paris, some for London. Almost all his notes and drawings were made in duplicate, which was amply justified when some of the shipments failed to reach their destinations.

The visit of Humboldt and Bonpland to the United States was a memorable occasion. They reached Philadelphia in May of 1804. There they visited the American Philosophical Society and then started for Washington by way of Baltimore. From June 1 to 13 they were in Washington, where Humboldt had many meetings with Thomas Jefferson, whose interest in geography has already been noted. Humboldt and Jefferson became close friends, and the great German scientist was given access to the White House without special invitation. Humboldt was greatly moved by the liberal ideas so eloquently expressed by the author of the Declaration of Independence, with whom he was thoroughly in accord. He and Bonpland finally set sail from Philadelphia for the return voyage to Bordeaux on June 30, 1804.[1]

In Paris

First Humboldt returned to Berlin. But especially after the defeat of the Prussians by Napoleon in the Battle of Jena in 1806, he found himself isolated from the world of science and scholarship. After a brief visit to Italy to observe an eruption of Vesuvius, he went to Paris on a diplomatic mission, but remained there for the next nineteen years.

It was in Paris that Humboldt wrote and published the thirty volumes in which the results of his American field studies were presented. In the French capital he was able to secure assistance from other scholars in putting his 60,000 plant specimens in order—specimens that included many species and genera never before known to Europeans. And in Paris he could work with skilled publishers and engravers. The thirty volumes are included under the general title: *Voyage aux régions équinoxiales du Nouveau Continent* (Humboldt, 1805—1834).[2]

[1] For a detailed account of Humboldt's stay in the United States, see Herman R. Friis in Schultze, 1959:142-195.

[2] The contents of the thirty volumes as originally published are as follows:

1-2 *Plantes équinoxiales . . .*, ed. A. Bonpland (Paris: Levrault et Schoell, 1808-1809). (143 plates.)

3-4 *Monographie des mélastomacées . . .*, ed. A. Bonpland (Paris: Librairie grecque-latine-allemande, 1816-23). (120 plates.)

5 *Monographie des mimoses et autres plantes légumineuses*, ed. C. S. Kunth (Paris: N. Maze, 1819-24). (60 plates.)

6-7 *Révision des graminées . . .*, introduction by C. S. Kunth (Paris: Gide fils, 1829-34). (220 plates.)

8 *Nova genera et species plantarum . . .*, with A. Bonpland, ed. C. S. Kunth (Paris: Librairie Schoell, 1815-25). (700 plates.)

15-16 *Vues des Cordillères et monuments des peuples indigènes de l'Amérique* (Paris: F. Schoell, 1810). (63 plates.)

17 *Atlas géographique et physique du Nouveau Continent . . .* (Paris: Dufour, 1814). (32 maps; with a supplement of an additional 7 maps published later.)

18 *Examen critique de l'histoire de la géographie du Nouveau Continent et des progrès de l'astronomie nautique aux 15ᵉ et 16ᵉ siècles* (Paris: Gide, 1814-34).

19 *Atlas géographique et physique du Royaume de la Nouvelle Espagne* (Paris: F. Schoell, 1811). (20 maps.)

The *Relation historique* (Volumes 28–30, of which the fourth volume was never published) made an enormous impact on the scholarly world. It was translated into many European languages, appearing in English in 1825 and in German in 1859–60 (London: trans. H. M. Williams, 1825; Berlin: trans. H. Hauff, 1859–60). In his *Ansichten der Natur* (Humboldt, 1808) he declared that his purpose was "to win the attention of educated but non-scientific readers for the fascination of the discovery of scientific truths" ((Kellner, 1963:75). Charles Darwin said later that he had read and reread this account of scientific travels and that it had changed the whole course of his life. There can be no doubt that numerous field studies in different parts of the world were stimulated by these volumes. Actually the *Relation historique* (or the *Personal Narrative,* as it was translated into English) dealt with Humboldt's own experiences and hardships very briefly but devoted most of its pages to a sober report on scientific problems that had been investigated and on the results achieved. Yet for a world emerging from the first shock produced by the impact of the discoveries, Humboldt's books were like a fresh breeze because they were filled not only with the excitement of travel in strange places, but also with the reports of careful scientific investigation, the seeking of answers to questions about the interconnections among the phenomena grouped together in rich diversity on the face of the earth. And as early as 1805 (Volume 27) he presented a synthesis of his detailed findings as a basis for the study of plant geography.

Another part of his great work that was widely influential was the *Essai politique sur le Royaume de la Nouvelle Espagne* (Volumes 25–26). This was one of the world's first regional economic geographies, dealing with the resources and products of a country in relation to population and political conditions. Humboldt was much impressed with the far greater prosperity he found in New Spain in comparison with the countries of northern South America, and he was curious about

20 *Tableau physique des Andes et pays voisins (Géographie des plantes equinoxiales)* (Paris: F. Schoell, 1805).

21–22 *Recueil d'observations astronomiques, d'opérations trigonometriques, et de mesures barométriques, faites pendant le cours d'un voyage aux régions équinoxiales du Nouveau Continent,* ed. J. Oltmanns (Paris: Schoell, Treuttel & Würtz, 1808–1820).

23–24 *Recueil d'observations de zoologie et d'anatomie comparée faites dans l'Océan Atlantique, dans l'intérieur du Nouveau Continent, et dans la Mer du Sud pendant les années 1799–1803,* in collaboration with Cuvier, Latreille et Valenciennes (Paris: Schoell & Dufour, 1805–1833). (54 plates.)

25–26 *Essai politique sur le Royaume de la Nouvelle Espagne* (Paris: F. Schoell, 1811). (In later editions with a supplement: *Essai politique sur l'ile de Cuba.*)

27 *Essai sur la géographie des plantes, accompagné d'un tableau physique des régions équinoxiales...* (Paris: F. Schoell, 1805). (One large, folded plate.)

28–30 *Relation historique du voyage aux régions équinoxiales du Nouveau Continent, faites en 1799–1804 par A. de Humboldt et A. Bonpland* (Paris: F. Schoell, 1814–25). (The account ends with the first part of the travels in Peru in 1801; a fourth volume was planned but never published.)

(All thirty volumes are being republished in facsimile, starting in 1970 by Theatrum Orbis Terrarum, Amsterdam).

the reasons for the difference. His interpretation was based on the theory that the only proper way to increase the general prosperity of a country was to make more effective use of natural resources, of which Mexico seemed to have an abundance. He supported his explanations with a wealth of statistical data he found in New Spain, organized and enriched on the basis of his own observations. One of the numerous digressions that interrupt the main theme of his work is the suggestion that a canal should be dug across the Isthmus of Central America and that the best place to do this would be in Panama.

In later editions of the *Essai politique* (after 1826) he included a supplement dealing with the Island of Cuba (*Essai politique sur l'ile de Cuba*). In this short essay he deplored the institution of slavery and outlined a procedure whereby slavery could be eliminated from Cuba without serious disruption of the economy.

During his stay in Paris, Humboldt enjoyed many stimulating meetings with the numerous scholars concentrated there. He became a close personal friend of the French physicist, François Arago, pioneer in the study of electromagnetism and in the wave theory of light. Humboldt enjoyed universal acclaim and was generally recognized as second only to Napoleon among famous Europeans. People came to visit him from all over the world, including the future leader of the independence movement in northern South America, Simón Bolívar, then an exile in Spain. Humboldt encouraged and actually aided many young scientists, including Louis Agassiz (the Swiss scholar who developed the hypothesis of universal glaciation and later taught at Harvard), Justus von Liebig (the German biochemist), Joseph Boussingault (the French geologist whose ascent of Mt. Chimborazo took the altitude record away from Humboldt), and many others.

In Berlin

In 1827 Humboldt moved back to Berlin. His own personal fortune had been exhausted by the cost of his travels and, especially, by the cost of printing his books. When he was offered a position as chamberlain in the court of the Prussian king, which included a steady income, he accepted. In 1829 at the invitation of the Russian czar he went to St. Petersburg and then on by horse and carriage into Siberia as far as the borders of China (Fig. 17). He visited the shores of the Caspian Sea. The whole trip was a triumphal tour, for as his carriage approached a village or town the inhabitants turned out to line the road and give their distinguished visitor an ovation.

On this trip Humboldt was impressed by his observations of temperature. He could see clearly that temperatures varied at the same latitude in accordance with distance from the ocean. On his return to St. Petersburg he urged the czar to set up a network of weather stations at which weather data could be recorded regularly and in accordance with standard procedures to make the results comparable. The czar promised to do this; by 1835 the Russian network of recording stations extended all the way from St. Petersburg to an island off the Alaskan mainland. From these stations Humboldt later received the data that permitted the construction of the first world map of average temperatures (Fig. 18). Following the example of Halley and Buache, who had used lines to connect points of equal value, Humboldt for the first time used lines to connect points of equal temperature (isotherms). Noting how the isotherms departed from the lines of latitude, Humboldt developed the concept of continentality: that continental climates are colder in winter and warmer in summer than places near the oceans at the same latitude.

On his Siberian trip Humboldt also observed and described the permanently frozen soil, which is now called *permafrost*. He saw the remains of a mastodon that had been frozen in the ground and preserved in this way. But he did not see any evidence of glaciation, and, therefore, he remained skeptical of the idea of a universal Ice Age then being advanced by the Swiss scholar, Louis Agassiz. Humboldt was right—large parts of Siberia were not covered by ice sheets during the Ice Age.

The Kosmos

In the winter of 1827-28 Humboldt offered a series of public lectures at the Royal Academy of Sciences in Berlin. His lectures had drawn such large and enthusiastic audiences that he had to repeat them in a larger room. In these lectures he not only made science interesting to the educated layman, but he also made it acceptable to the religious leaders of the time. Religions, he insisted, offer three different

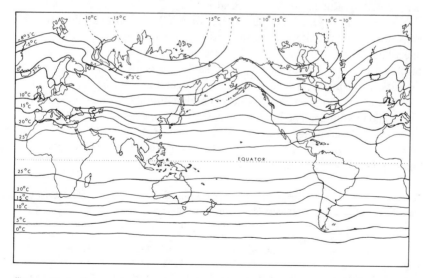

Figure 19. Isotherms of Average Annual Temperature, 1845, according to Humboldt.

things to mankind: a lofty moral idealism, which is common to all religions; a geological dream regarding the origin of the earth; and a legend concerning the origins of the religion. He always emphasized the unity and coherence of nature; although he made clear the wonder of the universe, some of his admirers complained that nowhere in his lectures or in his books did he mention God (Kellner, 1963).

For nearly fifty years he had been forming in his mind the plan of a book, or a series of books, that would

> ... give a scientifically accurate picture of the structure of the universe which would attract the general interest of the educated public and communicate some of the excitement of scientific study to the non-scientific mind. Since he saw nature as a whole and man as a part of nature, and therefore all intellectual and artistic activities as having a share in natural history, he linked his main theme to an exposition of its development through the centuries and to the history of landscape painting and descriptive poetry of nature . . . . The book, when it was finally completed, followed fairly faithfully the scheme of the course of lectures he had given in 1828 [Kellner, 1963:199].

He wrote the book, which he called the *Kosmos,* during the last years of his life. The first volume was published in 1845 when he was seventy-six; the fifth volume, published posthumously in 1862, was based on the copious notes he left.[3] Written in superb literary style, the *Kosmos* became the most prestigious scientific work ever produced up to that time. It was an immediate success. The first edition of Volume 1 was sold out in two months. Soon it had been translated into many languages, including almost all the European languages.

The *Kosmos* put together in one unified work all the various interests and discoveries of Humboldt's lifetime. In the first volume there is a general presentation of the whole picture of the universe. The second volume starts with a discussion of the portrayal of nature through the ages by landscape painters and by poets and then continues with a history of man's effort to discover and describe the earth since the time of the ancient Egyptians. Humboldt's enormous erudition becomes especially clear in this second volume. The third volume deals with the laws of celestial space, which we would call astronomy. The fourth volume deals with the earth, not only with geophysics but also with man. Here is what Humboldt had to say about man as a part of nature near the end of the first volume:

> The general picture of nature which I have endeavoured to delineate, would be incomplete, if I did not venture to trace a few of the most marked features of the human race, considered with reference to physical gradations—to the geographical distribution of

[3]*Kosmos, Entwurf einer physischen Weltbeschreibung,* 5 vols. (Stuttgart and Tübingen: J. G. Cotta Verlag, Vol. I, 1845; Vol. 2, 1847; Vol. 3, 1850; Vol. 4, 1858; Vol. 5, 1862). The best English translation is by E. C. Otté, *Cosmos, Sketch of a Physical Description of the Universe* (London: H. G. Bohn, 1849–58).

contemporaneous types, to the influence exercised upon man by the forces of nature, and the reciprocal, although weaker, action which he in his turn exercises on these natural forces. Dependent, although in a lesser degree than plants and animals, on the soil, and on the meteorological processes of the atmosphere with which he is surrounded—escaping more readily from the control of natural forces, by activity of mind, and the advance of intellectual cultivation, no less than by his wonderful capacity of adapting himself to all climates—man everywhere becomes most essentially associated with terrestrial life [Humboldt:1, 378; Otté translation:1: 360-361].

Humboldt believed that all the races of man had a common origin and that no race is necessarily inferior to the others: all races, he insists, are equally destined for freedom, individually or in groups.

Humboldt emphasized again and again the need for careful observation of nature in the field and for the careful and precise measurement of observations. Yet he was always seeking to formulate general concepts, or what we would now call abstract models or theory. However, he thought that observation had to come first. In Volume 1 he wrote:

> We are still very far from the time when it will be possible to reduce, by the operation of thought, all that we perceive by the senses to the unity of rational principle. On the other hand the exposition of mutually connected facts does not exclude the classification of phenomena according to their rational connections, the generalizing of many specialties in the great mass of observations, or the attempt to discover laws [Humboldt:1: 67-68; Otté translation:1: 58].

During Humboldt's long life the need for closer attention to the definition of fields of special study became important. By the time of Immanuel Kant, as we have seen, the course of lectures on physical geography started with a definition of the field. It was quite clear that history dealt with problems of chronology and that geography was concerned with problems of areal association and distribution. Also Kant's logical classification of knowledge made room for the specialist in the study of particular processes without reference to time or space. That this was a generally accepted division of the world of scholarship and not an invention of Kant is made clear by Humboldt's earliest studies of the subterranean plants in the mines at Freiberg. In the introduction to this monograph (1793:ix-x) Humboldt pointed out that he was not studying plants as such, but rather the plants in relation to their surroundings. Humboldt reprinted his earlier statement in a footnote (in Latin) in the *Kosmos* (1, 486-487). Hartshorne suggests that Humboldt was probably presenting ideas he had gained from his teacher, A. G. Werner (Hartshorne, 1958:100). In the introduction to the *Kosmos* Humboldt points out that

> ... the terms physiology, physics, natural history, geology, and geography were commonly used long before clear ideas were entertained of the diversity of objects embraced by these sciences, and consequently of their reciprocal limitation [Humboldt: 1:51; Otté

translation: 1:39]. Geography, which Humboldt called *Erdbeschreibung* (earth description), deals with the variety of different kinds of interrelated phenomena that exist together in areas or segments of earth space. This was essentially the same idea as that suggested by Kant, although there is no evidence that Humboldt was quoting Kant.

CARL RITTER

Carl Ritter was born in 1779, ten years after Humboldt. His father was a physician, and when he died the widow lacked any means of support for her family of five. At this time Carl, the youngest of the family, was only five years old. By the greatest good fortune in 1784 a German schoolmaster named Christian G. Salzmann was founding a new school to experiment with the radical innovations in educational methods then being proposed. He wanted a child who had never been exposed to the traditional methods and could be trained from the beginning by new pedagogical procedures. The child he selected for this purpose was Carl Ritter.

The Education of a Geographer

In the late eighteenth century in Germany and France the traditional educational methods were being vigorously challenged. It had long been customary to expect children to memorize selected passages from books, often in Latin or Greek, and then repeat them out loud. To repeat the words correctly was enough whether the passages were understood or not. In 1762 Jean Jacques Rousseau, in his novel *Emile,* outlined a new educational procedure that would end rote learning and would encourage a child to develop inborn potentialities. The Swiss educator, Johann Pestalozzi, further developed the ideas of Rousseau, insisting on the principles that clear thinking must be based on the careful observation of things and that words could have no meaning unless they were matched with perceptions. Salzmann was enthusiastic about these new suggestions and established his school at Schnepfenthal in the Thüringerwald to experiment with them.

The teacher who was selected to supervise the young Carl Ritter was a geographer named J. C. F. GutsMuths, whose own special field of interest was in the observation of natural features and who had already made some contributions to the teaching of geography by the new methods (Hartshorne, 1939:50–51). At an early age Ritter was able to observe the close involvement of man with the natural features of his surroundings. He was encouraged by his teachers to formulate for himself the concept of the unity of man and nature; and, from the richly varied landscapes of this region of hills and low mountains, he derived the idea of unity in diversity, which became a basic theme of his mature writings. He could only account for such unity as evidence of God's divine plan. Without Rousseau, Pestalozzi, Salzmann, and GutsMuths this kind of educational experience would not have been available; but, because of these men, Ritter received the best possible basic training for a career as a teacher of geography.

At the age of sixteen, when he was ready to go to a university, Ritter was again fortunate in finding financial support. A wealthy banker, Bethmann Hollweg, agreed to pay for Ritter's university expenses if Ritter, in turn, would agree to tutor the two Hollweg sons. At the University of Halle, with the commitment to become a teacher in mind, he took work with the famous educator, Professor Niemeyer. Ritter continued his own studies after he started tutoring the Hollweg children at their home in Frankfurt on the Oder. He undertook to learn Latin and Greek and to read widely in geography and history. With his pupils he made frequent field trips around Frankfurt, where in the process of teaching field observation he increased his own competence as an observer. He became a master of the art of landscape sketching, which, even after the age of photography, remains an effective way to preserve field observations for future study. Later he extended the range of his field trips to Switzerland and Italy, during the course of which he met many of the leading scholars of that period. In 1807 he met Humboldt and was deeply impressed by him. In 1811 he published a two-volume textbook on the geography of Europe, making use of previously prepared maps of the geographic features of that continent.[4]

After one of his pupils died, Ritter went to the University of Göttingen to accompany the other Hollweg son. At that university between 1814 and 1816 he studied geography, history, pedagogy, physics, chemistry, mineralogy, and botany.

Ritter as a Teacher and Lecturer

Unlike Humboldt, Ritter held several academic positions during his life. In 1819 he became professor of history at Frankfurt. He held this appointment only for one year, during which time he was married. In 1820 he was appointed to the first chair of geography established in Germany—at the University of Berlin; he continued to offer courses of lectures at this university until his death in 1859. During this time he held other positions. He lectured on military history at the Prussian military school and became the director of studies for the Corps of Cadets. He was appointed a member of a scientific commission on geography and history. He founded the *Gesellschaft für Erdkunde zu Berlin* (the Berlin Geographical Society). He was the private tutor for Prince Albert of Prussia. And in addition to these numerous undertakings, he continued to lead field trips each summer to various parts of Europe.

Ritter was a brilliant and influential lecturer. In interesting contrast to the obscurity of his style of writing, his lectures were clear and well organized. He was a master of the art of using the blackboard to illustrate his ideas. After the first two

---

[4]These maps, prepared between 1804 and 1806 (*Sechs Karten von Europa mit Erklärendem Text* [Schnepfenthal], were among the first to make use of hypsometric symbols to describe the shapes of surface features. The earliest such map was a world map by A. Zeune, 1804. See the discussion of the use of the contour method in Joseph Szaflarski, 1959: "A Map of the Tatra Mountains Drawn by George Wahlenberg in 1813 as a Prototype of the Contour-Line Map," *Geografiska Annaler*, 41:74–82.

or three lectures at the university, when he found his lecture room empty or with only a few students, his lectures became immensely popular and his lecture room was always full. Many were the young students whose enthusiasm for geography and for Ritter's interpretation of it was kindled by attendance at his lectures and who went forth to spread the word in other countries. Among his famous disciples were Elisée Reclus of France and Arnold Guyot, who became professor of physical geography and geology at the College of New Jersey (later Princeton) in 1854.

Ritter's public lectures were also highly successful. Some of his basic ideas concerning the influence of the earth's major features on the course of history were developed in lectures before the Royal Academy of Sciences in Berlin.[5]

Ritter's Geographical Ideas

Ritter emphasized repeatedly that he was teaching a "new scientific geography," in contrast to the traditional "lifeless summary of facts about countries and cities, mingled with all sorts of scientific incongruities" (Bögekamp, 1863:37). His scientific geography was based on the concept of unity in diversity, which he had developed for himself at an early age. His purpose was not just to make an inventory of the things that occupy segments of earth space, rather he sought to understand the interconnections, the causal interrelations, that make the areal associations cohesive. Again and again he used the German word *zusammenhang* (literally, hanging together) to refer to this quality of cohesion among diverse things.

To refer to the new scientific geography he made use of the word *Erdkunde,* or earth science. This he preferred to Humboldt's term *Erdeschreibung,* or earth description. *Erdkunde* is a German synonym for the Greek word *geography.* There was never any uncertainty in Ritter's mind that he was studying the earth as the home (*wohnort*) of man and, therefore, that he was dealing with the earth's surface. Later, as we will see, some German geographers took the word, *Erdkunde,* literally and focused their work on the whole body of the earth, not just its surface.

Ritter insisted that geography should be empirical, in the sense that the student should progress from observation to observation in the search for general laws and not from preconceived opinion, to hypothesis, to observation. The student should ask the earth itself for its laws (Ritter, 1822:1:23). By avoiding preconceptions and making his own empirical observations of the surface features of Europe, Ritter was among the first to point out the error in Buache's concept of continuous mountain chains.

Ritter's search for unity in diversity led him to make use of the regional

[5]Between 1826 and 1850 Ritter gave five lectures of great importance: "The Geographic Position and Horizontal Extension of the Continents," 1826; "Remarks on Form and Numbers as Auxillary in Representing Relations of Geographical Spaces," 1828; "The Historical Element in Geographical Science," 1833; "Nature and History as the Factors of Natural History, or Remarks on the Resources of the Earth," 1836; "The External Features of the Earth in Their Influence on the Course of History," 1850 (Gage, 1863).

approach to geography rather than the systematic study of individual features. Yet he realized the importance of systematic studies and acknowledged his indebtedness to Humboldt, whose general studies made Ritter's special studies of regions possible. For his larger regional units Ritter made use of the traditional continents and proceeded to formulate generalizations concerning the continents and their human inhabitants. The continued use of continents as major regional entities not only for the teaching of geography but also for the formulation of concepts has retarded the progress of geographical scholarship. Unfortunately, Ritter's identification of races by skin color and his identification of color by continents has produced only obscurity (Europe for white people; Africa for black people; Asia for yellow people; America for red people).

Ritter's concepts regarding the meaning of the observed geographical patterns on the earth were strongly teleological. Following the philosophers Immanuel Kant and Johann Gottfried von Herder, Ritter saw in all his geographical studies the evidence of God's plan. A Supreme Being, an all-wise Creator, was identified as the author of a plan for building the earth as the home of man, and all through Ritter's writings and lectures are words of praise for the divine creation. Even the arrangement of the continents Ritter saw as evidence of God's purpose. Asia, said Ritter, represents the sunrise—here the early civilizations of man originated. Africa represents the noon—because of the smoothness of outline as well as the uniformity of climate, the inhabitants are induced to slumber and to shun outside contacts. Europe is especially designed to bring out man's greatest accomplishments— because it represents the sunset, or the end of the day, the culmination of man's development is found there. But the discovery of America now suggests the approach of a new sunset and a new culmination toward which man continues to strive. The polar regions represent midnight, when land and people are locked in eternal sleep. Ritter enlarges on the concept of the land hemisphere, as suggested by Buache and developed by Malte-Brun, and points out that this, too, is a part of God's plan. Only in this central location among the earth's land areas can a world-conquering civilization arise.

Ritter's teleological ideas came under fire from his contemporaries. It was Julius Fröbel who said that one might, with equal truth, say that grass had been created as feed for cattle (Fröbel, 1831). Ritter replied that among all the creatures on the earth only man could comprehend the existence of a divine plan and so could adjust his life to it and make maximum use of God's gifts (Hartshorne, 1939:62).

The Erdkunde

Ritter, like Humboldt, produced one great work that represented his major scholarly achievement. This was *Die Erkunde*.[6] The translation of the full German title

---

[6]*Die Erdkunde, im Verhältniss zur Natur und zur Geschichte des Menschen, oder allgemeine vergleichende Geographie als sichere Grundlage des Studiums und Unterrichts in physikalischen und historischen Wissenschaften*, 19 vols. (Berlin: G. Reimer, 1817-18; 1822-59).

presents the basic purpose: *The Science of the Earth in Relation to Nature and
the History of Mankind; or, General Comparative Geography as the Solid Foundation of the Study of, and Instruction in, the Physical and Historical Sciences.* Before
Ritter became the professor of geography at Berlin in 1820, he was still thinking of
geography as the basis for the writing of history. The first two volumes of the
Erkunde (1817–18) were intended to be followed by a study of history. But when
Ritter went to Berlin he decided to devote himself to doing a more thorough piece of
work on the geography. In 1822 he published a second edition of Volume 1 and in
1832 a second edition of Volume II. But by this time he realized the magnitude of
the work he had started. After 1831 he gave up many of his positions so that he
could devote himself more fully to the completion of the *Erkunde.* Between 1832
and 1838 he completed six more volumes and between 1838 and 1859 eleven more
volumes. Yet the nineteen volumes of the *Erdkunde* Ritter actually finished only
covered Africa and a part of Asia.

Unlike Humboldt, Ritter's great work was largely put together on the basis of
other people's observations. He said that his field studies in Europe made it possible
for him to interpret what other people reported. Fritz Kramer comments on the
interesting point that Ritter's descriptions of places he had never seen were vivid
and accurate, whereas his descriptions of places he had seen often lacked zest
(Kramer, 1959).

In contrast to the clarity of his lectures, Ritter's published works are often
obscure. Scholars have struggled to find suitable translations for some of his passages that would make sense in another language yet not do violence to his ideas.[7] A
student of the German language can have endless fun with Ritter. But one must
arrive at the conclusion that Ritter himself was neither critical of his own ideas, nor
resolved about what he wanted to say. He was expressing a general feeling about
the subject, and many of his assertions of relationships have never been and could
never be subjected to rigorous verification. One may be pardoned the somewhat
irreverent observation that one way to gain the reputation for being a profound
thinker is to write obscurely.

RETROSPECT

So these two great scholars, who died in the same year in Berlin, each in his
own way attempted to establish a "new geography." Each tried to embrace the
knowledge of mankind concerning the earth as the home of man. Both of them saw
the field of geography as dealing with things and events of diverse origin that were
interconnected in segments of earth space, as did Kant and others. Both were
tireless workers, who wrote many books and whose influence on the scholarly world
was very great. Both recognized the need for seeking general concepts and both

[7]For example, see the discussion of the meaning of one of Ritter's frequently quoted statements that
geography is the study of *"der irdisch ertüllten Räume der Erdoberfläche"* (from Ritter, 1852) in
Hartshorne, 1939:57.

recognized that little progress toward higher theory could be made in their time. But both had confidence that continued use of proper geographical methods would eventually bring to light the inner meaning of the universe. Humboldt was an agnostic; Ritter once remarked that although the *Kosmos* was a magnificent piece of work, one found in it no single word of praise for the Creator. Ritter saw all of his studies of the earth and man as revealing more and more of God's plan.

Yet the two men were fundamentally different in their approach. Humboldt could not look at the world around him without finding innumerable questions demanding answers. He not only described what he saw with care and precision, but he also formulated hypotheses to account for the things he observed—and then he also subjected his hypotheses to the test of new observations. Ritter also had a vision of an ordered and harmonious universe, but instead of asking questions about it, he wanted to communicate to others the meaning he had found. As a teacher he wanted to make clear to his disciples how God's plan was revealed in the harmony of man and nature. Each in his own way was enormously successful, and each enjoyed wide personal prestige.

When Humboldt and Ritter died there was no one to replace them. Classical geography had come to an end—no individual scholar could hope any longer to master the world's knowledge about the earth. The specialization of subject-matter disciplines resulted in the development of new technical jargons, new paradigms of scientific behavior. As a result much that had been called geography was partitioned among a variety of logically defined fields. In Germany no one was appointed to fill Ritter's chair. Some decades later when geography was reestablished as a university study, the scholars invited to teach it had had no previous training in a field called by that name.

What do Humboldt and Ritter mean to us today? Ritter did influence his disciples to identify a new scientific geography based on the organic unity of man and nature (Guyot, 1860). But his teleology, reflecting the contemporary thinking of such philosophers as Kant and Herder, became outmoded and raised a barrier to the continued acceptance of this kind of new geography. Ritter's regional studies, moreover, deal for the most part with such large areas that the material he included had to be highly generalized. The interconnections he described could not be perceived by direct observation. Today Ritter's *Erdkunde* has chiefly an antiquarian interest. Humboldt's systematic studies are also outdated, although the methods he used represent important steps in the progress of geography. But Humboldt's regional studies "cannot become obsolete" (Hartshorne, 1939:82), especially his comparative studies of New Spain and Cuba, which provide invaluable material for studies in historical geography. Humboldt dealt with areas that were small enough so that he could discuss all the factors relevant to a problem that could be tested by direct observation—for example, the study of the Lake of Valencia Basin in Venezuela. These two great masters of the nineteenth century mark the culmination of thousands of years of effort to push knowledge out beyond the far horizon, and they both point to new horizons to be conquered along the road ahead.

PART TWO

MODERN

A major innovation in the world of scholarship took place in nineteenth-century Germany. The university as an institution first appeared in Medieval Europe when charters were issued by religious or secular authorities giving certain faculties the right to teach. The University of Paris in the twelfth and thirteenth centuries became the chief center (other than Rome) for the teaching of orthodox Christianity. But in 1809 Wilhelm von Humboldt, the brother of Alexander, founded the University of Berlin with the support of King Friedrich Wilhelm III of Prussia. For the first time anywhere the attachment of either faculty or students to any particular religious creed or school of thought was explicitly repudiated. Hitherto universities were places where the accepted dogma of state and church was taught to students. After 1809 the university as a free community of scholars began to appear.

Geography as a field of advanced study taught by professionally qualified individuals first appeared in Germany in 1874. Within a few decades geography departments offering graduate training leading to advanced degrees were established not only in Germany, France, and Britain but also all around the world. This was the "new geography," and it was guided for the first time in history by professional geographers. A profession had come into existence that could establish the paradigms of geographical study.

We date the Modern Period in the history of geographical ideas with the establishment of professional staffs in universities.

CHAPTER 7

What Was New?

The labors of Humboldt, of Ritter, of Guyot, and their followers have given the science of geography a more philosophical, and, at the same time, a more imaginative character than it had received from the hands of their predecessors. Perhaps the most interesting field of speculation, thrown open to the new school of the cultivators of this attractive study, is the inquiry: how far external physical conditions, and especially the configuration of the earth's surface, and the distribution, outline, and relative position of land and water, have influenced the social life and social progress of man.

But it is certain that man has done much to mould the form of the earth's surface, though we cannot always distinguish between the results of his action and the effects of purely geological causes; that the destruction of the forests, the drainage of lakes and marshes, and the operations of rural husbandry and industrial art have tended to produce great changes in the hygrometric, thermometric, electric, and chemical condition of the atmosphere, though we are not yet able to measure the force of the different elements of disturbance, or to say how far they have been compensated by each other, or by still obscurer influences; and, finally, that the myriad forms of animal and vegetable life, which covered the earth when man first entered upon the theater of a nature whose harmonies he was destined to derange, have been, through his action, greatly changed in numerical proportion, sometimes much modified in form and product, and sometimes entirely extirpated.

The quotation above is from G. P. Marsh, *Man and Nature, or Physical Geography as Modified by Human Action* (New York: Charles Scribner, 1864), pp. 8, 13-14.

Those scholars who from time immemorial have been seeking more and more useful knowledge concerning the face of the earth, including man's use of it, have always faced five basic problems, none of which has yielded to permanent solution. These problems, as suggested by Fred Lukermann, are: (1) What things in the universe should man select to observe and record? (2) What is the best way to observe them? (3) How can the resulting observations be generalized to reveal some kind of significant geometric arrangement on the earth? (4) How can the patterns of arrangement be explained or made plausible? (5) How can the results be communicated?*

The flood of new information that swept over the world of European scholarship as a result of voyages of discovery greatly complicated the search for answers to these questions. At first attention was focused on the marvels that were reported; and writers with vivid imaginations, like the author of *The Travels of Sir John Mandeville,* could scarcely be distinguished from sober reporters, like Marco Polo. The world revealed by the explorers was full of strange things, and there was no lack of subjects to be observed and recorded. Then little by little attention shifted from the marvels to things that formed some kind of pattern with familiar things at home. It became more important to report similarities than differences. Cluverius and Carpenter in the seventeenth century began omitting references to weird creatures and unusual natural phenomena, but not until 1761 did Jean B. B. d'Anville remove the drawings of strange creatures that hitherto had adorned the blank places on the maps.

The seventeenth century witnessed the beginning of the scientific revolution that led to the development of more useful ways of generalizing, explaining, and communicating. The effort to provide more exact descriptions of specific things was replaced by the effort to formulate general theory in relation to which specific things could be made significant. In the formulation and testing of theory and in communicating the findings, a step of major importance was the independent development of calculus by Newton and Leibniz. The use of mathematical procedures made the process of reasoning more precise and provided a universal language for the communication of the results. Most of the present fields of science had their roots in the eighteenth century, during which time the acceptable methods of study were being formulated and reliable procedures for the verification of hypotheses were being established. No longer could a hypothesis be supported by its plausibility, for scholars were learning that things perceived by the senses were not necessarily the outline of reality. Controlled experimentation began to bring spectacular results; and after the eighteenth century it would no longer be possible for a Montesquieu to draw conclusions from the study of a sheep's tongue.

By the end of the eighteenth century the ideas expressed by Kant had become generally accepted. As bodies of theory developed and proved useful, special fields

*In personal conversation.

of study appeared, each defined in terms of the segment of the universe being investigated. These new fields of study became what Kant called the logical division of knowledge—in contrast to the physical classification of knowledge in terms of time and space. Each logically defined field provided a method for describing and demonstrating the significance of a particular segment of human experience—and of creating new experience through use of the experimental method.

The last great figure who could claim universal scholarship was Humboldt. No student of the earth before or since has enjoyed such acclaim by his contemporaries. Ritter, too, attempted to embrace the whole of geographical knowledge concerning the earth and man, but was less successful.

The world of scholarship underwent a basic change during the nineteenth and twentieth centuries. Not only have the fields of learning—the academic disciplines—been greatly elaborated but also the total number of scholars has reached unprecedented size.[1] And the mere number of recorded facts about the earth has increased astronomically. A Humboldt could once master a very large part of the available knowledge concerning the earth. But no longer. The number of books and articles is causing libraries to bulge and researchers to seek new kinds of information retrieval systems. The computer was invented just in time to provide a mechanical means for data storage.

The question is: What is new about all this? Are the basic questions still the same? How much of what went on before the modern period is of any importance today—other than to satisfy the curiosity of historians of science?

THE LOGICAL SYSTEMS

The logical systems have become the familiar divisions of the academic curriculum. In the broadest sense these systems include the physical sciences, the biological sciences, the social sciences, and the humanities.

The Physical Sciences

The physical sciences, which appeared as separate disciplines earliest, are now the farthest advanced in the building of theory and in the continued testing of theory by controlled experiment. A particular physical process is artificially isolated in a laboratory and can then be observed free from the complications resulting from the presence of a great variety of logically unrelated processes in the total environments of particular places on the earth. General models can then be formulated to describe the observed sequences of events. What the physical scientist is trying to do is to

[1] For example, there were twelve professional geographers in Germany in 1880 (Wagner, 1880). In 1921 there were seventy (Joerg, *Geographical Review*, 1922, p.: 442). In 1964 the international directory, *Orbis Geographicus* (ed. E. Meynen. Wiesbaden: Franz Steiner) listed 546.

find order in human experience and to describe this order in the simplest possible terms. Ptolemy identified a kind of order in the movements of the celestial bodies; but Copernicus found Ptolemy's picture of celestial order too complicated and with too many motions unexplained. An example of the simplicity of order sought by the physical scientists is the law of gravitation formulated by Newton. This law states that every particle of matter in the universe is attracted to every other particle with a force proportional to the masses of the particles involved and inversely as the square of the distance between them. Albert Einstein showed that this simple statement of the law of gravitation only works with large numbers of particles and must be modified for studies of atomic physics.

Another major achievement of the late eighteenth century was the discovery by the French chemist, Antoine Laurent Lavoisier (1743-94), and by the English chemist, Henry Cavendish (1731-1810), that Aristotle's four basic substances (air, fire, earth, and water) were not really basic elements, no matter how plausible this might seem. In 1783 Lavoisier announced that water is made up of hydrogen and oxygen. Henry Cavendish in England anticipated Lavoisier by a few years, but his studies of water were not published until 1784-85. He also carried out the first measurements of the composition of the air, anticipating Gay Lussac. These scientific accomplishments required use of the experimental method.

A number of specialized fields of study emerged from the unspecialized cosmography to give attention to particular groups of processes on the face of the earth. The study of celestial bodies is now left to that branch of physics known as astronomy. The study of the interior of the earth is entrusted to the various branches of geophysics, including seismology (the study of earthquakes). The study of things and events on the surface of the earth involves geomorphology, geology, and the even more specialized mineralogy, petrography, and paleontology (where geology overlaps with biology). There is also the study of water on the land, hydrology. Oceanography became a separate discipline after Matthew Fontaine Maury started his collection of observations concerning winds and currents of the ocean. Climatology became the study of the average state of the atmosphere and meteorology the study of atmospheric processes that produce weather.

The Biological Sciences

The later voyages of discovery in the eighteenth century had a special impact on the development of biology as a separate field of study. Captain James Cook commanded the first expeditions on which scientifically trained people were included. Among those who sailed with Cook on his first voyage were Sir Joseph Banks and David C. Solander, a pupil of Linnaeus. On the second voyage Cook also took with him the two Forsters; Georg Forster was the one who focused Humboldt's interest on botanical observations. Humboldt himself brought back to Europe some 60,000 specimens of plants never before known to Europeans. When

Lamarck was appointed professor of zoology at Paris in 1793 he had access to a wealth of new plant and animal collections. The tropical parts of the world, long feared because of the persistence of ideas inherited from Aristotle, attracted much attention after they were vividly described by Forster and Humboldt. It was a copy of Humboldt's narrative that excited Charles Darwin to turn to the study of plants and animals.

Darwin was also greatly influenced by the uniformitarian ideas of Hutton and Sir Charles Lyell. Accepting Lamarck's concept of evolution, Darwin began looking for the processes of change in species that would explain the diversity of organic life on the earth. From December, 1831 to October, 1836 Darwin sailed around the world on H.M.S. *Beagle,* observing a great variety of physical and biotic processes, much as Humboldt had done. His concept of the stages in the transformation of coral reefs—from fringing reefs, through barrier reefs, to atolls—was published in 1842 (Darwin, 1842; Davis, 1928). It was on this voyage that Darwin formulated his hypothesis concerning the mechanism whereby random variations in plant and animal species would be selectively preserved and by inheritance lead to changes in species.

At about the same time another young scientist was traveling to the tropical parts of the world. This was Alfred Russel Wallace. In 1848 Wallace accompanied Henry W. Bates on a voyage up the Amazon River. Although his collections were lost when his ship burned on the return voyage, Wallace had been fascinated with the problem of how evolutionary changes in organisms could take place. From 1854 to 1862 he explored the Malay Archipelago, and in the course of this exploration he identified a sharp boundary that separated areas with very different kinds of native mammals. The line passed between Borneo and Bali on the west and Lombok and Celebes on the east. The more primitive animals farther east had been protected from competition with more advanced species farther west. The line between the two kinds of fauna was called Wallace's Line. In the process of plotting these geographic differences on a map Wallace, in a flash of intuition, saw the significance of what he called "natural selection." Influenced by the ideas of Malthus regarding the relation of population to food supply, he applied the same idea of the struggle for existence to animals. Wallace promptly dashed off a letter to Charles Darwin in England, presenting his hypothesis in brief form. The letter arrived just as Darwin was preparing to present a major paper to the Royal Society of London on exactly the same hypothesis, which Darwin had now verified by laboratory experiments. The idea was presented in a joint paper by Darwin and Wallace in 1858 entitled *On the Tendency of Species to Form Varieties, and on the Perpetuation of Varieties and Species by Natural Means of Selection.*

Darwin's *Origin of Species* was published in London in 1859, the year in which both Humboldt and Ritter died (Darwin, 1859). He demonstrated that evolutionary change in organisms was not the result of need or use, as Lamarck had thought. The giraffe did not get its long neck by stretching, rather the individual giraffes that were born with longer necks were better able to survive than their shorter necked relatives and so could pass on this characteristic to later generations. Darwin's contribution was to throw light on the mechanism whereby evolutionary change could take place. But he also produced clear evidence of the randomness of such variations. If evolutionary changes resulted from random variations, the teleological concept of a divine plan had to be abandoned. In spite of continued resistance by some biologists (such as Louis Agassiz) and the reluctance of Darwin himself to face the full implications of his conclusions, the world of science could never return to previously held belief.

The concept of evolutionary change was so stimulating that it was applied by analogy to many other fields beside biology. Applied to the study of landforms, it appeared as the theory of the cycle of erosion. Applied to soils, it was reflected in the concept of mature soils as developed from young or immature soils and parent materials. Applied to the survival of social groups because of the ability to adjust to environmental conditions, it became environmental determinism. As Stoddart points out, the geographers adopted the notion of evolutionary change as described by cause and effect sequences, but they overlooked the concept of random variations and failed to apply the theory of probabilities (Stoddart, 1966).

The Social Sciences

Among the logical systems aimed at the study of human group behavior, the first to develop as a special field was political economy, or economics, as it was renamed in the twentieth century. Some of the earliest attempts to formulate general theory regarding population and resources are associated with a group of eighteenth-century Scottish scholars at the University of Edinburgh. The group includes such people as David Hume (1711-76), Adam Ferguson (1723-1816), and Adam Smith (1723-90). Adam Smith's study, *An Inquiry into the Nature and Causes of the Wealth of Nations,* was published in 1776. The real source of a nation's wealth, he said, is its annual labor, its use of productive resources; and wealth can only be increased by making its use of resources more effective, by increasing the specialization of labor and accumulating profit in the form of capital. Money, he pointed out, was not wealth, but only the means of carrying on trade.

Adam Smith's work was followed by the essay on population by Malthus. It was Malthus who formulated into clear language the previously known law of diminishing returns from investments of capital and labor. David Ricardo in 1817 published a study, *Principles of Political Economy and Taxation,* in which he developed a theory of value. By 1830 political economy was a recognized field of study in most European universities, and in the course of more than a century it has led to the formulation of a large body of theory and methods of study.

During this time, economics has itself been further subdivided into distinct disciplines. There is a separation between economics as a pure science and economics as an applied science dealing with public and private problems of policy. Another distinct field is economic history; another is econometrics, based on the application of mathematical procedures to economic problems. Still another discipline embraces the study of economic theory.

What Happened to History?

History, like geography, did not fit easily into what Kant called the logical classification of knowledge. The various substantive fields into which the study of human behavior was partitioned were built around particular conceptual structures suitable for enlarging or testing these concepts. But traditionally history had dealt with whatever kinds of processes were necessary to understand the sequences of events—social processes, political processes, economic processes, military events, and especially the personalities who made an impact on the course of events. In a modern university, is history to be found among the humanities or among the social sciences? Actually it may be found in either.

Historians have long been concerned with the question about whether they should seek to identify general laws of human behavior or only to reach a more precise and "correct" knowledge of the sequence of events. There were many who felt that the verification of unique sequences of events constituted an ample justification for historical scholarship. There were others, however, who felt impelled to seek universal laws around which to arrange the historical facts. The earlier historians in Europe accepted the common belief that human behavior was a manifestation of the divine plan and that man was in process of development toward the perfect state that God had established as the goal. But during the eighteenth and nineteenth centuries historians gradually shifted away from this interpretation of history. The perfectibility of man, they said, was not demonstrated by empirical evidence but remained only an article of faith. What general laws of behavior then could be discerned from the historical record?

As the volume of historical data increased, historians had to find some way to specialize—to reduce the size and complexity of the questions being investigated. The last universal history was published in 1681 by J. B. Bossuet[2]; thereafter historians became specialists in particular countries or cultures and then just certain limited periods within those countries. Or perhaps they became specialists in the biography of a particular person. Specialization had gone so far by World War I that H. G. Wells sensed the great need for another universal history in which certain general and repeated historical trends could be noted for all of mankind. His book, which was criticized by some professional historians, took the chronological record back to the origin of the earth and the evolution of organic life.[3]

[2] J. B. Bossuet, *Discours sur l'histoire universelle* (Paris, 1681).
[3] H. G. Wells, *The Outline of History, Being a Plain History of Life and Mankind* (New York: Macmillan, 1920).

142   MODERN

The idea that historians should seek to formulate laws and models to explain the course of events became stronger as the teleological intrepretations were abandoned. In the early eighteenth century the Italian historian, Giambattista Vico, identified certain cycles that were repeated again and again in the history of different peoples. He agreed that historical law could not have the precision found in natural law but that at least the broad trends could be discovered.[4] The German scholar, Herder, presented the idea that to understand any sequence of events three interconnected factors had to be known: time, place, and national character.[5] Many other writers of history formulated general laws to explain the course of events.[6]

In Germany during the nineteenth century there was a similar discussion concerning the objectives of the new field of economics. Gustav Schmoller insisted that economics was only a branch of history and that the so-called laws then being formulated by economists could really only be applied to situations unique in time and place. On the other hand, Karl Menger said that the only purpose of economics was to identify the general and universal laws of man's economic behavior, not to record unique events. Menger's views came to be accepted as the pattern of economic scholarship, with the result that economics developed as a *nomothetic* field—that is, a field of study in which general laws are identified—rather than an *idiographic* field, or one that describes unique situations without reference to general laws. Economic history was left intermediate between history and economics. Generally speaking the parts of history in which bodies of theory have been formulated tended to develop as separate academic disciplines (such as sociology, anthropology, political science, or economics), whereas those parts of history where general laws seem to be less useful have been cultivated by scholars whose primary objectives were the discovery of new sources of information and the use of more precise methods for verifying the authenticity of the data (Shafer, 1969: 1–9, 37–42).

What Happened to Geography?

Essentially the elements that were included in what Varenius had called general geography were divided up among the separate disciplines, each with its own body of theory and its own methods of connecting observations with theory. Humboldt and Ritter recognized that when the newly emerging disciplines had divided up what used to be general geography there still remained an area of study not included in these substantive fields. Humboldt asked questions about the earth and man that were not asked by workers in any of these other fields—for example, he connected

[4]Glambattista Vico, *Scienza nuova* (Rome, 1725).
[5]J. G. von Herder, *Ideen zur Philosophie der Geschichte der Menschheit* (Berlin, 1784–91).
[6]Among these: Etienne Bonnot de Condillac (1715–80); the Marquis de Condorcet (1743–94); Auguste Comte (1798–1857); John Stuart Mill (1806–73); Karl Marx (1818–83); Friedrich Engels (1820–95); Oswald Spengler (1880–1936); Arnold J. Toynbee (1889–1975).

the vegetation cover on steep slopes with the water supply in the Lake of Valencia and with the economic and political conditions that led to the deforestation of the slopes. He wrote regional studies of Mexico and Cuba that were not just descriptions of unique places but which provided explanations in terms of general theory. Ritter in his regional studies of Africa and Asia did not just describe each element as a separate and distinct phenomenon—rather he sought the interconnections among things of diverse origin. These interconnections among the physical, biotic, and human features of the face of the earth Ritter identified as evidence of God's plan to lead mankind toward the state of perfection. This harmony of interconnected parts is what Ritter described by the expressive German word *zusammenhang*.

Meanwhile, the substantive fields of study included in the general categories of physical science, life science, and social science made spectacular progress by isolating the processes each examined and formulating an ideal or abstract model of how each process works in isolation. These sciences moved forward by specifically excluding the disturbing effect of *zusammenhang*. Chemistry and physics could isolate the processes they studied in laboratories. Biology set up experimental programs to test the validity of theory, also in isolation from the total environments of particular places on the earth. The social sciences had more difficulty in isolating the processes they studied; but economics especially established at least a symbolic isolation by the use of the phrase "other things being equal." The law of diminishing returns, for example, operates in undisturbed form only when the impact of irrelevant interconnections is eliminated. Now "other things" can be made equal by statistical procedures.

Geography was left with three major tasks. One was the continued collection of information about the still unknown or inadequately known parts of the earth and the presentation of this information in useful form. The second was the study of particular places in the world, whether for the purpose of throwing light on the processes at work in them or for the practical needs of government administrators, military commanders, or businessmen who needed clear descriptions of the facts and conditions relevant to particular problems. And the third task was the formulation of concepts: empirical generalizations, hypotheses, and perhaps even theory. There never was a time when geographers as a professional group were satisfied to describe unique situations without seeking to illuminate the geography of particular places by reference to generalizations or to seek explanations in terms of models.

The New Cartography

In all of these tasks new kinds of maps were needed to go along with the new geography. Pioneering work in making large-scale topographic maps had already been done by the Cassinis in France and by Nicolas Cruquius in the Netherlands, who made use of lines of equal elevation to show landforms in 1728 (Goode, 1927).

Improvements in topographic mapping had to await new methods of printing. The use of copper plates had started in 1493, but lithography was not invented until 1800. Electrotyping and photography were developed between 1840 and 1850. Only then could finely engraved details be reproduced with precision.

One of the earliest of the new cartographers was Adolf Stieler. Stieler had received a law degree from the University of Göttingen in 1797, but he had also developed a keen interest in geography and in the problem of representing geography on maps. He attended lectures by Gatterer (p. 109) and even taught geography in a girl's school at Gotha (which is 8 miles east northeast of Schnepfenthal, now called Waltershausen, where Ritter attended school). At Gotha Stieler found the old German publishing house of Justus Perthes, which he helped to convert into one of the world's leading centers of geographic study and cartography. In 1817 he published the first sheets of the *Stieler Handatlas;* in 1831, when the first edition of the Stieler atlas was complete, it contained seventy-five maps of the world as a whole and its different parts. Between 1829 and 1836 Justus Perthes published Stieler's map of Germany in twenty-five sheets.

Another German geographer-cartographer who contributed to the development of cartography and to the spread of German maps and atlases to other countries was Heinrich Berghaus. As a young man he was employed by the Prussian War Ministry in a field survey of Prussia, and from 1821 to 1855 he taught geometry and cartography at a school in Berlin. In nearby Potsdam he established a school of cartography at which several famous map-makers were trained between 1839 and 1848. Berghaus was one of the many scholars encouraged by Humboldt. Much of the information for his atlas maps came from Humboldt; and his *Berghaus Physikalischer Atlas,* 1837–48 (revised 1849–52), was intended to supplement the *Kosmos.* He included a great variety of thematic maps covering the latest information on climatology, hydrography, geology, earth magnetism, plant geography, zoogeography, anthropogeography, and ethnography–ninety-three maps in all (Beck, 1956). Berghaus also published a number of texts in geography as well as scholarly works. His six-volume *Allgemeine Länder-und Völkerkunde* (Stuttgart, 1837–43), and his five-part *Grundriss der Geographie* (Breslau, 1840–43) were widely read (Hartshorne, 1939:74).

Work on the *Berghaus Physikalischer Atlas* was carried on by Hermann Berghaus, nephew of Heinrich and one of those trained at the school in Potsdam. Hermann moved to Gotha in 1850 and remained there until his death in 1890. He compiled and edited a third edition of the *Berghaus Physikalischer Atlas,* which was published in 1883–91. Also at Gotha was Karl Vogel, who for many years kept successive editions of the *Stieler Handatlas* up to date. At Gotha, also, there was Eric von Sydow, who, in the 1830s, recognized the need for large wall maps for use in classrooms and who prepared the first series of such maps for Justus Perthes. Of this series the first, on Asia, was published in 1838. Sydow's *Schulatlas* (1847–49) not only made the new information about the earth available in the schools of

Germany and elsewhere but also set the standards for the use of blues, browns, and greens on hypsometric maps.[7]

Among the widely known scholars trained by Heinrich Berghaus, who later worked at Gotha was August Petermann. In 1845, at the age of twenty-three, Petermann went to Edinburgh to assist the Scottish map publisher Alexander Keith Johnston in bringing out an English edition of the *Berghaus Atlas*. He remained in Edinburgh and London until 1854, during which time he was appointed cartographer to the Queen and introduced numerous ideas and techniques from Germany into Britain. While he was in London he joined the controversy then raging regarding the existence of an ice-free polar sea. It is ice free, he insisted, because of the warming effect of the Gulf Stream, which flows into the Arctic Ocean. In 1854 he returned to Gotha; the next year he founded the famous geographical periodical, *Petermanns Geographische Mitteilungen,* which is still among the leading professional periodicals in the field of geography. Petermann's maps, for which the *Mitteilungen* became famous, set new standards in cartography. In the first twenty-four volumes and fifty-six supplements (*ergänzungshefte*) that he edited, he published a total of 850 maps. The topographic map, said Petermann, is the highest achievement of geography since it furnishes the most accurate reproduction of the earth's surface and thereby provides the best basis for all knowledge (*Petermanns Geographische Mitteilungen,* 1878:208).

When Petermann was in Edinburgh he was assisted by the young John Bartholomew, son of the director of the map-publishing firm of John Bartholomew & Son. When John became head of the firm in 1856 he was the fourth member of the family with the same name to occupy this position.[8] With his son, John George Bartholomew, he introduced the use of layer tints for hypsometric maps into the English-speaking world. These two Bartholomews also recognized the need for having a map-publishing firm closely supported by a geographic research center when they established the Edinburgh Geographic Institute. The next John Bartholomew, who died in 1962, was the editor of the world-famous *Times Survey Atlas of the World* in 1922, which was expanded into a new edition in 1955.

German cartographic ideas also had an important influence in the United States. Daniel Coit Gilman, who was professor of physical and political geography at the Sheffield Scientific School at Yale from 1863 to 1872, was in close touch with developments in Germany, especially with the mapping of statistical information by the Prussian statistical office. When the Ninth U.S. Census was being planned under the direction of the economist, Francis Amasa Walker, Gilman brought the German materials to Walker's attention. The result was the publication in 1874 of

---

[7]See the *Bulletin of the American Geographical Society,* 44 (1912): 846–848.

[8]The seventh Bartholomew is now head of the firm; John Bartholomew, 1831–93;John George Bartholomew, 1860–1929; John Bartholomew, 1890–1962; John C. Bartholomew, head of the firm since 1962.

the *Statistical Atlas of the United States* under Walker's direction. The atlas was an immediate success and in succeeding censuses the data on economic production and population were effectively presented on maps. The Tenth Census (1880), which Walker also directed, was published in twenty-two large volumes together with the atlas, which reviewed the changing geography of the United States since 1790. This census is a major source for the study of American historical geography. Neither Gilman nor Walker were primarily geographers, yet their influence on the development of the new geography in America was very great.[9]

THE LEGACIES OF HUMBOLDT AND RITTER

Humboldt and Ritter left quite different legacies for future generations. Humboldt sought answers to a great variety of specific questions (de Terra, 1955). For example, he attempted to develop a general picture of the distribution of average temperatures in the world in relation to the distribution of continents and oceans. With the assistance of the Russian network of weather stations he was able to do this. He attempted to define the effect of altitude in tropical mountains on plants, animals, and man. This he did on the basis of personal observation in tropical America. But Humboldt did not leave a school or disciples. The method of asking questions and seeking answers that he so effectively demostrated was not "rediscovered" until several decades after his death. Since he did not restrict himself to the study of physical, biotic, or cultural processes in isolation, his contributions to the substantive fields are usually considered minimal. With Georg Forster he laid the groundwork for the study of plant geography—but he is not listed by botanists or zoologists as a major contributor to their fields. He was, in fact, a geographer because he asked about the interconnections among things and events of diverse origin, sweeping aside the then-growing barriers among disciplines to get a last majestic view of the cosmos.

Ritter did found a school, in the sense that his enthusiastic teaching aroused a similar enthusiasm among his disciples (Sinnhuber, 1959). There were many who undertook to continue his plan of the *Erdkunde* in parts of the world he had been unable to complete. There were German studies of Australia and more detailed works dealing with parts of Europe, especially Germany. His most famous disciples, however, were Élisée Reclus and Arnold Guyot.[10]

[9]Francis Amasa Walker was an economist who taught at Yale from 1873 to 1881. Later he was president of the Massachusetts Institute of Technology. Daniel Coit Gilman became the president of the University of California in 1872 and of the newly organized Johns Hopkins University in 1876. At Johns Hopkins he introduced the German concept of the university as a community of free scholars. Advanced study for the degree of Doctor of Philosophy was made available and a faculty was selected on the basis of excellence in scholarship rather than ability to lecture to undergraduates (Wright, 1961).

[10]Another one of Ritter's students was the Russian geographer, Petr Petrovich Semenov Tyan-Shanski. The work of Semenov is discussed in Chapter 11.

Élisée Reclus

Élisée Reclus was a French geographer and anarchist who studied briefly under
Carl Ritter (Dunbar, 1978). Departing France following revolutionary activism in
1852, he spent the next five years traveling in Great Britain, the United States, and
Colombia. He returned to France in 1857, was imprisoned in April, 1871 and was
banished for ten years in February, 1872. In March, 1879 the ban was lifted. By
then he had become an ardent supporter of the Anti-Marriage movement and demonstrated his sincerity by permitting his two daughters to live with their husbands
without either civil or religious sanction. He was identified by the French government as one of the leading promoters of anarchism. Since he maintained his residence in Switzerland, he was not arrested. In 1892 he was appointed as professor of
comparative geography at the University of Brussels in Belgium. Because of his
continued activity as a revolutionary anarchist, his appointment was canceled. From
1894 until his death in 1905 he was director of the Institut géographique, which was
established at the New University of Brussels.

One reason for his immunity from any penalty other than banishment from
France was his standing as a scholar and the resulting efforts of European scholars to
give him protection. In 1867-68 he published *La terre,* a two-volume descriptive
systematic geography with preponderant emphasis on physical geography, which
was done in the Ritter manner (Reclus, 1867-68). But his major work was the
completion of the kind of universal geography that Ritter had started. His nineteen-
volume "new universal geography" was, in a sense, the last echo of the classical
period when one scholar could present all available knowledge about the earth as the
home of man (Reclus, 1876-94). Reclus took great care with the accuracy of his
sources and with the clarity of his writing. Unlike Ritter's *Erdkunde,* which
is well known for its numerous obscure passages, Reclus's work was easy to
read and understand. Also, he was one of the few disciples of Ritter who eliminated
the teleological element. His standing as a scholar was given further support by
numerous other writings, including a detailed description of the history of a stream
and a similar work on a mountain (Reclus, 1869, 1880).

Arnold Guyot

Arnold Guyot was born in Switzerland; in 1839 he was a colleague of Louis
Agassiz at the University of Neuchatel. Agassiz turned the attention of his younger
colleague to the study of glaciers and the effect of glacial action in producing
distinctive kinds of mountain landforms. Guyot also studied with Ritter and became
one of his most devoted disciples (Libbey, 1884). In 1848 Guyot came to the United
States, where he was invited to deliver a series of lectures at Harvard outlining the
"new geography." In 1849 his lectures were published in book form and served to
make Ritter's ideas known in America (Guyot, 1849). Guyot attacked the traditional

descriptive geography wherein encyclopedic collections of facts were given to students to be memorized (James, 1969). The "new geography" should not only describe but also compare and interpret: "it should rise to the how and wherefore of the phenomena that it describes" (Guyot, 1849:21). When the Massachusetts Board of Education asked Guyot to deliver a series of lectures on the "new geography" and the methods of teaching it, his influence on American schools spread rapidly; and for decades his textbooks set the standards for elementary and secondary classes in geography. He taught his pupils to observe their surroundings and to match their perceptions with the word symbols they used to describe them. When distant regions were studied, Guyot urged that pupils should become better acquainted with these places by the close examination of topographic maps.[11]

Guyot, who held the position of professor of physical geography and geology at the College of New Jersey (Princeton) from 1854 to 1880, remained a vigorous supporter of Ritter's ideas. Even as the widespread acceptance of the concepts of evolution as developed by Darwin, Wallace, and Huxley swept away the philosophical ideas of the teleologists, Guyot's stand remained unshaken. When he retired in 1880, the "new geography" he preached was not only old but its philosophical basis had been largely discredited. Here is what he had to say about the purpose of physical geography in 1873:

> The Earth, as an individual organization, with definite structure, character, and purpose, is the subject of geographical science. . . . A careful study of physical geography tends to lead the mind to the conclusion that the great geographical constituents of our planet—the solid land, the ocean, and the atmosphere—are mutually dependent and connected by incessant action and reaction upon one another; and hence, that the earth is really a wonderful mechanism all parts of which work together harmoniously to accomplish the purpose assigned to it by an all-wise Creator [Guyot, 1873; Davis, 1924: 165–169].

NEW APPROACHES TO GEOGRAPHY IN AMERICA

Not all the scholars who contributed to knowledge about the earth were Europeans or derived their ideas directly form European sources.[12] There were several who made notably original contributions to geographic knowledge. Among these

[11]In addition to Arnold Guyot, who taught at the College of New Jersey (Princeton) from 1854 to 1880, and Daniel Coit Gilman (see p. 145 and note), others who were important in the introduction of European ideas into American geography include: Jedidiah Morse (1761–1826); whose texts, *American Geography* and *American Universal Geography,* were published in 1789 and frequently revised thereafter; they were widely used in schools and read in American homes for many decades (James, 1969: 474–475); John Daniel Gross, who was professor of German and geography at Columbia College from 1784 to 1795; John Kemp, who was professor of geography at Columbia Gollege from 1795 to 1812; Louis Agassiz, who was professor of zoology at Harvard from 1848 to 1873 and introduced the ideas of natural history to America. He was strongly opposed to Darwin's concept of evolution through the survival of the fittest; his student, N. S. Shaler, was the teacher of William Morris Davis (Lurie, 1960).

[12]A more complete survey of the history of geographical ideas would include other eighteenth-

were George Perkins Marsh, Matthew Fontaine Maury, and the numerous out-doors men who participated in the exploration and survey of the American West (Colby, 1936; Curti, 1943; Glick, 1974).

George Perkins Marsh

Marsh was described by David Lowenthal as the "versatile Vermonter" (Lowenthal, 1958). Few men have excelled in a wider range of fields of interest. After graduating from Dartmouth in 1820 and passing examinations for the bar, he set up a law office in Burlington, Vermont, where for many years he carried on a small practice. In 1843 he was elected to Congress by the Whig Party; but when he was defeated for reelection in 1849 he was appointed the U.S. minister to Turkey. In 1861 President Lincoln named him minister plenipotentiary to the Kingdom of Italy, a post that he held until his death in 1882. During all these years he wrote on an amazing variety of scholarly questions. He was a master of the English language and in addition could read some twenty other languages. His books include a grammar of the Icelandic language, a treatise on the habits and uses of the camel (which he recommended for importation to the dry parts of America), and a book on the origin and history of the English language.

Marsh also occupies an important position in the history of geographical ideas. At an early age he began to focus his attention on the destructive effects of man's use of the land. In his wide reading, especially of the works of Humboldt, Ritter, Guyot, and Mary Somerville, he recognized that a "new geography" had appeared, focusing on the close interconnections between man and his natural surroundings. Marsh, who had never studied geography as such, developed a novel approach to the study of the relations of man to the land: he turned his attention to man's effect on nature, to the modifications of the organic and the inorganic parts of the habitat that resulted from human action. This was the point of view that Plato missed and that Buffon promoted. But Marsh began to seek examples of man's destructive use of land as a result of the damage he had seen done by widespread forest clearing in Vermont. His years of residence in Turkey and Italy gave him an opportunity to observe even more startling examples of the damage done by human action.

Many years of field observation and reading went into the preparation of Marsh's great works (Marsh, 1864, 1874). He had been working on the theme of the

---

century Americans. For example, Benjamin Franklin was a keen observer and, for his time, a careful scientist. His discovery of the nature of lightning and of electricity is well known. He also was the first to measure the temperature of the Gulf Stream. There was Hugh Williamson, who in 1760 observed that the warmer the waters of the Gulf Stream, the colder the weather that might be expected in New England. Lewis Evans, whose *Analysis of a Map of the Middle British Colonies of America* was published by Franklin in 1755, is credited with being the leading geographer of his time (Davis, 1924: 160-162; Brown, 1951: 192). Either Evans or Franklin first recognized that storms with northeast winds actually came from the southwest.

modification of nature by human action long before he left for Turkey. In 1847 he gave an address to the Agricultural Society of Rutland, Vermont, on "man's altera- tion of the landscape, intentional and unintentional, desirable and dangerous." He describes the objectives of his book, *Man and Nature,* as follows:

> To indicate the character and, approximately, the extent of the changes produced by human action in the physical conditions of the globe we inhabit; to point out the dangers of imprudence and the necessity of caution in all operations which, on a large scale, interfere with the spontaneous arrangements of the organic and of the inorganic worlds; to suggest the possibility and the importance of the restoration of disturbed harmonies and the material improvement of wasted and exhausted regions; and, incidentally, to illustrate the doctrine that man is, in both kind and degree, a power of a higher order than any of the other forms of animated life, which, like him are nourished at the table of bounteous nature [Marsh, 1864:iii].

Marsh's warning was sounded in a country with seemingly endless resources and at a time when the need for conservation programs had yet to be formulated. In Russia a similar note was sounded in 1901 by Alexander Ivanovich Voeikov, who was especially concerned about the destruction of the grasslands owing to overgraz- ing and about the supposed modifications of climate that resulted from changes in the cover of vegetation (Voeikov, 1901). In America it was not until 1905 that Nathaniel Southgate Shaler, professor of geology at Harvard and a student of Louis Agassiz, returned to the theme of man's destructive effect on earth resources. Shaler, however, was especially concerned about the depletion of mineral re- sources, whereas Marsh paid slight attention to minerals (Shaler, 1905). Marsh had to await rediscovery until the modern period when the destructive effect of human action is recognized as a major and very practical concern (Thomas, 1956).

Matthew Fontaine Maury

Another American who contributed major new concepts about the earth during the nineteenth century was Matthew Fontaine Maury, a Virginian by birth who was brought up in the backlands of rural Tennessee (Williams, 1963, Leighly, 1977), Maury received an appointment as midshipman in the U.S. Navy and was on the *Vincennes* when she became the first naval ship to be sailed all the way around the world—a voyage that lasted from 1826 to 1830. At an early age Maury had de- veloped an insatiable curiosity concerning all matters that lay beyond his immediate horizon. His voyage around the world left him with many unanswered questions concerning the characteristics of the oceans. In 1839 Maury was appointed director of the Navy Depot of Charts and Instruments (which later became the U.S. Naval Observatory and Hydrographic Office). He devised a blank form for ship's logs on which the captains could enter specific observations of winds and currents and other conditions of the sea. Each observation was located by latitude and longitude; when the logs were sent back to Washington, the data were plotted on maps. Maury also devised new instruments for sounding ocean depths and was, therefore, able to produce the first map of the floor of the North Atlantic Ocean—information of the highest practical value in planning the route of the first transatlantic cable. The wind and current data were plotted on charts that were published along with an explanatory text. On the basis of the new picture of winds and currents his data revealed, Maury was able to advise ships' captains concerning the best routes to follow. His sailing directions cut the trip from New York to Rio de Janeiro by ten days. The trip from New York to San Francisco, which used to take an average of 183 days, was cut to 135 days. Maury's sailing directions were fully as important as new rigging and new design in making possible the speed records of the clipper ships (Maury, 1851).

Maury was not satisfied by the mere collection of data. He sought to develop a generalized picture of the surface winds of the earth by identifying the prevailing winds and eliminating the temporary and local interruptions. His model of atmospheric circulation is shown in Figure 20 (Maury, 1850:137; 1855:75). Along the equator is a zone of equatorial calms, which became known as the doldrum belt. On either side of the equator as far as about latitude 30° are the trade wind zones, with prevailing winds from the northeast in the Northern Hemisphere and from the southeast in the Southern Hemisphere. At about latitude 30° in each hemisphere is a zone of middle latitude calms, which became known as the horse latitudes. In the middle latitudes, roughly between 30° and 60° in each hemisphere, are the prevailing westerlies. Maury showed the regions around both poles as zones of calms.

It is instructive to note the way Maury solved the problem of generalizing his information about wind directions. He recognized that there would be many interruptions of his simplified scheme owing to irregularities in the distribution of land and water. Major interruptions are the monsoons and the local land and sea breezes. He did not include these interruptions on his model. For example, his information about wind directions in the tropical South Atlantic (Fig. 21) showed that there was a high probability of encountering northeast winds rather than southeast winds along the coast of Brazil south of 10°S. In his sailing directions he was sufficiently realistic to suggest that captains should stay close to the coast on the way to Rio de Janeiro, and it was this recommendation that cut the sailing time from New York. On the other hand he did not modify his generalized model to take these northeast winds into consideration. His wind zones and belts of calms, running all around the earth, fitted so well into the traditional torrid, temperate, and frigid zones that Maury's model was widely accepted and taught in schools.

Maury also showed the wind directions aloft around the margin of his generalized model. If air moves steadily from latitude 30° toward the equator and if observations of air pressure show that pressure is high in the horse latitudes and low along the equator, then clearly there had to be a return current of air aloft moving in a direction opposite to the surface winds. If the surface winds in the middle latitudes

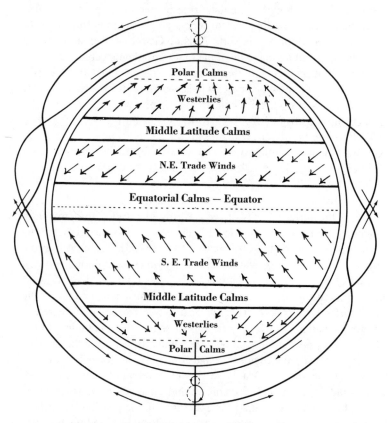

Figure 20. Maury's model of atmospheric circulation.

are from the southwest or the northwest, air must be moving in the opposite direction aloft unless the equilibrium of the world's atmosphere were to be destroyed. His vertical section shows winds crossing each other (Fig.20). Modified later to avoid this difficulty, diagrams showing the atmosphere in vertical section have been widely used in elementary texts.

    Having developed his model, Maury proceeded to make deductions from it. For example, the coastal desert of Peru, lying in the zone of the southeast trades, can be explained by the descent of air on the lee side of the Andes. This notion, that the Peruvian desert is a lee coast in the trade wind zone, is still found in many books. Yet it is an established fact, with which Maury was well acquainted, that the surface winds along the Peruvian coast come from the southwest. These interruptions of Maury's model should have suggested the need for developing a different model; yet because of the wide acceptance of Maury's concept, the error contained in the concept was permitted to persist (James, 1964).

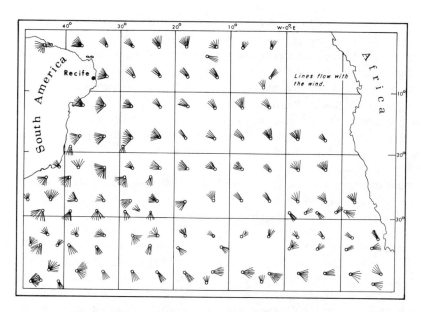

Figure 21. Maury's wind chart of the tropical Atlantic Ocean, 1859.

Maury, himself, was very explicit about his attempt to formulate a general picture of the movements of air and water. He said:

> I am wedded to no theories, and do not advocate the doctrines of any particular school. Truth is my object. Therefore, when the explanation which I may have at any time offered touching any facts fails to satisfy further developments, it is given up the moment one is suggested which will account for the new, and equally as well for the old system of facts. In every instance that theory is preferred which is reconcilable with the greatest number of known facts [Maury, 1855; quote from preface to the sixth edition, 1856:xv].

Maury also gave his support to an idea that was being hotly debated during the middle years of the nineteenth century—the concept of an open polar sea. This concept had had a long history and perhaps dated from 1527 when an English merchant suggested using the polar route to reach the Spice Islands (Wright, 1953). It was given strong support when a Dutch geographer in the seventeenth century provided a hypothetical explanation for the existence of a mild climate near the pole. Maury believed that the warm Gulf Stream flowed under colder surface water and emerged again at the surface in the vicinity of the pole. In 1853 when Elisha Kent Kane was preparing for a voyage to the far north, Maury explained his belief in the existence of such an open sea at great length. When Kane's ship was frozen in the ice with no open water in sight, he sent a sledge party farther to the north. At latitude 80°30′N, standing on a high cliff in northern Greenland, the leader of this party reported that "not a speck of ice was to be seen." The vast open sea extended northward beyond the horizon and was moved by the kind of huge swell that only could develop on a large expanse of open water. He provided the perfect demonstration of how a concept, vividly presented, can be so firmly implanted in the mind that evidence supporting it can be perceived and confidently reported. The idea of an open polar sea was not finally abandoned until the voyage of Fridtjof Nansen of the *Fram* (1893–96) who went to within a few degrees of the pole and did not find open water.

Maury sought to extend his collection of wind and current data by international cooperation. Enlarging on a suggestion made by the British, Maury promoted the idea of an international conference to which delegates from all the major maritime powers would be sent. He also wanted to include a system of land observations, but at this same time Joseph Henry, director of the Smithsonian Institution in Washington, was attempting to set up a network of weather-reporting stations in the United States and Canada for the study of storms. In 1835 the Russians had already set up such a network as a result of the visit of Humboldt. But Henry did not like the idea of cooperating with Maury. When the international conference was held in Brussels in 1853, the proposed cooperation was restricted to observations at sea. As a result of the conference, the flow of new data into Washington was greatly increased. Humboldt himself gave his support to Maury for establishing a new field of scholarship—the physical geography of the sea.[13]

Other American Climatologists

The conflict that developed between Maury and Joseph Henry is one of those sad episodes in the history of geography that have become more and more common during the past century. This episode offers an almost perfect model for all similar events. Maury, as we have seen, lacked an academic background. As a scientist he was self-taught, but he made up for his deficiency by a boundless enthusiasm and by a vast experience. He was a salesman, a promoter who sold the basic notion that knowledge of the ocean and of the atmosphere could only be brought together by widespread cooperation. His observations at sea had been carried out by sailors under proper direction. Why, said Maury, should not a similar system of weather

[13]Maury's career as a naval officer and as a geographer came to an abrupt end in 1861 when he resigned his commission to join the Confederacy. Born in Virginia, he felt a strong sense of loyalty to his native state. For the defense of the James River he invented the first torpedoes to be detonated by electricity. He went to England to purchase supplies and ships for the Confederacy. After the war he went to Mexico briefly to promote the settling of Virginian colonists who wanted to leave the United States. From 1868 to the time of his death in 1873 he was professor of physics at the Virginia Military Institute. His school textbooks in geography were widely adopted (Williams, 1963).

observations be extended to the land and be carried out by farmers? In the 1850s he worked steadily for the establishment of a national weather bureau, even to the extent of securing a leave of absence for the purpose of making a lecture tour to gain popular support.

Joseph Henry was a scientist. As a student of physics he had invented some of the basic devices required for electric motors and generators. He and Michael Faraday had developed these devices independently and at the same time, but Faraday had published his results sooner. In 1832 Henry had joined the faculty of the College of New Jersey (Princeton), where he taught physics and mathematics and a variety of other subjects, including geology and astronomy. In 1846 he was appointed the first secretary of the new Smithsonian Institution in Washington. Among the projects that Henry undertook at the Smithsonian was the establishment of a network of weather stations, each of which was to report to Washington by telegraph. Henry wanted these reports in order to chart the movements of winter storms.

Of course the two proposals met head on. Although Maury continued to insist that the proposals were not competitive, nevertheless, the fact is that Maury had access to government funds and Henry did not. Henry continued to oppose Maury's plan, insisting that Maury was not qualified to direct such a program or to make use of the results. Maury's attempted explanations of atmospheric circulation and his endorsement of the open polar sea only made Henry rate him still lower as a scientist. It was the old story of the enthusiastic promoter with little capacity for patient scientific work meeting the careful scientist. Henry was probably right: if Maury's plan had been approved by the Congress it would have submerged the Smithsonian project. So Henry with a small group of leading scientists saw to it that Maury's plan did not pass.

An official U.S. Weather Service was established in 1870 and assigned to the Signal Corps of the Army. In 1891 it was transferred to the Department of Agriculture, and in 1940 it was moved from Agriculture to the Department of Commerce.

Meanwhile, several scholars in America were working with climatic data (Leighly, 1949). Samuel Forry made one of the earliest maps of the United States showing the distribution of temperature by making use of Humboldt's isotherms. In 1853 Lorin Blodget used new climatic data to draw a temperature map, but, since he permitted his isotherms to end on the map and to split, his method was something less than expert. In 1854 James H. Coffin prepared a wind map of the Northern Hemisphere, and in 1875 he extended the map to the whole globe. His paper on the laws of atmospheric circulation went far ahead of Maury. In 1882 Elias Loomis of Yale compiled the first rainfall map of the earth (James and Jones, 1954: 334–361).

Meanwhile, a major contribution to the understanding of atmospheric circulation had been made by a high school teacher of mathematics in Nashville, Tennessee, William Ferrel. In 1856 he had read Hadley's explanation of the deflection of moving air by the earth's rotation and decided to work out the problem for himself. To Ferrell goes the credit for being first to give a mathematical explanation of the way moving bodies must behave on a rotating sphere.[14]

THE GREAT SURVEYS

In addition to the collection of new data on population or regarding the oceans or the world's climates, another kind of new geography appeared in the nineteenth century in the form of systematic surveys of unsettled territory. By midcentury the United States was in possession of a vast extent of thinly occupied land stretching roughly from the 100th meridian west to the Pacific Ocean. California was occupied, and the discovery of gold had resulted in a rush of new settlers. But what was the land like in between? This country had been penetrated by fur trappers, missionaries, and by a long list of explorers such as Zebulon M. Pike, Stephen H. Long, Lewis Cass, John C. Frémont, William H. Emory, and Isaac Stevens.[15]

Lewis and Clark

The Lewis and Clark Expedition of 1803-1806 was also primarily exploratory. The chief purpose was to explore the Missouri River and its tributaries and to find the best route to the headwaters of the Columbia River and thence to the Pacific Ocean. But the President, Thomas Jefferson, who was a keen student of geography, gave Lewis and Clark a careful and detailed directive. In addition to establishing the latitude and longitude of key places along the route, the expedition was to prepare a systematic record of observations concerning the nature of the country and its inhabitants. Jefferson specified that the reports should include observations on the number and characteristics of the Indians, their manner of making a living, their languages, and their relations to neighboring tribes. The reports should also provide information on the soil and face of the country, its vegetation, its animals, its minerals, its climate—including temperature; proportion of rainy, cloudy, and clear days; the occurrence of lightning, hail, snow, and ice; the prevailing wind directions; the dates of the first and last frosts; and the time of the year when particular

[14]William Ferrel, "The Motions of Fluids and Solids Relative to the Earth's Surface," *Mathematical Monthly,* 1 (1859): 140-148, 210-216, 300-307, 366-373, 397-406; 2 (1859-60): 89-97, 339-346, 374-390. There was also a brief report in *American Journal of Science and Arts,* 31 (1861): 27-50. For a review of the various scholars who have contributed to an understanding of the deflective force of the earth's rotation, including the work of G. G. Coriolis (1835), see C. L. Jordan, *"Bulletin of the American Meteorological Society,* 47 (1966): 401-403; see also 47 (1966): 887-891.

[15]Major Emory surveyed and prepared a geographic report on the United States-Mexican border, House Document 135, 34th Congress, 1st Session, 1857; also *Reports of Explorations and Surveys to Ascertain the Most Practical and Economical Route for a Railroad from the Mississippi River to the Pacific Ocean,* 1853-1855, Senate Document 46, 35th Congress, 2d Session, 1855. For Isaac Stevens, see Meinig, 1955.

plants lose their leaves—(Coues, 1893/1965). The expedition left St. Louis in May, 1804 and returned in September, 1806. The diaries and maps represent the most useful collection of data concerning a previously little-known land area ever assembled up to that date (Dillon, 1965).

Ferdinand V. Hayden

The term, Great Surveys, however, is usually applied to the expeditions specifically organized to map and make inventories of the western territories. There were four such expeditions working more or less simultaneously in the years 1866 to 1879 (Bartlett, 1962). The four leaders were Ferdinand V. Hayden, Clarence King, George M. Wheeler, and John Wesley Powell.

Hayden directed the U.S. Geological and Geographical Survey of the Territories, which was sent out by the Department of the Interior each year between 1869 and 1878. Hayden's men surveyed the mountain country of northern Colorado and Wyoming. He reported on the geysers and hot springs of Yellowstone and the spectacular scenery of the Teton Mountains. Hayden was one of the most effective proponents of the creation of a Yellowstone National Park. His reports, however, were hurried and were less than objective. He was enthusiastic about the possibilities of settlement and even defended the notion that the rainfall would increase with settlement. Nevertheless, his maps and the landscape sketches prepared by William H. Holmes remain major accomplishments.

Clarence King

Clarence King, who in 1863 was operating the California Geological Survey, was appointed director of the U.S. Geological Exploration of the Fortieth Parallel. The survey carried out a geological cross section of this western country along latitude 40°N between 1867 and 1872. He identified and named the now-dry lake basins of the Great Basin in Utah and Nevada (Lake Bonneville and Lake Lahontan). His studies of the Uinta Mountains and the Wasatch Mountains made it possible for him to correlate some of the fossils found there with fossils of the same geologic age in Europe. However, when King tried to convince the War Department that his survey should be broadened, he was not successful because it was felt that his approach was too narrowly geological.

In 1879 King was named the first director of the U.S. Geological Survey (which in that year replaced the separate surveys of the West). He resigned after one year.

George M. Wheeler

For many years before the Civil War (from 1813 to 1863), the U.S. Army Topographic Engineers had been engaged in making detailed maps of various parts of the United States. But during the war years this function was dropped. After the war the four surveys began once more to undertake the kind of work that had once kept the engineers busy in peacetime. General Humphreys of the Army Engineers felt that the surveys, especially those of Hayden and King, were too exclusively focused on geological problems and that any mapping accomplished was only incidental. For this reason he rejected King's bid to secure additional appropriations. But in 1871 he appointed Lt. George M. Wheeler, a West Point graduate, to head the U.S. Geographical Surveys West of the One Hundredth Meridian. Wheeler was directed not only to make topographic maps but also to undertake a land classification. The potential use of land was identified in four categories: (1) land suited for agriculture, (2) land suited for timber, (3) land suited for grazing, and (4) arid land. Between 1874, when the land classification work was added to the mapping, and 1879, when all the surveys were combined in the U.S. Geological Survey, some 175,000 square miles were classified. It is interesting that the army wanted not only maps showing landforms but also man-made features, such as mines, farms, roads, dams, and settlements. Wheeler was able to recruit some of the best qualified field scientists available, including Grove Karl Gilbert, who made some highly important original contributions to the understanding of landforms. In 1875 Gilbert resigned to join the Powell survey.

In 1881 Wheeler was sent to Venice to attend the Third International Geographical Congress. His report on the principal government land and marine surveys of the world is an important record of the progress of topographic mapping since the earliest mapmaking efforts of the Russians in 1720 under Peter the Great (Wheeler, 1885).

John Wesley Powell

The story of John Wesley Powell is so important for the history of geography that it needs to be told in somewhat more detail (James, 1979). Powell, like the other three leaders of Great Surveys, had not received any formal geographic training. From the enthusiasm of a high school teacher in Ohio, he developed a keen interest in natural history and a capacity to observe natural phenomena out-of-doors. In 1859 while he was a schoolteacher in Illinois, he became secretary of the Illinois Natural History Society. But when the Civil War began, he volunteered for the army. In the battle of Shiloh he was wounded and lost an arm, but he continued on active duty and served as engineer officer in the siege of Vicksburg. He left the army as major in 1864.

By this time Powell was fascinated by the unknown country to the west. He sought funds to organize an expedition but was repeatedly turned down in Washington. He did get some funds from several Illinois colleges, and in 1867 and 1868 he was able to travel to the Rocky Mountains and climb Pikes Peak and Longs Peak. Powell became a national figure in 1869 when, with a small party and four boats, he sailed down the Green River and the Colorado River through the Grand Canyon (Darrah, 1951; Stegner, 1954).

In 1870 the U.S. Geographical and Geological Survey of the Rocky Mountain Region was financed by an appropriation from Congress; it continued to examine the mountain country until 1879 when it was combined with three other surveys of the West. In 1871 Powell carried out a second canyon trip, but most of his attention was given to surveying the land, the resources, and the people. He made the first records of Indian customs and language. In 1875 he secured the service of Grove Karl Gilbert, whose *Report on the Geology of the Henry Mountains* (Gilbert, 1878) is a classic in the field of landform studies. Gilbert is credited with the first clear statement of the concept of grade—that is, the equilibrium reached between slope, volume of water, velocity of flow, and load of detritus. Powell, himself, during his first trip through the Uinta Mountains along the canyon of the Green River, saw for the first time just how a river could cut down through a mountain that was being raised up across its course. Fortunately, Powell had not read Buache and did not know that mountains were supposed to border drainage basins. He called a river that continues in its course through a rising mountain range an antecedent river.

Gilbert and Powell enjoyed two advantages not shared by the students of landforms in Europe. In the western part of North America there are large areas that are arid or semiarid, which means that the forms of the surface are not hidden under a thick cover of vegetation. Furthermore, these field men were not blinded by preconceptions based on earlier influential studies. Powell never had to consider the hypothesis set forth by Charles Lyell that landforms were chiefly the result of marine erosion: in the western part of the United States marine erosion did not seem plausible, and, clearly, the landforms had been sculptured by running water. Of course the Colorado River had cut its own canyon—the conclusion was inescapable when the canyon was viewed from a boat. Here in the western part of the United States a truly new geography was born.

Powell was also concerned about the human use of this country, but he did not share Hayden's enthusiasm about the possibilities of settlement. He could find no evidence to support the idea that the rainfall would increase if more of the area were settled. In fact, his own observations led him to insist that the only way to maintain the flow of water from the mountains to the bordering lowlands was to preserve the forest cover. Without ever having read Humboldt, he, nevertheless, predicted that the clearing of the forests on the watersheds would bring disaster. One of his projects was to produce a map showing the relatively small areas that could be supplied with enough water to support irrigated farms.

In 1880 Powell became the second director of the U.S. Geological Survey, following the resignation of King. The Bureau of Ethnology had been established the year before to carry on Powell's work with the Indian cultures. But the Geological Survey was specifically directed not to undertake geographical work (Powell, 1885). Why? Does it not seem clear that the Congress must have wanted information regarding the potential uses of land and about the recommended ways of maintaining the flow of water in the rivers? But no, the members of Congress were subjected to pressures from people who did not want Powell's information published. There were people who wanted to sell land to settlers: What of it if the land turned out to be in an area designated as arid by both Wheeler and Powell? There were also grazing interests that presented good arguments in favor of clearing the forests to enlarge the area of pasture. Neither of these groups wanted Powell's maps; and, in spite of vigorous efforts by Powell, himself, the Congress directed him to continue the search for mineral resources.[16]

The story of Powell's years as director of the Survey from 1880 to 1894 offers a fascinating study in how the director of a government agency can find ways to carry out the programs he feels are necessary in spite of opposition. Powell did study the arid lands, and his monograph is a major contribution to geography (Powell, 1878). He found the means to continue making topographic maps. But little by little his opponents gained in strength. He found opposition not only from the politicians and the groups with special interests in the West but also from scholars in the universities, whose opposition to Powell reminds us of the very similar treatment of Maury. From the scholars' point of view, the attacks on Powell—and men like him—were entirely justified. Here was a man who had never held a teaching post except in a high school many years before. Here was a man with no scholarly reputation—a tough field observer; a man with a brilliant mind, but lacking the patience to do the tedious job of testing hypotheses; a man who would jump ahead to explanations before accumulating the needed supporting evidence. A field of study progresses because both kinds of people contribute to it; but this does not make the apparently inevitable conflicts any easier to live with.

EXPEDITIONS AND GEOGRAPHICAL SOCIETIES

Gradually then the exploratory kind of expedition gave way to a new kind of expedition—a field survey. Of course exploration continued: in the Arctic and Antarctic both poles were reached on foot (Peary in 1909, Amundsen in 1911, and Scott in 1912) before airplanes could fly over them. Mountains were climbed (the summit of Mt. Everest was reached by Hillary and Tenzing in 1953) and deserts were crossed for the first time by Europeans. A series of famous sea voyages was carried out by specially equipped research vessels: the *Challenger*, the *Tuscarora*, the *Gazelle*, the *Valdivia*, the *Chun*, and the *Meteor*.

These and many other scientific expeditions were reported chiefly in the various geographical societies that were founded during the nineteenth century. The prototype of all the geographical societies was the Association for Promoting the

[16]Powell was also one of a small group of scholars in Washington who founded the Cosmos Club in 1878 (Stegner, 1954: 242).

Discovery of the Interior Parts of Africa, formed in London in 1788. This and the Palestine Association were merged in 1830 to form the Royal Geographical Society. The Société de géographie was founded in Paris in 1821, and in 1828 the Gesellschaft für Erdkunde zu Berlin was founded. The first geographical societies in America were established in Rio de Janeiro (Instituto Histórico e Geográfico Brasileiro) in 1838 and in Mexico City (Sociedad Mexicana de Geografía) in 1839. The Russian Geographical Society was founded in St. Petersburg in 1845 and the American Geographical Society of New York[17] was founded in 1851 (Wright, 1951, 1952). In 1888 the National Geographic Society was established in Washington, D.C., for the increase and diffusion of geographical knowledge. The first monograph of the society published in 1896, included a series of papers on geography, one of which was Powell's division of the United States into physiographic regions.

By 1875 there were twenty-eight geographical societies in Europe and one in Cairo (Ginsburg, 1972). The meetings of these societies and the periodicals each of them published were devoted to accounts of scientific expeditions in different parts of the world. The first International Geographical Congress was held in Antwerp in 1871, and congresses have been held approximately at four-year intervals since that time.[18] The International Geographical Union was formed in 1922 to tie all these and other geographical activities together and to coordinate their programs.

[17]Formed as the American Geographical and Statistical Society; the "and Statistical" was dropped in 1871 to shorten the name.

[18]International Geographical Congresses have been held at: Antwerp (1871); Paris (1875); Venice (1881); Paris (1889); Bern (1891); London (1895); Berlin (1899); Washington, D.C. (1904); Geneva (1908); Rome (1913); Cairo (1925); Cambridge, England (1928); Paris (1931); Warsaw (1934); Amsterdam (1938); Lisbon (1949); Washington, D.C. (1952); Rio de Janeiro (1956); Stockholm (1960); London (1964); New Delhi (1968); Montreal (1972); Moscow (1976); and Tokyo (1980) (Commission on History of Geographical Thought, 1972).

CHAPTER 8

The New Geography in Germany

If Galileo's famous experiment had merely demonstrated that when he, Galileo, dropped two specific objects of different weight from the Leaning Tower of Pisa they fell together at the same rate of speed, that fact would have found but a small place in scientific knowledge. Its great importance, of course, was that subsequent experiments showed that he had illustrated a universal, a relation that was true regardless of where, when, or by whom the weights were dropped. Few will question that it is an essential function of science to seek for such universals. On the other hand if later experiments had shown that the same results were not obtained elsewhere or at other times, the fact that they did take place on that one occasion, if substantiated as a fact, even though never explained, would represent a bit of scientific knowledge.... One might say that it is an axiomatic ideal of science to attain complete knowledge of reality—expressed as completely as possible in terms of universals, but in any case expressed in some way.

The nineteenth century witnessed a fundamental change in the role of the universities. The traditional university, as we have seen, was a place where students could be indoctrinated in the religious and political beliefs of the community and where studies were focused on Greek and Latin classics, on theology and law, or logic and rhetoric. But as the newly identified disciplines came into existence, the universities had to assume the responsibility for the advanced training of younger generations of scholars. The members of each professional field set the standards of

The quotation above is from R. Hartshorne, *The Nature of Geography* ... (1939), p. 378.

professional behavior (Kuhn, 1963). This kind of education required courses of advanced study in graduate schools.

The change started in Germany in 1809 with the establishment of the University of Berlin as a free community of scholars. Slowly at first, then more rapidly after the middle of the century, two principles came to be accepted regarding the new functions of universities. First, the students were freed from standard curricula and were permitted to select whatever course of study interested them. Second, the appointments to positions on the faculties were made on the basis of scholarly performance; and, after an appointment was made, the faculty member was granted the right to engage in research and to teach the results of his research free from any restraints except those established by professional opinion.

When these major innovations in the functions of universities began to appear, there were only a few universities in the world where a young scholar could be trained in the concepts and methods of a field called geography. Only a very few university teachers were conscious of being geographers, and most of these were pupils of Carl Ritter at Berlin.

Yet geographical research was being carried on. In America, as we have seen, the field men working on the Great Surveys were not trained as geographers. They were geologists, naturalists, engineers, or men trained for the military services. They were curious about the processes that produced the unfamiliar landscapes of the American West; but they were also faced with questions of pressing practical importance. There was an urgent need to know more about the resources and shortcomings of a region that was about to be occupied by waves of new settlers.

In Europe the situation was quite different. There were no nearby unoccupied regions where new pioneer settlement created an obvious need for more knowledge about the earth as the potential home of man, nor were there regions of unfamiliar appearance to challenge the curiosity. Men like Alexander von Humboldt had to travel to other continents to find such regions. In Germany there were centers of geographic research where the observations of scientific travelers could be collected and where exciting new maps were being devised to present this information in useful form. Geographical societies were organized so that the latest knowledge about the earth could be presented in lectures or in professional periodicals. But advanced instruction in universities was not available. In spite of Wilhelm von Humboldt's effort to improve the quality of instruction in the secondary schools, there were no places where teachers of geography could be trained. The result was a continued decline in the content of secondary-school geography; even a return to the kind of rote learning against which Pestalozzi and Ritter had waged so vigorous a campaign.

Geography as a field of advanced study in the universities appeared in Germany in the 1870s. From Germany the movement spread rapidly to France, to other European countries, and also to America. In most countries after 1870 it is possible to identify one outstanding pioneer who was chiefly responsible for establishing the scope and methods of the new geography at the university level. But at first the persons who were appointed to university positions in geography had had no previous training in this field. There was no professionally accepted paradigm to serve as a guide to the study of geography. The new appointees had been trained in history, geology, botany, zoology, mathematics, engineering, or journalism. In the absence of any guidelines regarding the field of geography, each new professor felt the need to set forth his own ideas concerning the scope of the field. Each tried to provide a definition of geography that would give it unity and that would establish its position among other academic disciplines. All around the world in the late nineteenth century the question echoed through the academic corridors: What is geography? In this chapter we will review the answers that were given in Germany (Hettner, 1927; Hartshorne, 1939:84–148; Van Valkenburg, 1951; Fischer, Campbell, and Miller, 1967: 81–174; Dickinson, 1969:51–185).

THE NEW GEOGRAPHY IN GERMANY

Geography in Germany after the death of Ritter lacked a focus to give it unity. Those who worked at Gotha were devoted to the careful plotting of information on maps and to the publication of clearer maps of finer design. There was no need at Gotha to ask what geography was—geography was anything that could be put on a map. Those who were engaged in military careers appreciated the practical need for more accurate and useful information about the earth as the scene of warfare. Those who were administrators of Germany's new colonial possessions and those who were engaged in doing business away from home also needed more knowledge of geography—not theoretical geography, but compilations of useful facts about places. The former pupils of Carl Ritter, who were giving lectures on geography in some of the German universities, were chiefly concerned to provide a background for the study of history.[1]

Oscar Peschel

Oscar Peschel's appointment as professor of geography at Leipzig in 1871 was the first new professorship created since the death of Ritter. Peschel was then forty-five years of age and had already established his reputation as an editor and

[1]Between 1859 and 1871 courses of lectures in geography were being given at three German universities. At Berlin, where Ritter had been professor of geography, he was replaced by a *dozent* (more or less equivalent to an associate professor). This was Heinrich Kiepert, a classical historian, who was promoted to professor in 1874. At Göttingen there was Johann Eduard Wappäus, who was appointed as *privat dozent* (assistant professor or instuctor) in 1838 and became a professor in 1854. At Breslau there was Karl J. L. Neumann, who had been appointed Professor of Geography and Ancient History in 1856. G. B. Mendelssohn had been at Bonn, but he died in 1857 and was not replaced for two decades (Wagner, 1880; Dickinson, 1969:51–55).

writer. He had been the assistant editor of the *Allgemeine Zeitschrift* at Augsburg and from 1854 to 1870 was the editor of *Das Ausland*, a periodical that printed articles about foreign countries and about problems of foreign affairs. He had written on history of ancient geography, and on this basis he was invited by the Historical Commission of the Royal Academy of Sciences to write a book on the history of geography to form a part of a series on science in Germany. Peschel's *Geschichte der Erdkunde* was published in 1865 (Peschel, 1865).

His studies of the history of geography led him to take issue with Ritter's method of making comparisons between regions, which Ritter called *vergleichende Erdkunde* (comparative geography). Peschel pointed out that Ritter made his comparisons between whole continents or major parts of continents and that such units were really composite concepts and were not properly comparable. He demonstrated how he would make comparative studies by focusing attention on a particular kind of landform, which he examined on the most detailed maps available to him. For example, he studied the much-indented fjorded coasts, which, he observed, occur on the western sides of continents in higher middle latitudes. He offered the hypothesis that the fjords were fissures in the earth's crust that had been occupied and gouged out by glaciers. His systematic studies of these features together with lakes, islands, valleys, and mountains were published in 1870 (Peschel, 1870).

Peschel is credited with being one of the founders of modern physical geography. He was not unmindful of the importance of showing the relation of the physical features of the earth to man's use of the earth; but in his systematic studies he concentrated attention on the earth's physical features and did not treat the elements of human geography in a similar systematic manner. He served for only four years at Leipzig before he died at the early age of forty-nine. His book on physical geography was published posthumously (Peschel, 1879).

New Professorships in Germany

In 1874 an event of major importance took place in Berlin. The Prussian government decided to establish a chair of geography (to be occupied by a scholar with the rank of professor) in each of the Prussian universities. Prussia was the largest and most influential of the separate political units that came together to form the German Empire in 1871 at the end of the Franco-Prussian War (1870–71). The action of Prussia was followed elsewhere in the newly unified Germany.

Why did Prussia take this step in 1874? The answer is not entirely clear. It is reasonable to assume that the Franco-Prussian War resulted (as all wars do) in a popular demand for the teaching of more geography. The officers of the army, many of whom had studied geography at Berlin under Ritter, wanted more and better geography taught in the schools and universities; and the new interest in colonial possessions in other parts of the world or in people of German origin who had settled outside of Germany after 1848 contributed to a recognition of the need to learn geography. It is also recorded that Hermann Wagner, then a teacher of geography at the Gymnasium (secondary school) in Gotha, was greatly concerned about the poor preparation of secondary-school teachers of geography. With Alfred Kirchhoff, who had lectured at the military academy in Berlin and had been appointed professor of geography at Halle in 1873, Wagner urged the Prussian government to provide advanced instruction in geography in the universities (Dickinson, 1969: 59). When the government took this action numerous openings for university professorships in geography appeared suddenly, and many of them were filled within a decade. By 1880 Hermann Wagner reported that there were professors of geography in ten of the Prussian universities, and vacancies in three more were about to be filled (Wagner, 1880:591).

Ferdinand von Richthofen

It was a geologist, Baron Ferdinand von Richthofen, who became the leading figure in the introduction of the new geography into the universities of Germany. He was an experienced field observer. As a young man he had carried on geological studies in the Alps and the Carpathians. In 1860 the Prussian government selected him to undertake an expedition to eastern Asia for the study of lands and resources. After working in China he sailed across the Pacific to California, where he spent six years in active geological studies. Still challenged to find out more about the resources of China, he was able to get the Bank of California to finance field work there with the agreement that he would report his findings to the Chamber of Commerce in Shanghai. In his subsequent survey of China he was the first to report and map the Chinese coal fields.

He did much more than locate minerals and fuels. He also sought to formulate hypotheses to explain China's surface features. He noted the presence of a fine, powdery material covering the land on the eastern side of the Gobi, and he was the first to identify this material as wind-blown dust, or *loess*. He also noted that in this part of Asia the loess, and also the stratified rocks, were deposited over a relatively level or gently rolling surface and that this more-or-less level surface cuts across ancient rocks of varying degrees of resistance to erosion. He concluded that the only force strong enough to level such resistant rocks must be the ocean. Ocean waves working on gradually sinking land could, he believed, produce the great, nearly level plains of the world. His studies of China were published in five volumes over the period from 1877 to 1912 (Richthofen, 1877–1912).

Richthofen's answer to the question, What is geography?, was enormously influential inside and outside of Germany. His was the pioneer statement of the scope and method of the new geography. His ideas were initially presented in the first volume of his study of China (Richthofen, 1877–1912:1, 729–732; Fischer, Campbell, and Miller, 1967:84–87) and then repeated in his inaugural lecture at Leipzig, delivered on April 27, 1883 (Richthofen, 1883; Fischer, Campbell, and

Miller, 1967:88-95). It was the distinctive purpose of geography, he said, to focus attention on the diverse phenomena that occur in interrelation on the face of the earth. To reach useful and reliable conclusions, he believed, a geographical study of any part of the face of the earth must start with a careful description of the physical features and then must move on to an examination of the relationships of other features of the earth's surface to the basic physical framework. He distinguishes the study of the processes creating the surface features (geology) from the description of the surface features themselves as the frame of reference to which other elements of the face of the earth (including the works of man) are to be related. The highest goal of geography is the exploration of the relationship of man to the physical earth and to the biotic features that are also associated with the physical features. This became the basic pattern for geographical studies, not only in Germany but also in other parts of the world.

Richthofen also thought carefully about two other problems that bothered all the newly appointed professors of geography. The first of these was similar to the questions being asked during the same period by historians and economists. Is geography concerned only to describe the unique features of particular regions or is it also concerned with the formulation of generalizations or theory? And the second problem had to do with the relationship of what Varenius had called *general geography* and *special geography*. Richthofen combined the answers to these two questions. In the first instance, he decided, the essential observations on which any framework of concepts must be built had to be made in the field in particular areas where the features are unique. This, he said, is *special geography*. Regional study first must be descriptive; but it must also go beyond the description of unique features to seek regularities of occurrence and to formulate hypotheses that explain the observed characteristics. Loess, for example, can be observed, measured, and carefully described; but a regional study must also look for the process by which loess is accumulated and for the consequences of the presence of loess in the cover of plants and in man's use of the region. In this procedure Richthofen was specifically following the lead of Humboldt as that master had carried out field studies in New Spain and Cuba.

Richthofen, being an experienced field observer, had a feeling for what could be learned from direct observation out-of-doors. The purpose of developing the general concepts regarding the world distribution of phenomena (general or topical geography) is to throw light on the causal interrelations among the diverse things in particular areas, to which he applied the term *chorology,* or regional study—a term already widely used in Germany (Hartshorne, 1939:92). He also realized that in addition to looking at the world as a whole, it was necessary to examine smaller and smaller segments of the earth's surface. He distinguished the different methods of study in areas of different size, which he named (in order of decreasing size): *Erdteile* (major divisions of the earth), *Länder* (major regions), *Landschaften* (landscapes or small regions), and *Örtlichkeiten* (localities).

Furthermore, Richthofen helped define the word *Erdkunde*. Ritter had preferred this term, which is of German origin, rather than the word *Geographie*, which is of Greek origin. Ritter intended to use *Erdkunde* as a synonym for *Geographie*, but some of the German writers took the word literally to mean the study of the whole earth body. Richthofen insisted that geography (and *Erdkunde*) must refer to a study of the surface of the earth (*Erdoberfläche*), where the lithosphere, the hydrosphere, the atmosphere, and the biosphere are in contact with each other.

Richthofen published another statement of the method of geographic study in 1886 when he accepted the professorship at Berlin.[2] He wrote a systematic outline of the processes shaping the surface of the earth that could serve as a guide to the field study of any region (Richthofen, 1886).

Friedrich Ratzel

Peschel and Richthofen had laid down the guidelines for the systematic study of the earth's physical features and at the same time had defined geography as a unified field of study by organizing the treatment of biotic and cultural features around the fundamental framework of the physical earth. It was Friedrich Ratzel who provided the guidelines for a comparable systematic study of human geography, or, to use the word he coined, *anthropogeographie*.

Ratzel, who was eleven years younger than Richthofen, came to geography from a quite different background. He completed his graduate study at Heidelberg in 1868, working in zoology, geology, and comparative anatomy. This was just that exciting period when Darwin's revolutionary concepts regarding the process of evolution were sweeping away old paradigms. Ratzel's dissertation dealt with the significance of Darwin's ideas (Wanklyn, 1961). But Ratzel was more interested in the observation of plants and animals out-of-doors than he was in laboratory studies. To study in the field required travel, and traveling cost money. He had his first taste of field study when he accompanied a French naturalist on a trip to the countries around the Mediterranean. When the Franco-Prussian War started in 1870, Ratzel joined the Prussian army and was wounded twice during the war.

The unification of Germany in 1871 was, for Ratzel, a very exciting occurrence. His strong sense of national pride drew his attention away from academic studies to field observation of the ways Germans lived and made use of resources. He wanted to continue his traveling, if possible, to seek out and describe the people

[2]Richthofen occupied several professorships in German universities. In 1875 he was offered the chair at Berlin and was immediately granted leave of absence to work on the manuscripts of his China study. Then in 1877 he accepted an appointment at Bonn. In 1883 he moved to Leipzig and then in 1886 he moved to Berlin. For a time he served as rector of the University of Berlin. Richthofen trained many of those who became leaders of the later generations of geographers in Germany and elsewhere. Among his famous students were Siegfried Passarge, Otto Schlüter, Alfred Philippson, Alfred Rühl, and Sven Hedin, the Swedish explorer of Central Asia (Dickinson, 1969).

of German origin now living outside of Germany. Some letters he wrote for a newspaper in Cologne so impressed the editor that the young Ratzel was employed as a roving reporter. Payment for articles written for the *Kölnische Zeitung* made it possible for him to extend his travels to more distant places. He visited Hungary and Transylvania, reporting on the German minorities in that part of eastern Europe. In 1872 he crossed the Alps for an extended visit to Italy.

Ratzel's visit to the United States and Mexico in 1874-75 was a turning point in his career (Sauer, 1971). Not only was he impressed by the contributions of people of German origin to American life in the Middle West and the Southwest but also his attention was focused on the success of other minority groups, such as the Indians, the Africans, and the Chinese in California. He began to formulate some general concepts regarding the geographic patterns resulting from contact between aggressive and expanding human groups and retreating groups. He recognized and described examples of the destructive use of land (to which the Germans attach the expressive term *Raubbau*), which he hoped was a characteristic of land settlement at an early stage and would be remedied as settlement matured. He later published his studies of the United States, based on his own keen observations and a remarkably penetrating study of earlier accounts, especially regarding the settlement of the West. It was this experience in the interpretation of an extensive region that turned his attention specifically toward the study of geography (Steinmetzler, 1956: 73-74; Speth, 1977).

Returning to Germany in 1875, he resigned his position with the newspaper and was appointed *privat dozent* in geography at the *Technische Hochschule* (Institute of Technology) in Munich. He was promoted to *dozent* in 1876 and to professor in 1880. In 1886 he accepted an offer to become the professor of geography at Leipzig, where he remained until his death in 1904.

It was at Munich that Ratzel first began to publish his ideas concerning the systematic study of human geography. In 1882 he published the first volume of *Anthropogeographie,* in which he traced the effects of different physical features on the course of history (Ratzel, 1882; Hassert, 1905; Steinmetzler, 1956; Wanklyn, 1961; Dickinson, 1969:64-72). Meanwhile, some other geographers, notably Kirchhoff, had used the opposite approach to the study of human geography. Instead of describing the influence of the physical earth on human affairs, they focused on man himself. Human societies were studied in relation to the physical features, but the most attention was given to the culture of human groups rather than to the physical earth. This was the method that Ratzel adopted in the second volume of *Anthropogeographie,* the first edition of which was published in 1891 (Hartshorne, 1939:91).

It is curious that Richthofen was much more influential than Ratzel among later generations of German geographers but that Ratzel had more influence than Richthofen outside Germany, especially in France and the United States. Ratzel, like Ritter, was a brilliant lecturer, and his classes at Leipzig were always full.

Sometimes as many as 500 people crowded into his lecture room. One of his most eloquent disciples was the American student, Ellen Churchill Semple, who studied at Leipzig in 1891–92 and again in 1895. Here is what she had to say about his ideas:

> Moreover the very fecundity of his ideas often left him no time to test the validity of his principles. He enunciates one brilliant generalization after another. Sometimes he reveals the mind of a seer or poet, throwing out conclusions which are highly suggestive, on the face of them convincing, but which on examination prove untenable, or at best must be set down as unproven or needing qualification. . . . He grew with his work, and his work and its problems grew with him. He took a mountain-top view of things, kept his eyes always on the far horizons, and in the splendid sweep of his scientific conceptions sometimes overlooked the details near at hand. Herein lay his greatness and his limitations [Semple, 1911:v–vi].

Among Ratzel's "brilliant generalizations" that led some of his disciples to carry his ideas further than he himself intended was the application of Darwin's biological concepts to human societies. This analogy suggested that groups of human beings must struggle to survive in particular environments much as plant and animal organisms must do. This is known as *Social Darwinism*.[3] Ratzel, like many other scholars in this period, was greatly impressed with the ideas of Herbert Spencer regarding the similarity of human societies to animal organisms. In his *Political Geography*, Ratzel compares the state to an organism: "Der Staat als Bodenständiger Organismus" ("the state as an organism attached to the land") (Ratzel, 1897: Chap. 1). However, he is very careful to point out that this analogy is not to be taken literally, for there are many ways in which human groups act quite differently from organisms. The comparison is not intended, he says, to be a scientific hypothesis, but only an illuminating remark (Ratzel, 1897; 2d edn., 1903: 13). Nevertheless, having said this, he proceeds to show that a state, like some simple organisms, must either grow or die and can never stand still. When a state extends its borders at the expense of another state this is a reflection of internal strength. Strong states must have room to grow in order to survive. This is only one small step removed from the concept of *Lebensraum* (living space) and the right of superior people to enlarge their living space at the expense of inferior neighbors. Ratzel himself never subscribed to the idea of superior and inferior races. But some later geographers made use of the *Lebensraum* concept as a pseudoscientific support for the national policy of the 1930s.

[3]Social Darwinism can be traced to the writings of the English philosopher, Herbert Spencer (1820–1903). Spencer pointed out the close resemblance of human societies to animal organisms. In both there are regulative systems (a ccentral nervous system and a system of government); in both there are systems for the production of energy (the digestive system and the economic system); in both there are systems of distribution (veins and arteries and roads and telegraphs) (Spencer, 1876–96). It was Spencer who coined the phrase "survival of the fittest" (Spencer, 1864:1:444).

Ratzel did not cease to be critical of his own generalizations (not a common trait among scholars). He made special note of cases where cultural differences were more important than differences in the physical character of the land. For example, in a regional geography of Germany (Ratzel, 1898) he pointed out the great contrast in the way people make use of the land in two places that are physically very much alike. The two places are the low mountain regions on either side of the middle Rhine Valley—the Vosges Mountains in France and the Schwarzwald in Germany. The differences to be observed in these two regions are related to the contrast between the French and German cultural traditions.

Ratzel's systematic studies of human geography led him to discard a number of ancient concepts. Ever since the time of Dicaearchus in the fourth century B.C., it had been accepted that man's use of the land had advanced through a succession of stages: first the primitive hunters, fishers, and collectors; then the pastoral nomads; then the agriculturists; and finally the horticulturists, such as the rice farmers of southern and eastern Asia. Humboldt had noted the absence of the stage of pastoral nomadism in America owing to the absence of domestic animals. Ratzel expressed doubt about the possibility, in every case, of distinguishing between herders and agriculturists since many people were a little of both (Ratzel, 1891: 2: 741).

Ratzel's criticism of the concept of stages was noted by Eduard Hahn, who in 1892 published a new map of the economic systems (*Wirtschaftsformen*) of the world. He distinguished six major kinds of rural economies: (1) hunting and fishing; (2) hoe culture (*hackbau*); (3) plantation culture where the land is worked with the plow (*ackerbau*); (4) European-West Asian agriculture, where farming and herding are combined; (5) unmixed herding; and (6) horticulture, as in China or Japan (Hahn, 1892). Hahn provided ample evidence to demonstrate that the three stages of man's use of the land had to be discarded; in fact, there was evidence that in some cases pastoral nomadism developed out of an earlier agricultural way of living. Hahn's work ended a period when man's relation to the physical earth could be described by deduction from a theory and when a new phase of inductive study of the origins of agriculture could be undertaken (Hahn, 1896, 1919; Kramer, 1967).

The Problem of Unity and Diversity

Although Ratzel has provided the guidelines for the systematic study of human geography and scholars, such as Hahn, had proceeded along the lines Ratzel suggested, many of the newly appointed professors of geography in the German universities were bothered by the diversity of the subject matter they had to command. The question could be put this way: Are the concepts and methods of physical geography so utterly different from those of human geography that these two branches cannot be properly included in a single discipline? And did Richthofen's restriction of geography to the study of the surface of the earth really provide sufficient unity or did it require those who hoped to become geographers to master the concepts and methods of many diverse fields?[4]

Numerous suggestions were made in the search for unity and coherence. At one extreme were those who wanted to exclude human geography entirely. Georg Gerland, who was appointed professor of geography at Strassburg in 1875, concluded that geography should be exactly what the word *Erdkunde* implied: the study of the whole earth body without reference to man. Physical science, said Gerland, can formulate exact laws, but no such laws can ever be formulated to account for the behavior of human groups (Hartshorne, 1939:89). Few of the geographers could accept such a radical break with the tradition; nor did Gerland himself follow his own prescription, for he always included the geography of man in his courses at Strassburg (Hartshorne, 1939:107).

The Concept of Chorology

It was Alfred Hettner[5] who elaborated Richthofen's concept of chorology. His earliest methodological statement was published in Volume 1 of the *Geographische Zeitschrift*, a professional periodical that he founded in 1895 and of which he was the editor until 1935. Hartshorne quotes Hettner's first statement of the nature of geography as follows:

> If we compare the different sciences we will find that while in many of them the unity lies in the materials of study, in others it lies in the method of study. Geography belongs in the latter group; its unity is in its method. As history and historical geology consider the development of the human race or of the earth in terms of time, so geography proceeds from the viewpoint of spatial variations [Hartshorne, 1939:97].

Hettner continued to develop his methodological ideas in published papers in the *Geographische Zeitschrift* during the next two decades. His more important

---

[4]Many of those who had been trained in geology were scornful of their colleagues who accepted appointment as geographers. One remark that was commonly repeated among the geologists at the turn of the century was a kind of pun: "Ein Geograph ist einer der die Erdoberfläche oberflächlich studiert" ("A geographer is one who studies the surface of the earth superficially"). Contributed by Dean Emeritus Edward H. Kraus of the University of Michigan, who studied mineralogy, chemistry, and geology at Munich in 1899–1901.

[5]Alfred Hettner was the first professor in a German university after Ritter who was trained as a geographer. He also studied philosophy. He began his advanced studies with Kirchhoff at Halle, worked with Fischer, Richthofen, and later with Ratzel, and completed his doctorate with Gerland at Strassburg. He wrote a dissertation on the climate of Chile and western Patagonia. Field studies in Germany and in the Andes of Colombia led to major publications in geomorphology. In 1897 he was appointed professor of geography at Tübingen, and he gave his inaugural lecture there in 1898. In 1899, however, he was appointed professor at Heidelberg, where he remained until his retirement in 1928. At Heidelberg he supervised more than thirty doctoral dissertations, and many of his students became leaders of the next generation in Germany—including Leo Waibel and Oskar Schmieder (Dickinson, 1969:113).

pronouncements were published in 1895 and 1905 (Hettner, 1895, 1905). In 1927 he brought all his methodological ideas together in one book, together with a history of geographic thought (Hettner, 1927). Hartshorne quotes his concept of chorology as presented in 1927:

> The goal of the chorological point of view is to know the character of regions and places through comprehension of the existence together and interrelations among the different realms of reality and their varied manifestations, and to comprehend the earth surface as a whole in its actual arrangement in continents, larger and smaller regions, and places [Hartshorne, 1959:13].

Geography as a chorological science was not a new concept when it was developed by Hettner. Hartshorne has traced the repeated appearances of this view of the place of geography among the fields of learning in the writings of different scholars since the eighteenth century (Hartshorne, 1958). In his lectures at the University of Königsberg, Immanuel Kant presented this view of geography as axiomatic—the view that geography was the field in which things were considered in their areal context on the face of the earth, much as history was the field in which things were considered in their time context. Humboldt expressed the same view of the field, but there is no evidence that he received the idea from Kant. In 1832 Julius Fröbel pointed out the similarity of the ideas of Kant and Humboldt, but Fröbel's writings were overlooked and only rediscovered many decades later. Richthofen accepted the same idea without finding it necessary to quote any earlier authority. When Hettner presented this view, it had apparently been accepted by so many generations of geographers that he felt no need to support it with references to earlier writings.

Hettner also faced and answered the question of whether geographers should be concerned only with the unique things and events that occur in a spatial context or whether geographers should seek to formulate general concepts. The arguments concerning the idiographic as opposed to the nomothetic approach to learning had been carried on for a long time, especially among historians and economists. The historians of nineteenth-century Germany conceived of their field as essentially idiographic; but the economists, after some discussion, decided that economics was a nomothetic field. But Hettner insisted that geography is both idiographic and nomothetic, as indeed almost all other fields of learning must be. If attention were paid only to universals, there would be a residue of information considered irrelevant because it could not be explained by existing models; if attention were paid only to unique things and events, the full richness of interpretation resulting from deduction from theory would be lost. But why is the question posed as a dichotomy? Geography, like other fields of learning, must deal both with unique things and with universals. Is the astronomer who maps the face of the moon less of an astronomer than one who studies planetary motions? Is the economist who describes the unique factors producing inflation in Brazil less of an economist than the one who formu-

Alfred Hettner

Alexander von Humboldt

Joseph Partsch

Albrecht Penck

Walther Penck

Friedrich Ratzel

Carl Ritter

Carl Troll

lates or tests the causes of inflation in general? Are not both considered to be valuable members of the profession? Hettner rejected the view that geography could be either idiographic or nomothetic but not both. As Varenius had decided more than two centuries earlier, special geography is concerned with many unique characteristics of places; yet this approach can only be effectively pursued when it is illuminated by general concepts that are necessary even to identify uniqueness. General geography is that aspect of geographic study in which general concepts are formulated and in which every unique element is a challenge that leads to the search for better theory. Since both Hettner and Hartshorne made this view of geography entirely clear (Hettner, 1927:221–224; Hartshorne, 1939:382–384), it is discouraging to find some writers who continue to accuse Hettner and his followers of defining geography as essentially idiographic (Schaefer, 1953; Harvey, 1969:50), thereby obscuring the underlying continuity of geographic thought.

Hettner's ideas concerning the organization of geographical studies influenced the course of German geography for decades. One result was the continued focus of attention on the theme of man's relation to his physical and biotic surroundings. The traditional outline of a regional study started with location or position and then proceeded through chapters on geology, surface features, climate, vegetation, natural resources, the course of settlement, the distribution of population, the forms of the economy, the routes of circulation, and the political divisions (Dickinson, 1969:122). The outline was based on the notion that this formed a kind of sequence of cause and effect; in dealing with each topic the relationships with the physical base were discussed, but not the relationships with other topics (Hettner, 1907; Gradmann, 1931a; Spethmann, 1931). In spite of the apparent rigidity of the outline, some highly effective regional studies were produced (Philippson, 1904; Gradmann, 1931b).

Geography as Landschaftskunde

Hettner's thought did not remain unchallenged. There were some geographers who were uneasy about the identification of geography as a chorological science and, hence, one that, like history, was defined by its method rather than its subject matter and its concepts. Others were also concerned about the overemphasis on the significance of the physical features of an area that resulted from following the *schema*. By tying everything back to the physical features, other important relationships were overlooked—such as the relation of population density to the economy, or the economy and the routes of circulation, or even the relation of all of these to the political units. Furthermore, many of the interrelations observed in regional studies were in the process of change through time; and only by examining the past geographies and the changes taking place could the idea of processes, or sequences of events, be introduced to geographical work (Troll, 1950).

As a means of reaching a more balanced treatment of the interrelations of things in particular areas, it was suggested that attention be given to the overall appearance of the landscape. The idea was first set forth by J. Wimmer in *Historische Landschaftskunde* in 1885 (Hartshorne, 1939:218). But the concept of geography as *Landschaftskunde* ("landscape science") was widely adopted in Germany after the inaugural address given at Munich in 1906 by Otto Schlüter[6] (Schlüter, 1906). Schlüter recommended that geographers look first at the things on the surface of the earth that could be perceived through the senses and at the totality of such perceptions—the landscape. He objected to the chorological definition of geography and noted that accepting the landscape as the subject matter of geography would give the field a logical definition, like that of any other academic field except history. The nonmaterial content of an area—such as political organization, religious beliefs, economic institutions, or even the statistical averages of climate—could not be considered primary objects of geographical study, although these could be introduced to explain the observable landscape.

Both Hettner and Schlüter were concerned about the variations in the character of the face of the earth—which became known later as *areal differentiation*. Both recognized that there were distinctly different kinds of areas on the earth that were distinguished from their surroundings and that showed a certain degree of homogeneity within boundaries that could be defined. But Hettner stressed the ways in which the features of a region reflected the basic patterns of the physical earth, whereas Schlüter focused attention on the interrelations of these features that gave the region its distinctive appearance. He used the method of historical geography in analyzing landscapes. First, he identified what he called the *Urlandschaft*—that is, the landscape that existed before major changes were introduced through the activities of man. He would then trace the sequences of change whereby an *Urlandschaft* was transformed into what he called a *Kulturlandschaft,* a landscape created by the human culture (Schlüter, 1920, 1928). To trace these changes, he said, was the major task of geography.

When Schlüter was over forty years of age he started a major research program to apply his ideas to a specific region. His problem was to reconstruct the *Urlandschaft* of Central Europe. He found that the approximate date when the major movement of new settlement into the forests began was A.D. 500. By using the evidence of place-names and reading the descriptions written by ancient Greek and Roman geographers, he reconstructed the pattern of forested land and open land for that date. He then traced the process of settlement that created the *Kulturlandschaft*. His three-volume work was published after his death in 1952 (Lautensach, 1952; Dickinson, 1969:126–136).

---

[6]Otto Schlüter began his academic training as a student of the German language and of history. At Halle he studied with Kirchhoff, who turned his interests toward geography. In 1895 he went to Berlin to study with Richthofen and to work as his assistant. He went to Munich briefly but was the professor of geography at Halle from 1911 until his retirement in 1938 (Dickinson, 1969:127).

The German geographers who followed Schlüter in identifying geography as *Landschaftskunde* surrounded the concept of landscape with a kind of mystical significance (Hartshorne, 1939:149–174). Unfortunately the word *Landschaft* has two distinct meanings. Those geographers who made use of the word in a technical sense were not always careful to distinguish these meanings. *Landschaft* can, and traditionally did, refer to an extent of territory distinguished by a more or less uniform aspect. The word carried this connotation for many centuries before it was used in the fifteenth century by artists to refer to the painting of scenery—the aspect of the face of the earth as seen in perspective but without connotation of areal extent (Schmithüsen, 1963: 17).[7] Since then, writers have used the word without being clear about its meaning. Humboldt used *Landschaft* to refer to the visual impression, the aesthetic appreciation of beauty in scenery. J. Wimmer also used it in this sense in 1885 (Hartshorne, 1939:97), and even in 1953 Ewald Banse used the word with this meaning (Fischer, Campbell, and Miller, 1967: 168–174). Meanwhile, the use of the word in the other sense—referring to an area of uniform aspect—never entirely disappeared.

But there have been other difficulties with the concept of *Landschaft*. Schlüter defined the landscape as the total impact of an area on man's senses—including such invisible phenomena as wind or temperature. Schlüter specifically included man as a part of the landscape. But some of his followers insisted on restricting the term only to those material objects that could be seen. On the other hand, there were other students of landscape who used the material features of the earth's surface only to indicate the existence of a region and who did not hesitate to include in their studies such nonmaterial elements as law, or religion or economic institutions.

Leo Waibel, in a paper discussing the meaning of *Landschaft* (Waibel, 1933), pointed out that the word came into common use in Germany just at the time when geographers were focusing their studies on smaller and smaller areas, for which the word *Landschaft* (with its connotation of a small region) seemed more appropriate than the word for a larger region (*Gebiet*). Furthermore, there was continued emphasis on Richthofen's concept of the interconnections among things associated in area, and the word *Landschaft* seemed to carry the connotation of harmony of related parts.

The uncertainty regarding the exact meaning of this word is responsible for a conceptual error that still persists. The sequence of ideas runs as follows: first, it is agreed that *Landschaft* is made up of concrete, observable freatures; second, it is agreed that the word *Landschaft* is a synonym for a homogeneous area, or region. Combining these meanings, it becomes possible to assert that a region is a concrete reality and not just an intellectual construct (Hartshorne, 1939:263).

In one way or another, the concept of lanscape as an area with a more-or-less

---

[7]Similarly, before A.D. 1000 the word *landscipe* in Old English referred to an extent of territory. The word *landscape* was reintroduced into English from the Dutch (*landschap*) in the early seventeenth century (James, 1934:78–79).

uniform appearance, the interpretation of which requires study of the nonmaterial as well as the material phenomena in an area, was adopted by most of the geographers of Germany before World War II (Krebs, 1923; Bobek and Schmithüsen, 1949). Lautensach, writing in 1952, reported that most German geographers followed Schlüter in identifying the study of landscape as the central purpose of geography (Lautensach, 1952:226). Schmithüsen, in 1963, says that "every landscape is a dynamic structure, a thing-area-time (*Sach-Raum-Zeit*) system of specified quality inside the whole geosphere (*Geosphäre*)." This is what we might describe as a spatial system. It is an open system, not one that is closed, like an organism (Schmithüsen, 1963:13).

The Concept of Chorology Applied to General Geography

In Germany, and also in other countries, it became common to accept the notion that general (or systematic) geography is necessarily analytic and makes use of general concepts, whereas regional geography is necessarily synthetic and deals with unique situations. This error, according to Hettner, can be blamed on Richthofen (Hettner, 1927:400). The concept of chorology, or the examination of the areal associations of things of diverse origin, can be applied to general geography as well as to studies of segments of the whole face of the earth. Before and during the spread of the concept of geography as *Landschaftskunde,* some German scholars were also looking at the whole surface of the earth.[8] One of the leading geographers of the late nineteenth and early twentieth century whose contributions were chiefly in the systematic physical aspects of geography was Albrecht Penck.

Penck—who is credited with the first use of the term *geomorphology* to refer to the origin and development of the earth's landforms—showed how the systematic study of physical features can be approached from the chorological point of view.[9] In 1910 Penck suggested the hypothesis that the climate of a region so impresses itself on the observable features of the landscape that a classification of climates can

---

[8]The whole surface of the earth is what Richthofen and others called the *Erdoberfläche.* Schmithüsen, in suggesting a number of new technical terms for the purpose of gaining greater precision, offers the word *Geosphäre,* or geosphere, to replace the older word (Schmithüsen, 1963:10).

[9]Albrecht Penck was professor of geography at Vienna from 1885 to 1906. In that year he succeeded Richthofen at Berlin, where he remained until his retirement in 1926. He was the rector of the university in 1917–18. In Vienna Penck collaborated with Eduard Brückner in the identification of four separate ice ages in the Alps, *Die Alpen im Eiszeitalter,* 3 vols. (1901–1909). He was associated at the university with Eduard Suess, who prepared maps of the major geological regions of the world, outlining the great crystalline shields of ancient rocks, *Das Antlitz der Erde* (1883–1908), translated by Emmanuel de Margerie as *La face de la terre* (1897–1918); it has also been translated into Italian and English. Also at Vienna was Julius von Hann, the famous climatologist whose work, *Handbuch der Klimatologie* (1883, 1897, 1908–1911) has been translated into English by Robert deC. Ward as *Handbook of Climatology* (1903). Penck's son, Walther Penck, was noted for his challenge to the Davis system of geomorphology (Martin, 1974).

be made even where instrumental records are lacking. He was the first to point out that the effective rainfall of a place is a balance between rainfall, runoff, and evaporation and that evaporation increases with higher temperatures. Although he started with the observable features of the landscape, he did not overlook the nonmaterial factors because of any rigid definition. Nor did he overlook man and the works of man as essential to an understanding of the varying characteristics of the face of the earth (Fischer, Campbell, and Miller, 1967:99–106).

Penck realized that no one geographer, and certainly no class or seminar, could actually see any considerable part of the earth. He realized, therefore, that the compilation of accurate maps, showing at least the major features that are associated in areas, was essential for the adequate study of geography. The maps and atlases that had been produced at Gotha and other places were excellent and informative, but it was not possible to show enough detail on the relatively small scales used in atlases. The large-scale topographic map, showing not only the shape of the land-forms but also water bodies, vegetation, the works of man, and other things, was the ideal way to bring the real surface of the earth down to a size that would permit study. But only a very small part of the world was covered by surveys detailed enough to permit the compilation of topographic-scale maps. Penck proposed a compromise. In 1891 (at the International Geographical Congress held in Bern, Switzerland) he suggested that the nations of the world should cooperate in the production of an International Map of the World at a scale 1/1,000,000 (or about 1 inch to 15.8 miles). He proposed that agreement be reached regarding the standards of accuracy, the categories of things to be included, the symbols to be used, and the projection. Not until 1913 was such a conference called together in Paris to reach the agreements Penck had suggested. Work on the "millionth map" started slowly; when the pressing need for such a map was underlined during World War II, it was still far from complete. In 1953 the Central Bureau (which had been established to coordinate efforts in different countries to facilitate the project) was placed under the Economic and Social Council of the United Nations. Another international confer-ence in 1962, attended by delegates from some forty nations, reviewed the proce-dures for compiling the map and the symbols to be used to take advantage of the new technology now available for map-making (Hard, 1969, 1970).[10]

Climate and Landscape

Penck's suggestion concerning the imprint of climate on the landscape inspired a number of studies at different scales to elaborate on the hypothesis. One of those who contributed to *Landschaftskunde*, not only in the detailed investigation of small

---

[10]The United Nations *Report* for 1966 (published 1968) shows that of the 975 sheets needed to cover the land areas of the earth, Europe, Asia, Australia, and South America were completely covered, Africa almost completely covered, and North America partially covered. Many sheets are old and must be redone to conform to new standards.

areas but also at the global scale, was Siegfried Passarge.[11] Passarge's field study of the landscapes of the Kalahari Desert was published in 1904. His interest in the treatment of landforms as a part of a broader geographical study of landscape led him to call for empirical landform description rather than the genetic method proposed by Davis (Passarge, 1919–20). This led to a vigorous reaction by Davis and his followers. Nevertheless, in the contemporary period the tendency is to describe the surface features of an area without attempting what Davis called "explanatory description."

Passarge rejected the description of landscapes as unique. He insisted that a landscape must be viewed as a type, and he gave some examples of what he had in mind. He saw a landscape type as what we would call a spatial system, an assemblage of interrelated elements. His way of studying landscapes led him to focus his attention on boundaries to be drawn around the area occupied by a type. In a discussion of the problem of defining the area of a landscape type (*landschaftsraum*), he postulated the existence of forested mountains rising above a grass-covered plain. The difficulty, as he saw it, was that, in wetter parts of the plain, the forest extended beyond the mountains. If the landscape is described as *Waldgebirge* (forested mountains), should the boundary of this landscape be drawn around the whole forested area, even where the forest extends onto the plain?[12] That such a problem could exist seems like a good example of a semantic trap that results from the nature of the German language rather than the nature of the German landscape.

When looking at the world on a global scale, Passarge suggested that the best indicator of the extension of landscape types would be the vegetation. His *Landschaftsgürtel der Erde* (*Landscape Zones of the Earth*) is based on major categories of vegetation. On the basis of his world map, it is possible to identify certain regularities in the arrangement of the world's major landscape zones in relation to latitude and position on the continents (Passarge, 1923; Fischer, Campbell, and Miller, 1967:143–154).

Penck's suggestion concerning a classification of climates based on observable features of the landscape was picked up by the Russian-born climatologist, Wladimir Köppen, who, from 1875 to 1919, was employed as a meteorologist in the German oceanographic office (Deutsche Seewarte) in Hamburg. Between 1884 and 1918 Köppen made several attempts to produce a satisfactory classification of climates. At first he used only temperature distinctions, attempting to select values that would correspond as closely as possible to categories of vegetation. But after

[11]Siegfried Passarge was trained as a medical doctor and a geologist, receiving a degree in both fields. He served as a medical doctor in World War I. His detailed field studies were in the Kalahari Desert, in Algeria, and in Venezuela. From 1908 to his retirement in 1936 he was at the Kolonial Institut in Hamburg (Dickinson, 1969:137–141).

[12]"Aus einer Steppenplatte erhebt sich ein Waldgebirge, an dessen Fuss aber der Wald über feuchte Teile der Ebene grieft. Wo soll man die Grenze ziehen?" (Passarge, 1930:34).

the publication of Penck's ideas concerning the relation of rainfall effectiveness to temperature and to seasonal changes, Köppen devised a new system of classification in which he made use, for the first time, of annual variations of temperature and rainfall. He published this new classification in 1918.[13] He continued to revise his definitions (Köppen, 1923), and his final version appeared in 1936 (Köppen, 1936).

Whenever climatic phenomena are plotted on world maps, certain regularities of pattern become evident, as Passarge had suggested in his highly generalized landscape zones. Humboldt had demonstrated this in his maps of temperature: because continental areas are colder in winter and hotter in summer than places exposed to the open oceans, any winter isotherm in crossing a land mass bends equatorward; any summer isotherm bends poleward. Similarly, deficiencies of moisture throughout the year are found on the west sides of continents between latitudes 20° and 30° in both hemispheres, owing to the cold water offshore. Areas of abundant moisture are found on continental west coasts poleward of about 40° and along all the east coast. Köppen demonstrated these regularities by drawing a geometric figure for a generalized continent and plotting on it the hypothetical positions of his categories of climate (Fig. 22). The generalized continent eliminates mountains and irregularities of coastline. The hypothetical pattern of climates shows how the climates would be arranged "other things being equal." It represents an imaginative new effort to identify certain of the regularities discovered on the face of the earth at a global scale.

The Geography of the Oceans

Regularities of pattern also exist in the oceans, and the first comprehensive treatment of this subject was by the German ocean geographer, Gerhard Schott, who from 1894 to 1933 worked in the Deutsche Seewarte in Hamburg (Schott, 1912, 1935). Schott's monumental work on the oceans deserves a place beside the similar treatments of the land by Eduard Suess and of the climates by Julius von Hann. The

---

[13]W. Köppen: "Die Wärmezonen der Erde, nach der Dauer der heissen, gemässigten und kalten Zeit, und nach der Wirkung der Wärme auf die organische Welt betrachtet," *Meteorologische Zeitschrift,* 1 (1884):215-226; "Versuch einer Klassification der Klimate, vorzugsweise nach ihren Beziehungen zur Pflanzenwelt," *Geographische Zeitschrilft,* 5 (1900):593-611; "Klassification der Klimate nach Temperatur, Niederschlag, und Jahreslauf," *Petermanns Geographische Mitteilungen,* 64 (1918):193-203, 243-248. See also R. deC. Ward, "A New Classification of Climates," *Geographical Review,* 8 (1918):188-191; and P. E. James, "Köppen's Classification of Climates: A Review," *Monthly Weather Review,* 50 (1922):69-72. A five-volume work on the climates of the world by various authors was planned but never completed. The first volume appeared in 1930, *Handbuch der Klimatologie,* W. Köppen and R. Geiger, eds. (Berlin: Gebrüder Borntraeger). Köppen's final definitions of his categories of climate are in Volume 1. These same definitions are in P. E. James, *A Geography of Man,* 3rd ed. (Waltham, Mass.: Blaisdell, 1966), pp. 478-487. These definitions differ from those used in *Goode's Atlas.*

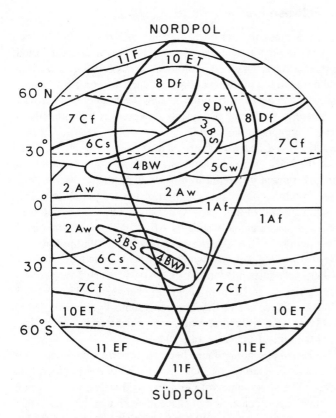

Figure 22. Köppen's generalized continent.

range of subject matter that Schott includes in his two volumes is impressive. He describes with great care the hydrological conditions (motion, temperature, salinity, color, and other elements), the climates over the oceans, the configuration and geological structure of the ocean basins, the marine organisms, the areas that support fishing or whaling industries, the routes of ocean commerce, and the air routes over the oceans; there are several chapters as well that deal with the history of discovery and exploration. Both volumes contain many maps of ocean features.

Schott produced a world map of oceanic regions. These are areas within which characteristic oceanic conditions are to be found, based on associations of hydrological conditions and marine organisms. This map, like the generalized maps of climate and landforms, reveals certain regularities of arrangement on the earth. It is possible to prepare a diagram of a generalized ocean similar to the generalized continent of Köppen (James, 1936:669).

German Geography Between the Two Wars

Samuel Van Valkenburg describes the period from 1905 to 1914 as the "golden age of German geography" (Van Valkenburg, 1951). This was a period of rapid growth and increasing productivity. The methodological discussions made working in this professional field not less, but more, exciting; and each new proposal regarding the scope and method of geography resulted in new field studies to provide examples. A number of new professional periodicals appeared as the influence of German geography spread throughout the world.

Then Germany was involved in two disastrous wars, and between 1933 and World War II the rise of the Nazi state made studies in objective scholarship more and more difficult. The government interfered frequently in academic affairs. Most of the geographers chose to remain discreetly silent on matters of policy, but there were some who came out in support of the "New Germany." Only those who were Jews were actually placed in confinement—such as Alfred Philippson, the distinguished author of the study of the Mediterranean region (Philippson, 19-04).[14] Hettner's *Geographische Zeitschrift,* which had been edited by Heinrich Schmitthenner after 1935, was suspended in 1943. In 1947 Carl Troll wrote a review of what had been going on in German geography between 1933 and 1945, which was published in the first volume of the new geographical periodical that he established, *Erdkunde* (Troll, 1947). Troll's review was translated in part by Eric Fischer and published in the United States (Troll, 1949).

In spite of the difficulties that scholars faced between the two wars, there were some notable contributions to geography. In 1931 Robert Gradmann published his work on south Germany.[15] The geographers of Germany paid little attention to the innovative study of the functional organization of space in south Germany published in 1933 by Gradmann's student at Erlangen, Walter Christaller (Christaller, 1933). Christaller's contribution to the development of central-place theory and the other German writings on the theory of location during this period will be discussed in the concluding chapter.

Troll does not mention Christaller, but he does point to some other positive advances made by German geographers during this period. There was a conservation movement, to which the geographers contributed field studies as a basis for landscape planning. There were studies of the relation of people to living space that were not politically oriented (Troll, 1949:115). Largely through the efforts of Max Eckert, German cartographers formed the *Deutsche Kartographische Gesellschaft*

[14]Philippson was not deported to Poland owing to the intervention of Sven Hedin of Sweden, who was moved to act in behalf of Philippson by Carl Troll.

[15]Robert Gradmann was trained as a botanist, but was directed toward geographical studies by Karl Sapper at Tübingen. In 1919 he became professor of geography at the University of Erlangen; while there, he completed the famous regional study of southern Germany, one of the outstanding examples of *Länderkunde* (Gradmann, 1931b). He also edited the German research series *Forschungen zur deutschen Landeskunde* (Dickinson, 1969:144–145).

in 1937 for the purpose of promoting cartography as a separate discipline (Troll, 1949:124). There was increased interest in the status and economic development of places all around the world occupied by Germans who had emigrated from Germany.

One of the deviant streams of geographic thought that developed in Germany betwen the wars was the application of geographic concepts to politics. This was given the German name *Geopolitik* (geopolitics). Geopolitics differs from political geography and is not to be confused with it. According to Karl Haushofer, *geopolitik* is the art of using geographical knowledge to give support and direction to the policy of a state. Haushofer himself published his *Geopolitik des Pazifischen Ozeans* in 1924, analyzing the significance of location around the Pacific Basin with reference to the various threatening conflicts. Haushofer drew his ideas from the Swedish political scientist, Rudolf Kjellén (1846–1922), who developed the basic ideas of *Geopolitik* by using Ratzel's analogy of the state as an organism that required room to grow. Haushofer was also influenced by the British geographer, H. J. Mackinder. At first there were numerous geographers writing papers for publication in the *Zeitschrift für Geopolitik* and dealing with problems in political geography. But as the periodical became the chief vehicle for the dissemination of geographic writings in support of Nazi policies and as the unscientific nature of the periodical became clear, many of the geographers withdrew their support from the movement. It seems that although Haushofer may have had some influence on Nazi policies his importance has been greatly exaggerated. With the collapse of Nazi Germany in 1945, Haushofer was put on trial at Nuremburg. In 1946 he committed suicide.

Postwar German Geography

In the aftermath of the disaster of World War II, German geography had to be rebuilt. Many universities and libraries were in ruins, and the funds for foreign field research were almost nonexistent. Yet by the 1960s German geography had regained its position of importance. In 1921, W. L. G. Joerg reported that there were twenty-three departments of geography and some seventy professional geographers in Germany (Joerg, 1922:441–449). The situation immediately after the war is described in some detail (Smith and Black, 1946; Troll, 1949; Van Valkenburg, 1951). In the early 1960s the International Directory of Gerographers, *Orbis Geographicus* (edited by E. Meynen, published by Franz Steiner at Wiesbaden), listed fifty-three departments of geography (forty-one in West Germany, twelve in East Germany), and 546 geographers (455 in West Germany, and 91 in East Germany).[16]

[16]In 1947 a new geographical periodical appeared: *Erdkunde: Archiv für Wissenschaftliche Geographie* (published by Ferd. Dümmler, Bonn; edited by Carl Troll and Herbert Lehmann). In 1949–50 the old *Zeitschrift der Gesellschaft für Erdkunde zu Berlin* resumed publication with Volume 1 under a new name: *Die Erde*, edited by W. Behrmann. In January, 1963 Hettner's *Geographische Zeitschrift* resumed publication with Gottfried Pfeifer as editor.

The German geographers in the 1960s went forward with the traditional concept of geography as *Landschaftskunde* but with the use of new and more precise methods of analysis. Hermann Lautensach's study of the Iberian Peninsula offers a new approach to the study of landscapes. Instead of seeking to define landscape regions as outlined by boundaries, Lautensach describes the variations from place to place as forming a continuum. The continuous variation of the landscape throughout the Iberian Peninsula is a function of four variables: latitude, elevation, distance from the ocean, and direction from the nearest coast (Lautensach, 1964). It is this kind of innovative approach that keeps regional geography (*Länderkunde*) as the central interest of the great majority of German geographers (Schmithüsen, 1963; Pfeifer, 1965; Schmieder, 1966; Dickinson, 1969:184–185).

Many decades ago the German geographers had reacted against the traditional physical determinism stemming from the teaching of Richthofen and Hettner. The reaction began when Schlüter focused attention on the historical changes of landscape resulting from human action. In the postwar period a kind of "cultural determinism" has replaced the earlier physical determinism. The new emphasis on culture is called *social geography* (*Sozialgeographie*) (Bobek, 1948; Schmithüsen, 1959). The purpose is to interpret the cultural landscape but with the clear recognition that the major force for landscape change is the human group—the "attitudes, objectives, and technical skills" that are parts of man's culture. Many collections of papers on the new social geography have been published recently (Hartke, 1960; Storkenbaum, 1967, 1969; Hajdu, 1968; see also comments on Hajdu's paper in the *Annals AAG,* 59 [1969]:596–599).

Substantively in post-war German geography several outstanding geographers contributed to both physical and human geography (Beck, 1957; Bartels and Peucker, 1969; Jager, 1972). These workers include particularly H. Bobek, H. Mortenson, H. Schmitthenner, and C. Troll. This is a significant matter when workers in a field of study may become so specialized that they can find themselves far removed from each other. Physical geography subdivided itself into geomorphology, geophysics, meteorology, climatology, oceanography, and so forth. Cultural geography also found itself the object of subdivision. An interest in the history of German geographers is revealed by Büttner (1978, 1979) and Büttner and Burmeister (1979). These subdivisions have been shared between geography and other disciplines. It is then that the regional concept and undertaking brings workers and their points of view closer to each other. The contributions of a variety of sciences and specializations are integrated for the purpose of understanding areal complexes. Perhaps the two strongest advocates of this, the regional undertaking, have been Carl Troll and Hermann Lautensach (Troll, 1966).

CHAPTER 9

The New Geography In France

Little by little, I have arrived at the idea that our doubts originated in the conflict between two concepts of geography: the traditional way of looking at it, which I have called classical, looking more toward the past and toward the delimitation of regions, and a forward looking interpretation which is not yet sure of its routes but which is playing an increasingly important role in contemporary studies. . . . Classical geography adopts the positions of Ritter. In spite of appearances, it owes little to Humboldt: from him it has borrowed the practical methods of symbolization, a taste for precise scientific description, but no general concepts. Humboldt's cosmologic vision of geography is lost in the classical period: geography is now concerned with small, local integrations, and has given up explaining distributions on a global scale. Modern geographers have returned to the ideas and preoccupations of Humboldt.

The "new geography" that appeared as a professional field in Germany after 1874 spread to other countries of the world with a lag of at least a decade. In each country the same kinds of philosophical and methodological questions were faced that had been faced in Germany by scholars who, like their German colleagues, had not been specifically trained in a field called geography. In each country somewhat different answers to these questions were found, depending on differences of language, difference of national scholarly traditions, and differences of earlier experi-

The quotation above is from Paul Claval, *Essai sur l'évolution de la géographie humaine* (1964), pp. 9-10.

ence with geographical study. In each country the "new geography" was introduced in the universities by one outstanding scholar who became the "grand old man" of that national school of geography.

The part played by the universities in the formation of geography as a professional field has been fundamental. To be sure, before the professors of geography were appointed at universities, there were those who, like Humboldt, wrote books and gave lectures on geography because there never has been a time when people were not curious about the world in which they lived. There were cartographic centers, where geographical information was compiled and transmitted on maps. But only through the training of younger generations in an accepted program of geographical study can a professional field make its appearance. As Paul Claval points out, it is through personal contacts with teachers in seminars that all the philosophical options involved in a methodological statement can be transmitted (Claval, 1964:20). The development of "schools of geography," each with a somewhat distinctive approach, appeared after chairs were established at the universities and after professional institutions had been provided, such as professional societies, periodicals, and texts. And only after a professional group comes into existence can a paradigm of scholarly work be progressively developed.

This chapter reviews the development of professional geography in France.

THE NEW GEOGRAPHY IN FRANCE

To understand the distinctive character of French geography it is necessary to recall what scholars had been writing and thinking about during the preceding century and a half. In 1752 Philippe Buache attacked the use of administrative divisions to provide the frame of organization in the presentation of geographic information. He proposed instead that geography should be organized by natural regions and that a river basin was the best kind of natural region. Buache believed that it was self-evident that high ground must form the water partings between different drainage basins, and "high ground" was easily translated into continuous ranges of hills or mountains. In fact, Buache presented the notion that the ranges of mountains continued under the oceans so that the whole surface of the earth was divided into naturally defined compartments. In spite of gradually accumulating observations that refuted Buache's concept, those who accepted his general picture of an earth sharply set off into compartments could easily overlook evidence by just not seeing it. This is another neat example of the way in which a mental image can determine the perception of so-called reality.

In France one of the earliest critics of the Buache concept was the government administrator, Charles Coquebert, baron de Montbret, who taught physical geography at the École de mines in Paris in 1796–97 (Gallois, 1908:21). When Baron Coquebert was made director of the French statistical office, he proposed a division of the national territory into natural regions and a provision for a concise description of each one. Statistical data would then be compiled by these regions instead of by the traditional administrative divisions. Unfortunately, the work of identifying these natural regions was not completed when Coquebert was assigned to another post. But the seed of interest in regional divisions had been planted; and Buache's hypothesis concerning river basins, like any challenging hypothesis, was beginning to promote the formulation of counterhypotheses.

In 1810 Malte-Brun in his *Précis de la géographie universelle* offered a carefully considered refutation of the idea that river basins are bordered by ranges of mountains or hills. But so strongly implanted was Buache's concept that Malte-Brun's refutation of it was overlooked. In 1823 a geologist, J. J. d'Omalius d'Halloy, prepared a geological map of France and the Low Countries accompanied by a description of the relation of landforms to the soils and underlying rocks. Even this clear contradiction of Buache's hypothesis failed to deter those who saw river basins as natural regions. In 1827 a French army officer published the first of two works on geographical methodology in which he showed all the drainage basins bordered by ridges or mountains, just as Buache had done (Denaix, 1827, 1841). Gallois remarked that ''jamais on n'a poussé plus loin le dédain des realités.''[1]

Meanwhile, the teaching of geography in the French schools suffered from a difficulty similar to that of Germany. There were no geographers in the universities and teachers' colleges: therefore the teachers in the schools had never been given any training in the concepts and methods of the field. There had been a chair of geography at the Sorbonne in Paris since 1809, but it was occupied by historians. The historian who taught geography was in the Faculty of Letters; in the Faculty of Sciences there was a geologist who offered courses in landforms.

As a result of the Franco-Prussian War of 1870–71, France lost the border provinces of Alsace and Lorraine to Germany, and experienced popular demand for better geography teaching in the schools. Donald V. McKay traces the development of France as a colonial power after 1871 to the influence of the French geographical societies, all of which urged France to bring the benefits of French civilization to the less developed parts of the world. There were two geographical societies in Paris (the Société de géographie at Paris had been founded in 1821), and one each in Bordeaux, Nancy, Lyons, Rochefort, Marseilles, Montpellier, and Douai. Their influence on French public opinion was widely disseminated (McKay, 1943).

Paul Vidal de la Blache

The one who led the development of the new geography in France was Paul Vidal de la Blache (Freeman, 1967:44–71). Vidal found his way into geography through his study of ancient history and classical literature. He became familiar with Greek geographical writings when he spent a year (1865) at the French School of

[1] ''Never has disdain for the realities been carried so far'' (Gallois, 1908:33).

Archeology in Athens. In 1866 he graduated from the École normale supérieure in Paris with high honors and completed work for the doctorate in 1872. For the next twenty-six years he devoted himself to improving the training of teachers of geography and to making up-to-date materials and ideas available to them.[2] During this time the chair of geography at the Sorbonne was occupied by Professor M. Himly, whose major interest was in the changing political boundaries of Europe. In 1898, when Himly retired, Vidal became the first geographer to be appointed to the chair of geography since the chair was established in 1809.

In his inaugural address at the Sorbonne (February 2, 1899), Vidal followed the custom of presenting his ideas concerning the scope and purpose of geography. There was need, he said, to focus attention on the close relationships between man and his immediate surroundings (*milieu*) by studying small homogeneous areas. In France such homogeneous areas are popularly recognized as *pays,* such as the *pays de Beauce* around its urban center of Chartres.[3] Vidal presented an effective refutation of the idea of environmental determinism. From Ratzel's second volume of *Anthropogeographie* (p. 169) he formulated the concept of *possibilism.* Nature, he insisted, set limits and offered possibilities for human settlement, but the way man reacts or adjusts to these given conditions depends on his own traditional way of living (Vidal, 1899).

The concept of a way of living (*genre de vie*) has been widely used in French geography. It refers to the inherited traits that members of a human group learn— what we may call a *culture,* borrowing the term from the anthropologists. The *genre de vie* stands for the complex of institutions, traditions, attitudes, purposes, and technical skills of a people. Vidal pointed out that the same environment has different meanings for people with different *genres de vie:* the *genre de vie* is a basic factor in determining which of the various possibilities offered by nature a particular human group will select (Buttimer, 1971:52-57).

Vidal supported the idea of studying small natural regions, but he was very much opposed to the definition of such regions in terms of drainage basins, as Buache had suggested. The use of drainage basins, he pointed out in 1888, would

---

[2]Vidal taught geography at the University of Nancy from 1872 to 1877 and then returned as professor to the École normale supérieure. In 1891, with the collaboration of Marcel Dubois, he founded a new professional periodical in which the best writings on geography could be published. In each number of the new *Annales de géographie* he included a bibliography of published materials where additional information and new geographical concepts could be found. This bibliography has since 1923 been published separately as the *Bibliographie géographique internationale,* compiled now each year with the collaboration of geographical societies throughout the world and issued by Armand Colin. In 1894 Vidal published the first edition of the *Atlas générale Vidal-Lablache* (revised editions in 1909, 1918, 1922, 1938, and 1951). He was the professor of geography at the Sorbonne from 1898 to the time of his death in 1918 (Dickinson, 1969:208-212).

[3]The French and German languages differ in the clarity of their word symbols. The French *pays* is roughly the equivalent of the connotation of *landschaft* as an extent of territory. The connotation of *landschaft* as aspect or scene is translated in French as *paysage.*

THE NEW GEOGRAPHY IN FRANCE

make it impossible even to recognize the existence of one of France's important natural regions, the *Massif Central*. This is the low mountain region lying west of the Rhône Valley, a region of massive crystalline rocks and radial drainage. If each river basin were identified as a natural region, the *Massif Central* itself could not be treated as a region. Vidal suggested that one of the major contributions of geographers would be the identification of useful natural regions, or *pays* (Vidal, 1903).

The idea of identifying the regions of France was picked up by one of Vidal's first students, Lucien Gallois,[4] whose dissertation—*Régions naturelles et noms de pays*—was published in 1908 (Gallois, 1908). In this study Gallois gives a useful review of the history of the regional idea in France.

In 1913 Vidal elaborated his ideas about the method of geographical study (Vidal, 1913). In a much-quoted passage he gave clear support to the concept of chorology as the study of things associated in area, mutually interacting, characterizing particular segments of earth space. In this paper he wrote:

> ... that which geography, in exchange for the help it has received from other sciences, can bring to the common treasury, is the capacity not to break apart what nature has assembled, to understand the correspondence and correlation of things, whether in the setting of the whole surface of the earth, or in the regional setting where things are localized [trans. Harrison-Church, 1951:73].[5]

When Vidal died suddenly in 1918 at the age of seventy-three, he was in the process of writing his definitive work, *Human Geography*. From the partially completed manuscripts and notes, Vidal's son-in-law, Emmanuel de Martonne, completed the book, which was published in 1921 (Vidal, 1921). The chapter and section headings in the *Principes de géographie humaine* give an idea of the breadth of Vidal's scholarship:

Introduction: The Sense and Object of Human Geography
  Critical examination of the concept of human geography
    The principle of terrestrial unity and the concept of *milieu*
    Man and the *milieu*

[4]Among the French scholars who were contemporaries of Vidal de la Blache, Lucien Gallois was of outstanding importance. Gallois graduated from the École normale supérieure in 1881. He was Vidal's first disciple and worked closely with Vidal for years. From 1898 to 1919 he was editor of the *Annales de géographie*. He was also the editor of the French geographical series *Géographie universelle*, which Vidal had proposed.

Other contemporaries included Élisée Reclus; Emmanuel de Margerie, translator of Eduard Suess's *Das Antlitz der Erde;* and Franz Schrader, the cartographer, who headed the map division of the Librairie Hachette, publishers.

[5]"... ce que la géographie, en échange du secours qu'elle reçoit des autres sciences, peut apporter au trésor commun, c'est l'aptitude à ne pas morceler ce que la nature rassemble, à comprendre la correspondence et la corrélation des faits, soit dans le milieu terrestre qui les enveloppe tous, soit dans les milieux régionnaux où ils se localisent" (Vidal, 1913:299).

Man as a geographic factor
I. The Distribution of Man over the Globe
    General View
    The formation of areas of population density [population clusters]
    The European agglomeration
    The Mediterranean regions
    Conclusions
II. The Patterns of Civilizations
    The relations of human groups to the *milieux*
    Tools and materials
    Foods
    Building materials
    The human establishments (*habitats*)
    The evolution of civilizations
III. Circulation
    The means of transport
    The route
    The railroads
    The sea
Fragments on which Vidal Was Working at the Time of His Death
The origin of races
The diffusion of inventions (examples: the plow, the wheel, and the draft
    animals)
Culture regions
Cities

La Tradition vidalienne

After the appointment of Vidal de la Blache to the chair of geography at the
Sorbonne, the number of professorships of geography in the French universities
increased rapidly. By 1921 there were departments of geography in almost all of the
sixteen French universities[6] (Joerg, 1922:438-441). In almost every case the
scholars appointed to teach geography were pupils of Vidal. In no other country,
said Joerg, not even in Germany around Richthofen, has the development of geography been so centered around one outstanding teacher. These disciples of the
master spread throughout France the point of view and method that came to be
known as *la tradition vidalienne* (Buttimer, 1971).

The disciple who not only elaborated Vidal's ideas about human geography
and spread them throughout France but also transmitted these ideas to other coun-

[6]Geography was also represented in 1921 in four universities in Belgium where French is used and
in seven Swiss bilingual universities where both French and German are used.

tries was Jean Brunhes.[7] Brunhes provided a classification of geographic facts that made Vidal's concepts easier to transmit in the classroom. Brunhes said that two world maps were of chief importance in the understanding of human geography: a map of water and a map of population. He divided the essential facts of human geography into three categories: (1) the facts of the unproductive occupation of the soil: houses and roads (including rural habitations, urban agglomerations, and circulation patterns); (2) the facts of plant and animal conquest: the cultivation of plants and the raising of animals; and (3) the facts of destructive exploitation: plant and animal devastation, mineral exploitation. He then illustrated the use of these categories by several small regional studies of sharply defined and distinctive associations of man and the land: the two Saharan oases—Suf in a sandy desert, Mzab in a rocky desert; the Fang people who carry on a destructive exploitation of a small area in present-day Gabon; and the seasonal seminomadism of the inhabitants of a single valley in the Alps (which was replaced in the English edition by Bowman's study of the Peruvian Andes). The last part of his book went "beyond the essential facts" to a discussion of different kinds of broader geographical analyses under the headings of human geography, regional geography, ethnographical geography, social geography, and political and historical geography.

The French school of geography under the leadership of Vidal achieved a notable balance between physical and human components. The French geographers of that period were not bothered by an apparent dichotomy between physical geography and human geography, the way the Germans were. The reason why this difficulty did not arise in France can be traced back to the work of two of Vidal's earliest disciples: Jean Brunhes, who led in the development of human geography; and Emmanuel de Martonne, who led the way in physical geography.[8] De Martonne

[7]Jean Brunhes had a broad training in history, natural science, law, finance, and geography. He taught geography at the University of Fribourg in Switzerland from 1896 to 1912. From 1912 until his death in 1930 he held a research professorship at the Sorbonne. In 1910 he published the first edition of his great work *La géographie humaine* (Brunhes, 1910), which went through several revised and enlarged editions in 1912, 1915, and 1934. The second edition was translated into English by I. C. LeCompte and edited by Isaiah Bowman and R. E. Dodge in 1920; a shortened English translation was published in 1952. His work had great impact in America because of the translations.

[8]Emmanuel de Martonne, who became Vidal's son-in-law, graduated from the École normale supérieure in 1899. He taught at the University of Rennes from 1899 to 1905 and at Lyons from 1905 to 1909. He was appointed to the Sorbonne in 1909 and remained there until his retirement in 1944. He founded the *Institut de géographie* at the Sorbonne and was its director from 1927 to 1944. This institute was set up in the Faculty of Letters with the close collaboration of the historians. As a result questions were never raised concerning the teaching of physical geography in a social science faculty. De Martonne became one of the world's leading physical geographers. He was secretary-general of the International Geographical Union from 1931 to 1938 and president from 1938 to 1949. His major work in physical geography, *Traité de géographie physique,* was first published in 1909 in one volume of 910 pages and was later expanded and revised (4th ed., 3 vols., 1925-27). He was also the author of *Europe Centrale* and "La France physique" in *Géographie universelle.*

Henri Baulig

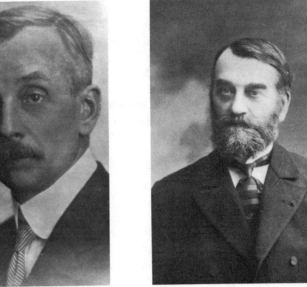

Paul Vidal de la Blache

Paul Claval

Lucien Gallois

Emmanuel de Margerie

Emmanuel de Martonne

Philippe Pinchemel

Elisée Reclus

combined the usual training in history and geography with a sound background in geology, geophysics, and biology. To him physical geography was an essential part of the whole geographical study of an area, as he first demonstrated in his regional monograph on the Wallachian Plain and his studies of the Carpathians (de Martonne, 1902, 1917). During all of his professional career he maintained a regional interest in central Europe, and a systematic interest in geomorphology and climatology (de Martonne, 1909, 1927). He became one of the leading physical geographers of the world and the most influential geographer in Europe between the two world wars. He was a strong supporter of Davis and made Davis's ideas known to the French-speaking world. Like Davis, he was a master of the art of landform description, including the drawing of eloquent landscape sketches with pen and ink (de Martonne, 1917:424). His work on the identification of arid regions through the use of an aridity index was a major contribution to the systematic study of climate (de Martonne, 1927).

An important part of *la tradition vidalienne* has been a relative freedom from concern about whether geography is one field or many. This dichotomy has worried German geographers for a long time; but French geographers stopped worrying about methodological questions of this sort after 1920. To a person who thinks in French, there is no real problem involved in recognizing that from one point of view geography is a unitary field, whereas from another it seems to tie together a variety of fields. Vallaux's book, *Les sciences géographiques* (Vallaux, 1925), expresses his understanding that geography is both a unitary and autonomous field of study and also an auxiliary aspect of many fields. Not only does geography have a philosophy of its own, he writes, but also "it is almost, in itself, a philosophy of the world of man" (Vallaux, 1925:viii). So it is that the French geographers can continue to make important contributions to systematic or topical studies and at the same time continue to produce regional monographs (Harrison-Church, 1951; L'Information géographique, 1957; Claval, 1964; Dickinson, 1969; Meynier, 1969; Beaujeu-Garnier, 1976).

But the scope and method of regional study have changed since the early part of the century, just as the scope and method of systematic studies have changed.[9] The earliest regional monographs (de Martonne, 1902; Demangeon, 1905; Blanchard, 1906; Vallaux, 1906) followed a more or less standard outline of topics, starting with the surface features and climate, advancing to the organic life in relation to the physical features, and then proceeding to the human inhabitants, looked at both as controlled by the environment and modifying the environment. The French region was a very similar concept to Schlüter's *Kulturlandschaft*. Yet by 1957, Roger Dion, writing on historical geography in the middle of the twentieth century (L'Information géographique, 1957:185), points to the many errors of in-

---

[9]By defining a regional study as one carried out in the manner suggested by Vidal in 1903, it is of course possible to prove that regional study must be dropped. Using the same reasoning, we could insist that systematic studies must be dropped.

terpretation that resulted from attempting to demonstrate strict controls. Man's economic life must be seen as in the process of transition and makes no sense at all unless it is viewed in historical perspective. René Musset, discussing recent regional studies in the same monograph (L'Information géographique, 1957:187-196), shows that the original idea of making a "complete" regional study had to be abandoned if only because of the vast increase of information available. For many decades regional studies have been tightly organized around single central themes or single problems, and all materials not relevant to such themes or problems have been omitted. Pierre Deffontaines's study of the Middle Garonne Valley in 1932 is organized in this way around the changing impact of human society on the landscape (Dickinson, 1969:217). Pierre Monbeig's regional monograph on the state of São Paulo in Brazil is focused on the contrast in settlement between the small pioneer farmers and the large coffee planters (Monbeig, 1952). Many decades ago the French geographers, like their colleagues in other countries, began experimenting with different methods of organizing regional studies (Meynier, 1969:113-119). Even Vidal in his last published work recognized the complexity of Alsace-Lorraine and the need for focusing attention on the changing significance of this border region between France and Germany (Vidal, 1917).

Géographie Universelle

Vidal thought that the field study of relatively small regions was the best possible way to train geographers. Many French geographers still think of the regional monograph as the best kind of doctoral dissertation. But Vidal also thought that regional studies could serve practical needs. He planned a series of books covering the whole of the earth's land areas to be carried out on a smaller scale than the French regional monographs to deal more broadly with the larger regions of the world. Vidal died before the plan could be carried out, but it was directed after his death by Lucien Gallois. The first volume was published in 1927; the whole series, except the volume on France in three parts, had been completed before the start of World War II. The last volume on France was published in 1948 (Vidal and Gallois, 1927-48). The whole series is a monument to the professional work of the first generation of French geographers after Vidal (Martin, 1964). Beautifully printed and illustrated with many detailed maps, the books include information never before available in one series. The parts dealing with economic conditions, population, and political boundaries are now out of date, but they are of inestimable value as the basis for studies in historical geography. The parts dealing with the physical earth and its cover of vegetation are as useful today as when they were written nearly half a century ago. Jean Gottman has written of the volume on La France physique by Emmanuel de Martonne:

> De Martonne's physical geography of France . . . gives him the opportunity of presenting a complete picture of the problems and ideas that have been his main concern

throughout his life. It is probably the most authoritative and best-written study ever produced on the physical aspects of any large section of Western Europe; it will long remain a classic [Gottman, 1946:82].

French Geography Since World War II

Dickinson quotes Philippe Pinchemel, a professor at the Sorbonne, regarding the status of geography in the contemporary period (Dickinson, 1969:262–266). He notes a great increase in the number of geographers and in the range of their interests. There has been a weakening of *la tradition vidalienne* and a loss of the feeling of coherence as French geography reacts to contemporary currents of thought that are international in range. The old unity of physical and human geography is being lost as geographers draw closer to scientific geomorphology on the one hand or to biology or sociology on the other (Gottmann, 1946). Regional theses of high quality have continued to appear—typified by studies of colonial Africa by Gilles Sautter (1966), Paul Pelissier (1966), and Jean Gallais (1967)—but systematic work has expanded impressively, particularly in urban geography and economic geography. This has both encouraged and been encouraged by what has been referred to as the quantitative revolution. In part this late arrival of numeracy owes to the grand literary tradition of French geography exemplified in regional studies, and in part to a disciplinally conservative epistemology. The fundamental mathematical techniques have now established themselves in most academic curricula, and this has manifested itself in the more recent publications of French geographers. Special mention should be made of the involvement of geographers in comprehensive planning processes, which are such a marked feature of the modern French scene. After much debate in the 1960s and early 1970s over the question of applied geography (la géographie appliquée) as a contradiction of traditional values in the profession (McDonald 1964, 1975), the younger generation has embraced it, both as theoretical stimulant and as a significant source of employment.

Interest in the history of geographical thought has also emerged in French geography with significant works by Paul Claval (1964, 1972, 1975), André Meynier (1952, 1969), the Comité National de Géographie (1972 and 1980), and Anne Buttimer (1971). Philippe Pinchemel continues as chairman of the Commission on the History of Geographical Thought of the International Geographical Union.

Finally, the 1970s have been marked by a continued diminution of the primacy of Paris, which, although still an attraction for French academics, has lost much of its dominance as the appeal of life in the metropolis has waned and the vitality (and employment opportunities) of numerous regional centers has increased. Strasbourg, Nancy, Grenoble, Lyon, Aix-Marseille, Rouen, Rennes, Lille, and other regional centers have witnessed an impressive expansion of geographic education and geographic involvement in practical regional affairs. The numerous fine regional geographical journals in which much of the best original work now appears are ample evidence of this trend (McDonald, 1965).

10

The New Geography in Great Britain

For the British the age of exploration did not come to an end with the voyages of Captain Cook. The extension of knowledge of the earth's surface through exploration and research has continued to be a major concern of British geographers and of the British geographical societies (Crone, 1964). A large proportion of the papers published in British geographical periodicals deal with reports of exploration of the still relatively unknown parts of the earth. Geography started its growth in the universities after the appointment of Halford J. Mackinder at Oxford in 1887, but the period of major expansion came after 1900 (Freeman, 1974, 1980).

The Nineteenth Century

In the nineteenth century British geography as taught in the schools was generally considered to be a dull and laborious subject. Uninspired and untrained teachers presented pupils with lists of places and products to be memorized. In the universities geography was offered by geologists, and lectures on geography as a background for understanding the course of history were given by historians.

Nineteenth-century Britain, however, had that remarkable lady—Mary Somerville (Baker, 1963:51-71). A self-made geographer, who read widely and who was in close touch with the leading scholars of her time, she was far in advance of her contemporaries in her understanding of the nature of geography as a field of study. After two earlier books on celestial mechanics and on the physical sciences, she started on her work, *Physical Geography,* in 1839. When it was ready for publication, the first volume of Humboldt's *Kosmos* appeared, and Somerville had to be persuaded by her friends to publish her own work. The first edition of *Physical Geography* came out in 1848.[1] In it she described the surface features of the land, the oceans, the atmosphere, plant and animal geography, and man as an agent of change of the physical features of the earth. She worked on many revisions during her long life (she died at the age of ninety-two in 1872), adding new materials as they became known to her, including materials contained in the *Physical Atlas* by Keith Johnston based on the *Berghaus Atlas* (p. 189) (Freeman, 1961). Yet her book seemingly made little impact on geography in Britain. Nevertheless, if Mary Somerville's work was eclipsed in Britain, it struck a responsive chord in faraway Vermont. George Perkins Marsh found her observations about man's destructive use of the earth very stimulating, and he made frequent references to her work.

Another British scholar who made important contributions to geography was Francis Galton, better known for his studies of heredity (Freeman, 1967:22–43). For Galton geography became a hobby, to which he devoted a considerable amount of time and thought. After traveling in South Africa, he served on the council of the Royal Geographical Society from 1854 to 1893. His interest in the study of British weather led him to make the first British weather map in 1861, based on reports from eighty stations. He was the first to point out the weather patterns that could be revealed by plotting lines of equal air pressure on a map (isobars), and he was also the first to recognize the nature of air circulation around a center of high pressure. The first weather map to be published in a newspaper was one that he prepared for the *Times* (April 1, 1875).

Galton's interest in plotting things on maps was bounded by no restrictive definition of the scope and nature of geography as a field of learning. He prepared a map of the world showing lines of equal travel time from London (an isochronous map) in 1881. He also made a map of female beauty in Great Britain, based on his own observations. He identified three classes—good, medium, and bad. On the resulting map the high point in female beauty was London and the low point was Aberdeen, Scotland (Freeman, 1967:41). Studying the hereditary genius among British scholars, he found that 92 percent of the scientists who became famous had been born in places located within only half of the country (Freeman, 1967: 36–37).

In 1885 Galton was the editor of a small pamphlet entitled, *Cambridge Essays,* to which numerous members of the faculty at Cambridge made contributions. He wrote a brief piece entitled, *Notes on Modern Geography.* He described geography as:

[1]The first three editions, published in 1848, 1849, and 1851, were in two volumes; later editions were in one volume—1858, 1862, 1870, and 1877. The last two editions were edited by Henry W. Bates (Baker, 1963:53).

a peculiarly liberalizing pursuit, which links the scattered sciences together and gives to each of them a meaning and significance of which they are barren when they stand alone [Galton, 1855:81].

At the time when Somerville and Galton were making their contributions there was no professional body of scholars to carry their ideas on because there were no clusters of geographers in the universities. Furthermore, in Great Britain the geologists included physical geography as a part of their own field and studied the influence of the physical features of a country on the people. Archibald Geikie wrote that the influence of physical features can be seen:

(1) in the distribution of migration of races; (2) in the historical development of a people; (3) in industrial and commercial progress; and (4) in national temperament and literature [Geikie, 1865].

The introduction of geography into the British universities resulted chiefly from the efforts of the Royal Geographical Society (Baker, 1963:64). In 1884 John Scott Keltie, then secretary of the Society, was asked to make a survey of the status of geography in Great Britain and to compare it with the position of geography in other countries. He reported that in the other countries of Europe and in America there were professors of geography in most of the universities and that Britain compared unfavorably with the rest of the world in this field. In 1886 the president of the Society wrote to the authorities at Oxford and Cambridge, pointing to the findings of the survey and urged that something be done about it. The result was the appointment of a geographer at Oxford in 1887, at Cambridge in 1888, and thereafter at almost all of the other British universities.

Halford J. Mackinder

Halford J. Mackinder became the grand old man of British geography who led the way in the expansion of geography offerings in the universities. He was appointed reader in geography at Oxford in 1887.[2] Mackinder's training was in natural science and history. He reached the conclusion that history without geography was mere narrative and that since every event occurred in a particular time at a particular place, history and geography, which deal respectively with time and place, should never be separated. In a lecture delivered before the Royal Geographical Society

---

[2]Mackinder was not the first to receive an appointment to teach geography in a British university. Baker observes that Baldwin Norton was a lecturer in geography at Oxford in 1540 and 1541. Richard Hakluyt was a lecturer in geography at Oxford after 1574. During the three centuries before Mackinder's appointment, lectures in geography were given by geologists and historians; but there was no place where future geographers could receive advanced training until the establishment of the School of Geography at Oxford in 1899, which Mackinder had recommended four years earlier (Baker, 1963:58-60, 119-129).

(Mackinder, 1887) he identified geography as the field that traces the interactions of man and his physical environment. He said, "we hold that no rational political geography can exist which is not built upon and subsequent to physical geography" (Mackinder, 1902; Fischer, Campbell, and Miller, 1967:258-261).

Mackinder was less interested in the details of man-land relations than he was in developing a world view.[3] His first major work, *Britain and the British Seas* (Mackinder, 1902), was an example of a regional study in a global context. Two years later in 1904 he gave the now-famous lecture at the Royal Geographical Society on "The Geographical Pivot of History" (Mackinder, 1904) in which he announced the heartland theory as a concept of global strategy. His warning regarding the challenge to sea power by land power fell on deaf ears in a Britain securely in control of the world's oceans. In 1919, however, people were ready to listen when he elaborated the same theme in *Democratic Ideals and Reality* (Mackinder, 1919/1942).

Mackinder's heartland theory was nothing less than a model to place the broad sweep of world history on the stage provided by global geography. He identified a world island consisting of the continents of Eurasia and Africa (Fig. 23). The most inaccessible part of the world island he called the heartland, throughout which the rivers flow either into inland seas, such as the Caspian, or into the frozen Arctic Ocean. And extending like a peninsula from one end of the heartland are the deserts of Arabia and the Sahara. In contrast to this curving area of generally thin population and difficult accessibility from the oceans are the coastlands on either side: the European coastland and the so-called monsoon coastland, both easily accessible from the sea. In these coastlands is found most of the world's population. Africa south of the Sahara, Mackinder identifies as a southern heartland, of slight strategic importance, but, like the interior of Eurasia, inaccessible from the sea (Teggart, 1919). As C. R. Dryer pointed out, Mackinder did not include the Americas except to name them as satellites out on the margins of things, along with Australia (Dryer, 1920). Mackinder's main theme has to do with the repeated invasions of the coastlands by conquerors coming from the heartland. He went back to the prehistoric migrations of mankind, spreading from the heartland in three directions: southeastward into the monsoon coastland and on to Australia; northeastward through Siberia and Alaska into the Americas; and westward into the European coastland and the southern heartland of Africa. Repeatedly throughout the course of history the earlier migrants along these routes were invaded and conquered by migrants who came

---

[3]Mackinder was both a scholar and a practical man of affairs. While he was at Oxford he also held the position of Principal of University College at Reading (1892-1903). In 1905 he was named director of the London School of Economics and Political Science. Between 1910 and 1922 he was a member of Parliament. From 1920 to 1945 he was Chairman of the Imperial Shipping Committee and from 1925 to 1930 Chairman of the Imperial Economic Committee. In 1922 he gave up his seat in Parliament. Incidentally, he was the first to climb to the summit of Mt. Kenya, 17,040 ft., which he did in 1899 (Freeman, 1961; Baker, 1963).

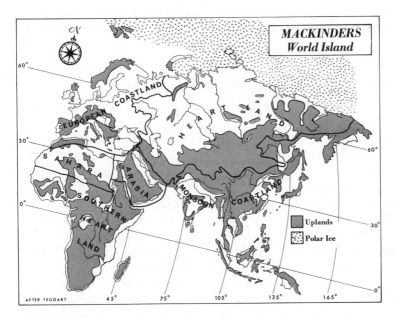

Figure 23. Mackinder's world island.

later. The coastlands, he insisted, had always proved vulnerable to attack from the heartland, and the heartland remained invulnerable because sea power could be denied access to it.

In 1919 after World War I Mackinder argued for the formation of a buffer zone of small states to keep Germany and Russia apart. He summarized his view of global strategy with the famous dictum:

> Who rules East Europe commands the Heartland;
> Who rules the Heartland commands the World Island;
> Who rules the World Island commands the World.

[Mackinder, 1919/1942:150.]

If Germany and Russia could form an alliance or if Germany could conquer Russia, the stage would be set for world conquest. These were the ideas that were embraced by the German geopolitician, Karl Haushofer, with the results previously discussed (p. 185).

We can understand now that, like all theoretical models that generalize geographic observations, Mackinder's heartland concept helps people to understand complex sequences of events by oversimplifying them. The model is based on the selection of a few facts of location and a few sequences of events and ignores complicating details. It cannot provide a precise blueprint of things to come, yet it cannot be entirely ignored. In terms of its premises, its deductions are startlingly clear.

The Development of British Geography to World War I

The paradigm that Mackinder set before the British geographers called for basic work in physical geography to provide a scientifically accurate description of the stage setting on which the human drama was to be played out. In this point of view he was close to the concepts of Richthofen and Hettner, but not Schlüter, and close to Vidal de la Blache and de Martonne, but not Brunhes. Mackinder himself was concerned with the global view, but most of his contemporaries and his followers turned their attention to the analysis of man-land relations in small areas.

Two of Mackinder's contemporaries who also contributed to the development of British geography were George G. Chisholm and Hugh Robert Mill (Freeman, 1977). Chisholm was a pioneer in the field of commercial geography, and his *Handbook of Commercial Geography* (Chisholm, 1889) became a classic that appeared in many revised editions. In it he brought together a vast amount of information about world trade, and he also formulated the basic theory describing such trade. The eleventh edition (edited by L. Dudley Stamp) starts as follows:

> The great geographical fact on which commerce depends is that different parts of the world yield different products, or furnish the same products under unequally favourable conditions. Hence there are two great results of commerce: the first, to increase the variety of commodities at any particular place; the second, to equalise more or less, according to the facilities for transport, the advantages for obtaining any particular commodity in different places between which commerce is carried on [Chisholm, 1889:1].

Hugh Robert Mill outlined the contrast between Mackinder and Chisholm in the following words:

> Mackinder was brilliant, brushing aside all irrelevancies, sketching the broad outlines of the science with a masterly hand, and by his gifts of generalization and exposition often suggesting new lines of research to the more pedestrian votaries of Geography.... Chisholm was profoundly learned, laborious, accurate, and meticulous in definition and safe-guarding of every detail [Mill, 1951:85].

Mill was not like either of the other two. Plagued by ill health for much of his long life, he became an avid reader and the master of many of the varied branches of geography. He had a fine feeling for words and a sense of the poetry that infuses all the aspects of physical and human geography. He sums up the major focus of his own research interests as follows:

> The study of the part played by water in the economy of the world through the action of solar heat and terrestrial gravitation raising vapour from the sea and carrying the condensed moisture back over the land, sustaining all forms of life and furnishing hydro-electric power, the only inexhaustible supply of energy [Wrigley, 1950:660].

Although the study of water was the major focus of his interest, he himself, describes his life as made up of nine interwoven strands. He never was able to take part in a polar exploring expedition, but polar exploration became one of his specialties. His accounts of polar exploration, including the biographies of famous explorers, were not only accurate recordings of events but were also literary achievements—among them, *The Siege of the South Pole* (1905) and *The Life of Ernest Shackleton* (1923). The study of water as part of the "realm of nature" had fascinated him as early as 1891 (Mill, 1891), but it became a major interest when he was named joint director of the British Rainfall Organization in 1900 and director the following year. Under his direction rainfall maps of Great Britain were prepared on the basis of fifty-year averages. Only approaching blindness made it necessary for him to retire from his post in 1919.

Mill recognized the need for detailed studies of British geography while he was working at the Royal Geographical Society. In 1896 he drew up a plan for using the sheets of the Ordnance Survey (1 inch to the mile) as bases on which to plot categories of land quality and land use for all of the British Isles (Mill, 1896). In 1900 he provided sample studies of two sheets to demonstrate the utility of such detailed field-mapping (Mill, 1900). Later he said that the neglect of his proposal was one of the greatest disappointments of his life. The idea was furthered by L. Dudley Stamp during the 1930s.

The geographers who followed Mackinder, Chisholm, and Mill gradually spread and strengthened British work in this professional field. A. J. Herbertson, who succeeded Mackinder as Director of the Oxford School of Geography in 1905, remained there until his death at the age of fifty in 1915. He was primarily interested in improving the teaching of geography, which for too long a time had been characterized by the study of encyclopedic collections of data organized by political units. Herbertson proposed a framework of natural regions for the study of world geography. On a global scale, he suggested, the great natural regions should be identified in terms of associations of surface features, climate, and vegetation. Here he was thinking along lines similar to Penck and Passarge rather than Vidal de la Blache and Gallois. For his categories of surface features he went back to the work of Eduard Suess and especially to the translation and amplification of Suess's work done by Emmanuel de Margerie. For his major divisions of climate he relied on

Supan. His fifteen major natural regions (Fig. 24) revealed the regularities of climate because the same regions appeared in similar positions on each of the continents (Herbertson, 1905; Dickinson and Howarth, 1933:238–239; Crone, 1964:201).

British Geography After World War I

World War I had the predictable result of stimulating the public demand for the more effective teaching of geography. After the war, departments of geography were established in almost all of the British universities, and the number of professional geographers soared.[4] As in other countries, many of those who were appointed to the new academic posts were not trained as geographers, and the result was a considerable amount of methodological discussion. The British geographers were greatly influenced by the French and Germans and also by the American geographer, William Morris Davis (pp. 281–294). Much discussion involved the relationship of physical geography and human geography. At first geomorphology, which was described as the last chapter of geological history, was accepted almost everywhere as a branch of geography and was a required part of every program of training. As a result, in the period since World War I a large number of studies in geomorphology by British authors have been published (Clayton, 1964). Only since World War II has a movement appeared to reduce the emphasis on geomorphology, but this has been resisted by those who received their training in earlier decades (Stamp and Wooldridge, 1951; Wooldridge and East, 1951; Wooldridge, 1956).

During the period of expansion following World War I, British geography developed five distinctive characteristics: (1) a continuing concern with exploration; (2) an emphasis on various kinds of regional studies; (3) the inclusion of field observation and map interpretation as essential parts of training programs; (4) an emphasis on studies in historical geography and a related concern with the history of geography; and (5) the study of geography because of its relevance to economic, social and political policy problems.

Exploration. The professional periodicals of Great Britain devote a large proportion of their pages to accounts of exploration. The earlier expeditions had the attainment of a destination as major objectives. Ernest Shackleton and Robert F. Scott in their expeditions to the Antarctic did a large amount of pioneering work in geology, meteorology, and biology, yet their essential purpose was to reach the South Pole.[5]

[4]In 1921 there were ten universities offering honors programs in geography (Keltie, 1921; Joerg, 1922), but by 1964 there were programs of advanced study in thirty-two universities (including seven separate colleges of the University of London). *Orbis Geographicus* for 1964–66 lists 464 professional geographers in Great Britain.

[5]In January, 1912 Scott reached the South Pole only to find that the Norwegian explorer, Roald Amundsen, had reached it a little more than a month earlier. The Scott party perished during the return to their base.

Similarly when H. J. Mackinder ascended Mt. Kenya in 1899 and when Sir Edmund Hillary and Tenzing attained the summit of Mt. Everest in 1953, it was the climb to the top that had been their objective. Of course behind the quest lay the epic theme of man versus nature and a desire to know the face of another part of the earth. The great advance of technology has robbed epic achievement of some of its meaning. Now to an ever increasing extent exploring expeditions are organized and financed for specific scientific purposes. Today people go exploring to study geology, botany, glaciology, geography, zoology, archeology, or other subjects. Between 1960 and 1964 the Royal Geographical Society financed 184 expeditions: of these 62 went to the Arctic or its fringe, 37 to Africa, 21 to South America, 19 to southwest Asia (which the British call the Middle East), and the remainder to a variety of previously unknown spots (Kirwan, 1964:223). There can be no doubt that the image of geography in Great Britain includes a continued concern with exploration[6] (Priestley, Adie, and Robin, 1964).

Regional Studies. There is fully as much confusion over the meaning of the words, regional studies, as there is over the German word *Landschaft* (Dickinson, 1976). There are at least three different meanings to regional studies as a characteristic of British geography: (1) There are regional studies the end purpose of which is to divide the surface of the earth into homogeneous areas or regions of varying size. (2) There are regional studies that are descriptions of segments of the earth surface. (3) There are regional studies produced by an individual geographer, who devotes a large part of a professional career to the continued study of different aspects of one part of the earth. And none of these three kinds of regional study is necessarily involved with the regional concept. In Chapter 16 an attempt is made to clear up the semantic complexity that surrounds these words and the ideas for which they stand. But it is important to appreciate that to a certain extent the obscurity is a result of British experience with regions.

Experiments with the classification of very general regions for use at the global scale were started when A. J. Herbertson proposed his scheme of major natural regions (Fig. 24) in 1905. These so-called natural regions, we may recall, were defined as associations of surface features, climate, and vegetation and were intended to provide an empirical generalization regarding the arrangement of these features for teaching purposes.

Following Herbertson, other British geographers experimented with a variety of different ways of classifying regions. In 1912 Marion I. Newbigin, editor of the *Scottish Geographical Magazine,* suggested an approach to the definition of regions that was a clear reflection of the ideas of Vidal de la Blache, Lucien Gallois, and Jean Brunhes. She started with a question: "Why is it easier for men to make their living at some places than at others?" (Dickinson, 1969:206). The answer, she

[6]The traditional concern with exploration at Cambridge is reflected in the foundation of the Scott Polar Research Institute at that university (Crone, 1964:206).

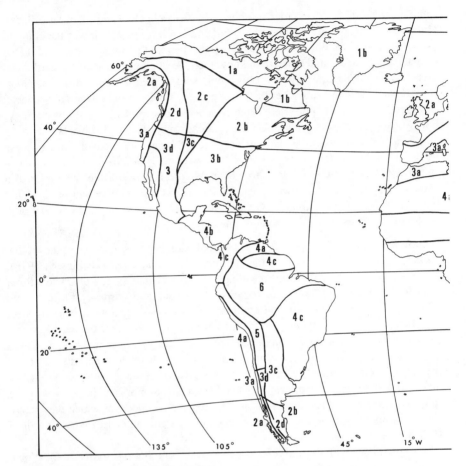

Figure 24. Herbertson's major natural regions.

believed, could be found by examining the relationship between the *genre de vie* and the productivity of the land. The classification of regions, therefore, should be based on the kinds of relationships observed between human communities and their natural surroundings. H. J. Fleure of Aberystwyth carried this idea forward by defining seven kinds of global scale regions.[7] He postulated that all human activities

[7]H. J. Fleure was trained as a zoologist and was elected to the Royal Society as an anthropologist. He insisted on the need for close cooperation between geography, history, and anthropology. He was greatly influenced by the ideas of the French sociologist, Frédéric Le Play, and by the Scottish regional planner, Patrick Geddes. With H. J. Peake he was the author of the series of books entitled, *Corridors of Time,* which demonstrated the method of treating the factors of time, type, and place together in one work.

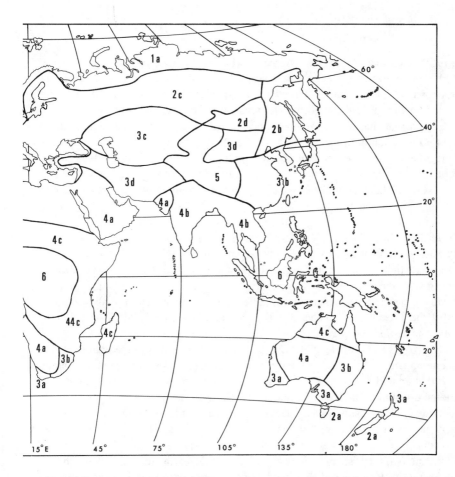

are primarily directed toward the accomplishment of three functions: nutrition, reproduction, and the increase of well-being. In seeking criteria for the delimitation of human regions, he eliminated the first two because without them "a race would perish." He then classified his regions according to the measure of the earth's response to man's efforts in the pursuit of well-being (Fleure, 1917, 1919). Fleure identified seven types of regions:

Regions of hunger
Regions of debilitation
Regions of increment
Regions of effort
Regions of difficulty

Regions of wandering

Industrialized regions

He recognized that regions of increment, while requiring effort by man, rewarded effort so liberally as to leave him a surplus of both food and leisure. But when certain advanced societies enlarged their expectations for the "good life," this resulted in dropping some former regions of increment to lower categories. He recognized that the new technology and finance of industrialized regions had so modified the relation of man to his environment that such regions could replace any of the others. The experiments by Newbigin and Fleure were imaginative efforts to formulate illuminating empirical generalizations; they were offered at a time when there were neither the statistical data nor the electronic equipment with which to store and process such data.

Another experimental regional scheme was proposed by J. F. Unstead in 1916 (Unstead, 1916). This was a classification of geographic regions in which the physical and human factors were to be given equal weight. Furthermore, Unstead recognized that regions could be defined at different degrees of generalization: starting with the immediately observable units of area, which he called *stows,* he moved on to somewhat larger regions called *tracts* (which were roughly the equivalent of the French *pays*). Then these were combined into *subregions, minor regions,* and *major regions.*

There were two difficulties with Unstead's scheme that in the course of time, became apparent. The first difficulty was the concept of a uniform unit of area, a stow, that could be used as the basic building block in erecting a structure of world regional divisions. The idea of a unit area so homogeneous that it cannot be further subdivided is found in many languages and appears, at least by implication, even in recent geographical writing. Yet it is clear that the stow is only indivisible because it is so conceived by the observer: actually it is necessary to start with the observable fact that no two microscopic points on the face of the earth are identical and that any area enclosed by a line and described as homogeneous is only homogeneous with respect to selected features. The face of the earth is an intricate system of interconnected features that forms a continuum of varying aspect. Regions are defined and drawn to illuminate some aspect of the problems geographers are concerned about; in fact, it would be impossible to reach any comprehension of the nature of the earth's surface without recognizing areas of partial homogeneity. The colors of the spectrum form a continuum, yet we find no intellectual difficulty with arbitrarily defining a certain segment of the spectrum and calling it red. The difficulty arises when the region is identified as a unit area, an indivisible segment of space.[8] If the geographer were reduced to the size of an ant, he would soon start subdividing the indivisible stow.

[8]See D. L. Linton, "The Delimitation of Morphological Regions," in Stamp and Wooldridge, 1951:199–217; ref. 209.

Robert P. Beckinsale

Richard J. Chorley

Henry C. Darby

Herbert J. Fleure

Thomas W. Freeman

Peter Haggett

Halford J. Mackinder

Hugh Robert Mill

Alan G. Ogilvie

L. Dudley Stamp

Michael J. Wise

Sidney W. Wooldridge

And the second difficulty with Unstead's scheme was his announced intention to recognize homogeneous associations of physical and human factors (Unstead, 1916:241). This idea is logically derived from the theory that the way people live is a reflection of their natural surroundings. According to this theory an area that is homogeneous with respect to its natural features will also be homogeneous in the ways people adjust to these features. The French geographers recognized these difficulties and, following Vidal, defined regions in terms of the way people live. Human use, said Vidal, creates homogeneity even where the natural features were not homogeneous. But the German geographers had difficulty in thinking of the different features associated in an area—in part because of the nature of the German language. As early as 1913 L. W. Lyde warned against attempts to make regions of human settlement fit exactly into the areas defined as natural regions (Lyde, 1913).

After Unstead the British geographers began to specify their criteria for defining regions more precisely and to avoid regional schemes that involved the associations of too many diverse factors. In 1919 C. B. Fawcett prepared a map of the service areas of the major cities of England (Fawcett, 1919). This was the first identification of functional regions (Fig. 25). In 1932 he mapped the continuously built-up urban areas of Britain (conurbations, as Patrick Geddes had called them). In 1937 a committee with Unstead as chairman (including Myres, Roxby, and Stamp) reviewed the numerous schemes for dividing the world into regions, not only in Britain but also in other countries (Unstead et al., 1937).

Meanwhile a very different kind of regional study was being produced. These were book-length treatments of specific parts of the world also stimulated by the similar studies by the German and French geographers. For the Twelfth International Geographical Congress which met in Cambridge, England, in 1928, the British geographers prepared a volume of regional essays, each dealing with a region of Great Britain. Twenty-four regions were marked off on the basis of surface features and underlying geological formations, but no map of regional boundaries was included in the hope that futile discussions about such boundaries could be avoided. "The purpose of regional geography," the committee decided, "is to describe the regions of the country as they are and to discover the causes that have made them what they are" (Ogilvie, 1928:1). We can understand today that what a region is depends on what concepts are in the mind of the observer, but in 1928 geographers still thought they were dealing with objective reality. For many years these essays stood as a model for this kind of writing. It is interesting to compare the essays edited by J. Wreford Watson and J. B. Sissons on the occasion of the Twentieth International Geographical Congress in 1964. At this time the geographical study of Great Britain was organized topically rather than regionally (Watson and Sissons, 1964).

Many regional monographs, some published since 1960, are included in the bibliography in Minshull's book, *Regional Geography* (Minshull, 1967). The tra-

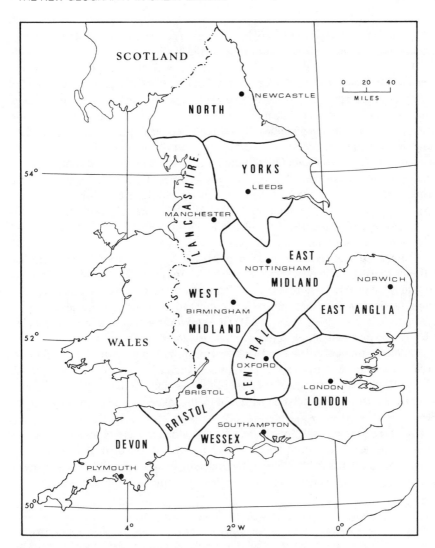

Figure 25. Fawcett's provinces of England.

ditional regional study follows a more-or-less standard outline of topics—like the early works in Germany and France. It begins with the bedrock geology, the surface features, the climate, and the vegetation and soils. The treatment of the stage setting is then followed by a history of the course of settlement, starting with the earliest human inhabitants. More recently, however, the authors first announce a general theme that characterizes the aspect of the region to be investigated (as indeed Fleure did in his chapter on Wales in the 1928 volume). The regional study is then tightly organized around materials relevant to the theme.[9]

The third kind of regional study is, perhaps, a form of applied geography. When an individual geographer devotes a large part of his professional life to studying different aspects of some one part of the world, he becomes known as a regional specialist; and his publications, each dealing topically with some aspect of the area, are called regional studies. There are many examples of such specialists among the British geographers, but perhaps we can illustrate this kind of work by two examples. One was David G. Hogarth, who devoted his life to the study of the people and problems of what the British called the Near East.[10] He began traveling in Turkey and Arabia in 1887, first as an archeologist, later as an observer of the people and their problems (Hogarth, 1902). The informed advice he gave to his government regarding the treatment of the Turks after World War I was, unfortunately, not acted on promptly.

Another regional specialist was Percy M. Roxby of Liverpool, who spent his life in the study of China[11] (Freeman, 1967:156–168). In 1912 he was awarded a fellowship that enabled him to travel to China among other places. He developed a lifelong fascination with the Chinese and published numerous papers dealing with that country (Roxby, 1916, 1925, 1938). During World War II he was employed by the British Naval Intelligence to prepare the *Handbook of China*, which was completed in three volumes in 1944 and 1945.[12] He was also one of Britain's important historical geographers. His treatment of East Anglia in the volume of regional essays in 1928 is an outstanding example of this method (Ogilvie, 1928:143–166). His published papers also include influential methodological studies (Roxby, 1926, 1930).

British geographers still contribute actively to the literature of regional studies, at least of the second and third types. The recognition of the difficulty of defining homogeneous areas has led to more sophisticated methods of identifying and analyzing regions (Haggett, 1966:241–263). In spite of the effective efforts of some of the

[9]Compare, for example, Freeman (1950,) or Monkhouse (1959,) with Cole (1960), Longrigg (1963), Harrison-Church et al. (1964), or Prothero, (1969).

[10]Hogarth was first an explorer and archeologist, but he gradually turned his attention to geography. He was at one time director of the British School of Archeology at Athens. From 1908 to 1927 he was the director of the Ashmolean Museum at Oxford. One of his famous students was Lawrence of Arabia, Thomas E. Lawrence.

[11]Roxby was a student of history at Oxford but never "bothered" to complete his graduate studies. In 1912–13 he received a fellowship to permit him to travel in the United States, China, and India. As a result he devoted his career to the study of China (and also of Chinese workers who were living in ghetto conditions around Liverpool). He was appointed to the staff at the University of Liverpool in 1904 and taught geography there until after World War II. Unfortunately, he died in China in 1947 before his intended definitive book on Chinese geography could be written.

[12]The *Handbook of China* included: Vol. 1, *Physical Geography, History, Peoples* (1944); Vol. 2, *Modern History and Administration* (1945); Vol. 3, *Economic Geography, Ports and Communications* (1945).

younger British geographers to formulate and establish a new paradigm of scientific geography (to be discussed in Chapter 18), there are numerous members of the profession who regard geography as a humanity, not a science. There are still papers and books in the literary tradition, seeking to present the "personality" of a region (Freeman, 1950; Gilbert, 1960; Minshull, 1967). There is even some eloquent support for the notion that the presumed dichotomy between science and art is a false one and that geography can continue to be treated as if it were both, or neither. And there are still some writers who are critical of the kinds of regional studies that were common in the 1920s and 1930s (Kimble, 1951; Wrigley, 1965).

Field Studies and Map Interpretation. The third distinctive characteristic of British geography is the continued requirement for training in field observation and map interpretation as essential parts of training programs at all levels. Even children in the elementary grades are expected to practice the observation of things out-of-doors, very much as recommended by Pestalozzi. There are exercises in the reading of topographic maps and the following of maps in the field, and then further training in the interpretation of detailed maps of unfamiliar areas. Perhaps it is this early familiarity with the use of maps and the resulting attention to the features of the surrounding landscape that results in the widespread British habit of taking long walks in the country or cycling to more distant places as a form of recreation.

This attention to field observation and map interpretation that is a part of the school experience of most of the British people, even those who live in cities, is carried on at a more sophisticated level in the training of geographers in the universities. It has long been customary to require theses dealing with the geography of small areas as part of undergraduate programs, and the more elaborate studies of regions have been the traditional subject matter for dissertations. Examinations for honors candidates regularly include map interpretation. Therefore, even in the contemporary period of increased use of electronic devices for gathering and analyzing information, the British geographers continue to make use of direct field observation as a basic method.

Historical Geography. The fourth of the distinctive characteristics of geography as it developed in Great Britain after World War I has been a continued emphasis on the historical method (Baker, 1972). This emphasis seems to have been derived quite directly from the geographic writings of some of the nineteenth-century historians. In 1838 Thomas Arnold published his famous *History of Rome* in which he included chapters on the natural surroundings.[13] He included also a map and discus-

[13]Thomas Arnold was appointed professor of modern history at Oxford in 1841, but died the next year. His pupils included A. P. Stanley and E. A. Freeman, who published on historical geography. Chapter 3 of the first volume of Thomas Macaulay's *History of England* (London, 1848) included a description of England in 1685 that Baker describes "as a model of what historical geography should be" (Baker, 1963:36). H. T. Buckle, however, went far beyond Montesquieu in relating the characteristics of people to climate. Bibliographies of the historical writings of this period are contained in Baker (1963:33-50) and Clark (1954:79-80).

sion of the area of occurrence of malaria in Roman times (Baker, 1963:33–50). Some of Arnold's pupils continued to insist on the importance of geography as the basis for understanding history. As a result, there were lectures on geography in British universities long before there were geographers on the faculties; some of these lectures were given by geologists and some by historians. Even now it is not uncommon to find books on historical geography written by anthropologists or economic historians (Fox, 1932; Beresford, 1954).

The development of historical geography by geographers came after Mackinder's work on the British Empire (Mackinder, 1902). In his training as an historian Mackinder learned to appreciate the need for looking at the story of man's settlement of the land from the perspective of time. He insisted that the geographer should attempt to re-create past geographies and show how sequences of change have led to the presently observable features (East, 1951:80); otherwise geography would become the mere description of contemporary features. Adding the time dimension permits the study of processes of change and reveals that the present geography is only the latest stage in a sequence of stages. This is the way the observed features of geography in an area could be explained.[14] After Mackinder most of the British geographers included accounts of historical geography in their regional studies. Among the many important contributions to historical geography, special mention should be made of Marion I. Newbigin (1926), E. G. R. Taylor (1930, 1934), E. W. Gilbert (1933), W. Gordon East (1935, 1951), and H. C. Darby (1936, 1940a, 1940b, 1951, 1952). Since then Darby has produced or edited *A New Historical Geography of England* (1973), the Domesday geography of all the main parts of England (1962–71), and *Domesday England* 1977. Possibly the best modern historical geography written by an English geographer is *An Historical Geography of Western Europe Before 1800* (1967) by C. T. Smith.

Related to this continued use of the historical method is the interest of British geographers in the history of geography as a field of learning. In the bibliographies of the early chapters of this book, which deal with ancient geography and with the progress of exploration, there are many references to British writings.[15] Even today interest in writing on different aspects of the history of geography continues among British scholarly works. A leading historian of geography in Britain was J. N. L. Baker.[16] The first parts of a massive history of the study of landforms have been

---

[14]For a long time there was confusion between historical geography and the history of geography, and, in fact, bibliographies used to list both of these categories under one heading. Historical geography can now be viewed as the re-creation of past landscapes and the tracing of geographic changes through time (Clark, 1954:72–73). Geographical history, then, is a study of the effect of geography on the course of history. The history of geography has to do with the development of geographic concepts and the progress of geographical studies.

[15]For example, see books by C. R. Beazley, E. H. Bunbury, Frank Debenham, Richard Hakluyt, G. H. T. Kimble, John Needham, Percy Sykes, E. G. R. Taylor, J. Oliver Thomson, and H. F. Tozer.

[16]J. N. L. Baker went to Oxford as assistant to the reader, H. O. Beckit, in 1923, after serving in the army in France in 1915–16 and in the army in India in 1918–19. He remained at Oxford until his retirement in 1962. His collected writings were published in a single volume in 1963 (Baker, 1963).

issued and will eventually comprise four volumes. Volume 1 deals with geomorphology before Davis (R. J. Chorley, A. J. Dunn, and R. P. Beckinsale, 1964), Volume 2 deals with W. M. Davis (R. J. Chorley, R. P. Beckinsale and A. J. Dunn, 1973), and the remaining two volumes, which are in preparation, with geomorphological thought since 1892. Other British writers on the history of geography include Dickinson and Howarth (1933), Mill (1951), Freeman (1961, 1967, 1980), Crone (1964), Kirwan (1964), and Dickinson (1969). Individual studies of geographers have been made also by Grigg (1977), Middleton (1977), Oughton (1978), and Jay (1979).

Applications of Geography. Geography in Britain has also included some notable contributions toward the solution of practical problems. Dickinson has traced the influence of the French sociologist Frédéric Le Play on the Scottish regional planner, Patrick Geddes (Dickinson, 1969:197–207). Geddes developed the concept of the regional survey of potential land quality and land use as the basis on which to draw up a plan for economic development. If one is inclined to wonder why someone did not set forth such an idea many decades earlier, it is well to remember the historical context. Before there could be regional surveys there had to be standard sets of detailed (topographic-scale) maps on which to plot the data. In 1896 Hugh Robert Mill suggested using the sheets of the Ordnance Survey (1 inch to the mile) and four years later provided an example of how such mapping of relevant information could be done and how it could be used (Mill, 1896, 1900). His suggestion was discussed, but no one did anything about it.

The British geographer who put these earlier ideas into practice was L. Dudley Stamp.[17] Returning in 1926 to the London School of Economics from some three years in Burma, he turned his attention to the study of Great Britain. Convinced that Mill and Geddes were right about the need for a survey of Britain as a basis for planning, he began searching for ways to carry out such a survey by plotting categories of land quality and land use on the Ordnance maps at 6 inches to the mile. The sheets of this map include the field boundaries, that greatly facilitated the work of plotting the data. He found that some surveys of this kind had already been attempted: for example, some maps of a parish in Leicestershire were done by school children under supervision of their teachers. Stamp recognized that quite aside from the practical importance a survey would have as a basis for planning, the work of making it would be an excellent educational experience for the children.

With great energy and patience, Stamp undertook to organize and direct what became known as the British Land Utilisation Survey (Stamp, 1947). With professional geographers as advisors, his first job was to define the categories to be plotted on the maps. Then it was necessary to "sell" the project to the Directors of

---

[17]L. Dudley Stamp completed two degree programs at King's College, University of London, in geology and geography. He then went to Burma for a petroleum company and from 1923 to 1926 was professor of geography and geology at the University of Rangoon. From 1926 until his retirement in 1958 he was at the London School of Economics and Political Science.

Education in the counties and then teach the field workers how to do the work. Some 22,000 volunteer school children were asked to give three or four days' time, after which the maps had to be checked. It took eight weeks for a skilled cartographer to redraw the maps on a scale of 1 inch to the mile and make them ready for publication. Geography students and staffs at various British universities cooperated in providing the professional personnel. Stamp reports that the hardest part of the project was securing funds to print the maps and to publish county reports explaining what the maps showed. The work was started in the summer of 1931, and by the end of 1935 the mapping was essentially complete.

When World War II began in 1939, the vital importance of the maps was quickly appreciated. Britain had to undertake a rapid program of agricultural expansion because there were not enough ships to bring in the usual supplies of food. The Ministry of Agriculture recommended that funds be appropriated to publish the maps as rapidly as possible, and this was done between 1939 and 1945. It would scarcely have been possible to increase the production of wheat so rapidly had it not been for the existence of these field studies. Many years later, in 1965, Stamp became Sir Dudley Stamp in recognition of his contribution to the survival of his country.[18]

The survey maps were used for a variety of purposes in addition to the planning of emergency crop expansion during the war. After the war they were the basis for the reconstruction of Britain. In the universities several studies were made of the historical geography of agriculture by reconstructing the crop patterns of former times and comparing them with the survey maps (Stamp, 1947).

Now a new survey of Britain is being carried out. In 1946 the Town and County Planning Act established a new survey to be done at a scale of 1/25,000 and to be kept continuously up to date (Coleman, 1961). In Lisbon at the International Geographical Congress in 1949, Stamp's proposal that a World Land Use Survey be established was adopted; and an international commission was set up to supervise the work. The maps are to be published at 1/1,000,000.[19]

British Geography in the Contemporary World

The waves of innovation in geographic study that distinguish what we call the contemporary period, and which will be discussed in Part Three of this book, reached Britain in the 1960s (Haggett and Chorley, 1967.) Not that the five distinc-

[18]On April 15–16 and May 10, 1941 air raids destroyed nearly 50,000 of the already printed sheets and also the set of plates from which they had been printed. Maps stored on a farm and at publishing houses survived.
[19]Another kind of applied geography involves the delimitation of political boundaries. Among the distinguished boundary specialists, the name of Thomas H. Holdich stands out. Holdich became involved in boundary surveys in 1884 when he was appointed chief of a survey team to demarcate the boundary between Russia and Afghanistan. From 1892 to 1898 he was superintendent of Frontier Surveys in India.

tive traits of British geography were swept away; rather, new traits have been added. R. J. Johnston suggests that overlapping the regionalism of the post World War II period came geography as spatial science, followed sequentially by overlays of the behavioral and radical-structural approaches (Johnston, 1976, 1978, 1979).

Origins of the spatial science movement seem to reside in the United States, but it was quickly the inspiration for work in Britain. Much attention was given to quantification, to statistical description of patterns, and to statistical manipulation and testing of hypotheses. The Madingley lectures for 1963, published as *Frontiers in Geographical Teaching* (Chorley and Haggett, 1965), demonstrated the application of quantitative methods to the understanding of geographical problems. Other books in this genre include *Locational Analysis in Human Geography* (1966) by P. Haggett and *Models in Geography* (Chorley, 1967). This movement was especially supported by some members of the younger generation who completed their graduate studies at Cambridge in the early 1960s.[20] Initially techniques had changed, but not objectives, and the idea of studying the distance variable in a multitude of contexts did not meet with universal approval.

One of the major objections to geography as spatial science was the assumption of rational economic behavior. Some geographers felt that behavioral-type geography should be based on business (and other) decisions already made, rather than on what decisions geographers think should be made.

Finally, since the late 1960s, yet another viewpoint intruded itself into professional thinking. A growing concern with relevance, with the inequalities within society, environmental degeneration, and the very structure of the capitalist system began to emerge. Geographers studied public policy, welfare, and the social good. This point of view was attacked, a new set of political values were postulated, and for the first time a political value-laden geography was postulated, albeit by a minority.

At the beginning of the twentieth century he was selected to establish the boundary between Argentina and Chile in Patagonia. His book, *The Countries of the King's Award*, was published in 1902. Another book, published in 1916, *Political Frontiers and Boundary Making*, summarizes the experiences of a lifetime in the use of geographical methods in the solution of a particular set of problems.

[20]Three British geographers, all of them graduates or faculty of Cambridge, were of chief importance as leaders of this new movement. Richard J. Chorley, a geomorphologist, was one of the first geographers to make use of general system theory in the study of landforms and also one of the first to point to the utility of such theory in human geography. David W. Harvey completed his graduate study at Cambridge in 1962, after spending the year 1960–61 studying with Hägerstrand at Lund. He was appointed assistant lecturer at Bristol University in 1961 and was promoted to lecturer in 1964. In 1965–66 he was at Pennsylvania State University and in 1969 joined the staff of the Johns Hopkins University. Peter Haggett completed graduate study at Cambridge in 1960. He taught at the University of London from 1955–57 and was appointed to the staff at Cambridge in 1957. In 1966 he was appointed to a second chair of geography at Bristol (where Ronald Peel had been Professor of Geography since 1957). One of the lecturers at Cambridge who has played a major role in stimulating the younger generation to experiment with innovations of method is Alfred Caesar, who from 1944 to 1946 worked in the Ministry of Town and Country Planning. He was appointed lecturer at Cambridge in 1949.

At least these were directions assumed by many of the younger geographers. Yet at the same time many of the traditional areas of the discipline continued as of old, seemingly shrugging off the new thrusts in direction as methodological additions to traditional directions of inquiry. Regional geography, historical geography, and human and physical geography remain strongly entrenched, and interest in the history of the discipline has perhaps been strengthened.(Gilbert, 1972; Chisholm, 1975) And an ecological approach was reintroduced by Stoddart (1965) and Chorley (1973b).

Geography gained ground especially from the mid-fifties to the mid-seventies (Robson and Cooke 1976). The number of institutions where geography was taught almost doubled. Total membership in the Geographical Association rose from about 5000 to about 8500; membership of the Institute of British Geographers simultaneously increased from about 600 to about 1550; and circulation of the *Geographical Magazine* reached 75,000 in the early 1970s.

11

The New Geography in the Soviet Union

Geographic science in the Soviet Union has traveled a prolonged and many-sided path of development. Arising after the Great October Socialist Revolution, Soviet geography received as a great and valuable heritage from prerevolutionary Russian geography not only an immense store of geographic facts, but also a whole system of fruitful progressive scientific traditions, schools, and concepts, many of which have become classic. During its existence, Soviet geography has repeatedly enlarged its scientific heritage. It has gathered new factual materials, continued and enriched the progressive classic scientific trends, and created new theoretical concepts which have developed on the basis of scientific Marxism-Leninism and in close connection with the practice of socialist construction.

The impact of German geographical ideas during the last quarter of the nineteenth century had quite different results in Russia than it had in France and Britain. By this time Russia had a long history of geographical work, including the production of maps and atlases and the writing of regional monographs. Although many explorers and scholars from Germany and other countries of western Europe lived and worked in Russia and were influential in promoting geographical studies, most of those who made the maps, gathered the statistical data, and wrote reports describing different parts of the national territory were Russians. Because of the language barrier, which is also an alphabet barrier, an understanding of the impor-

The quotation above is from I. P. Gerasimov, "Geography in the Soviet Union," *Soviet Geography...* , trans. L. Ecker, ed. C. D. Harris (1962), p. 1.

tance of the work of the Soviet geographers and of their predecessors in pre-1917 Russia was delayed in reaching the geographical scholars of western Europe and America for many decades.[1]

The exploration and mapping of the vast extent of land area that was eventually included in the Russian Empire were carried out mostly by Russians, but with considerable assistance from skilled map-makers from the west. Peter the Great, who ruled Russia from 1682 and 1725, recognized the vital importance of having accurate geographical information to guide the eastward expansion of the empire. He gave his support to exploring expeditions and to the publication of the results of their discoveries. The southern part of European Russia was surveyed in the late seventeenth century, and the resulting maps were published in Amsterdam (Bagrow and Skelton, 1964:170-176). In 1719 all official Russian map-making activities were placed under the direction of Ivan Kirilov, the first Russian to be named head of the Cartographic Office. With technical assistance from the French cartographers, Kirilov supervised the preparation of an atlas of Russia that was published in 1734 and frequently revised thereafter as new information became available.

At first the objectives of the expeditions was to establish the location of rivers, coasts, and mountains and to identify places where furs or precious metals could be found. But the great Russian encyclopedist M. V. Lomonosov urged that the exploring parties be charged with the systematic collection of information about the physical character of the land, the population, and the condition of the economy.[2] In 1758 he became head of the world's first officially named Department of Geography, which was in the Russian Academy of Sciences (Gerasimov, 1968b). But it was not until 1768, three years after the death of Lomonosov, that the Academy of Sciences sent out the first expedition for the specific purpose of gathering and reporting on the physical and economic geography of a part of the national territory (Nikitin, 1966:7).

The Academy of Sciences provided an institutional focus for the great variety of scholarly works, of which many were geographical in purpose. The Germans were very influential in this. Anton Friedrich Büsching was the pastor of a German Lutheran church in St. Petersburg from 1761 to 1765, and in 1766 the part of his *Neue Erdbeschreibung* dealing with Russia was translated into Russian. It was he who first described the division of European Russia into latitudinal zones of differing natural conditions—north, middle, and south. The Russian geographers quickly

---

[1]The transliteration of Cyrillic into Roman letters requires some arbitrary decisions regarding the spelling of names. This book follows the system recommended by Theodore Shabad, editor and translator of *Soviet Geography*.

[2]Lomonosov was a universal scholar in the classical sense. From 1736 to 1741 he studied at the university of Marburg, Germany. In 1755 he was one of a group of scholars who founded the Moscow State University. At that time Pushkin described Lomonosov as ''a university in himself.'' Before he was named head of the new department of geography in the Russian Academy of Sciences in 1758, he had been head of the department of chemistry. He was also famous as a linguistic scholar and as a poet.

adopted his suggestion that the national territory should be divided into natural regions for the practical purposes of administration.[3] At first most of the scholars named as academicians were foreigners who, like Büsching, were living in Russia. In the course of time the proportion of Russian academicians increased. Before 1800 there were already in existence numerous regional descriptions. Early in the nineteenth century two persistent characteristics of Russian geography had become established. One was the emphasis on regions as the basis of organization for geographical work and the insistence that regions are real entities that can be objectively defined. Between 1800 and 1861 there were fifteen different regional divisions of European Russia that were proposed and defended. The second persistent characteristic is that the study of these regions was undertaken for practical purposes. The Russian intellectuals of that period were disturbed by the poverty of the serfs and sought to undermine the system of bondage that tied most of the rural population to the aristocracy of landowners. For example, K. I. Arsenyev wrote a "Short Universal Geography" that was organized by regions and had a strong economic emphasis, reflecting the author's concern with improving the quality of life of the peasants. The book was first published in 1818 and went through twenty editions by 1848. In 1832 the same writer published a monograph on the historical geography of Russian cities in which he offered a classification of cities in terms of their economic functions.

Another distinctive characteristic of Russian geography has been the continued use of that name to cover a wide variety of specialties. While classical geography was undergoing analysis in Germany as each academic discipline sought to establish its separate existence, in Russia the tendency was for scholars with diverse interests to come together as geographers. In the 1840s this convergence of specialties on geography created a need for some kind of institution to provide a forum for the presentation and discussion of different kinds of studies dealing with the physical earth and its human inhabitants. In 1845 Arsenyev along with a number of foreign scholars founded the Imperial Russian Geographical Society. In addition to promoting geographical studies as such, the society also assumed responsibility for studies in geology, meteorology, hydrography, anthropology, and archeology. During the period from 1845 to 1917 the society published some 400 volumes of papers and monographs. The diverse specialties represented in the society were known collectively as "the geographical sciences" (Hooson, 1968).

GEOGRAPHY IN RUSSIA BEFORE 1917

In Germany after the deaths of Humboldt and Ritter there was a break in the continuity of geographical study until the new geographers, such as Richthofen, introduced the new geography. In Russia there was no such break. For this reason it

---

[3]Catherine II decreed in 1784 that officers and noblemen should wear uniforms of distinctive color in each of the three zones (Nikitin, 1966:6).

is difficult to select any one scholar as the grand old man of Russian geography. It would be more realistic to select four grand old men: one grandfather, Semenov Tyan-Shanski; and three fathers, Voeikov, Dokuchaiev, and Anuchin. These formed the nucleus of Russian geography before the October Revolution of 1917.

Petr Petrovich Semenov Tyan-Shanski

Petr Petrovich Semenov Tyan-Shanski provided the continuity between the scholars of the classical period, such as Lomonosov, Büsching, and Arsenyev, and the scholars of the modern period (after 1870), who could no longer claim competence in all branches of geographical study. In 1853–54 Semenov attended Ritter's lectures at Berlin and worked with Richthofen to prepare for exploring work in central Asia. In 1858 he explored the Dzungarian Basin and the bordering Altai Mountains to the north and the Tien Shan Mountains to the south. He was the first European to cross the latter range, and for this exploit the Tsar granted him and his family the right to add Tyan-Shanski as a suffix to his name. In 1888 he explored the desert of Turkestan east of the Caspian Sea. In the 1870s he was appointed director of the Geographical Society in St. Petersburg, a post he held for over forty years. Semenov was also a member of the Committee for the Emancipation of the Peasants from the Bonds of Serfdom (the serfs were emancipated in 1861).

Semenov was much more than an explorer. He had been turned toward the study of geography by his contacts with Ritter; but, like Reclus, he was not attracted by Ritter's teleological philosophy. He was much more concerned in using geography as a means for decreasing the poverty of the rural people. In other words he wanted to emphasize the practical importance of geographical study, or what we would call today its "social relevance." He, himself, wrote a number of regional monographs, including a five-volume work on Russia,[4] which has been described as a "perceptive blend of natural, historical, and economic phenomena" (Hooson, 1968:259). In 1871 he published a general work on the historical geography of Russian settlement. He was also a member of the committee that planned and directed the first Russian census of population in 1897. A man of diverse interests and competence, he was ideally suited to guide the fortunes of such a composite institution as the Geographical Society, and he was able to preserve its unity when in other hands it might well have fallen apart into separate specialties. When he died in 1914 he had set a distinctive stamp on Russian geography, giving it unity in spite of the variety of its parts and pointing it toward practical and remedial objectives.

The Followers of Semenov

The new geography came into Russia between 1880 and 1914. The ideas of Richthofen, Ratzel, and Hettner were familiar to the Russian geographers because many of them studied in Germany. The revolutionary ideas of Charles Darwin were,

---

[4]P. P. Semenov Tyan-Shanski, *Geographical-Statistical Dictionary of the Russian Empire*, 5 vols. (St. Petersburg, 1863–83).

perhaps, less intoxicating than they were in Britain because of the earlier studies of evolution by the Russian biologist, K. F. Rul'ye. In any case the Russians rejected the more extreme forms of environmental determinism stemming from Herbert Spencer and also the use of the biological analogy to describe sequences of land-forms as proposed by the American geographer, William Morris Davis. To be sure, some of the historians did support the ideas of climatic influence on national charac-ter or of the critical importance of the large Asian rivers in providing the setting for the development of early civilizations (Matley, 1966). But the geographers as a whole avoided these difficulties.

In the period between 1880 and 1914 there were three outstanding followers of Semenov who helped to give modern Russian geography its distinctive stamp. Two of these, A. I. Voeikov and V. V. Dokuchaiev, were primarily research workers whose innovative studies of climates and soils gave them international reputations. One, D. N. Anuchin, was primarily an educator, who established geography as a major university subject and who drew up the curricula for the primary and secon-dary schools (Eskov, 1978).[5]

Alexander Ivanovitch Voeikov,[6] 1842-1916, was a scholar who followed Semenov in the wide range of his interests (Fedosseyev, 1978). His doctoral disser-tation, "Direct Insolation in Various Parts of the Globe," was accepted at Göt-tingen University in 1865. His studies of the earth's heat and water balances con-tinued for the rest of his life. His studies in climatology were directed to the improvement of agriculture: hence, when he experimented with different ways to measure the depth of snow, he was partly concerned with the effect of snow cover on temperature and, to a very large extent, with the forecast of the crop yields of the following summer. He was the originator of what came to be called snow science. In the 1870s he traveled in the United States and Asia and thereafter was a regular correspondent with the Smithsonian Institution in Washington, D.C. In 1886 he published a discussion of James H. Coffin's *The Winds of the Globe* (Smithsonian Contributions to Knowledge, vol. 20 [Washington, D.C., 1875]). His concern with the improvement of Russian agriculture led him to compare the farm practices in places with climates similar to those of European Russia. These were the first systematic studies of climatic analogs. Based on his advice, tea was successfully planted in Georgia (east of the Black Sea), cotton in Turkestan, and wheat in Ukraine. His book, *The Climates of the World,* was published in Russian in 1884; but it became known to the climatologists of other parts of the world when it was translated into German and published in Germany in 1887. Similarly, a monograph,

"Distribution of Population on the Earth in Relation to Natural Conditions and Human Activities" (published in Russian in 1906), appeared in German in brief form that same year. He also became known to the rest of the world through papers and books published in France.[7]

One of Voeikov's most important contributions to international geography, however, was his insistence on the importance of studying the influence of man on the environment. He was one of the first Europeans to recognize and report on the destructive effects of man's use of the land. In fact, he criticized Richthofen for not calling attention to the man-caused gullying in the loess lands of China. In his opinion a variety of changes follow the removal of the cover of natural vegetation—in some places the changes are disastrous. He pointed to the overgrazing of some of the Russian steppes with a consequent acceleration of gully erosion. The clearing of the forests in the north could produce a change in the climate toward increasing drought. He was always enthusiastic about what irrigation could do to improve the productivity of arid or semiarid lands (Voeikov, 1901).[8]

V. V. Dokuchaiev, who became the first professor of geography at St. Petersburg in 1885 (the same year in which Voeikov was appointed as docent), was less known beyond Russia than Voeikov because his works appeared only in Russian. But Dokuchaiev deserves a major place among the world's geographers because of his innovative studies of soil. Looking more closely at the natural zones that Büsching had outlined, he was the first scientist to realize that soil is not just disintegrated and decomposed rock. The geographers of Germany, France, and Great Britain conceived of the soil as a faithful reflection of the underlying geological formations. So they spoke of pre-Cambrian soil, or Devonian soil, or glacial soil (derived from glacial deposits). But Dokuchaiev, working on the plains of European Russia, saw that the parent material only provided the substance in which a soil would form. He noted that different kinds of soil could be identified by looking closely at the layers or horizons, which differed because of differences in the soil-forming processes. Soil is formed by water percolating through the loose material at the surface and carrying away soluble minerals; and soil is also formed by the mixture of organic matter from plants and animals with the upper horizon. The soil, said Dokuchaiev, reflects the extraordinarily complex interaction of climate, slope, plants, and animals, with the parent material derived from the underlying geological formations (Gerasimov et al., 1962:111). A soil that had been exposed to all these

---

[7]These translations into German and French include: *Die Klimate der Erde* (Jena: H. Costenoble, 1887); "De l'influence de l'homme sur la terre," *Annales de géographie*, 10 (1901):97–114, 193–215; "Le groupement de la population rurale en Russie," *Annales de géographie*, 18 (1909):13–23; A. Voeikov, *Le Turkestan Russe* (Paris: Armand Colin, 1914). The brief summary of his study of world population is in *Petermanns Geographische Mitteilungen*, 52 (1906):241–251, 265–270.

[8]It is interesting that he made no mention of George Perkins Marsh in his writings. Yet Marsh's *Man and Nature* (1864) was published in Russian translation in 1866. Perhaps he came into contact with Marsh's ideas through Elisée Reclus.

conditions for a long time would more closely reflect the complex of climate and vegetation than it would the parent material.

Dokuchaiev's ideas made slight impact on geographers in western Europe not only because he wrote in Russian but also because there was little opportunity to observe such soils in the west. Especially in France and Britain it seemed quite clear that soils of different kinds were produced on rocks of different types. There were no large expanses of plain on which the arrangement of climate zones could produce observable soil differences. Dokuchaiev's doctoral dissertation, on the other hand, was a careful and detailed study of the Russian chernozem (or black earth), published in 1883. In 1889 he published his concept of soil-forming processes and of the natural zonation of soils according to climate. He recognized, as Voeikov did, that man was a major agent of change on the surface of the earth and that the transformation of natural zones into agricultural regions involved such factors as the attitudes of the people and their technical skills. His concept of natural zones transformed by man comes very close to Schlüter's concept of the landscape type. Dokuchaiev, in fact, described geography as "landscape science" (*landschaftovedenie*).

Dokuchaiev trained a number of students at St. Petersburg who continued to develop the master's ideas. L. I. Praslov became the editor-in-chief of a soil map of the Soviet Union, compiled from a variety of detailed studies on a scale of 1/1,000,000—a project that was later carried forward by I. P. Gerasimov (Gerasimov et al., 1962:113). N. M. Sibirtsev, who died in 1900 at the age of forty, contributed the concept of *zonal* soils as distinct from *azonal* soils—the former reflecting climatic patterns, the latter more closely related to underlying parent materials or such local conditions as poor drainage. The scholar who did the most to make Dokuchaiev's work known outside of Russia was K. D. Glinka. In 1908 he published a major work on the zonal soils of the world that extended the concepts of Dokuchaiev to types of zonal soils not found in Russia. This work was translated from the Russian into German and published in German in 1914; and it was from the German edition of Glinka that Curtis F. Marbut, an American soil geographer, published an edition in English in 1927.[9] Dokuchaiev's ideas about soils were enthusiastically received in the United States, where, unlike western Europe, there are large expanses of plain on which zonal soils can be observed.[10]

Meanwhile, geographical courses were also being offered at Moscow State University. In 1887 D. N. Anuchin was named as head of the new Department of Geography and Ethnography. Trained at Heidelberg in anthropology and anthropogeography, Anuchin's point of view in both writing and teaching was

---

[9]K. D. Glinka, *The Great Soil Groups of the World and Their Development*, translation by C. F. Marbut from *Die Typen der Bodenbildung* (Ann Arbor, Mich.: Edwards Bros., 1927).

[10]Before the unwary reader uses this as an example of environmental determinism, let us observe that neither the steppe nomads of Russia nor the plains Indians in America had the necessary technical skills and theoretical knowledge to recognize the existence of zonal soils.

D. N. Anuchin

V. A. Anuchin

Nikolai N. Baranskiy

Lev S. Berg

Vasily V. Dokuchaiev

Innokenti P. Gerasimov

P. P. Tyan-Shanski

Alexander I. Voeikov

strongly man oriented. His textbooks provided a new kind of geography to teach in the schools; after 1912 when the authorities recognized geography as a field in which students could major, Anuchin's students at Moscow State University were the ones who were appointed to most of the new positions throughout the country. L. S. Berg, for example, went to St. Petersburg, where he continued the development of Dokuchaiev's ideas concerning landscape science and where he remained active after 1917. There was also the geologically trained V. P. Semenov Tyan-Shanski (the son of Petr), who outlined a system of geomorphic regions of the Russian plains with special reference to man-land relations. His nineteen-volume semipopular regional geography of Russia was published in St. Petersburg between 1899 and 1914. In 1928 he contributed a report on the Russian census of 1926 to the *Geographical Review* (V. P. Semenov, 1928).

SOVIET GEOGRAPHY SINCE 1917

The October Revolution of 1917 affected all phases of Russian life, including the pursuit of geographical knowledge. Only those scholars could survive who were able to bring their ideas into line with the concepts of Marx, Engels, and Lenin. Marx was ambiguous concerning his geographical principles, and perhaps it is fortunate that passages from his writings can be found to support a great variety of points of view. Lenin, on the other hand, was much more specific. As an economic determinist he was strongly opposed to any suggestion that the natural environment could in any way control man's destiny. For him geography was the necessary foundation on which the design of a new kind of economy had to be based. The most important product of geographical study was the identification of rational regions within which the segments of the new national economy could be constructed. For him geography was an essentially practical subject; but it was so important that in 1921 he decreed that the subject should be taught in all the schools. Unfortunately, his efforts to provide for instruction in geography were not very successful for the usual reasons: the schoolteachers were not trained in the concepts of geography and there were no teacher-training institutions where they could study geography.

Yet in spite of the new directions of geography after 1917, most of the distinctive characteristics of Soviet geography can be traced back to the prerevolutionary period. The tendency to look at natural landscapes as systems of interrelated parts is typically Russian and derives from men like Semenov, Voeikov, and Dokuchaiev. The continued interest in such physical processes as the heat and water balances goes back to Voeikov. The preoccupation with the drawing of regional boundaries began in the eighteenth century, as did the concern with practical problems of economic development. And to find that geography is a focus of diverse specialties rather than a remnant left over from the separation of the disciplines is in line with long-standing Russian tradition.

The relations between the ruling Communist Party and the members of a profession such as geography are not always clear to people unfamiliar with the Soviet system. The kind of academic freedom that was developed in Germany in the nineteenth century and lost for a time in the 1930s does not exist in the Soviet Union. In the Stalin era scholars who ventured to express disagreement with party policies could be arrested and punished. But after the death of Stalin in 1953 there was more freedom to express critical views on policy questions. Disagreements within the profession concerning professional matters are given wide publicity, even in the public press. Discussion goes on until some kind of a consensus can be discerned; then the leaders of the Communist Party announce the decision. Disagreements on methodological questions involve the usual semantic traps when scholars sometimes mistake word symbols for reality and when certain words and phrases are repeated without clearly defined referents. Words in both Russian and English carry rich loads of connotations; when words are translated, many shades of meaning are lost. As a result the international discussion of theoretical or philosophical questions is not always fruitful. The best language for such discussion is mathematics.

Geography as a professional field is in a strong position in the Soviet Union in 1980. We shall attempt to trace the steps by which it reached this status.[11]

The Early Soviet Years

In the years immediately following the October Revolution the continuity of geographical teaching and research was greatly aided by Lenin himself. In 1918 he presented his "Draft of a Plan of Scientific and Technical Work." In this document he called on the Academy of Sciences to work on the development of a plan for:

> a rational location of industry in Russia from the point of view of proximity to raw materials and a minimal expenditure of labor from the processing of the raw materials through all subsequent stages of the processing of semifinished goods to the finished product [Saushkin, 1966:5].

The geographers in the Academy of Sciences were ready to make practical use of the large amount of information that had been gathered during the preceding half century under the guidance of Semenov and Anuchin. In 1918 the Commission for

---

[11]Such an attempt for those who do not read Russian becomes possible because of the English translations of Russian studies since 1959. There is the periodical, *Soviet Geography,* translated and edited by Theodore Shabad and published monthly (except July and August) by the American Geographical Society. There is also the volume translated by Lawrence Ecker and edited by Chauncy D. Harris, *Soviet Geography, Accomplishments and Tasks,* a symposium of fifty essays by fifty-six leading Soviet geographers edited by Gerasimov and others (Gerasimov et al., (1962). The issue of *Soviet Geography* for September, 1967 is a directory of Soviet geographers.

the Study of Natural Productive Forces was established in the Academy of Sciences, and one of its subdivisions was the Department of Industrial Geography. The first assigned task was to make an inventory of Russia's natural resources. At the same time, in 1918, the first graduate school in geography in the Soviet Union, the Institute of Geography at Leningrad University, was started by L. S. Berg, A. A. Grigoriev, and others.[12]

A geographer who appeared on the scene at this time in a key position was Nikolai N. Baranskiy.[13] As a student at Tomsk at the turn of the century, Baranskiy had been involved in underground activities against the government, and by 1917 he had established himself as one of the leading Bolsheviks. As Saushkin puts it:

> ... geography (and economic geography in particular) was for Baranskiy a logical continuation of his revolutionary party work. He saw in it one of the powerful tools for remaking the world, for building communism, for indoctrinating and educating the people. Baranskiy was clearly aware of the role that geography played in the apprehension of the universe, and of the strength and party orientation of its scientific ideas [Saushkin, 1966:20].

Furthermore, Baranskiy was a close personal friend of Lenin. For the development of geography in the Soviet Union here was the right man in the right place at the right time. It was he who was as much responsible as any one person for the development of Soviet economic geography.

Lenin was a strong supporter of the kind of economic regionalization proposed by Baranskiy. In 1920, in setting up the studies on which the construction of a network of electric power stations and transmission lines was to be based, Lenin specified that he wanted "a geography of each of the sections of the electrification plan" and that he expected to see a map of the main power stations. He insisted that

> ... the map should clearly show the regions to be served by the central power stations, the type of industry to be included, and everything associated with these regional stations [Saushkin, 1966:9].

The State Planning Commission (GOSPLAN) was established in 1921. After some difficulty in gaining general acceptance for any regional division of the Soviet

---

[12]The Institute of Geography was reorganized as the Faculty of Geography at Leningrad in 1925. The Commission for the Study of Natural Productive Forces of the Academy of Sciences in 1930 became the Institute of Geomorphology under Grigoriev; in 1934 it became the Institute of Physical Geography; and in 1936 it became the Institute of Geography.

[13]N. N. Baranskiy was born in Tomsk in 1881 and attended the Law School at the University of Tomsk. Like many young people in those days, Baranskiy was committed to the cause of social revolution. In 1901 he was expelled for Marxist activities, and before 1917 he had been arrested three times by the Tsarist police. At the university he had become interested in economic geography because of its relevance to social problems.

Union, a special commission on regionalization was appointed and given authority to recommend a rational plan for the division of the national territory into functional units. The commission formulated the concept of the economic region in 1922 as follows:

> A region should be a distinctive territory that would be economically as integrated as possible, and, thanks to a combination of natural characteristics, cultural accumulations of the past, and a population trained for productive activity, would represent one of the links in the entire chain of the national economy. This principle of economic integration makes it possible to construct further, on the basis of a properly selected complex of local resources, capital assets brought in from outside, new technology, and the national plan of economic development, a regional development plan that would make optimal use of all possibilities at minimal cost. This will also help achieve other important results: the regions will specialize to a certain extent in those activities that can be developed most fully in them, and exchange between regions will be limited to a strictly essential amount of purposefully directed goods. Regionalization will thus help establish a close link between natural resources, working skills of the population, and assets accumulated by previous cultures and new technology, and yield an optimal productive combination by insuring a division of labor among regions and, at the same time, organizaing each region as a major economic system, thus evidently insuring optimal results" [Saushkin, 1966:12].

GOSPLAN divided the Soviet Union into twenty-one such regions and then proceeded to a detailed study of each of them. The work was done by a large number of young men and women with a variety of professional backgrounds, including engineering, economics, and geography. Some of those who worked on the initial planning went on later to become leaders in economic geography. It was in 1920 that one of these newly created geographers, L. L. Nikitin, made his first investigation of the writings of the prerevolutionary geographers. The older data were combined with the latest information regarding natural resources, physical conditions, population, and types of economy in each of the regions. "Never before in Russian geography had such a vast, diversified body of material been generalized on a regional basis" (Saushkin, 1966:15).

Meanwhile, other geographers not directly employed by GOSPLAN were also working on problems of industrial location and resource development. Of special interest were the plans for the organization of interregional combines. N. N. Kolosovskiy was the author of the plans for the great Ural-Kuznetsk Industrial Combine. A theoretical model for an integrated industrial region, combining basic resources with steel production and with industries using the steel, had been drawn up in 1927 for the Dnieper Basin; but Kolosovskiy's plan called for the movement of raw materials and finished goods between regions. The plan was immediately attacked by other geographers on the basis of central-place theory; but in 1931 such controversy was dangerous because the loser could be arrested for anti-Soviet activities.

As might be expected, the geographers who worked on these practical problems of economic planning advanced professionally more rapidly than those who remained in the universities in teaching positions. The methodological and philosophical discussions among the university geographers seemed irrelevant to those who were using their knowledge of geography to find answers to "real" problems. In the universities there were even some who reverted to the traditional way of teaching economic geography by topics instead of by regions. Baranskiy, who headed the new section of economic geography in the Research Institute at Moscow, fought persistently for the regional approach because this seemed to him the only way that geography could make a useful contribution to practical questions of economic development.[14]

And some of the university geographers even continued to teach various forms of environmental determinism in spite of authoritative decisions outlawing such ideas. Lenin himself had taken vigorous exception to the notion that the steppes north of the Black Sea could never be used for agriculture because of climatic conditions. This point of view, he said, was based on the existing level of technology and ignored the inevitability of technological change (Nikitin, 1966:36).[15] Yet in 1923 A. A. Kruber then head of the Department of Geography at Moscow wrote in a textbook:

> Like all living things on earth, man is subject to the same forces of nature, which, with irresistible inevitability, determine both the conditions of settlement and the way of life of man [Saushkin, 1966:16].

There were also repeated efforts to separate physical geography from economic geography. The argument is the familiar one that the "laws" governing the physical world are not at all the same as those governing man's economic behavior and,

[14]At Moscow State University, where the Department of Geography and Ethnography had been established by D. N. Anuchin in 1887, a Department of Geography was established in 1919, headed by A. A. Kruber; and Anuchin became the head of a Department of Anthropology and Ethnography. These were still undergraduate departments primarily concerned with teaching. In 1923 a government decree (dated December, 1922) established the Scientific Research Institute of Geography, which was headed at different times by A. A. Kruber, A. A. Borzov, N. N. Kolosovskiy, and B. F. Kosov. In 1929 this institute was divided into two sections; one was to work on problems in physical geography; the other, under Baranskiy, was to work on economic geography. This was the first time that postgraduate training in economic geography as such became possible (Ryabchikov, 1968:348). The Department of Geography became the Faculty of Geography and Soils in 1933 and the Faculty of Geography in 1938, under A. A. Borzov. Borzov, a physical geographer, and Baranskiy, an economic geographer, led numerous field expeditions to various parts of the country; and their close and friendly cooperation was a major factor in assuring the unified approach of geography during the years before World War II (Ryabchikov, 1968:349–350).

[15]This statement by Lenin seems to give authoritative support for the principle that the significance of the physical and biotic features of man's natural surroundings is a function of the attitudes, objectives, and technical skills of man himself—which is a widely accepted bourgeois concept.

therefore, these two fields of study cannot logically, or even practically, be included in the same discipline. It was all very well for Borzov and Baranskiy to work together in harmony (much as Vidal de la Blache and de Martonne worked together at the Sorbonne), but their students could find little common ground between physical and economic studies. O. A. Konstantinov (who went on later to become one of the leading economic geographers of the Soviet Union) when he was a young man in 1926 wrote one of those statements that young people the world over often write—to their later embarrassment:

> We reject the possibility for economic geography to be simultaneously part of two quite different systems of science (the geographic and economic sciences). We hold that ours is not only an economic discipline, but a purely economic discipline. In other words we are for a complete break with geography, in the sense of complete outlawing of geographic approaches [Saushkin, 1966:23].

Baranskiy was the leader of the opposition to these challenges to the unity of geography. In the 1930s Stalin's policy of strong centralized control of the economy made regional study seem less relevant and seemed to negate the whole idea of an economic region as a "major territorial production complex with a specialization of national significance."

The Decisions of 1934

Baranskiy was victorious over his challengers. Little by little in the late 1920s and early 1930s the number of geographers who supported his regional approach and his belief in the unity of geography increased. In 1933 at Moscow State University, a Faculty of Soils and Geography was established.[16] Included in this faculty were departments covering both physical and economic geography (a good way of saying both yes and no to questions about the unity of geography).

Then in 1934 two decrees gave official recognition of the professional support that had been given to Baranskiy. The Council of People's Commissars and the Central Committee of the All-Union Communist Party issued a decree on May 16, 1934 concerning the teaching of geography in the elementary and secondary schools. The decree restored the teaching of physical geography, with emphasis on the use of maps. It also specified the teaching of economic geography. On July 14, 1934 the Presidium of the Committee on Higher Technical Education of the Central

---

[16]In a Soviet university a department is one unit organized to teach and do research in one specialty. A faculty includes several departments, and a school is still larger. The Faculty of Soils and Geography in 1933 included the following departments: soil science, physical geography (Borzov), economic geography (Baranskiy), physical regional geography, and cartography. In 1934 the Department of Economic Geography was split into a Department of the Economic Geography of the U.S.S.R. (Baranskiy) and a Department of the Economic Geography of the Capitalist Countries (Vitver). In 1938 the faculty was split into the Geography Faculty (Borzov, dean) and the Geology and Soils Faculty (Ryabchikov, 1968:349).

Executive Committee of the U.S.S.R. specified the kind of geography to be taught
in colleges and universities. Here is some of what was specified:

> ... economic-geography teaching should concentrate on a large body of concrete eco-
> nomic geography material, systematically represented on maps. The specifics of eco-
> nomic geography, the location of productive forces, and economic regionalization
> should be of central concern to teaching personnel. Most of the content of economic
> geography should be taught on a regional basis. In the course on the economic geog-
> raphy of the U.S.S.R., in particular, at least seventy percent of the time should be
> devoted to economic regions [quoted by Saushkin, 1966:30-31].

The result of the publication of these decisions was an accelerated growth of the
fields included in geography. Even the newspaper, *Pravda*, published an editorial
on September 10, 1937 entitled "Know Your Geography." In both Moscow and
Leningrad new departments and institutes were established, as they were also in
other universities throughout the Soviet Union. In 1938 the Faculty of Geography at
Moscow included departments covering general physical geography, physical geography of the U.S.S.R., physical geography of foreign countries, economic geography of the U.S.S.R, economic geography of capitalist countries, geodesy, and
cartography. But during and after World War II (by 1948), the Faculty of Geography included the following additional departments: geomorphology, meteorology
and climatology, hydrology, geography of the polar lands, geography of soils,
biogeography, paleogeography, and oceanography (Ryabchikov, 1968). Furthermore, studies of the geography of population had been included under economic
geography. Baranskiy himself offered a seminar on the economic geography of the
United States in which several specialists on this subject were trained (Saushkin,
1966:37).

Progress in Physical Geography

In 1963, I. P. Gerasimov, the director of the Institute of Geography at the
Academy of Sciences, himself an outstanding physical geographer specializing in
the study of soils, presented the following summary of the important progress being
made in the various specialties of physical geography and biogeography:

> The particular physical-geographic disciplines made extraordinarily rapid progress
> throughout the Soviet period, both in the development of theory and in the formation of
> new approaches and research methods. In climatology, for example, Soviet scholars
> developed the theoretical principles of forecasting and a typology of climatic
> phenomena based on dynamic meteorology; they developed the concept of an integrated
> climatography and, in recent years, gave rise to the promising study of the radiation
> budget and the moisture cycle and their role in the formation of climates. In the field of
> hydrology, Soviet scholars worked out the theory of the water budget, the relationships

between components (surface water, soil water, and groundwater) and methods of transforming one into the other. Glaciology saw the development of the physical theory of glaciation processes, based on the study of heat and mass exchange in various types of glaciers. Geomorphologists established the dynamic character of many exogenous processes (erosion, deflation, abrasion, etc.) and, on the basis of a general theory of external and internal forces and the study of recent crustal movements, developed the morphotectonic or morphostructural approach to geomorphology. Soil scientists identified many new soil types characteristic of taiga, desert, and mountain areas, and worked out new approaches (for example, physical and biogeochemical) to the study of the dynamics of soil-forming processes and the circulation of substances in the natural environment. Biogeographers gave emphasis to the ecological and biocoenotic approaches in the study of plant groupings and of animal populations; in recent years these methods have been enriched by an analysis of nutritional relationships and quantitative patterns of formation of biomass in various environments.

. . . Although the achievements of the particular disciplines provided the main elements of modern theory of physical geography, the formation of such general theory was also complicated by the growing differentiation of the physical-geographic disciplines, in which scholars concentrated their attention increasingly on specific components of the natural geographic environment [Gerasimov, 1968b:242; see also Kalesnik, 1958].

Philosophical Discussion

In 1960 many of the differences of opinion regarding the scope and nature of geography that had been smoldering for a long time in spite of official decisions were brought out into the open (Matley, 1966). The lively discussions of the 1960s were started with the publication of a book by V. A. Anuchin entitled *Theoretical Problems of Geography* in 1960. Hooson says that this book "for the first time in Soviet history, set out to investigate the theoretical basis of geography as a whole through historical and philosophical analysis" (Hooson, 1962:469). V. A. Anuchin (who is a distant relative of D. N. Anuchin), a vigorous supporter of the idea of a unified geography, attacks both an "inhuman" physical geography and an "unnatural" economic geography. He rejects the idea of geographical determinism, which he equates with bourgeois geography; and he also attacks the other extreme of indeterminism, which he associates with the name of the American geographer, Robert S. Platt. The geographical method is best illustrated, he says, in studies of territorial complexes (regions) in which the physical features, the history of settlement, the population, and the economy are in balance.[17]

V. A. Anuchin's book produced an immediate reaction among the Soviet geographers. It was warmly praised by Baranskiy and Saushkin and attacked by Gerasimov, Kalesnik, Konstantinov, and others. Konstantinov characterized it as

---

[17]In 1956 V. A. Anuchin published a regional study of Transcarpathia that is an excellent example of the balanced approach to a territorial complex (Hooson, 1959:79).

"unscientific and anti-Marxist," which is a curious echo of his earlier effort to remove economic geography entirely from geography. The methodological controversy continued for about a decade, much of it available for study in *Soviet Geography*.[18] Some geographers reacted with impatience. For example V. V. Volskiy protested that the Anuchin discussions took valuable time away from the important tasks of economic construction (Volskiy, 1963).

Constructive Geography

Constructive geography is geography applied to the practical purposes of building the socialist economy. Geographical concepts and methods have meaning in terms of what they contribute to economic development planning. As a result, the mathematical procedures as developed in regional science in the United States (Chapter 18) have been eagerly adopted in the Soviet Union. The term may be landscape science instead of regional science. For example, Isachenko makes use of new terms with old meanings to focus attention on applied geography as landscape science, which he relates to the Dokuchaiev "law of zonality" (Isachenko, 1968). The landscape, he points out, is a dynamic system in which matter and energy are circulating and in which there are rhythmic (seasonal) changes of heat and water balance and biological productivity. A classification of landscapes with a hierarchy of scales or degrees of generalization has been worked out for the whole Soviet Union to be mapped on a scale of 1/4,000,000.[19] These landscape definitions will provide an objective basis for defining physical regions. The chief purpose of doing this is not just to provide maps for teaching purposes, but as a basis for identifying regions useful for planning purposes.

Landscape science, says Gerasimov, goes back to the ideas of Humboldt and Dokuchaiev (but not Schlüter) that were brought together in the writings of L. S. Berg (Gerasimov, 1968a). A landscape is a combination of interrelated environmental components (local climate, landforms, soils, plants, and animals) occupying a discrete territory. It exists objectively in the natural environment. But the study of landscapes just for the purpose of describing them is not enough: constructive geography must use this knowledge for the effective transformation of nature. Gerasimov offers four examples of constructive geography:

1. The study of the geophysics of natural and cultural landscapes of the wooded steppe in the central chernozem zone. The purpose was to investi-

[18]V. A. Anuchin, who became a candidate in geography at Moscow in 1949 (that is, he completed all the work for the doctorate except the dissertation), submitted the book as his dissertation at Leningrad. In 1961 the faculty rejected it. It was similarly rejected at Moscow in 1962. On both occasions several hundred spectators had been present. A further two-day-long defense by Anuchin at Moscow University in 1964 resulted in bestowal of the doctorate.

[19]He suggests that a landscape is made up of discrete parts. For example, the slopes of a valley side or the flat valley bottom are units known as *facies;* the whole valley, combining two or more *facies*, makes up a *urochishche*. This seems to repeat Linton's notion of indivisible unit areas.

gate the heat and water budget on the earth's surface in virgin land and in cultivated land and then to experiment with various technical devices for controlling natural processes and so increase agricultural productivity.
2. The study of the irrigated lands of central Asia. The purpose was to find ways to control salt accumulation, to use water more efficiently, and to increase the irrigated area. Questions investigated include the fate of the Aral Sea and what the drying up of this body of water would mean for the total economy of the region.
3. The study of means of reclaiming the swamps of the Ob Valley by using properly placed dams and diversion canals. The study also includes the hydroelectric potential.
4. Studies of the water in Lake Baikal for the specific purposes of reducing pollution, regulating the flow through the Angara River and finding new ways of putting this natural resource to better use.

Urban Studies

Until after World War II the Soviet geographers had paid little attention to studies of cities. To be sure, V. P. Semenov Tyan-Shanski, as early as 1910, pointed to the need for classifying cities in terms of their economic functions. But the research studies of the 1920s and 1930s were mostly devoted to problems in physical or regional geography. In 1946 N. N. Baranskiy again suggested the need to develop a method of classifying cities and made reference to urban studies in the United States as examples of what might be accomplished. This time the suggestion was taken up, and a large number of urban studies resulted. In 1962 Y. G. Saushkin wrote that:

The geography of cities is the most rapidly developing branch of Soviet population geography. The literature on methodology, methods for studies of cities, systems of cities of the country as a whole and of its regions, as well as individual cities of the U.S.S.R. is abundant [Saushkin, 1962:34].

Most of these studies were undertaken for the practical purpose of providing the background for planning projects (Fuchs, 1964). Chauncy D. Harris in a monograph on Soviet cities also reviews the Soviet literature in this field. He reports that by 1970 some four hundred Soviet geographers had published studies in urban geography (Harris, 1970:28). He credits O. A. Konstantinov of Leningard with "playing a leading role in defining the methodology and philosophy of Soviet urban geography" (Harris, 1970:402). Harris summarizes his monograph as follows:

On the basis of data assembled from such diverse sources as the census publications, administrative handbooks, encyclopedias, gazetteers, atlases, maps, and scholarly publications on Soviet urban geography, 30 variables (characteristics) were recorded for

1,247 cities. These data were subjected to statistical analysis [earlier]. Three principal components were found to be highly significant in that they measured a high proportion of the variation for the thirty characteristics analyzed. The first factor to be extracted in the principal components analysis showed highest association with the logarithm of the population in 1959; it is called the *size factor*. The second exhibited the highest association with the logarithm of the urban population potential of each city within its major economic region; it is called the *density factor*. The third revealed highest association with the percentage increase in population 1926–1959; it is called the *growth factor* [Harris, 1970:403].

What Holds Geography Together?

In spite of the integrating effect of focusing geographical studies on practical problems, a tendency to fly apart into separate disciplines appears from time to time among academic people. Saushkin and Zvonkova identify recurring cycles with an amplitude of some twenty-five to thirty years during which centrifugal and centripetal tendencies can be observed. The 1970s, the authors believe, is the beginning of a period of synthesis and unity; but they foresee a return of the tendency to split apart before the year 2000 (Zvonkova and Saushkin, 1968). It seems that geographers are brought together because of their concern with the geographical landscape (variously defined) and with specific segments of earth space. The processes going on in a landscape are interconnected spatial systems, and the study of them separately is not the same as the study of them as systems. In the study of the economy of an area there is need for the full consideration of physical factors; and there is also need for the study of the changes in the landscape resulting from human action, whether planned or unplanned. The basis of physical geography is landscape science. The general and the particular complement and enrich each other. Especially in field study does the essential unity of the geographical approach become apparent (Ryabchikov, 1968:354–355).

The training of young people for advanced degrees in universities in the Soviet Union as elsewhere in the world perpetuates the paradigm of geography as a field of study. Within the system of higher education of the Soviet Union, geography is represented in 36 of the roughly 70 universities that train both research geographers and geography teachers as well as in 74 of the 185 teachers colleges. Economic geography is also taught in a few specialized schools of economics, such as the Plekhanov Institute of National Economy in Moscow and the Finance-Economics Institute in Leningrad.

Twenty of the thirty-six universities have separate faculties (schools) of geography consisting of two or more specialized geography departments; in other universities, geography may be combined with geology, botany, chemistry, geophysics, or other natural sciences. Soviet geographers say that the world's largest educational and research institution in geography is the Geography Faculty of Moscow University. In 1953 the imposing new buildings of the university were completed on Lenin

Hills overlooking the city of Moscow. The faculty occupies six floors and above them, in the central tower, is the Earth Science Museum, which is well worth an extended visit. In the mid-1970s, the faculty was broken down into fourteen departments and thirty teaching and research laboratories. It had a student body of 1700 and a teaching and research staff of 1780, of whom forty percent were employed on outside contracts.[20]

A New Paradigm for Soviet Geography?

We have already noted the inauguration of a searching philosophical discussion among Soviet geographers with the publication of V. A. Anuchin's *Theoretical Problems of Geography* in 1960 (Fuchs and Demko, 1977). That book seems to be the only one in Soviet history to have investigated the theoretical basis of geography. Anuchin elaborated the central concept of "the geographical environment," arguing for an integrated physical and economic geography. This contrasted starkly with the doctrine of the theoretical separation of physical and economic geography that had dominated in Stalinist times. Yet by the 1960s the environment was perceived as more than a physical entity (French, 1968, 1969). An ecological point of view had emerged in the Soviet Union and in the thinking of those populations living in the developed world. And so Anuchin's work came at a propitious time. He followed it in 1972 with a revision of his 1960 book, *Theoretical Foundations of*

---

[20]The institutional structure of geography in the Soviet Union is described in *Soviet Geography*, 18 (1977):540–556, following a useful directory of Soviet geographers, pp. 433–538.

According to this account, the staff includes 60 full professors (the academic rank usually associated with the Soviet doctorate, which is a higher degree than the Ph.D.) and 200 docents or senior research associates (with the lower candidate's degree). The student body is made up of 1500 undergraduates and 200 graduate students, who have completed the five-year course of studies and are working toward the candidate's degree.

The Geography Faculty's fourteen departments are: Physical Geography of the USSR; Economic Geography of the USSR; Cryolithology and Glaciology; Soil Geography and Landscape Geochemistry; World Physical Geography; Economic Geography of Socialist Countries; Economic Geography of Capitalist and Developing Countries; General Physical Geography and Paleogeography; Geomorphology; Hydrology; Oceanography; Meteorology and Climatology; Biogeography; Cartography and Geodesy.

Among the laboratories are problems laboratories, which are concerned with interdisciplinary problems, and specialized laboratories. The problems laboratories include complex maps and regional atlases as well as studies on avalanches and mudflows, recent sediments, reservoirs, soil erosion, Arctic problems, and remote sensing. Specialized laboratories are devoted to the study of soils, land-quality evaluation, landscape geochemistry, luminescence analysis, hydrochemistry, map compilation and publication, hydrology, geomorphology, and so forth.

For their fieldwork practice, students may be attached either to periodic field expeditions organized by the Geography Faculty under outside contracts with government agencies or to some of the field stations operated by the faculty, including stations in the Moscow area, in the Khibiny Mountains of the Kola Peninsula, the Elbrus station in the Caucasus, or a field station located in a preserved patch of virgin steppe in the central chernozem area.

Geography; then came *Fundamentals of the Management of Natural Resources* in 1978 (Hooson, in press). Anuchin resigned his professorship at Moscow University taking a position with the Council for the Study of Productive Forces to further his thinking on national planning. His aim was to devote more time to environmental and regional problems and to demonstrate the practical value of an integrated geographical approach in their resolution. All this has done much to establish a way of thinking in Soviet geography quite different from that of only two decades ago (Zvonkova and Saushkin, 1968).

12

Geography Around the World

Ever since the dimmest antiquity the spirit of man has felt the need of geographical, i.e., earth-describing knowledge. Acquaintance with one's own country has constantly stood out as necessary from a practical standpoint, and curiosity has been great with regard to foreign countries; but regarded as a science, geography was slow to raise itself above the primitive stage of collecting data. It was not until facts began to be systematically brought into relation with one another, and conclusions to be drawn therefrom, that geography became a true science. It was then, too, that there arose the question of its proper method and its limits in relation to older sciences. Now one side of geography, and now another, has primarily caught the interest of a generation; and the general conceptions of the essential character of the new science have varied accordingly.

Geography is the science of the present-day distribution of phenomena on the surface of the earth. . . .

All around the world during the past century some version of the new geography has made its appearance. Ideas similar to those we have discussed in the preceding four chapters have been presented in many different languages and in many different national settings. The scholar who could speak with authority on a wide variety of topics was gradually replaced by specialists of one kind or another; and one of the specialities that emerged had to do with problems of location. The emergence of a professional discipline had to be accompanied by the development

The quotation above is from S. De Geer, *Geografiska Annaler* (1923), p. 1.

of faculties and departments where advanced training could be offered. Following the example of Germany in 1874, chairs of geography were established in universities where such positions had never existed before.

It is amazing to discover how similar were the problems that resulted from all this in different countries. The same questions were asked: Can studies of physical geography be included in the same department or faculty with studies in human geography? How can the scope and methods of a field of study called geography be so stated that the underlying unity of the field becomes clearly apparent? Or, is geography a kind of federation of separate sciences? Do studies of particular places on the earth belong in a different academic discipline from studies of the laws governing the arrangement of particular phenomena over the whole earth? Or, is it true (as Varenius implied but did not live long enough to demonstrate) that general and special geography are essential to each other and cannot be separated without damage to both? Is the study of geography properly included in a Faculty of Science or in a Faculty of Philosophy? And, are all these apparent dichotomies irreconcilable or is the statement of them in word symbols just a semantic trap? All these and other questions that were first faced in nineteenth-century Germany have since puzzled scholars all around the world.

The new geography with its persistent problems has been transmitted directly or indirectly from Germany but not necessarily in the form developed by the Germans. Distinctive variations on the theme of the new geography were formulated in such national "schools" as those of France, Great Britain, the Soviet Union, and the United States. And other countries of the world have in many cases received the impact of the new geography from these secondary centers of innovation. Before examining the development of geographical ideas in the United States, we offer in this chapter a brief survey of what has been happening in other parts of the world.

THE GERMAN INFLUENCE

The new geography was transmitted directly from Germany to the Scandinavian countries and Finland, to the Netherlands, Austria, and German-speaking universities in Switzerland. In these countries one result has been the sharp separation of physical and human geography, each with quite separate curricula, even when they are located in the same department. More often, however, physical geography is in the Faculty of Science, and other aspects of geography are found in the Faculty of Philosophy or the Faculty of Letters.

Sweden

Sweden is notable as a country in which there has been a strong popular interest in geography for a long time. Sweden is not a large country, but the Swedes are inveterate travelers. As early as 1885 the popular demand for travel literature and guidance led to the organization of the Swedish Touring Club. It not only supplied the demand for information about foreign countries but also undertook to develop facilities, such as trails and rest houses, for the many Swedes who wanted to journey into the most remote parts of their own country. By 1900 the club had a membership of more than 25,000 (Anrick, 1923). In proportion to the population, also, Sweden has supplied more than its share of noted explorers. We think of Baron Adolf Erik Nordenskjöld, who in 1878–79 sailed his ship from Norway to Japan, thus for the first time navigating the Northeast Passage. Nordenskjöld's sons and nephew became field observers, two in anthropology, one in geography.[1] There was also the explorer of central Asia, Sven Hedin, who led numerous expeditions to this remote region between 1886 and 1934. In the same generation was Gunnar Andersson, who was professor of economic geography at the School of Economics in Stockholm. He also was an explorer, but for the purpose of discovering mineral and plant resources. He carried on a long-standing tradition in Sweden of field expeditions aimed at the careful mapping of resources, which apparently started in the mid-eighteenth century when the king commissioned the botanist Carolus Linnaeus to carry out the first such survey. In 1914 Gunnar Andersson was named as advisor to the Geological Survey of China. For some twenty years he was secretary of the Swedish Society of Anthropology and Geography. He was editor of the society's periodical, *Ymer,* and coeditor of the *Geografiska Annaler,* founded in 1919.

The Swedish scholar who brought this rather diffuse popular interest in travel and exploration to a sharp focus on a new kind of geography was Sten De Geer.[2] The course of Swedish geography thereafter was profoundly influenced by Sten De Geer's insistence on a quantitative approach. In the Swedish tradition, he was devoted to detailed regional studies; and he rejected any hint of environmental determinism. When he classified and mapped surface features, it was for the purpose of demonstrating the close interrelations between landforms and population. In 1908 he began experimenting with new methods of showing population by dots. He

[1] Baron Adolf Erik Nordenskjöid made many trips to the Arctic between 1858 and 1883. In the 1860s he taught chemistry and mineralogy at the Swedish Military Academy. Both of his sons became anthropologists. Gustav, who died in 1894 at the age of twenty-five, was the first to describe the Indian settlements of the Mesa Verde; Erland spent many years among the Indians of South and Central America and made a major contribution to knowledge of the origin and distribution of Indian civilizations. He held the chair of anthropology at the University of Göteborg. The nephew, Otto Nordenskjöld, became professor of geology and mineralogy at the University of Uppsala; but in 1905 when a chair of geography was established at Göteborg, he was appointed professor of geography. He explored the Arctic, including Greenland, and in 1920–21 traveled widely in little-known parts of Peru and Chile.

[2] Sten De Geer was the son of the distinguished glaciologist, Baron Gerard De Geer, who was the first to establish a date for the last retreat of the ice by counting the varves (bands of coarse and fine material) in clays deposited in glacial lakes. Sten De Geer taught geography at the Högskola in Stockholm from 1911 to 1928. He was appointed in that year to the chair of geography at the University of Göteborg, but, unfortunately, he occupied this chair for only four years due to his untimely death at the age of forty-seven in 1933.

made a detailed map of the island of Gotland, using a scale of one dot for 10 people (De Geer, 1908).[3] In 1919 he published an atlas of twelve maps showing the population of the whole of Sweden on the basis of the 1917 census. He used a scale of one dot for 100 people. The cultivated area of the country was indicated by a yellow coloring, and the places that remained empty and unused were left blank. The concentrations of urban people he showed by drawing globes with diameters proportional to the city sizes (De Geer, 1919, 1922a). This was the first time that a population map with this degree of detail had been prepared. In 1923 he published a translation in English of one of the first studies of urban patterns of land use—for Stockholm (De Geer, 1922b, 1923b). During the years 1926 to 1928 he published a series of writings on the identification of regions based on the synthesis of a variety of elements. One map outlined the racial character of the Swedish nation. On another map he carefully demonstrated the difference between regions based on the physical character of the earth's surface and regions based on human settlement (De Geer, 1928a). In the same year he published a wide-ranging study of ancient political geography in which he identified the core areas of states—that is, the small segment of territory in a state in which there is the greatest concentration of economic production and political power (De Geer, 1928b).[4]

In 1922 Sten De Geer spent the year traveling in the United States. One result of his American visit was a paper in which he made the first attempt to define the American manufacturing belt in quantitative terms (De Geer, 1927). He based his definition on the spacing of industrial cities (identified by the number of industrial workers as given in the census). He found that Detroit and Toledo are a little over fifty miles apart, as are also Cleveland and Ashtabula—all of which cities must be included in the manufacturing belt. His definition then included all the territory over which industrial cities are spaced no more than fifty-three miles apart. He also discussed the movement of the cotton textile industry from New England to the South. Like any good innovative study, De Geer's work provoked numerous critical reviews and led to an increase in urban studies during the following decade (Freeman, 1967:124-155).

De Geer was thirty-seven when he published his paper on the scope and method of geography (De Geer, 1923a). Geography he defined as the study of the present-day distribution of phenomena on the surface of the earth. He followed Richthofen in defining the surface of the earth as the zone of overlap between lithosphere, hydrosphere, atmosphere, biosphere, and anthroposphere. He recommended focusing attention on the present and using historical geography only insofar as it might be necessary to explain present patterns. Sten De Geer's paper, which was

[3]Apparently the first use of dots to show population on a map was in a paper by A. O. Kohlman, "Om naturliga omradero använding i statistiken," *Fennia* (Helsinki), No. 1 (1897-99):46-59 (Freeman, 1967:131).

[4]These papers in the *Geografiska Annaler* are written in French, German, or English as well as Swedish.

written in English, had a considerable impact on the development of geographic ideas in America in the 1920s.

A contemporary of Sten De Geer who aided in the development of Swedish geography in the 1920s was Hans W:son Ahlmann.[5] He occupied the chair of geography at the University of Stockholm for more than thirty years, during which time he trained many young geographers. One of his students, William William-Olsson, taught for a time at the university before accepting a position at the Stockholm School of Economics, where he became professor in 1946. William-Olsson turned his attention to the detailed study of the city of Stockholm, starting where De Geer had left off. A paper on Stockholm was published in the *Geographical Review* (William-Olsson, 1940), and an enlarged and beautifully illustrated monograph on the same city was one of the publications presented by the Swedish geographers to the Nineteenth International Geographical Congress in 1960.

Another contemporary of Sten De Geer who was directly responsible for the important present-day developments at the University of Lund was Helge Nelson.[6] Nelson was raised in the tradition of the field study of small areas. After the nationwide surveys of resources by Linnaeus and Andersson, the next step was to study specific regions more closely. Nelson published two detailed reports on the process of settlement in the central mining district of Sweden. His examination of the process of settlement turned his attention to the opposite movement—the emigration of Swedes to the United States. He carried out field studies in the United States identifying the links between certain Swedish communities and communities in the United States where the Swedes settled.

The appearance of Lund in the contemporary period as a major center of innovation in geographical studies can be largely credited to one of Nelson's students, Torsten Hägerstrand. Nelson assigned him a more-or-less traditional seminar paper, a detailed field study of a small area near Lund from which many emigrants had departed. The idea was to find out what happens in a community when it loses population in this way. Hägerstrand observed that the people who decided to emigrate were not scattered in random fashion over the area but were grouped in clusters. This set him to search for an explanation for the clustering.

[5]Hans W:son Ahlmann was professor of geography at Stockholm from 1919 to 1950. From 1956 to 1960 he was the Swedish ambassador to Norway and during this same time was the president of the International Geographical Union.

[6]Helge Nelson completed graduate study at the University of Uppsala in 1910. From 1916 until his retirement in 1947 he was the professor of human geography at the University of Lund. In the Swedish universities physical geography and human geography offer separate curricula, even when they are in the same department. At Lund the physical geographer since 1956 has been K. E. Bergsten. Helge Nelson's students in human geography now hold a large proportion of the professorships in this field: Sven Godlund is at Göteborg; Carl David Hannerberg is at Stockholm; and Torsten Hägerstrand is at Lund. In addition, Nelson's academic grandson is Gunnar Evald Tornqvist, a student of Godlund's, who is now a second professor of human geography at Lund.

Just at this time Edgar Kant arrived at Lund.[7] Kant is an imaginative and a widely informed geographer who had applied central-place concepts to the study of settlements in Estonia and who had become interested in the study of human migrations (Kant, 1953/1962). At Lund, Hägerstrand was greatly excited by Kant's use of a variety of mathematical procedures for studying settlement problems. He realized that the key to the arrangement of the clusters of emigrants was to be found by seeking some way to measure and predict the probability that specific individuals would receive information about America. By this time Hägerstrand was less concerned with the detailed description of the results of emigration in his area than he was in developing some kind of a theoretical model of information diffusion. Later in his book on innovation diffusion he wrote:

> This study is concerned with the analysis of a specific geographic area; its object is to deal with the diffusion of innovations as a spatial process. That the material used to throw light on the process relates to a single area should be regarded as a regrettable necessity rather than a methodological subtlety [Hägerstrand, 1953/1967:1].

At about this same time, Hägerstrand came across some studies in atomic physics in which the solutions of certain problems had been found through the use of game theory, the so-called Monte Carlo simulation involving the use of random samples from a known probability distribution. He was able then to distinguish between clusters that were randomly distributed and clusters that could be accounted for in terms of probability.[8]

As a result of all this, Lund has become a major center of geographic research in the contemporary period. Students from all over the world are attracted to Lund, and the Swedish geographers have established close ties with similar centers in the United States (Chapter 18).

Geography in Norway

In Norway work of a geographic character started with the Arctic explorer, Fridtjof Nansen, who was appointed professor of oceanography at Christiania University (Oslo) in 1908. The Norwegians have been leaders in studies of the oceans and the atmosphere. Among Nansen's distinguished successors was H. U. Sverdrup, who was director of the Scripps Institute of Oceanography at La Jolla,

---

[7]Edgar Kant received his doctorate at the University of Tartu in 1934 and was professor of economic geography there from 1936 to 1940, and from 1941 to 1944. He was also acting rector from 1941 to 1944. In 1944 he left Estonia and settled at Lund, where he lectured on central-place theory in 1945. At Lund he was associate professor of social and economic geography from 1950 to 1963 and professor of economic geography from 1963 to his retirement in 1970.

[8]The same kind of procedure had been used, it seems, by biologists seeking to imitate the migrations of animals. For this information concerning the various influences on Hägerstrand's thinking the authors are indebted to Peter R. Gould.

California, from 1936 to 1948. Norway also produced the famous Bjerknes family, whose studies of physical meteorology provided the basis for the modern approach to weather forecasting.[9] At the University of Oslo there is now a well-balanced department of geography in which physical geography, including oceanography, is matched with professors of various aspects of human geography.

Finland

Scholarly interest in geography has also been apparent for a long time in Finland. A department of geography was established at Helsinki in 1893, but not until more than a decade later was the department headed by a professor. In 1907 J. G. Granö was appointed to direct advanced study in geography. Granö followed Otto Schlüter in identifying the landscape as that part of the natural surroundings of man that can be perceived by the senses. He illustrated his methodological study with examples from field studies in Finland and Estonia (Granö, 1929).[10]

Denmark

In Denmark the new geography was introduced by H. P. Steensby, who became the professor of geography and head of the newly established department at the University of Copenhagen in 1911. Steensby is noted for his studies of Eskimo culture. He not only contributed a detailed report on the close interrelations between the Eskimos and their natural surroundings but he also offered the hypothesis that the Eskimos had originated as northward-migrating Indians—a hypothesis with which the American anthropologist, Clark Wissler, did not agree (Wissler, 1920). Steensby also investigated the routes of the Norse explorers and the possible locations of their settlements in North America.[11]

The Netherlands

German geography also spread directly to the Netherlands. A chair of geography was established at the University of Amsterdam in 1877; but, as elsewhere, there were no geographers to appoint to the position. Professor C. M. Kan occupied the chair for thirty years, from 1877 to 1907. When he retired in 1907 the university

[9]Vilhelm F. K. Bjerknes (son of the mathematician, C. A. Bjerknes) in 1897 formulated a new model of atmospheric circulation. Between 1895 and 1932 he taught at various universities, including Stockholm, Oslo, Leipzig, and Bergen. His son, Jakob Bjerknes, was the first to show how cyclones developed on sloping frontal surfaces between air masses—in 1919.

[10]In 1970 the professor of geography at the University of Helsinki was Leo Aario, who has written on the connection between the Viking voyages to America and the climatic conditions (a warm, dry period) of Europe. Granö's son, Olavi Johannes Granö, became the professor of geography at the University of Turku in 1962.

[11]In 1970 the professor of geography at the University of Copenhagen was Axel Schou.

established two chairs: one in physical geography in the Faculty of Science and one in political geography, dealing especially with the geography and ethnography of the Dutch East Indies, in the Faculty of Letters and Philosophy. The same separation of these aspects of geography was solidified at the University of Utrecht, where a professor of physical geography was appointed in the Faculty of Science and the appointee in the Faculty of Letters was expected to cover "political, economic, and general geography" (Joerg, 1922:460). As a result any possibility of developing geography as a unitary field was lost. Furthermore, the physical geography became a well-developed science and political, economic, and general geography became a collection of uncoordinated elements. But the demand in the Netherlands for knowledge of people and products all around the world was as strong as it had been in the seventeenth century when Varenius wrote about Japan and Siam for the merchants of Amsterdam.

The demand for more training in economic geography was met chiefly at the School of Commerce in Rotterdam. The economic geographer at the University of Utrecht, J. F. Niermeyer, responded to the demand for better training for young people who were to be employed in the widespread trading activities of the Dutch by commuting to Rotterdam to offer a course of lectures. Eventually, other economic geographers were appointed to the School of Commerce, among them Hendrik Blink, who in 1910 started a professional periodical devoted entirely to economic geography—the *Tijdschrift voor Economische Geographie.*

The concept of chorology as the solution for the unfortunate separation of the physical and human aspects of geography was strongly advocated after World War II. One of the most extended discussions of the content of chorology was presented by G. De Jong (1962). De Jong, who was on the staff at the University of Amsterdam from 1955 until his death in 1968, was the coeditor of the *Tidschrift voor Economische en Socialische Geographie,* as that periodical was renamed to indicate that its scope had been broadened to include what the Germans call social geography. De Jong suggested that the emphasis on the physical aspects of geography derived from the German influence should be modified by following the ideas of Vidal de la Blache and Brunhes, by whom chorology had been revised to give greater emphasis to the human group.

German Universities in Switzerland and Austria

In Switzerland, where some of the universities are modeled after those of Germany and others after those of France, the latter have had less of a problem with the separation of the physical and human aspects of geography. Zurich has an outstanding Geographisches Institut, founded in 1895, which for many years was directed by Hans H. Boesch. The fact that the institute is in the Faculty of Science means that advancement is difficult for those geographers who have never published papers in physical geography, no matter how well qualified they might be in other aspects of the field, such as the study of transportation or population.

At Zurich, also, is the famous Kartographisches Institut, located in the Technische Hochschule (Institute of Technology), at which Eduard Imhof was a leading figure until his retirement in 1965. Some of the spectacular new map-making techniques that have made Swiss maps outstanding and that have given the map-publishing house of Kummerly and Frey an international reputation were first developed at the institute.

But while the geographical institutes at Zurich and Bern were organized on the German model, the study of geography at the University of Fribourg, which is bilingual, has followed the French plan. It is the genius of the Swiss that these diversities lead to great strength rather than enfeebling conflict.

The universities in Austria are traditionally interconnected with those of Germany. As Joerg pointed out after World War I, professorships in these universities, like the German ones in Switzerland, are interchangable with positions in the universities of Germany. Many of the geographers discussed in Chapter 8 spent part or all of their academic careers in Vienna or Zurich (Joerg, 1922:464).

THE FRENCH INFLUENCE

The new geography was transmitted through France during the time when Vidal de la Blache, Emmanuel de Martonne, and Jean Brunhes and their students were influential; the result was the development of geography as a coherent discipline. It was less of a problem to include physical geography and human geography in the same curriculum than was the case where German influence predominated. The countries most directly influenced by France include Belgium, Italy, Spain, Portugal, the Latin American countries, and French-speaking Canada.

Belgium

The new geography was brought to Belgium from France by Elisée Reclus, the French geographer. In 1892 he was invited to come to the University of Brussels, but his political activities by this time were so well known that he was not appointed. Instead he went to the Université nouvelle at Brussels, where he established the Institut géographique in 1898. Even as late as the 1920s at the University of Brussels, geography was a part of history; but, elsewhere in Belgium, geography was accepted much earlier as a subject for advanced study. A leading university was at Liége, where J. Halkin, a student of Richthofen and Ratzel, was professor; but when Halkin retired in 1933 he was succeeded by Professor Omer Tulippe, who had studied with Demangeon at the Sorbonne. Tulippe introduced the French version of the new geography. In his earlier years he focused attention on *l'habitat rurale,* as Demangeon had done; but in his later years he was concerned with using geographical methods and concepts to aid in the solution of practical problems. He searched for ways in which the physical environment, the rural settlements, the distribution of population, and the economic conditions might be modified and improved. When he retired in 1966 his students organized a collection of original papers by geographers from all over the world. The 109 contributions are arranged under five major headings: physical geography, human geography, economic geography, applied geography, and regionalization and theory. The two volumes offer a revealing cross-section of the field of geography as it was actually cultivated in the 1960s (Sporck, 1967).

Meanwhile, the chair of geography at the University of Louvain was occupied by Paul Michotte, who made use of geographical ideas derived from both Hettner and Brunhes. In 1921 at the age of forty-five he wrote a methodological paper describing what he called the new orientation of geography and introducing ideas derived both from the Germans and the French (Michotte, 1921). Geography cannot be a general science of the earth, for knowledge has become too specialized, but it can be restricted to phenomena of the earth's surface as Richthofen had first suggested. He was critical of Brunhes for giving too much emphasis to the connectivity among things on the face of the earth. The proper objective of geography, he decided, is chorologic. He concludes:

> The surface of the globe comprises a mosaic of patches, different in color, pattern, and physiognomy. One can understand them by a process of mental abstraction. . . . Delimiting these areas, describing the landscapes and explaining their character, finally classifying them in spaces of an hierarchical order of larger and larger extent, is the whole objective of geography . . . [Fischer, Campbell, and Miller, 1967:289].

Michotte made it clear that there was nothing new in his conclusion but that he was only presenting the generally accepted point of view for the benefit of his Belgian colleagues.

Italy

In Italy the founder of modern geography was Giuseppe Dalla Vedova, who was appointed to one of the first chairs of geography outside of Germany—at Rome in 1875. In 1881 he published a paper contrasting the popular idea of the content of geography with the scientific objectives of the field. He identified geography as chorology (Dalla Vedova, 1881). Most of the geographers who were appointed to chairs of geography in Italian universities during the 1890s were Dalla Vedova's students. In 1908, to celebrate the fiftieth anniversary of the beginning of his career as a teacher, some of his students put together a collection of his many writings (Joerg, 1922:450).

Dalla Vedova was strongly supported by a younger contemporary, Giovanni Marinelli, who was appointed to the new chair of geography at the University of Padua in 1879. Marinelli is noted for the publication of a compendium of physical and human geography of the world in eight volumes, which reminds one of the similar works by Reclus and Malte-Brun.[12] But he was also an effective promoter of the new geography in Italy. He was founder (and until his death in 1900 the editor) of the professional periodical *Rivista Geografica Italiana*. Giovanni's son, Olinto Marinelli, was appointed to the chair of geography at the University of Florence in 1902 and occupied it until his death in 1926. He also took over the editorship of the *Rivista*. Olinto Marinelli was a productive scholar who published an amazing variety of studies. His papers include works on mountains, glaciers, karst landforms, rivers, deltas, population, settlements, urban areas, cartography, the historical geography of Italy, and the history of geography (Marinelli, 1919, 1922; Marinelli and Dainelli, 1912; also Dainelli, 1929).

Another distinguished Italian geographer who was a student of Dalla Vedova was Roberto Almagià, who was appointed to the University of Padua in 1911 and occupied the chair of geography at Rome from 1915 to 1959. Like Marinelli he was a prolific writer on a variety of topics. His greatest work was the history of cartography, but he also wrote on the historical geography of Italy and the history of exploration (Almagià, 1929, 1959).

The Italians were not troubled by most of the methodological questions that bothered the Germans. Accepting the idea of geography as the study of the arrangement of things that gives character to places on the face of the earth, they proceeded in the French tradition to study their own country and the parts of Africa in which they were at one time interested as a colonial power (Milone, 1955).

Spain and Portugal

The new geography has only recently reached Spain and Portugal. There were individual scholars studying and writing about the various parts of their own countries. After the founding of geographical societies (at Lisbon in 1875, and at Madrid in 1876), there were regular meetings at which papers of geographical interest were read and discussed. There were also government surveys of climate, surface features, and other aspects of the resource base. In 1939 the Instituto Juan Sebastián Elcano started the publication of a research series called *Estudios Geográficos*. But departments of geography in the universities headed by professors only appeared at Coimbra in 1942 and at Lisbon in 1943; and in Spain at Madrid, Granada, Barcelona, and Murcia during the 1950s.

[12]*La Terra* (Milan, 1883–1901).

Brazil

Brazil is the Latin American country into which the new geography has made the deepest penetration.[13] The sequence of events in Brazil is perhaps representative of what may yet happen in other parts of Latin America. The start of the flow of new geographical ideas took place in the 1930s with the arrival of the French professor, Pierre Deffontaines, at the University of São Paulo. France has been the chief source for the new geography, but in the contemporary period the Brazilian geographers have reached out to the world's major centers of geographical innovation and have made use of ideas from a great variety of sources.

Before the decade of the 1930s the geography that was taught in Brazilian schools reminds one of the situation in Europe against which Rousseau and Pestalozzi started a revolt. School children were expected to memorize information collected by political divisions, including state capitals and chief cities, products, the names of physical features, and other uncoordinated elements. There were no colleges or universities where teachers could be trained. Furthermore, in Brazil this situation applied not only to geography teachers but teachers of all subjects. Geiger remarks that secondary-school teachers were mostly frustrated engineers, lawyers, or doctors. Engineers who could not meet the requirements in engineering schools became teachers of mathematics and physics; doctors turned to the teaching of science; lawyers taught history; and geography was taught by lawyers, doctors, or even by unsuccessful poets.[14]

In the 1930s there were many influential Brazilians who recognized the need to train teachers in the universities. The result was the organization of Faculties of Philosophy on the French model. The assistance of the French Government was requested to provide scholars in a variety of academic fields. As part of this program, with costs partly covered by France, the French geographer Pierre Deffontaines came to Brazil in 1934 as professor of geography at the University of São Paulo. At São Paulo he organized the Instituto de Geografia (a department offering both undergraduate and graduate training). Later he organized departments at Rio de Janeiro and at Belo Horizonte, and he founded a professional association for geographers and a professional periodical.[15] When he left Brazil in 1939, he had laid the

[13]For information concerning the development of geography in Brazil the authors are indebted to Carlos Delgado de Carvalho, Pierre Deffontaines, and Pedro Pinchas Geiger.

[14]One of the first Brazilians to deplore this situation was Carlos Delgado de Carvalho, the son of one of Brazil's career diplomats, who had been educated at the École des sciences politiques in Paris, where he studied in 1906-1908. There he became acquainted with the school texts and atlases of Vidal de la Blache and other French geographers. Returning to his native country, he found no school texts and no atlases or wall maps to portray Brazil. So he undertook to prepare such books and maps. Later, in 1919-20, he studied at the London School of Economics with H. J. Mackinder. In Rio de Janeiro during the 1920s and 1930s there was one Brazilian scholar thoroughly aware of the kind of geography being studied and taught in other parts of the world.

[15]In 1936 he founded the Associação dos Geografos Brasileiros, a professional society organized on the model of the Association des geographes français, which continues to hold periodic meetings

groundwork for the introduction of the new geography and for the creation of a body of trained professional geographers. He had also taken the opportunity to study in the field and to gather materials for a number of papers on the geography of Brazil (Deffontaines, 1938). His successor, Pierre Monbeig, also contributed to the geographical study of Brazil.

The transmission of French geographical ideas to Brazil was given additional support by the visit to Brazil of Emmanuel de Martonne after World War II. De Martonne introduced physical geography to the Brazilian universities as part of the training in geography, and he also carried on field studies of geomorphology. On his return to Paris he formed two institutes for the study of Latin American problems to which many students from Brazil and other Latin American countries have come for advanced training.[16]

A very important part of Brazilian geographical progress has resulted from the organization in 1936 of a government agency attached to the census, which became known in 1970 as the Fundação Instituto Brasileiro de Geografia (IBG).[17] In the course of time the IBG grew to employ about 100 professional geographers, mostly trained in the Brazilian universities. These geographers were bombarded by new ideas regarding geographical purposes and methods as a result of visits to Brazil for varying periods of time by a succession of foreign geographers. From Canada and France came the concepts of bioclimatic geomorphology; from Germany, by way of the United States, Leo Waibel brought a new approach to settlement problems, replacing the idea of response to environment with studies of agricultural systems in which the physical earth constitutes one element (Waibel, 1948, 1950). The Portuguese soil specialist, Luis Bramão, introduced modern ideas about soil science,

---

attended by professional geographers from all over Brazil. In that same year he started the *Revista Brasileira de Geografia* under the editorship of Christovam Leite de Castro. At São Paulo in 1935 he was succeeded by another French geographer, Pierre Monbeig. Departments of geography were also formed at the universities in Recife, Salvador, Fortaleza, Curitiba, Pôrto Alegre, and Rio Claro. At these universities a large number of young Brazilian geographers were given professional training in succeeding years.

[16]The institut des hautes études de l'Amerique Latine at Paris and the institut de géographie tropicale at Bordeaux.

[17]Originally, the agency was called the Conselho Nacional de Geografia (CNG) and formed one part of the Instituto Brasileiro de Geografia e Estatistica, which is charged with carrying out the decennial census. The CNG carried on geographical studies and prepared maps as part of the continued program of the census. Later the CNG was renamed the Instituto Brasileiro de Geografia (IBG); in 1967 it was made a part of a federal interagency planning organization, in which capacity it is known as the Fundação Instituto Brasileiro de Geografia. The IBG was designated as the adhering agency in the International Geographical Union and was in charge of the organization of the Twentieth International Geographical Congress which met at Rio de Janeiro in 1956. The IBG is also the adhering agency to the Instituto Panamericano de Geografia e Historia (Pan American Institute of Geography and History—PAIGH), a specialized agency of the Organization of American States. The Commission on Geography of the PAIGH has its central office in Rio de Janeiro and is directed by the Brazilian geographer, Nilo Bernardes.

which were carried on at the Institute of Agronomy at Campinas by José Setzer. Ideas current in North American geography were introduced by several geographers from the United States.[18] In 1970 the Brazilians were able to announce the first publication of a study using the newest mathematical procedures. In recent years, moreover, the Brazilian geographers, like their colleagues elsewhere in the world, have found themselves working on interdisciplinary teams seeking answers to complex problems of development planning.

Other Latin American Countries

Other Latin American countries are traveling along a similar route toward the development of a new approach to geography. A department of geography was organized in each of the Argentine universities (at Buenos Aires in 1917). Similar departments offering advanced work are to be found in Chile, Colombia, and Mexico. In all of these the influence of the French school is predominant. In many of the other Latin American countries there are individual scholars offering courses in geography and carrying on research studies.

The New Geography in French Canada

The French-speaking part of Canada also received its new geography from France. As early as 1910 a Belgian geographer, trained in France, occupied a chair of geography (but with no department) in the École des hautes études commerciales at Montreal. In the schools of Quebec, economic and commercial geography had been offered since the 1930s, but no facilities were available for the advanced training of teachers. Although departments of geography were not established at the Université Laval in Quebec until 1946 and at the Université de Montréal until 1947, courses in geography had been offered at these universities for many years. Two distinguished French geographers lectured at Montreal: from 1925 to 1927 Jean Brunhes offered work in geography; from 1927 until the start of World War II Raoul Blanchard paid an annual visit to Montreal to offer a series of lectures. Both of these scholars encouraged many young Canadians to seek advanced training in France at Paris, Grenoble, or Strassburg.

THE BRITISH INFLUENCE

The new geography was transmitted from Great Britain to the former British colonies and dominions. In fact, many of the universities in these areas have close administrative connections with a university in Britain. Many of the distinctive

[18]Among those who have carried on field studies or taught in Brazilian universities are Mark Jefferson, Robert S. Platt, Clarence F. Jones, Raymond Crist, Preston E. James, John Augelli, and Richard P. Momsen. In recent years the methods of mathematical geography and regional science have been introduced by Brian Berry, Howard Gauthier, and from Britain, J. P. Cole and Peter Haggett.

characteristics of British geography were passed on to universities throughout the Commonwealth. But no one scholar did more to spread the British style of geography outside of Great Britain than Griffith Taylor, who occupied the first chairs of geography in both Australia and Canada.

Australia and New Zealand

Like many British geographers before World War II, Griffith Taylor began his career as a geologist and explorer.[19] After serving on the Scott Antarctic Expedition from 1910 to 1913 and completing his work for the Sc.D. degree, he began to lecture on geology at Melbourne. But then in the period after World War I he found himself embroiled in a controversy over national policy. The Australians at this time were seeking immigrants from Britain, hoping to fill the vast empty spaces on their continent with new settlers. The more enthusiastic journalists and political leaders, looking at the large parts of Australia with less than two people per square mile, were convinced that modern engineering could provide the necessary water to make the dry lands productive. But Taylor's studies of the physical character of the continent told him otherwise. The potential sources of water were, in fact, much more limited than was popularly believed. Taylor estimated that instead of providing a home for more than 100 million people, the maximum population that could be supported with a high standard of living was perhaps less than 30 million. Taylor did not hesitate to speak out in the public press and from the university platform. His interest as a scholar was shifted from the study of glacial landforms to the meaning of the physical features for man, which he identified as geography. The obvious "relevance" of this kind of study to practical problems of policy led to the formation of the first department of geography at Sydney in 1920, and Taylor was selected as the first professor of geography. His stand on the problem of potential settlement, however, was very unpopular with the general public, and Taylor became the object of abuse. Some of his books were even banned (Taylor, 1926). Life in Australia became so unpleasant for him that in 1928 he accepted a position at the University

[19]Grifith Taylor was born near London in 1880. His father was a mining engineer who in 1892 moved his family to New South Wales. In 1904 Taylor graduated from the University of Sydney in geology and was employed by the Australian government as a physiographer. He surveyed the new federal territory and suggested the name that was later adopted—Canberra. In 1910 he went to Cambridge to start graduate work in geomorphology, but that same year he was selected from among a large number of applicants to serve as senior geologist on the Scott Antarctic Expedition. In 1916 he received the Sc.D. from Sydney for a thesis on Antarctic geology. His first academic appointment was at Melbourne to teach geology. He began to lecture on the relation of settlement to geology and climate in 1918. In 1920 he was appointed professor of geography at Sydney and chairman of Australia's first department of geography. He was at Sydney from 1920 to 1928; from 1928 to 1935 he was a member of the Department of Geography at the University of Chicago; and from 1935 until his retirement in 1951 he was professor of geography at the University of Toronto. He was president of the Association of American Geographers in 1941. During his professional career he was the author or coauthor of some forty-three books and a large number of professional articles (*Annals AAG*, 54 (1964):622–629; *Geographical Review*, 54 (1964):427–429).

of Chicago. In 1935 he once again became a pioneer when he was appointed to the first chair of geography in Canada, after which he was again involved in discussions of potential settlement of Canada's great northland.

Taylor became a vehement and eloquent spokesman for the concept of what he called "stop-and-go determinism." It may be, he said, that the well endowed parts of the world offer a number of different possibilities for making a living, but in some nine-tenths of the earth's land area nature speaks out clearly—"this land is too dry, or too cold, or too wet, or too rugged" (Taylor, 1951:11). Any settlers who fail to heed this nature-given limitation must face disaster. He insisted that he was not an "old-fashioned" determinist, such as Montesquieu or Buckle, who made extreme statements about the effect of climate on man but who never subjected their statements to scientific examination. Here is what he had to say, with considerable satisfaction, about the outcome of his arguments in Australia concerning potential settlement:

> The modern scientific determinist has an entirely different technique, and he knows his environment. Thirty years ago I predicted the future settlement-pattern in Australia. At Canberra [in 1948] it was very gratifying to be assured by the various members of the scientific research groups there, that my deductions (based purely on the environment) were completely justified. This aspect of geography is *Scientific Determinism* [Taylor, 1951:7].[20]

When Taylor retired from Toronto at the age of seventy, he returned to Sydney. In Australia he was welcomed as a national hero, for by that time the general public had discovered that he had been right.

Since World War II most of the universities of Australia have formed departments of geography offering graduate study. For nearly three decades the University of Sydney had the only department, but in 1947 a college of the University of Sydney was founded at Armidale in New South Wales—the University of New England. In 1954 this university became autonomous. Its Department of Geography, headed by G. J. Butland, had a staff of fifteen in 1966. A department was formed at Queensland in 1950, and O. H. K. Spate became the professor at the Australian National University at Canberra in 1951. Since then four more university departments offering graduate study have been organized. When these departments were first established, they were headed by geographers from Great Britain. Later, Australians were appointed to university posts after graduate study either in Great Britain or the United States. Now it is possible to work for advanced degrees in Australia.

The scholar who introduced geography into New Zealand was George Job-

---

[20]For discussions of the determinism-possibilism problem, see the comments by George Tatham of Canada (Tatham, 1951); O. H. K. Spate, a British geographer now in Australia (Spate, 1958); and Gordon R. Lewthwaite, a New Zealander now at California State University at Northridge (Lewthwaite, 1966).

berns.[21] Like Griffith Taylor he started as a geologist and later found himself more and more concerned about the relations of the physical features to the pattern of settlement. He described himself as "a geologist gone wrong." Although Jobberns introduced a distinctively British style of geography to New Zealand, he influenced his students to seek advanced degrees in the United States. He frequently appointed geographers from the United States as visiting professors at Canterbury, including Andrew H. Clark, who during his stay in New Zealand completed a study in the historical geography of South Island (Clark, 1949).

Departments of geography offering graduate study are now established in six New Zealand universities. At Auckland the first professor of geography was Kenneth B. Cumberland in 1949, who introduced to New Zealand the ideas of Carl O. Sauer and Richard Hartshorne, including Sauer's ideas concerning land-use surveys. At Victoria University in Wellington, Keith Buchanan became the first professor of geography in 1953. At these and the other New Zealand universities the geographical concepts and methods from Great Britain are still influential, but they are blended in a distinctive way with influences from the United States.

English-Speaking Canada

Before Griffith Taylor came to Toronto in 1935 there had been courses in geography offered in Canadian universities for many years. As early as 1915 physical geography was listed at the University of British Columbia, and in 1922 the Department of Geology and Geography was established there (Robinson, 1967:216–217). At the University of Toronto a course on commercial geography was offered in the Department of Political Economy; and by 1920 similar courses were offered at Queens, Western Ontario, and McMaster. The economic historian, Harold A. Innis, was appointed assistant professor of economic geography at Toronto in 1928.[22] He was a strong advocate of the importance of geography as a basis for the understanding of the limits and possibilities of settlement.

When the Department of Geography was established at Toronto in 1935, the courses were immediately popular. Taylor added to his staff Donald F. Putnam and George Tatham, and the three lectured to classes of 600 or more students each year (Robinson, 1967:217). During the 1940s and 1950s, as in French Canada, depart-

[21]George Jobberns received the M.A. from the University of Canterbury (Christchurch) in 1921 in geology. He was on the geology staff at Canterbury until 1936 when he completed the Sc.D. In 1937 he was named to head the first department of geography in New Zealand at that same university, and in 1942 he was appointed professor. He retired in 1960.

For information concerning the development of geography in New Zealand the author is indebted to Evelyn M. Stokes of the University of Waikato. She became editor of the *New Zealand Journal of Geography,* a professional periodical for teachers started in 1969.

[22]Harold A. Innis received the Ph.D. in economics at the University of Chicago in 1920. His studies of historical geography and economic history involved work with documentary sources and also much actual field observation. He contributed to Bowman's pioneer zone studies. Innis was one of those who helped to establish Canada's first chair of geography at the University of Toronto in 1935.

Gunnar Andersson

Eduard Brückner

Eugene de Cholnoky

Charles A. Cotton

Sten De Geer

Torsten Hägerstrand

Akinlawan L. Mabogunje

Giovanni Marinelli

Eugene Oberhummer

Eugeniusz Romer

T. Griffith Taylor

Paul Teleki

ments of geography were organized in all the major universities. It is no longer necessary for Canadians to leave Canada to complete graduate study.

In 1947 the Canadian government established the Geographic Branch in the Department of Mines and Technical Surveys, headed from 1949 to 1954 by J. Wreford Watson. The geographers employed there were responsible for preparing reports regarding the resources and physical character of the regions of Arctic Canada and of other land areas bordering the Arctic Ocean. They prepared maps of land quality and land use of certain critical regions. The Geographic Branch also carried out a systematic air survey of ice distribution in the Gulf of St. Lawrence and the Canadian Arctic Archipelago. The branch contributed maps of land use to the World Land Use Survey of the International Geographical Union. A national atlas of Canada was produced. In 1966, however, the Department of Mines and Technical Surveys was reorganized as the Department of Energy, Mines, and Resources. At the same time several new branches were set up to recommend and implement government policy with respect to the use of natural resources, especially water. It was decided that the geographers could be employed to greater advantage by reassigning them to these other specialized branches. In 1968 the Geographic Branch was discontinued and the geographers were transferred to the Geological Survey or to the Policy and Planning Branch, where they are grouped in a new Economic Geography Section in the Resources Research Center (Fraser, 1967).

India and Pakistan

The study of geography at the university level appeared late in India and Pakistan. In India the first college-level teacher was at Aligarh Muslim University in 1931, where I. R. Khan, who was trained at the University of London, offered courses primarily for teachers. A teacher-training course was set up at Madras in 1932 as a result of the efforts of N. Subrahmaniam. During the 1930s, when geography was required in all the high schools, the demand for teacher-training programs resulted in numerous one-year courses in many universities; but the staffing of these universities with scholars competent to offer post graduate work was delayed for many years. The first independent departments offering graduate work were at Aligarh in 1936, Calcutta in 1941 (with S. P. Chatterjee), and Madras in 1948 (with George Kuriyan). By 1965 there were twenty-two universities with separate departments offering graduate study.[23] In Pakistan the development of advanced study was similarly delayed, with departments at the University of the

[23]In 1968, when the Twenty-first International Geographical Congress was held in New Delhi, S. P. Chatterjee was president. He had prepared a report on geography in India in 1963, and in 1968 he brought this report up to date (Chatterjee, 1964, 1968). There is also a report on Geography in Indian Universities, published by the Universities Grants Commission (New Delhi, 1968). Chatterjee reported that the Indian geographers had concentrated their work on geomorphology, agricultural geography, soils, urban geography, and cartography. In 1956 a government agency, the National Atlas Organization, started work on the atlas of India.

Punjab (Lahore) in 1944 and at Dacca in East Pakistan in 1947. Since then, there has been a considerable expansion of geography in other universities of Pakistan.

Egypt

Modern geography also spread to Egypt from British sources.[24] To be sure, in the latter part of the eighteenth century a major survey of Egypt was carried out by the *Expédition Française,* but this was done in the classical tradition to gather information for a variety of purposes. In the second half of the nineteenth century Egypt became a major focus of interest among the British explorers and archeologists. British and Egyptian expeditions were sent out to map the sources of the Nile and to gather geological and hydrographic information concerning the Nile Valley. A widespread popular interest in exploration and archeology led to the foundation of the Egyptian Geographical Society in 1875. After the British occupation of Egypt in 1882, a Department of Surveys and Mines was set up to carry on the work of exploration and mapping. British and Egyptian scholars made detailed studies of the geology and physiography of the country and also of the flow of Nile water and the rates of evaporation. The patterns of land use were examined in relation to the resource base. An additional impetus was given to geographical studies when the International Geographical Congress was held in Cairo in 1925. One result of all these efforts was that before World War II Egypt was completely covered by published maps at various scales.

The first university department of geography was established at Cairo University in 1925, and thereafter more and more Egyptians were attracted to geography as a professional field. The well-known leaders of this first generation of Egyptian geographers were Mohammed Awad and Soliman Huzayyin, both of whom received their professional training in Britain.[25] Before World War II most of the students of Awad and Huzayyin went to British universities to complete their graduate work. These are the Egyptian geographers who were appointed to head up the new departments established at Alexandria University (formerly Farouk University) in 1942 and at Ain Shams University (formerly Ibrahim University) in Cairo in 1950. At Cairo there was also an Institute of African Studies.

Later generations of Egyptian geographers have received their advanced training in these Egyptian universities, where work to the Ph.D. degree is now offered.

[24]The authors are indebted to Professor Farouk M. El Gammal for information concerning the emergence of modern geography in Egypt.

[25]Mohammed Awad was introduced to British geography through field studies in the Department of Surveys and Mines. He studied for advanced degrees in Great Britain. When he died in 1967 he was generally recognized as the grand old man of Egyptian geography (Awad, 1954). Soliman Huzayyin graduated from Cairo University in 1929, and from 1930 to 1935 was carrying on advanced study in England, France, and Austria. In 1936–37 he made field studies in Egypt and Arabia. He has contributed numerous papers on Arab geography, and on the physical and human geography of Egypt and Arabia (Huzayyin, 1956).

Only a few continue to go abroad for graduate study, chiefly to Great Britain or the United States. The declining influence of Britain is reflected in a decrease of interest in the study of geographical problems in the field.

Africa and the West Indies

After World War II the British government adopted the policy of preparing the former colonies for independence as rapidly as possible. A plan was worked out whereby British universities, usually the University of London, would assist in the establishment of new universities overseas. The faculties were made up mostly of British scholars and the course offerings were patterned after those in Britain. Universities that had this special relationship with the University of London gave their graduates London degrees. Many new universities were founded in this way after the war; and in all of them there were departments of geography.[26]

Geography in these former British colonies is in a strong position in the schools. In popularity it rivals history, mathematics, and English. In the secondary schools most of the work deals with local geography and world regional geography, but there is an increasing attention to systematic studies. The teachers are expected to be qualified for their jobs by university-level training in the subject. The method and content of school geography were derived from the ideas brought from Britain by scholars who had graduated from British universities. During the 1970s some of the newer appointees to university positions began to introduce the use of quantitative techniques, following the pattern developed at places like Bristol. During the 1960s, also, there were a few African geographers who had gone to Great Britain for graduate study and then returned to take positions in the African universities.[27]

[26]University colleges were founded as follows: Makerere at Kampala in Uganda, 1946; University of Khartoum in Sudan, 1947; University of Ghana in (Gold Coast), 1948; University of Ibadan, Nigeria, 1949; University of Sierra Leone at Freetown, 1953; University of Nigeria at Nsukka, 1961; Ahmadu Bello University at Zaria (Northern Nigeria), 1962. In 1963 Makerere joined the Royal College at Nairobi in Kenya and the University College at Dar es Salaam in Tanzania to form the University of East Africa, but in 1970 this effort at federation was abandoned. In 1971 there was the University of Nairobi, Makerere University, and the University of Dar es Salaam.

Most of the graduate work offered in geography in the universities of South Africa was started in the 1950s. The three major universities where geographers can be trained are: University of South Africa at Pretoria; University of Witwatersrand at Johannesburg; and Rhodes University at Grahamstown. The one-time University College of Rhodesia at Salisbury retained ties with the University of London. There is a teacher-training college at Roma in Lesotho.

[27]We offer a selection to illustrate these trends. S. J. K. Baker had been a lecturer in geography at the University of Liverpool for sixteen years when he came to Makerere to start the new department there in 1946. When he retired in 1969 he was succeeded by B. W. Langlands, a London graduate. K. M. Barbour, who founded the department at the University of Khartoum in 1948, was an Oxford graduate. He became the head of the department of Ibadan in 1956. Barry N. Floyd, who started the department at the University of Nigeria in Nsukka, was a Cambridge graduate who earned his Ph.D. at Syracuse University and taught for two years at Dartmouth before going to Africa. In 1966 Floyd was named head of the department of geography at the University of the West Indies in Jamaica.

THE SOVIET INFLUENCE

The distinctive style of geography developed in the Soviet Union, with its clustering of specialized fields around a core of geography, its development of research programs in Institutes of Geography in the Academies of Science separate from the Institutes and Departments in the Universities, and its emphasis on practical contributions to the construction of a socialist society, has been exported to other countries of the communist world. In Eastern Europe the new geography was introduced mostly during the first two decades of the twentieth century from both Germany and France; but, whatever the first source, the geographical concepts and methods that prevail today have come from the Soviet influence. The story is similar in all the countries of Eastern Europe, but geography has had its greatest development in Poland.

Poland

In Poland, the grand old man who trained most of the Polish geographers of the post-World War I period was Eugeniusz Romer (Joerg, 1922:475–477). Romer had studied with Albrecht Penck in Vienna; and, when he was appointed professor of geography at the University of Lwów (now Lvov in the Soviet Union) in 1911, he was the first to offer courses at the university level in Poland. The expansion of geography in the Polish universities after World War I was carried out by Romer's students. Romer himself took issue with the popular notion that rivers should be used as boundaries between countries. Rivers, he said, are axes of settlement, not boundaries—which is an echo of the idea first set forth by Herodotus many thousands of years before. Romer was a strong Polish nationalist and pointed out that Poland was destined for leadership in this part of Europe owing to its strategic location. Romer retired in 1931.

After World War II Poland was physically and culturally rebuilt. Some of Romer's students were able to adjust to the new ways of living, but the new Soviet style of geography was soon dominant. In 1972 there were seven universities listed in *Orbis Geographicus* with institutes of geography or departments with associated chairs. In either case these geographical groupings included specialized subgroups covering such fields as physical geography, economic geography, meteorology and climatology, pedology, hydrography, agricultural geography, and regional geog-

---

African geographers are becoming more numerous. Three outstanding African scholars in geography are: Akinlawon L. Mabogunje—who received the Ph.D. from the University of London in 1961 after studying at Northwestern University and who then took a post at Ibadan in Nigeria—is a specialist on urban geography (Mabogunje, 1968) and editor of the *Nigerian Geographical Journal;* E. A. Boateng, who was trained at Oxford, went to the University College at Ghana and then became the principal of Cape Coast University; and Simeon Ominde, with a Ph.D. from London in 1963, became professor of geography and dean of the Faculty of Arts at Nairobi.

raphy. Many of the same scholars who hold positions in the universities are also in the Polish Academy of Science (Polska Akademia Nauk, Instytut Geografi), which was founded in 1953. In 1971 the chairman of the Institute of Geography at the University of Warsaw was also the chairman of the Institute in the Academy of Science—Professor Stanislaw M. Leszczycki. The chairman of the Department of Physical Geography, both at the University and the Academy, was Professor Jerzy Kondracki. Both Leszczycki and Kondracki completed their graduate studies in Poland during the 1930s. Altogether there are some 200 geographers in Poland.

As in the Soviet Union, there is an eager acceptance by the younger geographers of the new quantitative and theoretical procedures that originated in the United States and at Lund in Sweden. The new concepts of regional science are identified as setting a new paradigm for geographical study and highly relevant to the problems of economic development (Chojnicki, 1970).

Hungary

In Hungary, Count Paul Teleki was the leading geographer between the two world wars (Joerg, 1922:479). Trained as a political scientist, he became deeply interested in geographical field studies. After World War I he prepared maps of the "nationalities" of Hungary on the basis of which the new boundary lines were drawn at the Paris Peace Conference. He became the foreign minister of Hungary and three times was named prime minister. When he was unable to keep his country free from Nazi domination in 1941, he took his own life.

In 1971 there were institutes of geography at four universities and one Institute in the academy of sciences in Budapest. There were chairs of physical and economic geography and a separate chair of cartography at Budapest. At the Karl Marx University of Economic Science, also in Budapest, there was a chair of economic geography.

Czechoslovakia

Before World War I there were two universities in Prague, both offering advanced studies in geography. One was a German university staffed by geographers trained in Germany or at Vienna with Albrecht Penck. The other was Charles University (Czech), in which there were two chairs of geography (Joerg, 1922:477). One was occupied by V. Švambera, widely known for his studies of man and the land in the Congo. He had been appointed to his chair in 1902 when the first Czech geographer, Jan Palacký, retired. On the occasion of Švambera's seventieth birthday in 1936, the Czech geographers honored him with a symposium at which his many students presented the results of their geographical investigations. The second chair at Charles University was occupied by J. Daneš, a physical geographer who specialized in the study of karst landforms.

In 1972 *Orbis Geographicus* listed chairs of geography in four of the universities of Czechoslovakia. At the Charles University, where Palacký had first introduced geography in 1891, there was a geographical institute with chairs of cartography and physical geography and of economic and regional geography. There were similar clusters of geographers at Bratislava, Brno, and Olomouc (where the Palacký University was named after the first Czech geographer). There were also three Academies of Science: in Prague, Bratislava, and Brno.

Romania

In Romania, German and French influences introduced the new geography. At Bucharest students who worked with Ratzel began to offer anthropogeography in 1900; but at Cluj, physical geography was stressed by students of de Martonne, starting in 1918.

In 1972 Romanian geography still shows evidence of its German and French sources. At the University of Bucharest geography is in the Faculty of Geology and Geography and also in the Institute of Economic Sciences. At Jassy geography is in the Faculty of Natural Science and Geography.

Bulgaria

Geography was started at the University of Sofia in 1898 by a student of Ratzel. At this university in 1966 there was a Faculty of Geology and Geography with professors in the usual range of specialized fields.

Yugoslavia

Geography was introduced into what is now Yugoslavia by another grand old man, Jovan Cvijić, who was appointed professor of geography at the University of Belgrade in Serbia in 1893 (Joerg, 1922:482). Cvijić was influenced by both Ratzel and Brunhes, but he felt that both of these masters had emphasized forms of settlement rather than the people involved. Like de Martonne, he was primarily a physical geographer; and his major contribution in this field was the understanding of the processes whereby limestone is dissolved to form karst features (Cvijić, 1918a; Sanders, 1921). But he was also a careful observer of the inhabitants of the Balkans. Every year between 1888 and 1915 he spent at least a month on walking trips. He made his way to the most remote and isolated parts of the Balkans, observing both physical features and people and recording his observations with great care. He prepared maps showing the areas occupied by the various ethnic groups in this rugged region and their languages, religions, customs, and attitudes toward political rule from outside. In 1918 he published a regional monograph on the Balkans that was of the greatest practical importance in outlining the boundaries of the newly created country that later was called Yugoslavia (Cvijic, 1918b). Although there was a great difference in traditions and attitudes between the Serbs and the Croatians, he recognized that if a new state was to survive it must have an outlet to the sea. Therefore, he worked tirelessly to convince both Serbs and Croatians that they must learn to live together. His regional monograph on the Balkans is an example of one of the most successful studies carried out anywhere in *la tradition vidalienne* (Freeman, 1967:72–100).

There are now five institutes or departments in Yugoslavian universities in which advanced studies in geography are offered. At each of these centers the curriculum includes the usual variety of subspecialities, ranging across both physical and economic geography.

German Democratic Republic

The separation of Germany into two parts after World War II offers an opportunity to watch the interplay of German and Soviet geographical ideas. The present leading scholars in East Germany were trained in the traditional geography as developed in Germany between the two world wars. They took part in the methodological discussions that characterized this period. Most of them have rejected the efforts of Hettner to define geography as a unitary field on the basis of the chorological concept. Many of them were closely associated with Otto Schlüter at Halle and have gone forward with the study of landscapes. But they have also rejected Schlüter's historical approach to landscape interpretation, convinced that studies of physical processes do not belong in the same discipline as studies of social or economic processes. Physical geography and economic geography, they insist, are separate disciplines (Kazakova, 1966:43–44). The Soviet geographers, as we have seen, were for a time sharply divided on this issue; but the matter has been officially resolved in favor of the traditional Russian view of geography that embraces a variety of subspecialties, ranging across the physical and economic segments. At present in the twelve universities of East Germany there are institutes of geography; each includes the usual range of special fields found throughout the Soviet world.

The East German geographers have made important contributions to the methods of identifying and interpreting landscapes. In 1955 a government commission on regions (*Regionalkommission*) published a monograph on the landscapes of East Germany based on their natural conditions (Schultze, 1955), including a map of the regions on a scale of 1/1,000,000. Ernst Neef has edited a collection of studies of the physical geography of the earth by the Leipzig geographers (Neef, 1956, 1967). Landscape science is a rapidly developing field in which the Soviet geographers and the geographers of East Germany work closely together.[28]

---

[28]Two of the professional periodicals that have long been published in Germany are *Petermanns Geographische Mitteilungen* (still published at Gotha) and the former *Zeitschrift der Gesellshaft für Erdkunde zu Berlin,* now renamed *Die Erde* (edited by J. H. Schultze and published in Berlin).

GEOGRAPHY IN CHINA

China is remote from the sources of the new geography that originated in Germany. But, as we have seen, China had developed indigenous geographical ideas at the time when the Greeks were developing the basic concepts of the western world. The writing of local historical-geographical gazetteers, which is an ancient Chinese tradition, was continued during the nineteenth and twentieth centuries and forms a part of geographic work in the contemporary period.

China's first steps leading toward the formation of a group of professional geographers were taken in 1912, after the revolution that ended the Manchu dynasty. To be sure there had been European scholars carrying on field studies in China long before the date. Ferdinand von Richthofen had been on the Prussian expedition to map the resources of China in 1860; and after working in California, he had returned to study China's physical characteristics (p. 166).

British and American Influence in China

The introduction of geography into the Chinese universities may be credited to two Chinese scholars trained abroad. One was Wen-kiang Ting, a geologist trained in Scotland; the other was Co-ching Chu, a meteorologist and climatologist trained at Harvard.[29]

Wen-kiang Ting brought back from Britain the idea of a government agency devoted to the field survey of resources. The training he received in Edinburgh, Cambridge, and Glasgow gave him a broad approach to geology and zoology, including the study of the human use of physical and biotic resources. Like the British geologists, he accepted the idea that the geography of settlement and population was only the last chapter of the geological record. In fact, during the years 1913–14 when he was doing field work in Yünnan, he wrote reports not only on the tin ores of that province but also on the distribution of racial minorities in relation to the physical features. Ting trained many young Chinese scholars at Peking to work

[29]Wen-kiang Ting studied at Edinburgh, Cambridge, and Glasgow between 1904 and 1911. In 1912 he was appointed chief of the geology branch in the Ministry of Industry and Commerce in Peking and established a training institute to prepare young Chinese scholars for employment in field studies of resources. The Institute became the Department of Geology at the Peking University in 1916. From 1931 to 1934 Ting was professor of geology at Peking, and from 1934 to his death in 1936 he was secretary-general of the Academia Sinica in Peking (predecessor of the Academy of Sciences).

Co-ching Chu received his Ph.D. in meteorology and climatology at Harvard under Rober DeC. Ward. In 1922 he became professor of geography at the National Southeastern University (now National Central University) in Nanking. He was the first director of the Institute of Meteorology in the Academia Sinica. In 1936 he became president of the National Chekiang University at Hangchow and established the Department of History and Geography there. In the communist period Chu became the vice-president of the Academia Sinica in Peking. His students have carried on the studies of physical geography in the contemporary period.

Information supplied by Chiao-min Hsieh.

as geologists and physical geographers. In the 1920s and 1930s many of this younger generation were sent to British universities for part of their graduate study.

Co-Ching Chu introduced ideas from the United States. From Robert DeC. Ward at Harvard he absorbed the ideas of descriptive climatology that Ward had received from Julius von Hann; he also received from Ward a concern about the effects of climatic conditions on man. When he became professor of geography at Nanking in 1922, he aroused among his students an awareness of the meaning of the great climatic contrasts within China with regard to the process of settlement. Stimulated by the concepts of climatic cycles developed by Ellsworth Huntington, he prepared a synopsis of the cycles of floods and droughts in China based on the very long records kept by the Chinese (Chu, 1926), however, modern climatic data were largely lacking. When Chu became the director of the Institute of Meteorology at Peking, he started the monumental task of organizing a network of properly equipped weather-reporting stations to cover the vast area of China. The students trained by Chu (either in China or in foreign countries) together with Ting's students became the physical geographers of the contemporary period.[30]

Geography in Communist China

In 1949 mainland China was taken over by the communists and became the People's Republic of China. This meant that the study of geography was reformed to fit the Soviet model. Geography became the core subject around which a variety of specialities were grouped; but geography itself was described as physical geography or economic geography (Wiens, 1961). The physical geographers of pre-communist China could adjust their work without too much strain to the requirements established by the Communist Party, but economic geographers either had to accept a wholly new set of premises or escape from the country. Many of the physical geographers of today were the students of Ting or Chu; but today's economic geographers come almost entirely from the People's University in Peking.

There have been four fundamental requirements for continued work as a geog-

[30]Among the many British and American geographers who also worked in China, three deserve special mention here. One was the British geographer, Percy M. Roxby of Liverpool, who began his studies in China in 1912-13 and thereafter devoted his career to work with the Chinese, both in China and in Britain. Robert S. Platt taught geology in the Yale Collegiate School in Changsha in 1914-15. When he led field parties into the countryside, he found himself more and more concerned with the interrelations between the underlying geology and the patterns of settlement and land use. This led him later to become a geographer. George B. Cressey taught geology at Shanghai between 1923 and 1929, and during this time he carried out field studies throughout China. The main theme of his book on China had to do with people using resources (Cressey, 1934). In a later book he described the results of population pressure on a limited land base (Cressey, 1955). Cressey earned a Ph.D. in geology at Chicago in 1923 and in geography at Clark University (Mass.) in 1931. In 1931 he became chairman of the Department of Geology and Geography at Syracuse University and in 1945 was named chairman of the separate Department of Geography. From 1949 to 1952 he was president of the International Geographical Union.

rapher. First, research projects must be relevant to problems of building a productive socialist society, mere intellectual curiosity is ruled out. This does not mean that innovative and theoretical studies are not permitted, but only that the ultimate justification for the continued support of a scholar as a geographer depends on the extent to which his work contributes to the communist objectives. Second, every scholar must accept the Marxist-Leninist doctrine. Third, in China individual scholarship becomes increasingly difficult. Instead, groups of scholars are formed to work together. For most published monographs or professional articles numerous coauthors are listed. And fourth, evidence of "right thinking" consists of denouncing the "bourgeois" geographers. This thinking was greatly modified in the 1970's.

The attacks on the geographical ideas of noncommunist geographers were usually based on one of two basic themes. One is that geography in Germany, France, Great Britain, and especially the United States is an instrument of imperialist aggression. Wiens points out that the Chinese geographers identify geography as a fundamental science "because every kind of economic construction requires fundamental knowledge in geography." Geography is given a prominent place in the twelve-year plan for the development of science and technology. Even Chu is quoted as saying that "in an imperialist society, geography is used as a tool for aggression, hence it could not be termed 'fundamental science'" (Wiens, 1961:414). Cressey was vigorously attacked for his suggestion that China's population is too large (Hsieh, 1959:542). Beneath the party line attacks on Cressey's motives (that Americans want to decimate China's population because they fear the power of so many people), there is the logic of one of the basic elements of communist doctrine. As Mao Tse-tung said: "China's big population is a very good thing." Wiens describes the basic difference in attitude as follows:

> The crux of this theoretical battle lies in the Communist contention that the population problem lies not in any limits set by the natural resources and environment, but in the retarded technological development which results in low productivity. The power of productivity lies in people, and thus, the more people, the better, provided, of course, that this latent productivity is released through the wisdom and guidance of the Communist Party. This socio-political dictum, therefore, is another of the difficult premises under which the economic geographer is required to labor in Communist China. Should he question it, he would be faced with a barrage of abuse, ridicule, and condemnation, and with demands for thought reform until he would be driven to a humiliating recantation and repeated confessions of conversion to orthodoxy [Wiens, 1961:416].

The policy of giving new emphasis to geography as a professional field has introduced some important changes. Although Chu was able to set up only about a hundred weather stations in pre-Communist China, by 1959 there were 2400 well-equipped stations plus some 30,000 cooperative observers. A major study of the climate of Tibet has revealed that this highland area is not the cause of a split in the jet stream, for similar splits are observed in similar latitudes where there are no highlands. All these and other study projects require the services of a vastly increased manpower. In 1956 it was estimated that China had between 90 and 120 professors of geography and some 6700 students in 23 cities (Wiens, 1961:421). Field studies in selected problem areas involve a large number of individuals; for example a study of the Ordos Desert in 1959 was done by some 3000 people, including many geographers.

The idea of a resource inventory was not new to China. But it was given new impetus when, after the communist takeover, several of the physical geographers who had studied in Britain reported on L. Dudley Stamp's survey of land and land use. Stamp made use of volunteer secondary-school pupils under professional supervision. In China a similar inventory was started in the 1950s, to be done by vast numbers of individuals who were assigned to the work. In 1958 soil surveys in Szechuan involved 140,000 people, and in Kwangtung more than 170,000 (Wiens, 1961:428).

In addition to these surveys of relatively large areas, the Chinese geographers have also carried out detailed topical studies of small, critical areas. There has been much attention to the identification of microclimates, where measurements of heat and water balance are correlated with crop yields. In mountainous areas of the south many enclaves of tropical microclimates have been discovered in which such tropical crops as coffee, cacao, and rubber can be produced (Kikolski, 1964). Defining such microclimates is not enough: the means of changing them for increased productivity are also sought. The river basins have been studied in detail, and geographers have been at work on preliminary studies of a huge new dam in the Yangtze gorge. A project was surveyed to bring water from the Yangtze system to the Yellow River basin.

All these group research projects are coordinated, implemented, and supervised by the Institute of Geography of the Academy of Sciences in Peking. According to a report by the Polish geographer, Stanislaw M. Leszcycki, in 1958 there were some 500 workers at the institute, where there are well-equipped laboratory facilities and an excellent library (Leszczycki, 1963). There were departments of geography organized on the Soviet model in seven universities, in seven teacher-training universities, and in eighteen pedagogic institutes—all of which are involved in the kinds of research projects described above as well as many others. Out of all this activity has come a greatly increased volume of professional publication.[31]

GEOGRAPHY IN JAPAN

During the two and a half centuries when Japan was sealed off from the outside world, interest in the study of geography was effectively stifled.[32] But after the

[31] See the report on studies of the physical geography of China in *The Physical Geography of China*, Institute of Geography, U.S.S.R. Academy of Sciences, translated by U.S. Government Joint Publications Research Service (New York: Praeger, 1969).

[32] In the preparation of this section the authors are indebted to Professor Akira Watanabe of Ochanomizu University in Tokyo.

Meiji Restoration in 1868 there was a sudden and widespread concern with world geography.[33] Interest in geography, not only among scholars but also among the educated people throughout Japan, increased steadily. Since World War II geography has been made a compulsory subject in elementary schools and a large number of teacher-training institutions have been set up to provide qualified teachers. In the universities and in government agencies professional geographers have found many opportunities for employment, especially in various aspects of economic planning.

When a demand for the services of trained geographers began, there were no geographers and no backlog of geographical writings. In 1869, however, a compilation of information concerning the countries of the world was published and gained immediate popularity. The writer was a Japanese scholar named Yukichi Fukuzawa, and his book on world geography was very much in the tradition of the "universal geographies" of the western world. Its impact on the Japanese was enormous, bringing as it did the first glimpse of what things were like in the world outside. In a decade, interest in geography was sufficient to support the organization of the Tokyo Geographical Society (1879).

Geography as a professional field in the universities owes its first development to geologists, historians, and agricultural specialists who became interested in the study of the interrelations between human settlement and the physical earth. The geologist who led the way was Bunjiro Koto, the founder of the Department of Geology at the Imperial University of Tokyo. Koto had been a student at Tokyo; after completing an undergraduate course in geology he had been sent at government expense to study geology in Germany—at Munich and Leipzig. After his return to Japan in 1890 he began to offer courses in geography in the new Department of Geology at Tokyo. The first scholars to be appointed as professors of geography in Japan were Koto's students, who had taken geology at Tokyo and then had studied geography in Germany.

The first professor of geography was appointed in 1907 in the Institute of History at the University of Kyoto. This was Takuji Ogawa,[34] whose geographical ideas were largely derived from the teaching of Eduard Suess at Vienna. Appointed with him as an assistant professor was a young historian named Goro Ishibashi. Ogawa and Ishibashi pioneered in the study and teaching of historical and regional

---

[33]From 1637 until 1853 the only outside contacts that were permitted were made through a Dutch trading post on an island in the harbor of Nagasaki. In 1853 Commodore Matthew Perry of the U.S. Navy sailed into Yokohama harbor and thereafter outside contacts became more and more frequent. With the Meiji Restoration of 1868, which returned the emperor to the throne, a policy of rapid modernization was instituted.

[34]Takuji Ogawa graduated in geology from the University of Tokyo in 1896. He was then sent to Vienna for graduate study where he worked with Eduard Suess. Ogawa was a very broadly trained scholar. Not only was he thoroughly qualified in geology and physical geography but he also had a profound knowledge of the Chinese classics and was fluent in several Western languages.

geography. Ogawa's own writings included a regional study of the Kinki District (where Osaka, Kyoto, and Kobe are located) and several works on the historical geography of China. He never lost his concern with the physical earth, and, when an Institute of Geology was established at Kyoto, he was appointed to it, leaving Ishibashi to head the work in geography.

Another of Koto's students was Naomasa Yamasaki, who played a major role not only in providing trained geography teachers for the secondary schools but also in the development of geography as a professional field requiring graduate training in the universities.[35] Yamasaki had derived his geographical ideas from Albrecht Penck in Vienna, and the record of his studies and his teaching reflects the wide range of interests of his teacher. He investigated the glacial landforms of the high mountains of Japan and the landforms produced along fault lines. But he also made important contributions to regional geography and studies of settlement. In 1925 he was the founder of the Association of Japanese Geographers and was the only geographer at that time in the Imperial Academy of Science.

The next generation of Japanese geographers, who rose to prominence in the 1930s and 1940s, were trained as geographers from the start in Japanese universities. Most of them came from the Higher Normal School in Tokyo, the Higher Normal School of Hiroshima, or from the departments of geography at the University of Tokyo or the University of Kyoto. Some of this younger generation were given the opportunity to do graduate work in Germany, France, or the United States. But the writings of the leaders in the western world were made available through excellent Japanese translations. Students in Japan were thoroughly familiar with the ideas of Hettner, Schlüter, and Ratzel in Germany; Reclus, Vidal, Brunhes, and Demangeon in France; and Davis, Semple, and Huntington in the United States. In the research studies published before World War II there was a strong emphasis on physical geography (Shimomura, 1926-27; Tanaka, 1927; Seki, 1930; Fukui, 1933), but studies of regional and historical geography were not neglected.[36]

[35]Naomasa Yamasaki graduated in geology from the University of Tokyo in 1895, one year earlier than Ogawa. After teaching in the undergraduate program at Sendai (Miyagi prefecture) for three years he was sent to study in Germany and Austria. During the years 1898 to 1901 he studied with Rein at Bonn and with Albrecht Penck at Vienna. In 1902 when he returned to Japan he was appointed to teach at the Higher Normal School in Tokyo (later Bunrika University, now Kyoiku University). Here he was a major factor in supplying well-trained teachers for the Japanese secondary schools. He continued to teach at the Higher Normal School even after he was appointed to teach economic geography in the Faculty of Law at the University of Tokyo in 1908. He became the first professor of geography at Tokyo in 1911 when he was appointed to this chair in the Institute of Geology. He was transferred to the new separate Department of Geography when it was created in 1919.

[36]Prominent geographers of the generation following Ogawa and Yamasaki include: Togo Yoshida, a student of historical geography and a specialist in Japanese place-names; Shigetaka Shiga, whose interest in forestry and land use led him to a career in politics; Taro Tsujimura, who introduced the Davis system of geomorphology into Tokyo; Keiji Tanaka, who taught regional geography at Bunrika University until his retirement about 1950; Saneshige Komaki, a political geographer and archeologist, who was president of Shiga University from 1959 to 1965; Ryuziro Isida, one of the leading economic geographers;

After World War II there was another period of rapid expansion in Japanese geography. In 1945 the Geographical Survey Bureau (in 1970 the Geographical Survey Institute) was established to carry on a varied program of mapping at different scales. The bureau not only produced topographic maps but also many thematic maps of such things as land classification and land use, transportation, population, and various kinds of economic data. The application of geographical methods to the study of practical problems of urban and regional planning was a notable development—notable because before the war very few professional geographers were in any way involved in planning operations. But in the contemporary period geographers are employed in almost every city and prefectural planning commission and in the planning work of the larger metropolitan areas. There is also a private agency—the Japan Center for Area Development Research—that undertakes studies relevant to the policy problems of Japan as a whole.

In addition to such applications of geography to practical problems, there is also an increasing volume of research on various aspects of regional geography. There is a study of Japan's agricultural regions (Birukawa, 1950) and another of Japanese land use (Ogasawara, 1950). Akira Watanabe has published a monograph on the regional divisions of Japan in which he summarizes the various regional schemes proposed at different times by Japanese geographers and attempts a synthesis (Watanabe, 1970).

There can be no doubt that among all the various parts of the world to which the new geography, in one form or another, has been spreading, Japan is among the most active in the production of scholarly works and in effectively improving the quality of geography teaching.[37]

---

and Fumio Tada, who was a professor at Tokyo from 1926 to 1961. Akira Watanabe, who supplied much of this information and who taught at Ochanomizu University from 1958 to 1970, was a leader in the organization of the Geographical Survey Bureau.

Among the American geographers who specialized in the study of Japan and spent many years in that country is Robert B. Hall of the University of Michigan; most recently, he was the representative of the Asia Foundation. His study of the road connecting Yedo (Tokyo) and Kyoto and of the region that it ties together remains a masterpiece of historical geography (Hall, 1937).

[37]See the summary by John D. Eyre in the *Geographical Review*, 45 (1955):120–121.

13

The New Geography in the United States Before World War I

Reputations for originality are more often made by giving names to ideas already coming into circulation than by inventing new ideas. In this way the fruits of many men's thinking are appropriated by the more lucid and articulate among them.

When the new geography arrived in North America about a decade after its appearance in Germany, there was already a long record of interest in geographical studies and in the teaching of geography in schools and colleges, (Warntz, forthcoming; Aay, forthcoming). As in other parts of the world, work that can be identified as geographical in nature was contributed by scholars (e.g., Benjamin Franklin and Thomas Jefferson) for whom geographical studies constituted only one of many intellectual interests. There were pioneers, such as George Perkins Marsh and Matthew Fontaine Maury, who brought new understanding to the study of the earth as the home of man. European geographical ideas were brought to America by such scholars as Louis Agassiz at Harvard and Arnold Guyot at Princeton. Guyot's pupil, William Libbey, Jr., who succeeded Guyot as professor of physical geography, carried on the tradition of individual scholarship in his studies of oceanography, especially of the relationship between the Gulf Stream and the Labrador Current. During the nineteenth century, too, major advances in thematic mapping were introduced from Europe through the efforts of such men as Lorin Blodget, Joseph C. G. Kennedy, Daniel Coit Gilman, and Francis A. Walker.

The quotation above is from John P. Plamanatz, "Rousseau," *Encyclopaedia Britannica,* 1969 ed., s.v.

An important part of the background for the development of the new geography in America was the tradition of the field survey and the resulting emphasis on induction from observations rather than deduction from theory. In the 1880s the Great Surveys of the West had just been combined in the U. S. Geological Survey. [For geographers at work in the federal government, see Friis (forthcoming); for Amerindian antecendents, see Lewis (forthcoming).] The men who worked on these surveys had not received previous training in the concepts and methods of geography; and they had to find their own answers to the five questions listed at the beginning of Chapter 7 (what to observe, how to observe, how to generalize, how to explain, and how to communicate). Since these fieldmen had not been indoctrinated with Lyell's ideas about marine planation or Werner's ideas about the origin of the earth, they were able to observe landforms and the processes that produced them without strong preconceptions. There was a practical motivation for their work. They distrusted theory and the findings of scholars that had been deduced from theory. Grove Karl Gilbert wrote:

> In the testing of hypotheses lies the prime difference between the investigator and theorist. The one seeks diligently for the facts which may overthrow his tentative theory, the other closes his eyes to these and searches only for those what will sustain it [Gilbert, 1886].

By 1880 the stage was set for the appearance of what we have called the new geography. As we have seen in Chapters 7 and 8, the formation of a professional field requires the existence of clusters of scholars working closely with graduate students in universities (Koelsch, forthcoming). The concept of the university as a community of scholars first appeared in America in 1876 when Daniel Coit Gilman became president of the newly founded Johns Hopkins University; but thereafter the idea spread rapidly to other established universities. For the first time faculties qualified by advanced training and continued activity in research were selected to guide the training of younger generations. For the first time a professional group could lead and direct scholarly performance in each discipline, free from external interference.

Geographers began to participate in this new kind of university in Germany in 1874, and thereafter the innovation spread around the world. The pioneer who introduced the new geography in America was the geologist William Morris Davis, who had been appointed instructor of physical geography in the geology department at Harvard in 1878 (Beckinsale, forthcoming). Davis set the early paradigm for geographical study and helped found some of the professional institutions. At first geography was usually associated with geology, and soon departments offering six geography courses emerged at Columbia University (1899); Cornell University (1902); University of California (1903); University of Chicago (1903); University of Nebraska (1905); Miami University, Ohio (1906); University of Minnesota (1910);

University of Pittsburgh (1910); Nebraska Wesleyan University (1911); University of Wisconsin (1911); Harvard University (1911); University of Pennsylvania (1913); New York University (1913); Yale University (1914); and Denison University (1914).

Of these universities, Harvard, Yale, Pennsylvania, and Chicago were perhaps the major sources of the ideas that were involved in the scholarly competitive discussion from 1904 to 1914. At Harvard William Morris Davis was developing physical geography, though finding a place for man in his construct. At Yale was H. E. Gregory (one of Davis's own pupils) who developed a fine departmental offering in human geography—notably owing to E. Huntington and I. Bowman. At the University of Pennsylvania E. R. Johnson and J. R. Smith developed economic and commercial geography. At the University of Chicago in 1903 the first geography department in the United States was founded, offering advanced study to the doctorate.[1] The Chicago program included strengths in physical, human, and economic geography. (For sketches of other individual geographers, see Dunbar 1978; Koelsch 1979a, 1979b; Sherwood 1977.)

WILLIAM MORRIS DAVIS

William Morris Davis[2] spent his formative years as a scholar at Harvard with Nathaniel Southgate Shaler. He learned three distinctive habits of thought from Shaler. First, he developed the habit of careful field observation and made use of it in logical and impersonal argument. Second, he acquired the habit of seeing man and his works as part of the landscape, not separate from it. And third, he gained a clear appreciation of the importance of processes of change in explaining the varied

[1] Of the Ph.D. degrees listed by Whittlesey in 1935 (Whittlesey, 1935), there were nineteen granted before 1916: five at the University of Pennsylvania (in economics); five at Chicago (starting in 1907); three at Johns Hopkins (in meteorology and geomorphology); two at Cornell; two at Yale; and two at Harvard.

[2] William Morris Davis was born of Quaker parents in Philadelphia in 1850. He graduated from Harvard in 1869 and a year later received the degree of Master of Engineering. From 1870 to 1873 Davis worked as an assistant at the Argentine Meteorological Observatory in Cordoba, Argentina. Returning to Harvard for further study in geology and physical geography, he was appointed assistant to N. S. Shaler in 1876 and was promoted to instructor in physical geography in 1878. In 1885 he was appointed assistant professor of physical geography and later promoted to professor. In 1899 he was named the Sturgis Hooper Professor of Geology at Harvard, which chair he continued to occupy until his retirement in 1912. In 1909 he was visiting professor at the University of Berlin and from 1911 to 1912 at the Sorbonne. After his retirement from Harvard he held temporary appointments at the universities of Oregon, California, Arizona, Stanford University, and the California Institute of Technology. He was one of the founders of the Association of American Geographers in 1904 and was its president three times—in 1904, 1905, and 1909. He was also president of the Geological Society of America and the Harvard Travelers Club. He never received the Ph.D. degree, but he was honorary doctor of many universities. He was a Chevalier de la Légion d'Honneur and received medals from many geographical societies of the world (Bryan, 1935).

features associated on the face of the earth. How were these habits of thought transmitted from Shaler to his young assistant?

The habit of careful observation and logical argument was derived, it seems, from working with Shaler in the field. Davis found out from personal experience that just going out to see "what an area is like" is less productive of useful results than going out to find the answers to questions. This was just the time when a major question among students of the earth had to do with the origin of the sand and gravel deposits that are so widespread in New England. Were they laid down during the flood, as described in the Bible—which was the traditional explanation—or were they the results of the melting of the great ice sheets, as Agassiz insisted? Shaler had given new support for the Agassiz hypothesis as a result of his studies of the glacial history of Cape Cod. Davis, working with Shaler, looked for evidence for or against the glacial hypothesis; and he learned to marshal his observations in terms of the questions he was asking. The results of field study had to be put together in logical form and communicated in scholarly—that is, impersonal—argument.

Shaler also transmitted to Davis a vision of the earth as the resource base on which the human inhabitants were dependent. Shaler was the first scholar since George Perkins Marsh to point out how man's activities changed the earth, especially through the depletion of unrenewable resources. Shaler has been described as a geologist by training but a geographer by instinct in that he was always conscious of studying the earth as the home of man.

And the young Davis learned to think in terms of evolutionary change as a basis for scientific inquiry in the best possible way—by participating in the replacement of one hypothesis by a new one. Louis Agassiz was a persuasive and influential lecturer with a large following among the educated people of New England, who supported his work with substantial endowments. But at the same time there was a rising tide of opposition to the notion that organisms, once created, remained without change. Asa Gray (1810–1888), a botanist at Harvard, was amassing carefully evaluated evidence in support of the concept of organic evolution, which to the lay public was unpopular. Shaler, who had learned much from Agassiz about the observation of landforms, nevertheless joined those who supported the doctrine of evolution. Davis took part in the discussions in that exciting period when a widely accepted explanation was being rejected on the basis of careful scientific procedures.

Davis did not absorb and digest these ideas all at once. When he became Shaler's assistant in charge of field work in 1876 his teaching was not stimulating. His largely empirical approach to landforms left his students struggling with unordered detail (Davis and Daly, 1930:314–315). As a result, when the time came for his reappointment as an instructor in 1882 he received the following letter from President Charles W. Eliot, dated June 1st:

> The Corporation offer you a reappointment as instructor in geology at a salary of $1200 a year. . . . The Corporation are quite aware that this position is not suitable for you as a permanency; but it is all that they are able to offer you now, with their present re-

sources, and all that they expect to be able to afford for some time to come. In considering whether it is your interest to accept this offer temporarily, I hope that you will look in the face of the fact, that the chances of advancement for you are by no means good, although the Corporation have every reason to be satisfied with your work as a teacher.[3]

At this point another member of the Harvard faculty in geology, Raphael Pumpelly, who knew Davis as a student, offered him a position on the survey he was then conducting in Montana to identify the resources along the route of the Northern Pacific Railroad. Davis was assigned the task of describing the Montana coal measures. While this work was in progress, he began to visualize the outlines of a "cycle of erosion." He noted the existence of certain terraces above the Missouri River, which he interpreted as the result of the removal of an "unknown thickness of overlying strata," and the reduction of an earlier surface close to the baselevel of the drainage. The baselevel concept was derived from Powell, and other insights into the process of river erosion came from Gilbert and Dutton. But Davis, observing the landforms of Montana, began to formulate a theoretical model that would describe all such processes and surfaces (Chorley, Dunn, and Beckinsale, 1964:622). The concept of the cycle of erosion, first announced in 1884, was presented in revised form in 1899 (Davis, 1899a). But it is important to see what this did to Davis's teaching: the observed details of landforms in particular places, which had failed to interest his students a few years earlier, could now be related to a generalized model of landform development anywhere. In 1885, only three years after President Eliot's letter, Davis was appointed assistant professor of physical geography at Harvard.

There were two sides to Davis's work that can be examined separately, although of course he carried them on simultaneously. First, we may consider his contributions to geomorphology, especially his model of the cycle of erosion; second, we may look at his promotion of geography as a field of study in the schools, colleges, and graduate school.

Contributions to Geomorphology

Among the many contributions Davis made to geology and geomorphology, the one that was central to all the others was the concept of the cycle of erosion, which he called the "geographical cycle" (Davis, 1899a).[4] This was in the nature of a model, an ideal sequence of landforms that would take place during the erosion by running water of an upraised portion of the earth's crust. In his model Davis postulated that after the upheaval no further up-or-down movements would take

---

[3]Quoted from Chorley, Dunn, and Beckinsale, 1964:623. The letter from Eliot was copied by Davis and kept in his personal files. See also Davis and Daly, 1930:314–315.

[4]Twenty-six of Davis's papers were reprinted in 1909 in *Geographical Essays,* edited by D. W. Johnson (Davis, 1909).

place and that during the resulting cycle there would be no essential change of climate. When an initial surface is raised, rivers at once begin the work of erosion. The surface is cut by narrow V-shaped valleys that are extended headward as more and more of the initial surface is consumed. But rivers cannot cut down their valleys indefinitely. As Powell has pointed out, there is a baselevel below which rivers cannot cut—a level determined by the surface of the body of water into which a stream flows. Furthermore, before a valley is cut down all the way to baselevel, the river establishes a slope that is enough to permit the river to continue to flow. This is known as *grade*—an idea originally presented by Gilbert. Grade is an equilibrium between slope, volume of water, and the amount of load being carried. After the valley bottoms reach this graded condition, the rivers begin to widen their valleys and the high country between the valleys is gradually reduced.

Importantly, Davis provided a terminology. When the initial surface is still undissected between the valleys, when the valleys are V-shaped, and when the rivers descend through them turbulently—this is a stage described as *youth*. The greatest amount of relief occurs when the last remnant of the initial surface is dissected. Then the surface is gradually reduced and the valleys begin to widen. This is the stage Davis described as *maturity*. When the rivers meander across wide valleys and the land between the valleys has been reduced to gently rounded slopes—this he called *old age*. The upheaved block of the earth's crust is worn down almost to a level plain, which Davis called a *peneplain*. The whole cycle, Davis pointed out, could start again with another uplift, resulting in *rejuvenation*.

Davis offered a formula for the description of landforms. These could be understood in terms of the interaction of three factors; *structure*, or the character and position of the underlying rock; *process*, or the combination of agents of erosion, such as running water, soil creep, underground solution, or ice; and *stage*, or the point in the sequence of landforms that had been reached at a particular time.

Davis insisted that this ideal sequence of landforms was not to be considered as rigid but rather that it could provide a theoretical framework in reference to which actually observed landforms could be described. This is what he called the "explanatory description of landforms." He understood clearly that there would be an infinite variety of disturbances to the ideal sequence because almost every region observed in specific detail would constitute a special case. This is exemplification of the question of whether to describe the features of a particular place as unique or of describing them in terms of regularities or similarities among many places. Davis himself developed sequences for a variety of special conditions: for surfaces cutting across the upturned edges of tilted rock strata; for blocks bordered by fault scarps; for places where the climate was arid. He pointed out that only rarely, if ever, would uplift take place quickly and be followed by no further movements, as postulated in his model. He showed how the theoretical sequence would be modified if further uplift and rejuvenation should take place at any stage in the cycle, and he provided innumerable examples of how the ideal cycle had been changed in specific circumstances. He also applied the evolutionary concept to the ideal sequence of forms in mountain regions sculptured by glaciers, in islands bordered by coral reefs (Davis, 1928), and in a limestone region with solution caverns (Davis, 1930a). Here is the way he saw the utility of his scheme:

> In the scheme of the cycle of erosion . . . a mental counterpart for every landform is developed in terms of its understructure, of the erosional process that has acted upon it, and the stage reached by such action stated in terms of the whole sequence of stages from the initiation of a cycle of erosion by upheaval or other deformation of an area of the earth's crust, to its close when the work of erosion has been completed: and the observed landform is then described not in terms of its directly visible features, but in terms of its inferred mental counterpart. The essence of the scheme is simple, and easily understood; yet it is so elastic and so easily expanded or elaborated, that it can provide counterparts for landforms of the most complicated structure and the most involved history [Davis, 1899a].

Not only did Davis provide names for the stages of his cycle but he also suggested technical terms for the various landforms, each term with an exact definition. He adopted Powell's three types of rivers: consequent, antecedent, and superimposed; and to these he added subsequent, obsequent, and resequent (Davis, 1909:483 513). For the low mountains that stand above the general level of a peneplain, he gave the name *monadnock,* from Mt. Monadnock in New Hampshire, which stands above the New England peneplain (Davis, 1909:362, 591). He demonstrated again that when a name is given to a feature of any kind, students of the face of the earth at once begin to perceive it. This is an interesting example of the principle that percepts are closely related to concepts.

Davis demonstrated and defended the cycle of erosion and his terminology with such vigor that they were largely accepted throughout the world. He was invited to lecture in many countries, and his ideas were translated into several languages. In fact, the only complete statement of the cycle of erosion and its variations is contained in a German translation of his lectures at Berlin by Alfred Rühl (Davis, 1912). One of the most lucid presentations of Davis's ideas was written by the French physical geographer, Emmanuel de Martonne (de Martonne, 1909; see also Baulig, 1950). Davis visited many countries where he made specific applications of his method of explanatory description to the landforms of particular places. He also experimented with methods of geographical writing (Davis, 1910), and in his famous paper on the Colorado Front Range he accompanied his description of the landforms with a commentary on the method he was using (Davis, 1911). In 1915 he published a lengthy paper on the principles of geographical writing that is important reading even these many years later (Davis, 1915).

Of course the scheme was attacked, as all such hypothetical models should be and as Davis expected it to be. It is interesting that in Germany, where the only complete statement of the model was published by Davis, there was the largest resistance to his ideas. Hettner's disagreement with Davis over his cycle of erosion was that a theoretical model was too rigid and too specific to fit real world conditions as they exist in unique situations (particular places). Davis had offered a theoretical model and Hettner pointed out that such models were misleading. The discussion that ensued, when examined closely by scholars familiar with both German and English, turns out to hinge on misunderstandings of word meanings. Passarge, who wanted to describe landforms as a basic part of the study of landscape, was opposed to the Davis method of explanatory description. He called instead for a purely empirical treatment of the landform base. Davis took issue with Passarge's attempt to treat landforms empirically, pointing out the numerous inconsistencies that resulted (Davis, 1919). It is now clear, however, that the Passarge-Davis controversy became heated because neither quite understood the basic objectives of the other. For Davis the natural history of landforms was the core of geographical work to which other elements of landscape could be related; for Passarge the goal was the treatment of landscape, including the many elements of diverse origin associated in area of which the landforms were basic but not necessarily the most important. Davis himself had good reason to recall his own experience with the empirical approach to landforms and his own success with description based on an explanatory model.[5]

[5]Many attacks on the hypothesis of the cycle rather than the explanatory method were made in the period after World War I. Walther Penck, the son of Albrecht Penck, suggested that the initial steep slope on the margin of an upraised block would retreat parallel to its initial position, maintaining its steepness rather than flattening. Successive uplifts would be recorded by a series of scarps, each in process of retreat; and the rivers would pass over a series of "nick-points," each marking a change of baselevel and each in process of retreat upstream. More recently L. C. King, the South African geomorphologist, has demonstrated the existence of a series of erosion levels bordered by scarps in Brazil and in South Africa. He describes the flattish erosion levels as *pediments* (Walther Penck, *Die Morphologische Analyse* [Berlin, 1922], translated as *Morphological Analysis of Landforms* [London, 1953]; L. C. King, "Cannons of Landscape Evolution," *Bulletin of the Geological Society of America*, 64 [1953]:721-753; idem, *The Morphology of the Earth* [New York, 1962]; see also P. E. James, "The Geomorphology of Eastern Brazil as interpreted by Lester C. King," *Geographical Review*, 49 [1959]:240-246).

The part of the Davis scheme most subject to attack is the postulate of a single uplift. It is now understood that when a foot of rock material is removed from an area of the earth's surface this results in an isostatic rise of the rock column beneath the surface by some 9 to 11 feet. This means that no surface could stand still long enough to permit the formation of a peneplain and explains why no peneplains are found undissected. As early as 1878 G. K. Gilbert wrote that the reduction of a drainage basin to a plain would demand a uniformity of conditions that nowhere exists (Gilbert, 1878). In modern times A. N. Strahler has expressed the opinion that "the cycle concept of Davis does not seem well adapted to express the dynamics of the erosion process. Instead, the concept of a steady state in an open system seems a logical replacement of the idea of 'maturity' while the stage of 'old age' may well be abandoned" (Strahler, 1950). (For other critical comments on the Davis system, see Chorley, 1965.)

It is important, also, to understand that Davis himself wrote critical reviews of his own scheme during the late years of his life and that he also contributed many additional original works (Davis, 1922, 1928, 1930a, 1930b).

Contributions to Geographic Education

Davis also attempted to rescue the teaching of geography from too much attention to factual knowledge and not enough use of general concepts around which to organize the facts. This is precisely what Ritter attempted to change in 1817 and what Guyot pointed out to the Massachusetts Board of Education in 1848. In spite of these earlier efforts, Davis found the same emphasis on factual knowledge in the 1880s, and he proposed to do something specific about it. In 1932, looking back over a lifetime spent in the effort to improve the teaching of geography, he wrote:

> No geographer, therefore, need feel himself unfortunate because of the great diversity of facts that his composite subject requires him to study; for in the progress of his work he may discover relations and principles which bind his facts together in a thoroughly reasonable manner, and he may then concern himself, especially in his teaching, largely with those relations and principles, and introduce items of fact chiefly to illustrate the principles. Unhappily, geographers are often so impressed with the innummerable facts of their subject that much of their attention is given to individual occurrences in specified localities rather than to principles which the occurrences exemplify; and this is regrettable. Yet practically the same comment may be made on History, in which the mere sequence of events, or, still worse, the mere occurrence of events is often given greater emphasis than their inherent importance warrants . . . [Davis, 1932:214-215].

In a lecture delivered before the Scientific Association of Johns Hopkins University in 1889, Davis outlined a relatively simple model of landform development that a teacher could use to replace the bewildering description of detail then frequently attempted (Davis, 1909:193-209). This was essentially his concept of the cycle of erosion adapted to elementary and secondary schools. But in this address Davis was beginning to conceive of the study of geography as a means for introducing many kinds of physical science in a simple coherent framework. He was formulating the idea of a general earth science organized in terms of a dynamic model of earth-forming processes. About this same time Davis, who had learned from Shaler to see organic life, including man, as a part of the whole physical landscape, began to seek an even larger conceptual structure for geography. He began to seek cause and effect generalizations, "usually between some element of inorganic control and some element of organic response" (Davis, 1903:3-22). In 1906 he specified that:

> . . . any statement is of geographical quality if it contains a reasonable relation between some inorganic element of the earth on which we live, acting as a control, and some element of the existence or growth or behavior or distribution of the earth's organic inhabitants, serving as a response.
> . . . There is, indeed, in this idea of a causal or explanatory relationship the most definite, if not the only unifying principle that I can find in geography [Davis, 1906; quoted from Davis, 1909:8].

Albert P. Brigham

William Morris Davis

Grove K. Gilbert

Mark S. W. Jefferson

Emory R. Johnson

John Wesley Powell

Rollin D. Salisbury

Ellen C. Semple

Davis's attempt to bring cohesion to geography by introducing the causal notion had some less fortunate consequences. To this day it is not uncommon to find nongeographers—and even some geographers—using the words, geographic factor, to refer to a condition of the physical earth that influences human activities. Professional geographers would now use these words, if at all, to refer to some factor of location. Two famous Harvard scholars, a psychologist and a philosopher—Charles S. Peirce (1839-1914) and William James (1842-1910)—were laying the groundwork of pragmatism in the 1870s. They demonstrated that simple cause and effect relations were of questionable value as explanations of events because the existence of complex systems of functionally related parts made any simple explanation impossible. Pragmatism, as Peirce and James formulated it, was a method of determining the meaning of intellectual concepts; the meaning of any idea, they insisted, must be revealed in the practical consequences of the idea. The teaching of physical cause and human response was already outmoded before 1890. Furthermore, the pragmatic approach to knowledge was being introduced into the elementary and secondary schools through the work of John Dewey.

By 1892 there was such a ferment of new educational ideas that the National Educational Association appointed a Committee of Ten, headed by President Eliot of Harvard, to study the related problems of the content of precollege-level school programs and college entrance requirements. The committee organized nine different conferences, each to consider a specific field of study. One of the nine was directed to consider the content of courses in geography. The chairman of the Conference on Geography was T. C. Chamberlin, former president of the University of Wisconsin and in 1892 the chairman of the Department of Geology in the newly organized University of Chicago. The other members of the conference were geologists, meteorologists, and college or high school teachers of physical geography or natural history.[6] The member of this conference who had the chief hand in the resulting report was Davis. The recommendation that was passed and referred to the Committee of Ten concluded that

> Physical geography should include elements of botany, zoology, astronomy, commerce, government, and ethnology, and . . . it should take on a more advanced form and should relate more specifically to the features of the earth's surface, the agencies that produce or destroy them, the environmental conditions under which these act, and the physical influences by which man and the creatures of the earth are so profoundly affected [Mayo, 1965:20-21].

[6]The Conference on Geography included: T. C. Chamberlin, University of Chicago, chairman; George L. Collie, Beloit College; William Morris Davis, Harvard University; Delwyn A. Hamlin, Rice Training School, Boston; Mark W. Harrington, Weather Bureau, Washington, D.C.; Edwin J. Houston, Philadelphia; Charles F. King, Dearborn School, Boston; Francis W. Parker, County Normal School, Chicago; Israel C. Russell, University of Michigan.

The Committee of Ten expressed surprise when it received the report of the Conference on Geography which presented many more far-reaching changes in the secondary-school curricula than were presented by any of the other eight conferences. The committee wrote:

> Considering that geography has been a subject of recognized value in the elementary schools . . . , and that a considerable proportion of the whole school time of children has long been devoted to a study called by this name, it is somewhat startling to find that the report of the Conference on Geography . . . exhibits more dissatisfaction with prevailing methods . . . and makes the most revolutionary suggestions.[7]

Nevertheless, the report was adopted, and its radical proposals to raise geography from pure memory work to the status of a general science were recommended to the schools. Many schools introduced physical geography, or physiography as it came to be called, and many new textbooks were written to include Davis's ideas.[8] But there were also many secondary-school teachers who were opposed to the new materials and there were many more who were quite unprepared to teach them. Few indeed were the secondary-school teachers who could identify a landform out-of-doors or who could make Davis's theoretical models come alive. Lacking proper training in the concepts of the new geography, they fall back again on recitations from memory. Within ten years the new physiography was described as an uninteresting subject. Thereafter it was gradually pushed out by general science, social studies, and commercial geography.

Davis himself was a remarkable teacher. He was master of the lecture method and could capture the imagination of popular as well as scientific audiences. His pen and ink sketches of landscapes were superb. (Fig. 26). In the field with a group of students, he could arouse a deep interest in deciphering the sequences of events that produced contemporary landscapes. But with his students he was sharply critical of inferior performance and discouraged many of his more sensitive students from further work in geography. Only the best students could "take it." Mark Jefferson, one of Davis's most effective disciples, had this to say about the master's teaching:

> Davis' teaching was the most interesting thing I ever met. Confronted with the world of out-of-doors his formulae proved up. I took all his courses at Harvard, a summer school

---

[7]National Education Association, *Report of the Committee of Ten on Secondary School Studies* (New York: American Book Co., 1894), pp. 32–33.

[8]The "new geography" with its emphasis on the physical earth appeared in a series of school texts, such as A. E. Frye, *Elements of Geography* (1895); R. S. Tarr, *Elementary Physical Geography* (1895); J. W. Redway and R. Hinman, *Natural Advanced Geography* (1897); R. S. Tarr and F. M. McMurry, a series of geographies starting in 1900. See also Mark Jefferson's course in geography for some 1300 Cuban schoolteachers in the summer of 1900 at Harvard (reported in Martin, 1968). For a discussion of the teaching of geography in America up to 1924, see Dryer, 1924.

Figure 26. *A landscape sketch by W. M. Davis: "Looking eastward down the normal early-mature valley of Fourmile Creek." (From Davis, 1911:81. Reprinted by permission of the Association of American Geographers.)*

with him and two other students in the Rocky Mountains. . . . The more you checked his teaching against the out-of-doors the sounder you found it. But he was not always easy to take. His was a school of intellectual hard knocks [Martin, 1968:41].

What Davis Started

We must give Davis credit for his tireless devotion to the advancement of what he thought of as geography. His students were among the outstanding geomorphologists and human geographers of the early twentieth century.[9] They were appointed to positions in many of the older eastern universities and colleges and to posts in the U. S. Geological Survey and the Soil Survey (see Krug-Genthe, 1903).

[9]There were six outstanding graduate students at Harvard with Shaler and Davis in 1891–92: A. P. Brigham, former minister in Utica who became a disciple of Davis and taught at Colgate University from 1892 to 1925; Richard E. Dodge, who taught at Teachers College, Columbia, from 1897 to 1916 and at Connecticut State College, Storrs, from 1920 to 1938, and founded the *Journal of School Geography* (later the *Journal of Geography*) in 1897; Curtis F. Marbut, who taught at Missouri from 1895 to 1910 and was in charge of soil surveys for the Department of Agriculture from 1910 to 1935; Ralph S. Tarr, who taught at Cornell from 1892 to 1912; Robert DeC. Ward, the climatologist, who taught at Harvard from 1890 to 1930; and Lewis G. Westgate, who carried out surveys for the U.S. Geological Survey chiefly in the West. Later distinguished students who worked with Davis at Harvard include: A. H. Brooks, who after 1903 was in charge of U. S. Geological Survey work in Alaska; Ellsworth Huntington, author of numerous books on climate and man and research associate at Yale from 1919 to 1945; Mark Jefferson, who inspired many young scholars to become geographers when he taught at Michigan State Normal College at Ypsilanti from 1901 to 1939; Isaiah Bowman, director of the American Geographical Society from 1915 to 1935 and president of Johns Hopkins from 1935 to 1949; Douglas W. Johnson, geologist at Columbia; and J. W. Goldthwait, geologist at Dartmouth.

Davis recognized that if geography was to become established as a professional field it would be necessary to organize a professional society in which members could present their ideas. When he was vice-president of the American Association for the Advancement of Science (1904) for Section E—Geology and Geography, he used the occasion of the vice-president's address to point out that although geology and geography had been granted equal status in the association for the previous twenty years, no vice-president of Section E had ever spoken concerning geography. "He then proceeded to fling geography under the eyes or into the ears of the assembled geologists" (Brigham, 1924). He said that the study of geography could lead to no professional career outside of schoolteaching and that there were few opportunities in the universities. There was no organized body of mature scholars in geography and, therefore, no opportunity for the mutual encouragement that comes from professional fellowship. He proposed that geographers should organize a professional society similar to the Geological Society of America, "with criteria of expert training and ample publication as a basis for membership." He went on to point out where a nucleus of members could be found among teachers of geography, members of national or state weather services, members of many government agencies dealing with geology, hydrography, biology, ethnography, or statistical studies. During the following months of 1904, A. P. Brigham called a group of interested people together, and from this meeting there emerged a plan for the organization of such a society.

The first meeting and the act of incorporation took place in Philadelphia in December, 1904.[10] Davis, as president, spoke about the objectives and opportunites for the new association (Davis, 1905). He was reelected president for 1905 (and for a third term in that office in 1909). It was at the 1905 meeting of the association that

[10]The charter members of the Association of American Geographers [from P. E. James and G. J. Martin, *The Association of American Geographers: The First Seventy-Five Years, 1904-1979* (Washington D.C., AAG, 1979), pp. 36-37.]:

C. Abbe, Jr.	R. E. Dodge	W. Libbey, Jr.
Charles C. Adams	C. R. Dryer	G. W. Littlehales
Cyrus C. Adams	N. M. Fenneman	C. F. Marbut
O. P. Austin	H. Gannett	F. E. Matthes
R. L. Barrett	G. K. Gilbert	W J McGee
L. A. Bauer	J. P. Goode	C. H. Merriam
A. P. Brigham	H. E. Gregory	R. W. Pumpelly
A. H. Brooks	F. P. Gulliver	H. F. Reid
H. G. Bryant	C. W. Hall	W. W. Rockhill
M. R. Campbell	R. A. Harris	R. D. Salisbury
F. E. Clements	A. Heilprin	E. C. Semple
H. C. Cowles	R. T. Hill	G. B. Shattuck
J. F. Crowell	E. Huntington	L. Stejneger
R. A. Daly	M. S. W. Jefferson	R. S. Tarr
N. H. Darton	E. R. Johnson	R. DeC. Ward
W. M. Davis	M. Krug-Genthe	B. Willis

he gave his presidential address, "An Inductive Study of the Content of Geography" (Davis, 1906), in which he identified geography as the study of the relationship between inorganic controls and organic responses.

Another of Davis's accomplishments in behalf of the profession was the Transcontinental Excursion of the American Geographical Society in 1912. As a result of successful experiences in Europe in directing field excursions, he suggested to some colleagues that a very ambitious project would bring a number of European geographers to the United States. Davis was able to secure financial help and the cooperation of railroads, universities, chambers of commerce, university clubs, newspapers, scientific societies, government agencies, and business organizations all across the United States. The excursion included 43 European geographers from thirteen different countries (see photograph of some of the participants). About 100 American geographers accompaned the Europeans for at least part of the excursion. The party left New York on a special train on August 22 and returned to New York on October 17, after covering 12,965 miles from coast to coast. Many professional papers were written on the basis of notes made during the excursion and published in many languages; but the greatest benefit was derived from the close personal friendships that developed among leaders of the geographic profession in America and Europe and from numerous unhurried professional discussions that were carried on. There has never been anything quite like this excursion.[11]

Geography Moves On

During the early years of the twentieth century the movement to introduce a professionally acceptable kind of geography into the schools, colleges, and universities gradually gathered strength. At the end of the nineteenth century there were only three professors of geography in American universities: in addition to Davis at Harvard, there were Ralph S. Tarr at Cornell and William Libbey, Jr., the successor to Guyot, at Princeton. There were many teachers' colleges around the country that offered courses in geography, but there were only a few where these courses were taught by persons with some kind of geographical training.[12] Outstanding among the teachers' colleges was the Michigan State Normal College at Ypsilanti, where Charles T. McFarlane taught geography between 1892 and 1900 and was succeeded in 1901 by Mark Jefferson.

[11]See the *Memorial Volume of the Trancontinental Excursion of 1912 of the American Geographical Society* (New York: American Geographical Society, 1915). See also Wright, 1952:158–166. The excursion was made possible by a substantial gift from Archer M. Huntington.

[12]C. E. Cooper, "The Status of Geography in the Normal Schools of the Far West," *Journal of Geography*, 18 (1919):300–305; idem, "The Status of Geography in the Normal Schools of the Middle States," ibid., 19 (1920):211–222; idem, "The Status of Geography in the Normal Schools of the Eastern States," ibid., 20 (1921):217–224. See also Dryer, 1924.

The Transcontinental Excursion, 1912.

Geographical research studies, as reported to the meetings of the Association of American Geographers, increased in number and began to depart from the restricted paradigm of physical condition and human response set forth by Davis. The charter members of the association were geologists, climatologists, botanists, sociologists, teachers—but only a few of them had ever had any advanced training in geography (Brigham, 1924). Some, like Douglas W. Johnson, who made important contributions to the evolution of shorelines, nevertheless, remained close enough to the newly emerging field of geography to aid in its development. Curtis F. Marbut, who became one of the world's leading authorities on soils, made use of Davis's ideas when he described soils as young or mature. His work in making the Russian soil studies known in the English-speaking world has been mentioned previously. Robert DeC. Ward, who taught climatology at Harvard for forty years, was the president of the Association of American Geographers in 1917.[13] Many geographers who began their careers before World War I continued to work actively in the field in the period between the two world wars. Some of them require special attention both in this chapter and in Chapter 14. We will discuss at greater length the work of Mark Jefferson, Isaiah Bowman, Ellsworth Huntington, Ellen Churchill

[13]R. DeC. Ward taught at Harvard from 1890 to 1930. In 1903 he published a translation of the *Handbuch der Klimatologie* by the Austrian scholar Julius Hann (Hann, 1903). In 1908 he published his book on the effect of different kinds of climate on human life (Ward, 1908). Thereafter most of his efforts were devoted to the organization of material on regional climatology (Ward, 1925; Ward and Brooks, 1936).

Semple, and Albert P. Brigham. We will also reyiew the efforts made to define and outline physiographic regions for the United States. Then we will look at the beginnings of commercial and economic geography at the University of Pennsylvania.

Mark Jefferson

None of Davis's students did more to promote and improve the teaching of geography in the United States than Mark Jefferson, who was professor of geography at the Michigan State Normal College in Ypsilanti for thirty-eight years from 1901 to 1939. Jefferson deserves a special place in the history of geography not only because of the enthusiasm he kindled in his students but also for the many contributions to the conceptual structure of geography that came from his pen.[14]

The Michigan State Normal College at Ypsilanti was already famous for its geography teaching before Jefferson went there in 1901. Since 1853 lectures on geography had been offered—after 1860 by John Goodison. When Goodison died in 1892, the vacancy was filled by Charles T. McFarlane, then twenty-one years old and newly graduated from the New York State Normal College. McFarlane turned out to be an excellent selection. Two years later his courses were described as scientific, partaking of "the close reasoning of physics and mathematics, and the rich insights into the ground of history and social life" (Martin, 1968:78). McFarlane's course of study in geography made use of the recommendations of the Committee of Ten.[15]

When McFarlane resigned from Ypsilanti in 1900, Davis recommended Jefferson as a replacement. Jefferson started teaching at Ypsilanti in June, 1901 and only retired in 1939 at the age of seventy-six after the Michigan Board of Education had set an age limit for active teachers. During his thirty-eight years in Michigan he had a hand in the training of a large number of teachers of geography, some of whom

[14]Before Mark Jefferson graduated from Boston University in the class of 1884, he left to accept a position at the observatory in Cordoba Argentina. He was in Argentina from 1884 to 1889, serving for two years as submanager of a sugarcane plantation near Tucuman. He then returned to Massachusetts, where he held various administrative posts in secondary schools and where he taught geography. In 1897–98 he studied geography at Harvard with Davis and was greatly stimulated by this experience. In the summer of 1900, when Harvard received a group of some 1300 Cuban schoolteachers seeking to find out about the new geography, Jefferson was appointed to lecture to them in Spanish. He gave eighteen lectures and led twelve field trips. Davis was greatly impressed with Jefferson's ability as a teacher (Martin, 1968).

[15]McFarlane went to Vienna on sabbatical leave in 1898 and studied with Albrecht Penck. In 1901 he became principal of the New York State Normal College at Brockport. In 1910, however, he went to the Teachers College at Columbia as comptroller, a post he held until his retirement in 1927. He also taught courses on the teaching of geography during that time. One of McFarlane's outstanding students at Ypsilanti was H. H. Barrows, who finished his work there in 1896. Barrows, teaching at Ferris Institute in Big Rapids, Michigan, had a student who later became a leader in the profession—Isaiah Bowman. Bowman had been inspired to seek a career in geography when he heard one of McFarlane's lectures.

were to make important contributions to the profession.[16] Jefferson, who was a great admirer of Davis, nevertheless, took issue with many of his teacher's ideas. He never accepted the concept of determinism. Furthermore, he disagreed strongly with the recommendations of the Committee of Ten regarding the content of school geography. Jefferson insisted that the focus of geography instruction should be "man on the earth," in that order—not "the earth and man." Nor was he willing to omit the teaching of the geographic conditions in certain particular countries, which we might describe as regional geography. He wanted to make these countries seem real to grade-school pupils—real in landscape, life, and institutions (Martin, 1968:343–344). This kind of geography was not part of the recommended systematic approach that Davis favored. In 1904 Jefferson surveyed the geography instruction in Michigan schools, and among the 129 largest high schools he found less than a dozen following the Committee of Ten program.

Jefferson remained aloof from the endless discussion of the question: "What is geography?" He never began a course with a definition of the field but rather let the scope of geographical study emerge from the materials he covered. No one definition, he believed, could be more than partially inclusive. Here is what he wrote in 1931 in answer to a questionnaire:

> Some one has said that anything that you can put on a map is subject matter of geography. That is what I would call locational or distributional geography. . . . But geographers are contemplative persons who cannot be satisfied with so meagre an account of the subject. . . . The nature of geography is the fact that there are discoverable causes of distributions and relations between distributions. We study geography when we seek to discover them. . . . But there is an art of geography—the delineation of the earth's features and inhabitants on maps—cartography, and a science of geography, which contemplates the fact delineated and seeks out causes of the form taken by each distribution and its relationship to others [Martin, 1968:319–321].

Jefferson was the perfect example to demonstrate that vitality and effectiveness in teaching are related to research and to the communication of the results of research in scholarly publication. Although his teaching "load" in many semesters was as much as six courses (eighteen semester hours), he, nevertheless, established an outstanding reputation as a productive scholar. In the period between 1909 and 1941 he had thirty-one papers published in the *Bulletin of the American Geographical Society* and its successor, *The Geographical Review* (Wright, 1952:294). This is by far the largest number of professional articles published in these prestigious periodicals during this time by any one scholar. These were not trivial pieces written

---

[16]Among Jefferson's outstanding students were Isaiah Bowman (who came to Ypsilanti in 1901 to find that McFarlane had gone and who remained to find great stimulation from the teaching of Jefferson), Charles C. Colby, D. H. Davis, W. M. Gregory, George J. Miller, A. E. Parkins, and Raye R. Platt (see also Dryer, 1924).

for the purpose of building a list of titles. Some of them were major contributions to the concepts of geography (Jefferson 1909, 1915, 1928, 1939). As one of the first American students of population distribution and of urban structures, he was an innovator and trailblazer.

Isaiah Bowman

One of Jefferson's outstanding students was Isaiah Bowman.[17] After completing his undergraduate work at Harvard with Davis in 1905, Bowman received an appointment in the Department of Geology at Yale, where the geologist H. E. Gregory (one of the charter members of the Association of American Geographers and its president in 1920) was gathering together a vigorous group of young scholars in geography. Bowman was called on to teach a course in geography at the Yale Forestry School; and out of his notes concerning surface features and soils related to forest cover, he wrote the first book to give systematic coverage of the physical characteristics of the regions of the United States (Bowman, 1911).

Bowman spent three field seasons studying the Andes of Peru, Bolivia, and northern Chile. In 1907 he landed at Iquique, Chile, and made his way across the Atacama Desert to the Bolivian Altiplano and thence to the forested eastern slopes of the Andes. He returned by way of Peru, through Cuzco, Arequipa, and Mollendo. From this field expedition he wrote his doctoral dissertation, *The Geography of the Central Andes*. In 1911 he was the geologist-geographer on the Yale Peruvian Expedition headed by Hiram Bingham during which Bingham rediscovered the lost Inca fortress of Machu Picchu. Here is the way Bowman described his role in the 1911 expedition:

The geographic work of the Yale Peruvian Expedition of 1911 was essentially a reconnaissance of the Peruvian Andes along the 73rd meridian. The route led from the tropical plains of the lower Urubamba southward over lofty snow-covered passes to the desert coast at Camaná. The strong climatic and topographic contrasts and the varied human life which the region contains are of geographic interest chiefly because they

[17]Bowman was brought up on a farm in Michigan, where his mother stimulated his early interest in natural history. When his first effort at drawing a map in school was graded *A*, he had had the "success experience" that turned him toward a career in geography. Teaching in a country school and attending summer institutes to improve his knowledge, he was greatly inspired by hearing one of McFarlane's lectures. Bowman was at the Michigan Normal College in 1901-1902. He then studied with Davis at Harvard in 1902-1903 and returned in the next academic year to teach at Ypsilanti under Jefferson's supervision. In 1904-1905 he returned to Harvard and completed his B.S. degree there in 1905. He was then appointed as instructor at Yale under H. E. Gregory and received his Ph.D. from Yale in 1909. He remained at Yale until 1915 and from 1915 to 1935 was the director of the American Geographical Society. From 1935 to 1948 he was president of Johns Hopkins University (Carter, 1950; Wrigley, 1951; Martin, 1977, 1980).

present so many and such clear cases of environmental control within short distances.
... My division of the Expedition undertook to make a contour map of the two-hundred mile stretch of mountain country between Abancay and the Pacific Coast...
[Bowman, 1916:vii].

In 1913 Bowman received a grant from the American Geographical Society to permit him to return to Peru for the third time. The results were published in two important books (Bowman, 1916, 1924). Clearly at this time in his life he was still following closely the paradigm of geographical study formulated by Davis, but in the course of time and field experience he became more cautious about so-called environmental controls. He was also seeking an effective way to generalize the many detailed observations regarding the terrain that he had gathered in three expeditions. His imaginative innovation, which he used in both books, was his use of "regional diagrams." He recognized in the high Andes of southern Peru six kinds of what he called "topographic types":

1. An extensive system of high-level, well-graded, mature slopes . . . below which are:
2. Deep canyons with steep, in places cliffed, sides and narrow floors, above which are:
3. Lofty residual mountains composed of resistant, highly deformed rock, now sculptured into a maze of serrate ridges and sharp commanding peaks.
4. Among the forms of high importance, yet causally unrelated to the other closely associated types, are volcanic cones and plateaus of the Western Cordillera.
5. At the valley heads are a full complement of glacial features, such as cirques, hanging valleys, reversed slopes, terminal moraines, and valley trains.
6. Finally there is in all the valley bottoms a deep alluvial fill formed during the glacial period and now in process of dissection.
[Bowman, 1916:185-186.]

The topographic maps show all these features in their complex arrangement and with many variations that make each view unique. The regional diagram, on the other hand, shows these various types in their characteristic arrangement, simplified and compressed within small rectangles (Figs. 27 and 28). These diagrams are what we would now call "empirical generalizations." Here is what Bowman said about them:

This compression, though great, respects all essential relations. For example, every location on these diagrams has a concrete illustration but the accidental relations of the field have been omitted; the essential relations are preserved. Each diagram is, therefore, a kind of generalized type map [Bowman, 1916:15].

Some of Bowman's most important work was in the application of goegraphical methods to the study of practical problems during and after World War I. We will return to this aspect of his career in Chapter 15.

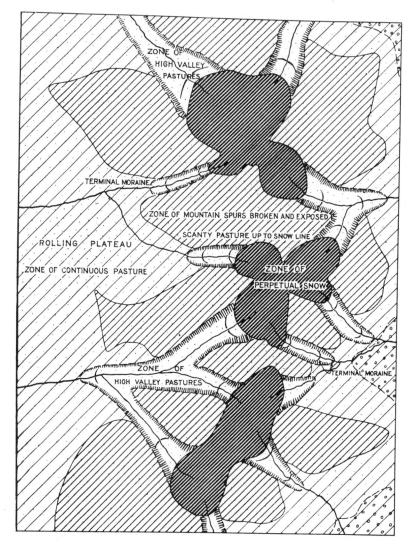

Figure 27. Regional diagram in the eastern Cordillera of Peru. (From Bowman, 1916:65. Reprinted by permission of the American Geographical Society.)

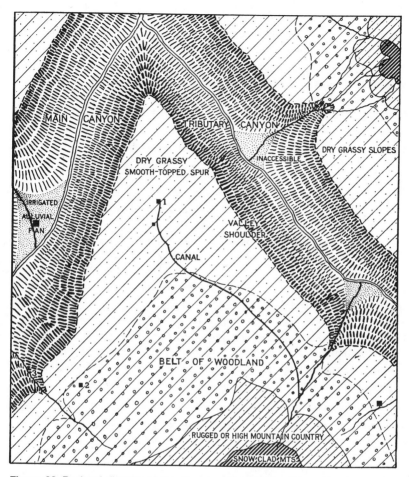

Figure 28. Regional diagram of the deep canyon regions of the Apurimac. (From Bowman, 1916:58. Reprinted by permission of the American Geographical Society.)

Ellsworth Huntington

Another scholar who studied with Davis at Harvard and was associated with Bowman at Yale was Ellsworth Huntington.[18] Huntington was a creative thinker, a prolific writer, and an imaginative interpreter of the effects of climate on human life. During his field studies in Asia between 1903 and 1906 he found much evidence to support the idea that there had been a worldwide progressive desiccation

---

[18]Ellsworth Huntington graduated from Beloit College in 1897 and was given an appointment as assistant to the president of Euphrates College in Harput (Turkey). He was in Turkey from 1897 to 1901. He took advantage of every opportunity to travel in different parts of the country, including a trip through the gorge of the Euphrates River during which he kept copious notes on the character of the land, the

since the glacial period. But on his return from Asia during the early part of 1906, the idea came to him that instead of progressive change toward warm and dry conditions, there were cycles of cool-wet and hot-dry periods of varying length. Correlating the periods of drought with historical dates, he developed the hypothesis that the great outpourings of nomadic peoples from central Asia, which led to the Mongol conquests of India and China and the invasions of eastern Europe in the thirteenth century, could be explained by the drying up of the pastures on which the nomads were dependent. This thesis was presented in his book, *The Pulse of Asia* (Huntington, 1907), which started him on the road to fame as a student of climatic influences on human society. In 1915 he published *Civilization and Climate* (Huntington, 1915) in which he developed the hypothesis that man's civilizations could only develop in regions of stimulating climate and that the monotonous heat of the tropics would forbid attainment of the higher levels of civilization. His book, *Principles of Human Geography* (Huntington and Cushing, 1920), which was written as a college textbook, organized a picture of world geography in terms of human activities with the "explanatory description" of the physical earth omitted or greatly reduced.[19]

Huntington's books have enjoyed recognition not only among geographers but also among historians, sociologists, and medical scholars. His generalizations about climate and man remain thought provoking. His vivid descriptions of places are among the most effective examples of geographical writing from any age. Yet in the period when Huntington was making his studies, the quantitative data on which such work had to be based were not in existence (Butzer, 1964:437). The identification of climatic cycles was based on scattered evidence, including the study of

---

climate, and the people. In 1901 he received a scholarship to study at Harvard with Davis, and in 1902 he completed the work for the M.A. degree. He started further graduate study leading toward the doctorate, but when the opportunity came to return to Asia for field study he left Harvard. In 1903-1904 he was a member of Raphael Pumpelly's expedition to central Asia. In 1905-1906 he visited northern India and then went across the Tarim Basin to the Lop Nor, returning by way of Siberia. In 1907 he joined the Yale faculty as instructor in geology. Yale granted him the Ph.D. degree in 1909 on the basis of some of his published works, and he was promoted to assistant professor in 1910. In 1916, however, he resigned from Yale when his request for promotion to professor was turned down. The reason given was his lack of success as a teacher of undergraduate courses. He had to support himself by writing textbooks. After serving in the army in the field of military intelligence in 1918 and 1919, he returned to Yale as a research associate with the rank of professor, but with a token salary (at the start) of $200 a year. He continued as a research scholar at Yale, supervising dissertations and offering graduate courses, until his retirement in 1945. He was president of the Ecological Society of America in 1917, president of the Association of American Geographers in 1923, and president of the American Eugenics Society from 1934 to 1938. He was author or coauthor of twenty-eight books and part author of thirty others. He published approximately 240 professional and popular articles (from G. Martin, *Ellsworth Huntington: His Life and Thought*, 1973).

[19]This book was severely criticized by H. H. Barrows in a review published in the *Geographical Review*, 12 (1922):157-160, which many geographers at that time thought was unnecessarily harsh. Huntington published a revised edition in 1922 in which many of Barrows' criticisms were met.

growth rings on trees, the bands of clay in drained lake beds, or even the scattered references to floods and droughts in the literature (Chappell, 1970). Modern studies of growth rings and clay bands are much more reliable than the information he was able to bring together. His maps of degrees of civilization were based on the opinions of people with whom he corresponded. Since people commonly rate their own country as the most civilized and since most of Huntington's correspondents lived in the northeastern United States, western Europe, or eastern Asia, he found these regions most highly civilized (Huntington, 1915:291–314). In his earlier years he could write passages such as this:

> The geographical distribution of health depends on climate and weather more than on any other single factor [Huntington and Cushing, 1920:248].

But as time went on he came to realize that things were much more complex than he had at first believed. In his last book he even suggested that diet was as important as climate as an explanation of human energy (Huntington, 1945:417). Huntington worked on subjects for which objectively defined data were lacking and in a period before the methods of collecting such data had been worked out. D. H. K. Lee writes of Huntington's work as follows:

> Huntington's brilliant generalizations covering such a wide range of relationships to climate are worth reading for two reasons: first they are thought-provoking, and not all of them have been disproved; and second, as a demonstration of effective presentation they are unequalled [Lee, 1954:473].

Regional Geography and the Delimitation of Regions

In the early years of the twentieth century those geographers who followed Davis in seeking to define a cohesive field of study including physical and human elements were attracted by the British approach to the delimitation of regions. Many concluded that the highest expression of geographic research was regional geography. Here, within limited confines, the student could go back to causes and forward to consequences without loss of confidence in the results (Fenneman, 1919). But how should one identify and delimit a region?

The first attempt to divide the territory of the forty-eight conterminous states and territories into physiographic regions was done by John Wesley Powell in 1896 (Powell, 1896). He divided the national territory into sixteen regions, some with several subdivisions. Davis, himself, made a similar map in 1899 (Davis, 1899b;719). By 1914 a considerable number of regional divisions of the United States and of North America had been published, each differing in minor ways from the others. In 1914 W. L. G. Joerg reviewed twenty-one such maps and drew one of his own that combined the best features of the others. Joerg used the term "natural region" and defined it as "any portion of the earth's surface whose physical conditions are homogeneous." The concept of scale or degree of generalization in de-limiting regions had not at that time been widely appreciated. Joerg's regions were highly generalized and of course were homogeneous only by definition. N. M. Fenneman also published a study of physiographic boundaries in the United States in 1914; and his resulting map incorporated new features not found on the twenty-one maps that Joerg compared—for example, the separation of the Southern Rockies from the Northern Rockies by the Wyoming Basin (Fenneman, 1914). At the Chicago meeting of the Association of American Geographers in December, 1914 one session was devoted to a conference on regions. A committee was appointed to draw a map of physiographic regions on the basis of the criteria accepted by the conference, and Fenneman was named as chairman. The map of the physiographic regions with a detailed statement of the characteristics of each region was published in 1916 and included a folded map on the scale of 1/7,000,000 (Fenneman, 1916).

The concept of physical control and human response entered into the drawing of regions with C. R. Dryer's paper in 1915 (Dryer, 1915). Dryer proposed that the best way to identify natural regions is to measure the economic functions of each. He called his divisions of the U.S. "natural economic regions." The "solid, liquid, gaseous, and biological phenomena fit into workable living combinations," he wrote (in an amazing return to the natural elements of Aristotle). Economic activi-ties fit into these same regions, and a study of the economic activities provides the best guide to the delimitation of natural regions. This reflection of Unstead's syn-thetic regions appeared again and again in later years in school and college textbooks that defined a region as homogeneous in its natural characteristics and, therefore, homogeneous in its economic functions. Once the map was drawn, the environmental control was demonstrated. Later we will see what Brigham had to say about this kind of reasoning.

Ellen Churchill Semple

Another pioneer geographer of this period was Ellen Churchill Semple. After graduating from Vassar in 1882 she taught for a few years in her native Louisville, Kentucky. She received a master's degree from Vassar in 1891, earned externally on the basis of a two-year program of readings in sociology and economics, a final written exam, and a thesis, "Slavery: A Study in Sociology." From friends she heard enthusiastic reports about a professor in Germany, at Leipzig, whose lectures were bringing new worlds into view. She went to Leipzig; and in spite of difficulties placed in the way of women who wanted to undertake graduate work, she studied with Ratzel in 1891–92 and again in 1895. She returned to the United States greatly stimulated by Ratzel's new approach to anthropogeography and his interpretations of so-called geographic influences on the course of history. She rejected his ideas about the state as an organism—which he had derived from Herbert Spencer. Semple's aim then was to present Ratzel's ideas in English, but clarified and reorganized with many new illustrations drawn from different parts of the world. In 1897 she published her first professional article dealing with the Appalachian Barrier in American history (Semple, 1897), and in 1901 she published a paper based on her own field observations on the highlands of eastern Kentucky regarding the results of isolation on the settlers of that area (Semple, 1901). This second paper started her along the road to fame, but her professional status was confirmed in 1903 with the publication of her first book, *American History and its Geographic Conditions* (Semple, 1903).

She presented her version of the first volume of Ratzel's *Anthropogeographie* in her great book *Influences of Geographic Environment,* which was published in 1911. Here is what she had to say about her method:

> The writer's own method of research has been to compare typical peoples of all stages of cultural development, living under similar geographic conditions. If these peoples of different ethnic stocks but similar environments manifested similar or related social, economic, or historical development, it was reasonable to infer that such similarities were due to environment and not to race. Thus by extensive comparison, the race factor in these problems of two unknown quantities was eliminated for certain large classes of social and historical phenomena [Semple, 1911:vii].

And here is another quotation from the opening paragraph of her book:

> Man is a product of the earth's surface. This means not merely that he is child of the earth, dust of her dust; but that the earth has mothered him, fed him, set him tasks, directed his thoughts, confronted him with difficulties that have strengthened his body and sharpened his wits, given him his problems of navigation or irrigation, and at the same time whispered hints for their solution. . . . On the mountains she has given him leg muscles of iron to climb the slope; along the coast she has left these weak and flabby, but given him instead vigorous development of chest and arm to handle his paddle or oar. In the river valley she attaches him to the fertile soil, circumscribes his ideas and ambitions by a dull round of calm, exacting duties, narrows his outlook to the cramped horizons of his farm. Up on the wind-swept plateaus, in the boundless stretch of grasslands and the waterless tracts of the desert, where he roams with his flocks from pasture to pasture and oasis to oasis, where life knows much hardship but escapes the grind of drudgery, where the watching of the grazing herd gives him leisure for contemplation, and the wide-ranging life of a big horizon, his ideas take on a certain gigantic simplicity; religion becomes monotheism, God becomes one, unrivalled like the sand of the desert and the grass of the steppe, stretching on and on without break or change. Chewing over and over the cud of his simple belief as the one food of his unfed mind, his faith becomes fanaticism; his big spacial ideas, born of that ceaseless regular wandering, outgrow the land that bred them and bear their legitimate fruit in wide imperial conquests [Semple, 1911:1–2].

These quotations suggest two things: first, that her style of writing has a certain literary quality that makes reading it a delight, yet which might—and sometimes does—carry the theme beyond what sober judgment would permit; second, that the concept of the earth as the controlling factor in human life is carried beyond the possibility of objective verification. It is true that in combing the writings of all nations for examples to illustrate her principles, she fell into an error not uncommon when deductive reasoning is followed—she failed to look carefully for examples that contradicted her principles. Is it not possible to find examples of people who worship one God yet are not pastoral nomads? And, are there no examples of inhabitants of the boundless steppes who are pantheists? People who live in pass routes, she wrote, tend to become robbers of passing travelers. Then she presented case after case of people in pass routes who rob for a living. But she did not look for people in pass routes who do not rob, nor did she seek an explanation for robbers who do not live in pass routes.

Nevertheless, two other observations must be made. First, she was very careful to make the point that the environment does not control human action: only that under certain circumstances there is a tendency for people to behave in predictable ways—which is a verbal approach to probability theory. Second, there are some brillant passages where her insight is even now thoroughly relevant. Her "islands of ethnic expansion and islands of ethnic retreat" offer an important modification for the contemporary theory of innovation dispersal (Semple, 1911:204-228).

Allen Bushong[20] has suggested that the much quoted first paragraph of Semple's book was written as an "attention getter" and that the second paragraph of the same book reveals a more objective and mature Semple:

> Man can no more be scientifically studied apart from the ground which he tills, or the lands over which he travels, or the seas over which he trades, than polar bear or desert cactus can be understood apart from its habitat. Man's relations to his environment are infinitely more numerous and complex than those of the most highly organized plant or animal. So complex are they that they constitute a legitimate and necessary object of special study. The investigation which they receive in anthropology, ethnology, sociology and history is piecemeal and partial, limited as to the race, cultural development, epoch, country or variety of geographic conditions taken into account. Hence all these sciences, together with history so far as history undertakes to explain the causes of events, fail to reach a satisfactory solution of their problems largely because the geographic factor which enters into them all has not been thoroughly analyzed. Man has been so noisy about the way he has "conquered Nature," and Nature has been so silent in her persistent influence over man, that the geographic factor in the equation of human development has been overlooked.

Ellen Semple was an enormously persuasive teacher. Generations of American geographers were brought up to believe these teachings. During the time that she lectured at Chicago and later at Clark University, a large number of future geographers passed through her classrooms. It is easy to condemn her for presenting

[20]Allen D. Bushong to G. J. Martin, letter of February 4, 1980.

concepts that have not withstood the test of time; but she must be appreciated for kindling among her students an enthusiasm for the broad view of the earth as the home of man.[21]

In 1911 Semple started to work on the geography of the Mediterranean region. Over a period of twenty years she was a frequent visitor in the countries bordering the Mediterranean, including the parts of Asia to the east. She did a vast amount of reading in the literature—both ancient and modern—concerning the Mediterranean countries. In 1915 she published the first of numerous articles on different aspects of the region—this one on the mountain barriers and the breaches through them as factors in the history of the region (Semple, 1915). Her papers dealt with Mediterranean agriculture, the relation of forests to climate, the relation of climate to religion, and the geographic basis of Mediterranean trade. One of the most delightful of these papers dealt with the ''Templed Promontories'' where the gods were asked to watch over the seafarers who had to sail around dangerous capes (Semple, 1927). All of this work was brought together in her last great book on which she was at work when stricken in 1929. With great courage she persisted, able to work not more than two hours each day, until the work was completed and published only a few months before her death (Semple, 1931).

Albert Perry Brigham

One of the ''outstanding graduate students'' who were at Harvard working with Shaler and Davis in 1891-92 was Albert Perry Brigham (James, 1978).[22] Brigham, who was on the faculty of Colgate University for many years, was one of the major supporters of W. M. Davis in the effort to establish geography as a professional field in the United States. Brigham's book, *Geographic Influences in American History,* was published in 1903, the same year in which Ellen Semple's book on the same subject appeared. Brigham's book (Brigham, 1903) placed heavy emphasis on the origin of what he called ''geographic conditions'' but was relatively light on history, as the historians were not slow to point out. Although

[21]Ellen Churchill Semple was a visiting lecturer at Chicago between 1906 and 1924. In 1912 and again in 1922 she lectured at Oxford in England. She was visiting lecturer at Wellesley College in the fall of 1914, at the University of Colorado in the summer of 1915, at Western Kentucky University in the summer of 1917, Columbia University in the summer of 1918, and at the University of California at Los Angeles in 1925. From 1921 until she was stricken with a heart attack in 1929, she was a member of the staff of the Graduate School of Geography at Clark University. In 1921 she was the president of the Association of American Geographers (Colby, 1933).

[22]A. P. Brigham was the minister of a large church in Utica, N.Y., when he took a summer field course in geology at Harvard in 1889. Two years later he resigned his ministry and went to Harvard for advanced study. He received the M.A. degree at Harvard in 1892, and from then until his retirement in 1925 he was on the faculty of Colgate University. In 1904 he was one of the charter members of the Association of American Geographers and was its secretary-tresurer from 1904 to 1913. In 1914 he was president of the Association [see the *Annals AAG,* 20 (1930):55-104; and 23 (1933):27-32].

somewhat different in emphasis, the points of view of Semple and Brigham were very similar.

In the course of time Brigham took a more and more vigorous stand against the way geologically trained geographers asserted the existence of "responses" to environmental conditions without attaching precise meaning to their words and without showing what influences really do and how they do it. He was familiar with the work of the European geographers, especially Ratzel and Brunhes. He insisted that Ratzel was a pioneer, and really could not be blamed for not exploring all the parameters of man-land relations. In his presidential address before the Association of American Geographers in 1914 (Brigham, 1915) he specified that it was the geographer's task to provide a careful and scientific description of the physical environment, but that geographers should use caution and common sense in asserting the existence of influences and that every possible test should be made to ascertain the validity of any general principles that were suggested. He wrote of the relation of specific factual information to general concepts:

> Our goal is broad generalization. But the formulation of general laws is difficult, and the results insecure until we have a body of concrete and detailed observations.... Detailed investigation of single problems, in small and seemingly unimportant fields, must for a long time prepare the way for the formulation of richer and more fundamental conclusions and general principles than we have yet been able to achieve [Brigham, 1915:24–25].

Brigham was especially critical of generalizations concerning the influence of climate. He observed that "perhaps there is no subject, unless it be politics, on which men say so much and know so little as about climate." He was especially disturbed by vague and unproved assertions of climatic influence on racial character, skin color, or man's institutions. The infinitely variable factors of the total environment, he insisted, produce infinitely diverse results upon body and mind.

During the period from the 1890s to World War I, the point of view toward geography that was almost universally accepted in the United States was that of the search for environmental influences. Furthermore, the meaning was the same whether or not the word, "influences," was replaced by "responses" or "adjustments." And in spite of Brigham's warning concerning the need for careful use of words and patient testing of alleged influences, many geographers continued to draw plausible, but unverified, conclusions from their studies (Hartshorne, 1939:23).

Emory R. Johnson and J. Russell Smith

While anthropogeography was being developed by people trained for the ministry or in history, and physical geography was being developed by people trained in geology, the advanced study in economic and commercial geography was being started by two scholars trained in economics. The individuals who led the way in this aspect of geography were Emory R. Johnson and J. Russell Smith, both in the Wharton School of Finance and Commerce at the University of Pennsylvania.[23] In 1899 when Congress set up the Isthmian Canal Commission to recommend the best route for an interocean canal, Johnson, who was a specialist on the geography of transportation, was appointed to make a cost-benefit study of alternative routes. He selected his graduate student, J. Russell Smith, as his assistant. Although he had been teaching courses on "The Theory and Geography of Commerce" and on "Physical and Economic Geography," he was not experienced in the actual methods of geographical study. He and Smith had to work out their own procedures. Here is the way Smith described the situation:

> We came out of that job with a firm conviction that the American educational field needed geography in the colleges—quick. Because of our helplessness in the face of a concrete problem, it convinced us that it was extremely important [Rowley, 1964:22–23].

In 1919 J. Russell Smith was invited to the newly established School of Business at Columbia University to develop a curriculum in geography leading to advanced degrees. He set himself to remedy what he felt was a serious lack of good textbooks in economic geography and in 1913 published his influential *Industrial and Commercial Geography* (Smith, 1913). The book was revised several times and provided the basic text for courses in this kind of geography for perhaps fifty years. The last edition was published in 1955 with the aid of two coauthors, one of whom was his son, Thomas R. Smith. In addition to this book, Smith published twenty-nine other books for use at various grade levels. He was active in the conservation movement, strongly recommending that steeply sloping land should be planted with tree crops to protect it from erosion. Here is the way Virginia M. Rowley summarized his career:

[23]Emory R. Johnson received the Ph.D. in economics for a dissertation, "Inland Waterways: Their Relation to Transportation," at the Wharton School in 1893. Whittlesey lists this as the first doctoral dissertation in geography in the United States (Whittlesey, 1935:213). On the staff of the Wharton School from 1893 until his retirement in 1933, Johnson offered courses in economic geography. He was dean of the Wharton School from 1919 to 1933. Students of Johnson who received the Ph.D. from Wharton School include J. Paul Goode, 1901, J. Russell Smith, 1903, and Walter S. Tower, 1906. Johnson was a charter member of the Association of American Geographers.

J. Russell Smith graduated from the University of Pennsylvania in economics in 1898 and started graduate work with E. R. Johnson. Johnson and Smith worked on the Isthmian Canal Commission in 1899–1901. In 1901–1902 Smith studied anthropogeography with Ratzel at Leipzig and then returned to the Wharton School, where he completed the Ph.D. in 1903. On the faculty of the Wharton School from 1903 to 1919, he became chairman of the new Department of Geography and Industry, which he founded. From 1919 until his retirement in 1944, he was chairman of the Department of Geography in the School of Business at Columbia University. He was president of the Association of American Geographers in 1943 (Rowley, 1964).

J. Russell Smith is thus a unique man, of unusual energy and versatility. His restless, probing, creative mind caused him to go beyond the narrow subject bounds of a single academic discipline and to view knowledge as a related whole. Some may criticize Smith for his occasional inaccuracies, his untested theories, or, at times, his de-emphasis of specific details. These criticisms are sometimes justified and when detrimental to truth and objectivity, reflect definite weaknesses which, as a professional geographer, Smith should have eliminated. On the other hand, it must be remembered that Smith's goal was different from that of the pure research specialist. To him an idea had worth only if it were put to work. We must see Smith the academic geographer, the generalist, as well as synthesizer and experimenter, as many others have seen Shaler, Smith's "unseen master," as more concerned with awakening minds than with imparting specific information [Rowley, 1964:200–201].

GEOGRAPHY AT THE UNIVERSITY OF CHICAGO

The pioneering works of Davis and his disciples, of Ellen C. Semple, and of Emory R. Johnson and J. Russell Smith were of major importance in the introduction of a revitalized geography into the United States. But geography as a professional field with departments staffed by scholars trained in geography could not appear until there was a department in a university staffed by scholars, all of whom were devoted to this kind of study. This began at the University of Chicago in 1903, when the first separate department offering advanced graduate study was formed.

The University of Chicago was founded on July 1, 1891. John D. Rockefeller had provided the financial support to organize a new university to be staffed by professionally qualified scholars. William Rainey Harper took office as the first president. Harper was building another university similar to Johns Hopkins, which had been modeled on the German concept of a society of scholars by Daniel Coit Gilman. When the first classes were held at Chicago on October 1, 1892, Harper had assembled a faculty of high quality. There were 103 members on the new faculty, and of these eight were former university presidents. Outstanding scholars included Thorstein Veblen in economics, Albert Michelson in physics, and Thomas C. Chamberlin in geology. Chamberlin, who resigned as president of the University of Wisconsin to accept the appointment of Chicago, brought with him his associate Rollin D. Salisbury.[24] In 1892 H. J. Mackinder visited Chicago and was urged by Harper to join the faculty. Mackinder did not accept the invitation.

[24]T. C. Chamberlin was the son of a pioneer farmer in Wisconsin. He graduated from Beloit College in 1866 and returned there as professor of natural history in 1873. He became professor of geology in 1880. From 1882 to 1887 he worked full time on the U. S. Geological Survey. As a geologist he is noted for his demonstration of the occurrence of several advances and retreats of the ice in Wisconsin and for his formulation of the planetesimal hypothesis regarding the origin of the earth. In the field of scientific methodology he is also known for his paper on the need for using multiple working hypotheses (Chamberlin, 1897). From 1887 to 1892 he was president of the University of Wisconsin. From 1892 until his retirement in 1919 he was chairman of the Department of Geology at the University of Chicago.

R. D. Salisbury graduated from Beloit College in 1881 and succeeded Chamberlin there in 1882. In

Rollin D. Salisbury

Rollin D. Salisbury, who was the chairman of the Department of Geography at Chicago from 1903 to 1919, was a major force in the development of professional geography in the United States (Pattison, 1979). He influenced a large number of students and was generally recognized as the best teacher in the university. His freshman course in physiography was always filled. Whereas Davis was noted for his polished lectures, Salisbury was a master of the art of stimulating and directing class discussions. By skillful questioning he insured student participation. But when any student attempted to obscure a lack of preparation behind a screen of generalities, Salisbury would remark: "perfectly true, perfectly general, perfectly meaningless." He insisted that a student learn to express himself clearly—"not so that he could be understood, but so that he could not be misunderstood" (Chamberlin, 1931:128). His classroom was no place for the dull student; but the better ones were enormously stimulated, especially by participation in his advanced seminars. Students who attended one of his seminars, even for a short period in the summer session, went out to establish similar seminars in other schools and colleges. In 1913 he started a regular weekly meeting of staff and graduate students in which geographical questions and problems were discussed.[25] In the give and take of scholarly discussion those in attendance not only clarified in their own minds the methodology and scope of geography but they also learned a valuable lesson about the way to carry on such a discussion with their peers, to accept criticism without emotional reaction, and to respect the words of others even if they were in disagreement. Colby said that these seminars did more to establish high standards of work and thought at Chicago and elsewhere than any other single part of the program (Colby, 1955).

Salisbury was himself concerned with that part of geography that he called physiography (physical geography). He was hopeful, but skeptical, that workers in anthropogeography might develop that part of geography also on a scientific basis. He is not properly described as "a follower of Davis," although he did make use of the terminology that Davis had proposed. But in Salisbury's teaching the cycle of erosion was given a minor place. Salisbury also rejected the idea of a simple cause and effect relation between the physical earth and the human response. Physiography, for Salisbury, was the scientific study of the stage setting on which the

---

1891 he came to the University of Wisconsin but resigned the next year to accept an appointment with Chamberlin as professor of geographic geology at Chicago. In 1894 he was named dean of the Ogden School of Science at Chicago, which position he held until his death in 1922. Salisbury was named chairman of the new Department of Geography in 1903. In 1919, when Chamberlin retired, Salisbury became chairman of the Department of Geology; and Harlan H. Barrows replaced him in geography.

[25]The first such seminar was held in the fall of 1913 and was attended by Salisbury, Barrows, Tower, and Goode from the staff; and by graduate students, Charles C. Colby, Wellington D. Jones, A. E. Parkins, William Haas, S. S. Visher, and Mary Lanier (Colby, 1955).

human drama unfolded. But the relation of the stage setting to human action was not a causal one.

Salisbury published his ideas in his *Physiography* (Salisbury, 1907), which was widely used throughout the United States and went through several editions. With Wallace W. Atwood[26] he made a selection of topographic maps from the U. S. Geological Survey, parts of which were reproduced along with notes interpreting the origin of the landforms (Salisbury and Atwood, 1908). With other members of the geography staff, Salisbury collaborated in a college text presenting a basic course in geography, which for many years was the standard college text in America (Salisbury, Barrows, and Tower, 1912). He also demonstrated his method and point of view in studying physical geography in a monograph on the stage setting of Chicago (Salisbury and Alden, 1899).

Building the Department

When the new Department of Geography was established in 1903, Salisbury immediately recruited two young scholars to form his staff. One was J. Paul Goode, who had studied geology at Chicago in 1896–97, but who had gone to the University of Pennsylvania to study geography with Johnson.[27] When Salisbury found himself responsible for developing a program of study in geography, he remembered the young student who had so impressed him with sound ideas about geography some six years earlier. He asked Goode to prepare a proposed program of courses for the new department. Goode answered in detail and most of his suggestions were adopted. In 1903 Salisbury invited Goode to become a member of the geography staff.

The other young man recruited at that time was Harlan H. Barrows, who had just received the B.S. degree in geology at Chicago, but who had already taught geography for more than five years.[28] Barrows was promoted rapidly, and in 1919,

[26]Wallace W. Atwood received his B.A. degree in geology at Chicago in 1897. On a field trip to the Devil's Lake Driftless Area of Wisconsin with Salisbury, he developed a keen interest in the interpretation of landforms. He received the Ph.D. in geology at Chicago in 1903, and from 1903 to 1913 he was on the staff of the Department of Geology there. For his later career at Harvard and Clark, see p. 318.

[27]J. Paul Goode received the B.S. degree from the University of Minnesota in 1889 and from then until 1898 held the position of professor of natural science at the Minnesota State Normal College at Moorhead, Minnesota. In 1894 he attended a summer session at Harvard and worked with Davis. In 1896–97 he was a "fellow in geology" at Chicago. In 1899 he was appointed professor of physical science and geography at Eastern Illinois State Normal College. Finding no place where he could pursue advanced graduate study in geography, he went to the University of Pennsylvania to study with E. R. Johnson and received the Ph.D. in economics in 1901. He was an instructor in geography at Pennsylvania when, in 1903, he was offered a position at Chicago.

[28]Harlan H. Barrows completed an undergraduate program at Michigan State Normal College under Charles T. McFarlane in 1896. He had been teaching at the Ferris Industrial School (later Ferris Institute) in Big Rapids, Michigan, when he decided to take additional undergraduate work in geology at Chicago. He received the B.S. degree in geology in 1903. He was such an outstanding student that he was

when Salisbury moved back to geology, he was named chairman of the Department of Geography.

Geology and geography at Chicago remained very closely associated during this period. When Chamberlin was away Salisbury acted as chairman of both departments. Students took courses in both departments. Geology and geography shared space in the same building and made use of the map collection. Salisbury and Goode undertook to set up no new courses that would duplicate work already being offered in geology. But meteorology and climatology, which had been taught in the Department of Geology, were transferred to geography, where they were taught by Goode. The new courses in geography were planned to occupy the great uncultivated field between geology and climatology on the one hand, and biology, history, sociology, economics, anthropology, and political science on the other. The first Ph.D. granted by the new department went to F. V. Emerson in 1907 for what was one of the first American urban studies by a geographer—"A Geographic interpretation of New York City." (Emerson, 1908–1909)[29]

Within a decade the new department had already established its preeminent position in the training of the younger generation of geographers. In addition to Salisbury, whose chief concern was with the physical earth, and Goode, who offered work not only in meteorology and climatology but also in the economic and commercial geography of Europe and the history of geographic thought, Barrows began to develop his ideas on the historical geography of the United States, which came into full flower after World War I (Barrows, 1962). His course was very popular with the undergraduates. A questionnaire circulated among Chicago alumni some years later listed Barrows's course on historical geography as one of the most worthwhile courses in the whole undergraduate program of the university (Colby, 1955). During his career Barrows offered some twenty-five different courses (Koelsch, 1976).

Walter S. Tower was added to the department in 1911 to offer courses in the economic geography of South America and in political geography (Tower, 1910).[30] Tower was one of the earliest "regional specialists" in the Latin American field to be appointed to a university post in the United States. Others who taught occasionally at Chicago were Ellen C. Semple, in alternate years between 1906 and 1924, and Bailey Willis, a geologist from Stanford University.

---

immediately appointed as Salisbury's assistant. He was promoted to instructor in geography in 1907, to assistant professor in 1908 to associate professor in 1910, and to professor in 1914. From 1919 until the time of his retirement in 1942 he was chairman of the department (for other aspects of his career, see Chapter 14).

[29]F. V. Emerson studied at Edinboro (Pennsylvania), Colgate, Cornell, and Harvard before completing his graduate studies at Chicago. He then worked with C. F. Marbut at the University of Missouri and from 1913 to his death in 1919 was professor of geology at Louisiana State University and also director of the Soil Survey of Louisiana.

[30]For a biographical sketch of Walter S. Tower, see p. 344.

The graduate students who worked in the department during the first decade included many who became leaders of the profession in the period after World War I.[31] When the government of Argentina planned the construction of railroads westward across Patagonia after 1902, it looked to the example of railroad surveys undertaken earlier in the American West and requested assistance from the U.S. Geological Survey. Bailey Willis was appointed to organize and operate this survey. When he found that he needed someone trained in economic geography, he turned to Salisbury for a recommendation. The young man who worked with Bailey Willis in Patagonia in 1912 was Wellington D. Jones. It was this experience in Patagonia that inspired Jones, along with his fellow graduate student Carl Sauer, to suggest the importance of detailed field-mapping of agricultural areas. For the first time it was suggested that maps of land use should be prepared at the same scale and the same degree of detail as the maps of the physical environment. The paper that Jones and Sauer published in 1915 had been thoroughly discussed at the Chicago seminar (Jones and Sauer, 1915).

Colby identifies three things in pre-1917 America that caused the rapid increase in the teaching and writing of geography. We have already mentioned the critical importance of the surveys of the American West and especially the pioneer work of Gilbert, Powell, Wheeler, and Hayden. Also in the decade before World War I there was a great increase in overseas commerce that was causing the public to demand the teaching of commercial and economic geography—just as the traders of Amsterdam had made similar demands in the seventeenth century and were answered by Varenius. The University of Pennsylvania responded to this demand in the United States through the activities of the economists and others in the Wharton School. And the third influence that Colby listed was the rapid opening up of new natural resources, including oil. This drew the attention of educators to the need for teaching the geography of resources and the methods of conserving them. All these matters were discussed at length by the participants in the Chicago seminar.[32]

Another distinctive characteristic of the Chicago group was the emphasis they placed on field studies. In the tradition of the exploring expeditions of the West, all graduate students were expected to examine the character of the landscapes and to identify geographical problems from direct observation. In September, 1913 Tower led a party of six students in a traverse of the Northern Appalachians from Pittsburgh to Harrisburg. In September, 1914 Barrows led a much larger group on foot across the Southern Appalachians. In 1915 Goode conducted a trip to the West,

[31]These students included Charles C. Colby, Wellington D. Jones, Stephen S. Visher, V. C. Finch, Carl O. Sauer, Mary Lanier, Mary Dopp, Mabel C. Stark, L. P. Denoyer.

[32]The four members of the department staff made distinctive contributions to the seminar discussions. Salisbury contributed the sure touch of the master in directing these discussions. Barrows had the keenest mind: he had a prodigious memory and was a strict logician. Tower introduced challenging and original ideas. And Goode never failed to insist that the best way to communicate geographical ideas was through the expert use of maps (Colby, 1955).

. visiting ranches, mines, and irrigated areas, and including a visit to the Panama Pacific Exposition in San Francisco. This kind of field course became a distinctive feature of Chicago before World War I. But, as we will see, students like Wellington D. Jones, Carl O. Sauer, and K. C. McMurry began to visualize a quite different kind of field experience in which students would not be taken on a conducted tour but would be set to work in a restricted area to identify problems and demonstrate their ability to find answers.

It is important, also, that in their field studies, the geographers learned to cooperate with scholars in other disciplines. For example, H. C. Cowles, professor of botany at Chicago, was interested in plant ecology. He involved geographers in his studies of plant succession on the Indiana Dunes. Cowles was one of the founders of the Association of American Geographers (1904) and its president in 1910.

MODERN GEOGRAPHY IN 1914

In 1914 George B. Roorbach, who was assistant professor of economic geography at the University of Pennsylvania, published the results of a questionnaire he had sent out to people who called themselves geographers. He found that in a seminar discussion of the scope and method of geography almost everyone had his own definition. As in Germany four decades earlier, there were very few people teaching geography who had been formally educated as geographers. Therefore, each new geographer felt impelled to answer the question: "What is geography?" And true to the nature of most scholars, it would not do to accept any other scholar's definition of the field. Consequently, there was little resolve concerning the nature of geography. Roorbach asked for a listing of the most important tasks to be undertaken by geographers. He received twenty-nine replies, all but four of which were from scholars in the United States. The four others were well-known British geographers (Roorbach, 1914).

He found an almost unanimous agreement that geography was the study of the relationship between the earth and life—which was essentially the idea proposed by Davis. The respondents then listed the following tasks as important in the order given:

1. The exact determination of the influence of geographic environment. This was placed first by twenty-two out of the twenty-nine respondents.
2. Regional studies of selected areas. There were some British suggestions that a major task would be dividing the world into its major natural regions.
3. The definition and organization of geographical material.
4. The improvement of the teaching of geography.
5. The study of the influence of geographic factors on history.

6. The exploration of unknown or little-known places (suggested by the British geographer Scott-Keltie and by Robert E. Peary).

7. The study of physical geography.

So it was in 1914. Geographers did not realize that culture, not nature, determined the significance of environment, site, and natural resources, in spite of the critique of environmentalism advanced by ethnologists since before 1900. Moreover, the interuniversity conferences between Columbia-trained ethnologists and Yale geographers, which were arranged by Franz Boas in 1913 for exploring the problem of environmental conditioning of society, failed to inoculate geographers against the naturalistic assumption. The principle that culture is the fundamental extra-environmental factor in the derivation of human activities did not penetrate geography more generally until after World War II (Speth 1978: 10-11).

14

The New Geography in the United States—World War I to Midcentury

Scarcely was physical geography established, or perhaps I should say rejuvenated and reestablished, before an insistent demand arose that it be "humanized." This demand met with prompt response, and the center of gravity within the geographic field has shifted steadily from the extreme physical side toward the human side, until geographers in increasing numbers define their subject as dealing solely with the mutual relations between man and his natural environment. By "natural environment" they of course mean the combined physical and biological environments.

The period from World War I to the decade of the 1950s was transitional as the paradigm of acceptable geographical study was reformulated. Trained geographers began to emerge from graduate departments of geography and enter the profession with the result that the traditionally close ties with geology were gradually loosened (Harris, 1979; Trewartha, 1979). In the course of time the focus of geographical inquiry shifted toward social science and away from exclusive concern with earth science. Indeed there were many who were deeply disturbed by the growing neglect of the methods and concepts derived from geology and by the tendency to relinquish the study of physical geography to other disciplines. The period has been incorrectly described as one in which geographers devoted themselves to the "mere description of unique places" without any effort to formulate general concepts. Such a characterization seems unwarranted. Much attention was given to the information and use

The quotation above is from Harlan H. Barrows in his presidential address to the Association of American Geographers, 1922 (Barrows, 1923:3).

of concepts and models, and many principles and ideas current in the 1970s can be traced back to their early appearance in the 1920s and 1930s.

As the entry of the United States into World War I approached, the ideas of William Morris Davis were almost unchallenged in geomorphology and were only beginning to be challenged in human geography. With the benefit of hindsight, we can now see that the careful observation and measurement of physical processes were neglected in favor of qualitative studies of natural history. In the field of human geography, social Darwinism was under attack, and indeed most of the historians and other social scientists had already rejected it (Hayes, 1908; Barnes, 1925). Many geographers, too, were ready to follow A. P. Brigham in rejecting strict environmental determinism and R. D. Salisbury in avoiding simple cause and effect explanations for complex associations of things on the earth's surface. But not all the geographers were aware of the validity of the criticisms of Davis's scheme of human response to physical controls. The persuasive teaching of Ellen Semple, the creative work of Ellsworth Huntington, and to a lesser extent the work of Whitbeck (1926) continued to gain support for some kind of environmental control of human behavior (Huntington, 1924). Long after the physical cause and human response paradigm had been dropped, some geographers continued to use the language of "geographic factor" and "environmental control" (Baker, 1921; Peattie, 1929, 1940; Whitbeck and Thomas, 1932; Atwood, 1935; Martin, A. F., 1951; Lewthwaite, 1966.)

The tradition established at Harvard was carried on after the retirement of Davis in 1913 by Wallace W. Atwood (Bushong, forthcoming).[1] As professor of physiography at Harvard, Atwood attracted many students who were excited by his teaching and by his leadership in field studies. After 1921, when the Clark Graduate School of Geography was established with Atwood as director, students came not only from the United States but also from many foreign countries. Atwood's school texts were very popular, departing from the traditional organization by political units and adopting one based on natural regions. It has been said that "no American has ever brought geography to so many people." Unfortunately, the geographical ideas he taught were already disputed by his colleagues when he reached the peak of his influence—much as Davis's ideas of the causal notion were already outmoded when he used them as the organizing principle of the "new geography."[2]

---

[1] Wallace W. Atwood was on the staff of the Department of Geology at Chicago when he was selected to succeed Davis at Harvard. At Harvard Atwood continued his interest in field studies in geomorphology and in the teaching of geography in elementary and secondary schools. His study of the San Juan Mountains of Colorado (Atwood and Mather, 1932) is a classic of its kind. The last chapter deals with "The Utilization of the San Juan Region by Man." In 1920 he became president of Clark University and in 1921 the director of the Graduate School of Geography. In 1925 he founded the periodical, *Economic Geography*. He was president of the Association of American Geographers in 1934. He retired in 1946.

[2] Another brilliant teacher who supported the ideas of environmental determinism was Griffith Taylor. He was on the staff of the Department of Geography at Chicago from 1928 to 1935. Taylor's work in Australia and in Canada is discussed on pp. 259–261.

CHANGING CONCEPTS

The period after World War I witnessed the gradual erosion of concepts of physical controls and human responses and a vigorous competition among proposals for new approaches to geographical inquiry (Brunhes, 1925). There is always a certain lag in such changes, a regrettable persistence of traditional error (Jastrow, 1936; James, 1967). But such a period of change is an exciting one because a variety of new ideas are used experimentally (Popper, 1959; Wright, 1966).

There were four main currents of geographic thought to examine. One proposal was that the scope of geographical study whould be narrowed to focus on the adjustments made by man to his physical and biotic environment. This was the proposal that geography should be described as *human ecology*. A second proposal was that geographers should focus on the identification and explanation of observed differences from place to place on the face of the earth. Such studies are included in *chorology*, or the study of places or regions. But chorology was to be more than descriptive. The search for explanations that would make sense out of observed diversity took two chief directions: one was to seek genetic explanation in terms of processes of change acting through time, leading to *historical geography* and its specialized offshoot *sequent occupance;* the other was to seek functional explanations, leading to the concept of the *functional organization of space*. These explanatory procedures were applied in various topical fields.[3] Meanwhile, the decade after World War I also saw a notable shift of professional attention from academic studies to the use of geographic concepts and methods in the study of practical economic, social, and political questions. *Applied geography*, as it developed in the period between World War I and the decade of the 1950s, is the subject of Chapter 15.

Human Ecology

That geography should be focused on the study of human ecology, or the adjustment of man to his natural surroundings, was presented by Harlan H. Barrows in his presidential address before the Association of American Geographers in 1922 (Barrows, 1923). Adjustment, as Barrows used the word, was not caused by the physical environment but was a matter of human choice. Barrows felt, however, that although the subject matter of geography had been partly lost to other disciplines, it was still too broad and that such specialities as geomorphology, climatology, and biogeography should be relinquished. Like others before him, he sought a unifying theme that would bring coherence to the study of geography. The unifying theme, he argued, could be provided by restricting attention to human ecology. He continued:

---

[3]For full summaries of the contributions made in the various fields of geography in the United States up to 1954 together with extensive references to published materials, see James and Jones, 1954. See also Colby, 1936 and Whitaker, 1954.

I believe that those relationships between man and the earth which result from his efforts to get a living are in general the most direct and intimate; that most other relationships are established through these; that, accordingly, the further development of economic regional geography should be promoted assiduously, and that upon economic geography for the most part other divisions of the subject must be based. . . . I believe that geography has been too much a library subject, and too little a field subject. I hold that the field is the geographer's laboratory. I believe that we have made only a beginning in the development of rigorous, scientific methods of field work in physiography and geology, and that the development of a thoroughly effective technique in field work is perhaps our greatest immediate need. Since most of us are "rebuilt geologists" do we not, in general, study the geological items and merely observe, in more or less haphazard fashion, the geographical items? Precisely how should one study in the field those relationships which are truly geographic? . . .

I believe that much of our so-called geographical exposition is something else, that to be truly geographic a discussion must involve from beginning to end an explanatory treatment in orderly sequence of human relationships, and that the development of a satisfactory technique of exposition is only less important than the perfection of field methods . . . [Barrows, 1923:13–14].

But geographers still had to examine skillfully two or more different sets of factors. To be sure, Barrows insisted that the physical conditions should only be studied in relation to man, but this proved to be more easily said than done. Although Barrows's paper has often been quoted and assigned as reading for graduate students, it did not provide guidelines for a new orientation of the field (Hartshorne, 1939:123).

Chorology

Some sturdy chorologic inquiry was rendered by M. Jefferson (1917) and W. L. Joerg (1914, 1936). But a much greater impact on the development of geography in the United States resulted from Carl O. Sauer's study entitled "The Morphology of Landscape" (1925) (Also see Leighly, 1976; Stanislawski, 1975). This was written shortly after Sauer became chairman of the Department of Geography at the University of California (Berkeley) in 1923 and was intended as a kind of inaugural lecture—a declaration outlining his concept of the field of geography to his colleagues in other departments of the university. Such a declaration was deemed necessary because of the common and uncritical acceptance of the earlier definitions of geography solely in terms of environmental influences. Sauer insisted that no field of study can be defined in terms of a single causal hypothesis that would commit the student to a particular outcome of an investigation in advance (Sauer, 1927:173). To go into the field to look for influences or evidences of control exerted by the physical conditions, is to accept a single dogma. Sauer did not deny the possibility of environmental determinism in specific cases but insisted that the concept of influences should be exposed to objective testing. Sauer referred to Siegfried Passarge, who recommended that the first step in any geographic study must be to determine the facts by describing the visible characteristics of an area without attempting to explain them in advance.

Sauer went back to the writings of Humboldt and Hettner who supported the so-called chorological concept of the nature of geography. Geography, he pointed out, is concerned with the study of things associated in area on the earth's surface and with the differences from place to place—both physical and cultural. Man, behaving in accordance with the norms of his culture, performs work on the physical and biotic features of his natural surroundings and transforms them into the cultural landscape.

> The design of the landscape includes (1) the features of the natural area and (2) the forms superimposed on the physical landscape by the activities of man, the cultural landscape. Man is the latest agent in fashioning of the landscape. The study of geography begins therefore with physical geography, but—coasts are marked by ports; mountains have flung over them the trails and workings of man. A phrase that has been much used in German literature, unknown to me as to origin, characterizes the purpose perfectly: "the development of the cultural out of the natural landscape." This is the newer orientation that continues the traditional position [Sauer, 1927:186–187].

This, Sauer suggested, is what geography is all about. It is the study of areas, not to describe them as unique occurrences—for there is no such thing as an idiographic science—but rather to identify the regularities and recurrences from place to place that permit the formulation of generalizations. To understand the changes man has made on the face of the earth, it is necessary to go back far enough in time to establish the nature of the processes. Geography as chorology, or the study of the associations and interconnections of things in areas or regions, is what Sauer calls a "naïvely given section of reality,"—that is, a division of knowledge that is accepted as axiomatic. He concludes his paper with these remarks:

> Our naïvely given section of reality, the landscape, is undergoing manifold change. This contact of man with his changeful home, as expressed through the cultural landscape, is our field of work. We are concerned with the importance of site to man, and also with his transformation of the site. Altogether we deal with the interrelation of group, or culture, and site, as expressed in the various landscapes of the world [Sauer, 1925:53].[4]

Sauer's purpose was to make a clean break with the traditional geography inherited from the period before World War I. He might have discussed "the

---

[4]For Hartshorne's criticism of the use of the word, landscape, and the justification for its use presented by Schmithusen, see pp. 177–178. See also Broek, 1938.

morphology of regions, or areas.'' But the word, region, in 1925 was encrusted with more ambiguities that the word, landscape, including the notion of the uniform physiographic region that was also uniform in human response. The word, area, was even more ambiguous. As a result of these difficulties with confused word meanings, discussions of the nature of geography that followed not infrequently descended to controversy over the meanings of words.

Sauer's paper won widespread acceptance among the younger members of the profession, most of whom had completed their graduate training since 1920 and had recently been appointed to one of the several new geography staffs then being formed.[5] The new geographers had been raised on the search for geographic influences, but by 1925 there was enough skepticism concerning the content or method they had been taught to make the younger generation ready to accept a change of paradigm. With enthusiasm they turned to the study of landscapes, or regions, seeking the kind of interacting systems among diverse phenomena that gave character to particular places and tracing the changes introduced by the human settlement back to origins (Dodge, 1932; Broek, 1932). Here is what Norton Ginsburg wrote many years later in a position paper for the Commission on College Geography:

> Theirs was above all a "scientific" geography, concerned with regions as systems, and with the comparative method as a device for developing hypotheses concerning areal relations and processes. The use of statistics was simple and even primitive, to be sure, but their concerns were far from trivial, and the problems with which they dealt were of—to use a somewhat abused word—"overriding importance," at least to the development of geographic discipline.

The younger generation developed new jargon, including the use of the symbols of the Köppen classification of climates, and proceeded to reject the older generation of seekers after environmental influence. Since most of these younger geographers had taken at least some of their graduate work at Chicago—where they had been participants in Salisbury's famous seminar—they spread to other universities the idea of regular staff-student discussions of philosophical or methodological questions.

Yet there is a curious fact about the impact of Sauer's paper. These things had all been said before. In 1924 Sauer himself had published a paper in the *Annals*

[5] Sauer had been appointed to the newly renamed Department of Geology and Geography at Michigan in 1915. The chairman was the geologist, William H. Hobbs. In 1921 Wallace W. Atwood became chairman of the newly founded Graduate School of Geography at Clark University. In 1923, when the Department of Geography was established in the Social Science Division at Michigan, K. C. McMurry became its chairman. Sauer became chairman of the department at the University of California in Berkeley. New separate departments were established in 1925 at Minnesota and in 1928 at Wisconsin. Meanwhile, there were many positions to be filled in departments of geology and geography. The number of new Ph.D.'s increased rapidly: ten in 1916–20; 32 in 1921–25; sixty-six in 1926–30; and fifty-one in 1931–35 (the period of the Great Depression) (Whittlesey, 1935; see also Hewes, 1946 and Browning, 1970).

AAG attacking the study of influences and advocating the field survey of the "areal expression of man's activities" (Sauer, 1924). Instead of going into the field with a set of a priori principles concerning the effect of the physical environment on man, one should seek to observe the facts and then draw conclusions from them. This part of Sauer's proposal drew immediate criticism from some of the older generation. As Dryer pointed out, no one could actually observe anything or describe anything without some kind of working hypothesis, conscious or unconscious. There would be no way to select things to record and describe. If anyone does try to do what Passarge and Sauer recommend, he wrote, "the result is likely to be a catalogue half rubbish, like a child's collection from a dump heap, and wholly unscientific."[6]

Dryer, himself, in his presidential address to the Association of American Geographers in 1919, had presented the chorological concept, but not by that name:

> It seems clear and beyond question that the psychological foundation of the geographic concept is the sense of distribution in terrestrial space. We must concede the pertinence of the doctrine of Kant that "geography is a narration of occurrences which are co-existent in space." The idea, more sharply put by Bain[7] in the statement that "the foundation of geography is the conception of occupied space," fits and includes every work generally recognized as geography from Strabo to Ritter and Reclus. With various additions and qualifications, it forms the essence of most of the current and accepted definitions of geography, of which quotation is unnecessary [Dryer, 1920:5-6].

Furthermore, N. M. Fenneman made almost the same point in "The Circumference of Geography," which was his presidential address to the Association of American Geogeophers in 1918 (Fenneman, 1919).

Dryer's paper seems to have had slight impact on his fellow geographers. Nor, for that matter, was Alexander Bain's very modern-sounding idea of geography as dealing with "the conception of occupied space" (1879) given any attention. Sauer, who was present at the St. Louis meeting of the association in 1919 when Dryer gave his paper, makes no reference to it in "The Morphology of Landscape." The report on the St. Louis meeting in the *Geographical Review* has the following to say about Dryer's address:

> President Dryer's address on "Genetic Geography: The Development of the Geographic Sense and Concept" was scholarly to a high degree and will rank among the finest presidential addresses that have been presented before the Association. It ought to be given a much wider circulation than it will receive if its publication is confined to the Association's *Annals* [*Geographical Review*, 9 (1920):139].

[6]See Dryer in *Geographical Reyiew*, 16 (1926):348-350.
[7]Alexander Bain (1818-1903), a Scottish philosopher, professor of logic and English at Aberdeen from 1860 to 1880, in *Education as a Science* (London, 1879), p. 272.

The report on the meeting goes on to say that the average attendance at the sessions was about thirty-five, half of whom were members, and that only three of the members were from eastern colleges. The large number of younger people about to enter the profession had not started in 1919.

After 1925, when a new generation of younger geographers began to emerge, it became common for geographers to report on situations where the physical features of an area were not of major importance. While some of the older geographers and a few of the younger ones continued to report on responses or influences, many of the younger ones took delight in describing cases where other factors were more significant than the physical ones. Richard Hartshorne presented a paper to the association in 1926 concerning the location factor in geography with special reference to manufacturing industries (Hartshorne, 1927). Location relative to the sources of raw materials, markets, power, and labor was more important than location relative to such features as relief, drainage, soil, or climate. For those who had been "explaining" the concentration of cotton textile factories in New England by the humidity of the climate (which permitted the spinning of thread without snarling due to static electricity), this reference to relative location with no mention of the elements of the physical environment came as an innovation. People who came to such conclusions were accused of leaving the "ge" out of geography.

Historical Geography

Those who adopted the chorological theme were never content merely to describe the content of an area in static terms. Attention was necessarily focused on the processes, or sequences of events that provided an explanation of the observed landscapes. To explain is to make sense out of apparently endless diversity. Of course the study of sequences of events gave a dynamic quality to regional studies that purely contemporary description could not provide. Andrew H. Clark explains it as follows:

> The genetic approach to geographical study inevitably leads to an examination of the past. This does not mean that one is to seek simple causes in the past to account for contemporary conditions, but rather that the conditions observed at any period of time are to be understood as momentary states in continuing and complex processes of change. Simple cause and effect relations are elusive, for no matter how far back a scholar may penetrate there is always a more distant past calling for further investigation. The genetic approach focuses attention on processes, for whatever interests us in the contemporary scene is to be understood only in terms of the processes at work to produce it. It is not, therefore, a search for origins in any ultimate sense, but rather views the present, or any particular time, as a point in a long continuum [Clark, 1954:71].

It is important to understand that the new approach to historical geography that appeared in America in the 1920s was not at all like that of Brigham and Semple in

1903 or like Barrows's course on "The Influence of Geography on American History." Barrows had been greatly influenced in his early years by Ellen Semple's interpretation of Ratzel and by the historian Frederick Jackson Turner, who in 1893 gave his famous lecture, "The Significance of the Frontier in American History" (Koelsch, 1969:634). Turner was an eloquent speaker for geographical influences on history. But at some time between 1920 and 1922 Barrows changed his basic approach to this topic. In 1923 he renamed his course "Historical Geography of the United States," and he focused his attention on examples of "creative human adjustments to a passive natural environment" (Koelsch, 1969:637). We should note, also, that Ellen Semple's book on the Mediterranean, published in 1931, offers outstanding examples of the method of historical geography.

The course Barrows gave at Chicago made a lasting impression on his students. Many of the graduate students wrote dissertations that can be classified as historical geography (e.g., Parkins, 1918); dissertations in other universities in this period were also contributions to historical geography (Clark, 1954:84–85). Yet it seems that not many of the new geographers trained in this way continued to produce studies in historical geography as such—although most of them made use of genetic explanations that involved some attention to the time dimension. Some of the most important studies in historical geography were written by nongeographers.[8] And one study that described in detail the movement of a former hill town down into the valley to locate at a new site on the railroad was written by the gologist, J. W. Goldthwait (the story of Lyme, New Hampshire, in Goldthwait, 1927). This paper, which was published in the *Geographical Review*, was regarded for many years as a model of its kind.

During the period we are discussing, there were two American geographers who are recognized as the chief innovators in historical geography (Clark, 1954). One was Ralph H. Brown, the author of *Mirror for Americans* (Brown, 1943). In this study Brown undertook to write a geography of the eastern seaboard of North America as portrayed about 1810 in the writings of the previous two decades or so. This imaginative approach to the re-creation of a past geography as perceived by scholars of the time foreshadows the modern attention to environmental perception. Brown then published a second book, *Historical Geography of the United States* (Brown, 1948), in which he traced the geographical changes during the course of settlement. Unfortunately, the career of this outstanding innovator was cut short by his untimely death at the age of fifty.[9]

The other major source of inspiration in historical geography was Carl O. Sauer (Speth, forthcoming). At Berkeley, Sauer formed close intellectual ties with

[8]For example: Allan C. Bogue, *From Prairie to Corn Belt* (Chicago, 1963); Bernard DeVoto, *The Course of Empire* (Boston, 1950); H. A. Innis, *The Fur Trade in Canada* (New Haven, 1930); J. C. Malin, *The Grassland of North America: Prolegomena to Its History* (Lawrence, Kans., 1947); W. P. Webb, *The Great Plains* (New York, 1927).

[9]See the list of his numerous articles in a brief obituary by Stanley D. Dodge in the *Annals AAG*, 38 (1948):305–309.

two other workers in allied fields: Herbert E. Bolton, historian, and A. L. Kroeber, anthropologist. These outstanding scholars, each bringing a different background to his studies, came together around the problems of the interpretation of Latin America. The combination proved enormously stimulating not only to these three men but to the many graduate students in all three fields. The first of many monographic studies that Sauer wrote with a graduate student as coauthor described the prehistoric Indian frontier of settlement on the Pacific coast of Mexico (Sauer and Brand, 1932). Sauer, himself, undertook to locate the old colonial highway from Guadalajara to Tucson on the basis of field study (Sauer, 1932). Additional samples from these studies in historical geography include: Kniffen, 1931; Broek, 1932; Meigs, 1935; Spencer, 1939; Carter, 1945; Parsons, 1949; Clark, 1949; Hewes, 1950; and West, 1952. Sauer extended his works on historical geography to cover a wide variety of topics (Sauer, 1952, 1956, 1966b, and his presidential address before the Association of American Geographers, 1941).

Out of these studies of sequences of settlement certain principles began to emerge. One was the principle that the same physical conditions of the land could have quite different meanings for people with different attitudes toward their environment, different objectives in making use of it, and different levels of technological skills. In agricultural areas it was clear that slope had one meaning for the man with a hoe and quite another for the man with a tractor-drawn plow. It might be that the introduction of machinery could reduce the arable area of a country or change the kind of soil considered desirable. People with one kind of culture might concentrate their settlements on flattish uplands, whereas another people in the same area might concentrate in the valleys. Water power sites that were useful for the location of industries before the advent of steam lost that attraction when power came from other sources.

One of the early studies in which the changing significance of the land was traced through a sequence of periods with different cultures dealt with an arbitrarily outlined area on either side of the Blackstone Valley, extending from the outskirts of Worcester, Massachusetts, to those of Providence, Rhode Island. The study, published in 1929 was summarized as follows:

> Thus the landscapes of the Blackstone area are made up of a complex of cultural impressions set one upon the other. The three chief cultures, the native Indian, the rural European, and the urban manufacturing, have each modified the natural setting in a unique and characteristic way. Forms developed by the Indian culture are visible, even today, in the shell mounds, the deposits of chipped stones and broken utensils, or the scarcely discernable trails. The forms of the rural European culture are visible on every side, some of them continuing without change of function to the present, others significantly modified in their use, and others remaining as weather-beaten ruins or brush-entangled fields to tell of a period which exists no more.... Finally the urban landscape, in spite of its relatively small area, has come to occupy the position of commanding importance around which the economy of the region is oriented [James, 1929:108].

In that same year Derwent Whittlesey gave studies of this sort a name. He described the studies of the processes of change in the occupance of an area as *sequent occupance*. Referring especially to New England, he wrote:

> . . . each generation of human occupance is linked to its forbear and to its offspring, and each exhibits an individuality expressive of mutations in some elements of its natural and cultural characteristics. Moreover, the life history of each discloses the inevitability of the transformation from stage to stage [Whittlesey, 1929:163]

Studies in sequent occupance represent the antithesis of environmental determinism. In a sense they represent a form of cultural determinism, for it is recognized that with any significant change in the attitudes, objectives, or technical skills of the inhabitants of a region, the significance of the resource base must be reappraised. A large number of studies published during the 1920s and 1930s made use of the method of sequent occupance, whether or not that term was adopted (e.g., Colby, 1924; James, 1927, 1931; Platt, 1928, 1933).

This was not a period when general concepts were neglected. O. E. Baker made effective use of economic principles to explain the development of American agriculture (Baker, 1921, 1923), and H. H. McCarty used general concepts to interpret economic conditions and population regions in America (McCarty, 1940, 1942). Both Whittlesey and Hartshorne formulated theoretical structures to enlarge the reach of political geography (Whittlesey, 1939: Hartshorne, 1950). In 1939 Mark Jefferson wrote, "A country's leading city is always disproportionately large and exceptionally expressive of national capacity and feeling"—which he called the "Law of the Primate City" (Jefferson, 1939:231). As early as 1921 Marcel Aurousseau, an Australian geographer working in Washington, D. C., investigated the world distribution of population and sought to quantify the "expansion ratios" of already occupied regions (Aurousseau, 1921). In 1932 Stanley D. Dodge proposed that studies of population could be related to a statistically normal growth curve, a portion of a sine curve (Dodge, 1933). Studying population changes in Vermont and New Hampshire, he classified each minor civil division in terms of its position on the curve of growth or decline. Applying the concept to the whole of New England, he revealed a new pattern of population regions by plotting his results on a map (Dodge, 1935).

One of the most imaginative geographers of the period was Robert S. Platt,[10] who was a member of the Department of Geography at Chicago from 1919 to 1957.

[10]Robert S. Platt graduated from Yale in 1914 in history and philosophy. He taught for a year in the Yale Collegiate School at Changsha, China, and then returned for graduate study in geography at Chicago. He was appointed instructor in geography in 1919 and received the Ph.D. in 1920. He was a professor by 1939 and from 1949 until his retirement in 1957 he was chairman of the Department of Geography. When he returned to Chicago after teaching in China, he found that Wellington Jones was already teaching a course on Asia and that Walter S. Tower, who had taught a course on South America, had left the university; so he turned his attention to Latin America as his region of special interest. He was president of the Association of American Geographers in 1945.

Wallace W. Atwood

Harlan H. Barrows

Isaiah Bowman

J. Paul Goode

Richard Hartshorne

Ellsworth Huntington

Carl O. Sauer

John K. Wright

On his first field trip to the Antilles in 1922, he discarded the ideas of environmental determinism and became one of the most eloquent adversaries of those who continued to speak of responses or controls (Platt, 1946, 1948). It was Platt, who in 1928 in a report on a field study of a small Wisconsin community, first formulated the concept of a hierarchy of central places. "The radius of the community," he wrote, "is measured by the reach of the village institutions" (Platt, 1928: 92). He noted that the individual farmer looked to the smallest village for those immediate services that had to be close enough for daily contacts. For his larger needs, which did not require such frequent visits, the farmer looked to the larger towns. The hierarchy that Platt identified started with the individual farmer, proceeded to the village of Newport, then Ellison Bay and Sister Bay, to Sturgeon Bay (the county seat), to Green Bay (the regional center), and to Chicago (the metropolis). If Platt were studying Ellison Bay today, he would have described it as a spatial system and he would have used quantitative techniques to make his observations of the functional relationships more precise. But the ideas were all there in 1928 (Thomas, 1979).

STUDIES OF SCOPE AND METHOD

The geographers in North America, like their colleagues in other parts of the world, had to clarify their own ideas regarding scope and method. Geography in America was nurtured in its early years by geologists, and most of the first generation of scholars in geography had a common background in geology. The few who were not geologists were mostly meteorologists or botanists. The direction offered by Davis in the 1880s and 1980s had led to the causal notion becoming the first paradigm in American geography soon after the turn of the century. As departments of geography were established in the universities and as graduate students trained in geography began to enter the profession, there was a close examination of the spirit and purpose of geography (Johnson, 1929; Finch, 1939). Initially, the main objective was to establish geograpy as an independent field of study. There was concern to define limits that would separate geography from other fields. This drive toward disciplinal independence probably retarded the development of geographic ideas because the workers in any field of learning must be in close contact with ideas being generated in other fields.

The habit of discussing philosophical and methodological questions was supported in at least three ways. One was the nature of Salisbury's seminars and many others that were patterned on the Chicago example. Then there were the many opportunities for such discussion offered by the annual meetings of the Association of American Geographers. By long-standing tradition the presidents make use of their presidential addresses to set forth their own ideas concerning the scope of geography. And of no small importance were the foreign visiting lectures who were invited to American universities for periods ranging from a summer session to a whole academic year.[11]

Moreover, there was a widely based tradition in America of studying geography out-of-doors, and discussions concerning geographical ideas and procedures were more vigorous in the field than within the walls of the seminar rooms. Few indeed were the young geographers of that period who were content to give only a verbal definition of geography in logical terms. In the field there was no difficulty in moving the discussion promptly to the search for an operational definition—that is, in reaching an agreement about what must be done to identify a geographic idea or how a geographic idea could be used to increase knowledge of the earth as the home of man. In the field the concept of the region came alive. Symptomatic of this operational approach to the definition of geography was the oft-quoted remark that "geography is what geographers do."[12]

These two traditions—the habit of discussion and familiarity with field study—resulted in the organization early in the 1920s of an annual spring field conference (James and Mather, 1977). The first such conference was held in 1923 in the Indiana Dunes south of Lake Michigan. The participants were former students at Chicago who held positions in several midwestern universities.[13] A somewhat enlarged group met in May, 1924 at Bagley, Wisconsin, and in May, 1925 at Hennepin, Illinois. In 1925 a report representing the joint conclusions reached at these conferences was published, entitled "Detailed Field Mapping in the Study of the Economic Geography of an Agricultural Area" (Jones and Finch, 1925). Thereafter field conferences were held almost every spring and included approximately the same people. In 1926 a second conference was organized consisting of junior scholars. In 1935 both groups were combined, meeting that spring in Menominee,

[11]For example, the French geographer Raoul Blanchard lectured at Harvard in 1917, at Columbia in 1922, at Chicago in 1927, at Berkeley in 1932. He returned to give a lecture series at Harvard every year between 1928 and 1936. In addition to Blanchard in 1927, other Europeans invited to Chicago included James Fairgrieve in 1920, Sten De Geer in 1922, Ernest Young in 1924, L. Rodwell Jones in 1925, Helge Nelson in 1926, and Patrick Bryan in the summers of 1928 and 1929.

[12]Credited to A. E. Parkins (see J. Russell Whitaker in the *Annals AAG,* 31 [1941]:48).

[13]The following geographers participated in these conferences:

From Chicago: C. C. Colby, W. D. Jones, R. S. Platt, D. S. Whittlesey, and C. O. Sauer in 1923 only (Whittlesey continued to attend after he went to Harvard in 1928).

From Wisconsin: V. C. Finch, A. K. Lobeck (until he went to Columbia in 1929).

From Minnesota: D. H. Davis

From Northwestern: W. H. Haas

From Michigan: K. C. McMurry

From George Peabody: A. E. Parkins

The junior group that met first in 1926 included:

From Michigan: S. D. Dodge, R. B. Hall, P. E. James

From Minnesota: R. H. Brown, R. Hartshorne

From Wisconsin: L. Durand, G. T. Trewartha, J. R. Whitaker

From Chicago: H. M. Leppard

Michigan, and in the spring of 1936 in Pokagan State Park in Indiana. In 1938, the last spring conference was held in the Muskingum Watershed in Ohio, where the Soil Conservation Service was carrying out a program of erosion control based on detailed studies of rainfall and runoff. In 1940 at Pokagan State Park proposed field studies were discussed. During the 1920s and 1930s many papers that originated in conference discussions were published (Whittlesey, 1925, 1927; Jones, 1930; James, 1931; Platt, 1931; Hartshorne, 1932; Finch, 1933; Hall, 1934; James, Jones, and Finch, 1934; Platt, 1935).[14]

Experiments in Method

The field conferences had the effect of focusing attention on problems of methodology. How were geographical problems to be identified in the field? How were the necessary data to be collected in useful form? W. D. Jones had returned from Patagonia many years earlier with an appreciation of the need to prepare maps of land use on the same scale and degree of generalization as the traditional maps of the physical features. Jones and Sauer had jointly published a paper on this in 1915. The field conference discussions of field methods began where the earlier discussions had ceased.

The basic problem involved in the study of agricultural areas was how to identify and plot on maps the significant units of area relevant to the understanding and guidance of land use. At first it was proposed that several separate maps all on the same scale should be prepared to show the critical elements in land use problems. The next step was to reduce the number of such maps to two: one to show the conditions of the land, including units of soil, slope, drainage, and cover of wild plants; the other to show the categories of land use. When these two maps were superimposed, the precise covariance of the physical features and the land use could be examined in detail. Later the suggestion was made that all this information could be plotted on one map by making use of a fractional code symbol. The denominator

[14]It should be recorded that the members of these two field conference groups were the ones who shaped the policies of the profession in the 1920s and 1930s. In those days membership in the Association of American Geographers was by election only, based on "an original contribution to some branch of geography beyond the doctoral dissertation." A nominating committee each year selected a single slate of officers. The president and vice-president held office for one year only, but the officers whose terms ran for several years naturally had the greatest influence on association policy. These were the secretary, the treasurer, and the editor of the *Annals*. At least two of these were always members of the conference group. Often at the spring conferences one evening was devoted to a discussion of association problems. Actually the conference membership was not exclusive because the group was always looking for younger men with both energy and ability. Nevertheless, when the new constitution of the association was adopted after World War II, which opened the membership to any interested person, the existence of a "clique" was severely criticized by the flood of new geographers then emerging from the graduate schools. The association is now democratically operated under the management of a paid executive director. In 1941 the membership was 167; in 1980 it was approximately 6000 (James and Martin, 1979).

of the fraction was made up of digits representing categories of soil, slope, and drainage; the numerator comprised digits representing types of land use or wild vegetation. In the course of field-mapping, when the observer noted any difference of the physical land or the land use, a boundary had to be drawn and a new fractional code symbol shown. In each unit area, represented by one fraction, there was a single, uniform association of land quality and land use. Since every microscopic point on the face of the earth differs from all other points, these unit areas were, in fact, generalizations. Within each unit area there was a certain range of diversity that had to be small enough to be acceptable—and this of course depended on the proposed use for the resulting map.

The fractional code system was tested by V. C. Finch with a group of graduate students from the University of Wisconsin, and the results were published in the so-called Montfort Study in 1933 (Finch, 1933). The map was published on a scale of approximately 1/15,000 (Fig. 29) and provided an extraordinary amount of detailed information. From this map it was possible to prepare a number of thematic maps of individual elements, such as slope, soil, or land used for specific purposes. But the field-mapping had required so much time that Finch was unable to recommend the method as useful. In Chapter 15 we will look at a new method of random sampling applied experimentally to a part of the Montfort area.

There was a very great increase in the number of books and articles by American geographers after the 1920s—so great an increase, in fact, that it becomes impossible to do more than offer a few selected examples. Work was done in all the topical fields of geography: in population and settlement studies, in urban geography, transportation and other aspects of economic geography; and also in the various fields of physical geography and biogeography. All this work is reported at length in *American Geography: Inventory and Prospect* (James and Jones, 1954).

The innovative papers in the field of population geography may be singled out as examples of the many studies in the 1930s, 1940s, and 1950s. The experiments with the mapping of population were greatly stimulated by the work of the Swedish geographer, Sten De Geer. Various kinds of dot maps and density maps were prepared. But one basic problem continued to bother the workers in this field. The census data are summed up within enumeration areas that are seldom relevant to the kinds of problems geographers want to study. In 1936 John K. Wright published a short paper on the mapping of population, using Cape Cod as an example.[15] He showed that quite different patterns of population are brought out by using enumera-

[15]John K. Wright was graduated from Harvard in history in 1913 and received the M.A. in 1914 and the Ph.D. in 1922. After serving in World War I, he was appointed librarian at the American Geographical Society. At the society he devised a new research catalog for use by geographers, with books classified both topically and regionally. He edited many of the society publications. From 1938 to 1949 he was the director of the American Geographical Society, and in 1946 he was president of the Association of American Geographers. Some of his more important writings are included in a book entitled *Human Nature in Geography* (Wright, 1966; also Wright, 1952).

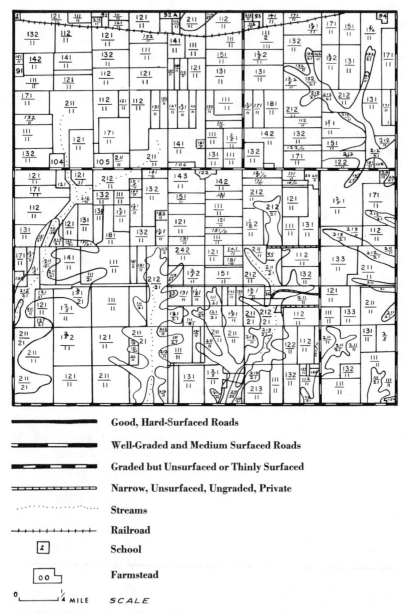

▬▬▬▬▬▬▬ Good, Hard-Surfaced Roads

▬▬ ▬ ▬ Well-Graded and Medium Surfaced Roads

▬ ▬ ▬ Graded but Unsurfaced or Thinly Surfaced

═══════ Narrow, Unsurfaced, Ungraded, Private

· · · · · · Streams

+++++++ Railroad

School

Farmstead

0 ⌞___⌟ ¼ MILE SCALE

Figure 29. A portion of the Montfort area. (From Finch, 1933.)

NUMERATOR

Left-hand Digit: Major Use Type	Second Digit: Specific Crop or Use Type	Third Digit: Condition of Crop
1. Tilled land	1. Corn (maize, 2. Oats 3. Hay in rotation 4. Pasture in rotation 5. Barley 6. Wheat 7. Peas (mainly for canning) 8. Soy beans 9. Potatoes T. Tobacco X. Sudan grass $^2/_5$ Oats and barley mixed	1. Good 2. Medium 3. Poor
2. Permanent grassland	1. Open grass pasture 2. Pasture with scattered trees or brush 3. Wooded pasture 4. Permanent grass cut for hay	1. Good 2. Medium 3. Poor
3. Timber land	1. Pastured 2. Not pastured	1. Good 2. Medium 3. Poor
4. Idle land	1. Is capable of use	

DENOMINATOR

Left-hand Digit: Slope of Land	Second Digit: Soil Type (Wis. Soil Survey Terminology)	Letter x: Condition of Drainage
1. Level, 0°–3° 2. Rolling, 3°–9° 3. Rough, 9°–15° 4. Steep, over 15°	1. Marshall silt loam 2. Knox silt loam 3. Knox silt loam (steep phase) 4. Lintonia silt loam 5. Wabash silt loam 6. Rough, stony land	X Poor XX Very Poor

tion areas of different sizes and shapes. He described a quantitative method for distributing the densities within a large enumeration area, such as a township. He described his method as follows:

> Assume, for example, a township with a known average density of 100 persons to the square mile. Assume, further, that examination of topographic maps and consideration of other evidence have shown that this township may be divided into two parts, m, comprising 0.8 of the entire area of the township and having a relatively sparse population, and n, comprising the remaining 0.2 of the township and having a relatively dense population. If, then, we estimate that the density of population in *m* is 10 persons to the square mile, a density of 460 to the square mile must be assigned to *n* in order that the estimated densities of *m* and *n* may be consistent with 100, the average density for the township as a whole [Wright, 1936:107].

He provided a table to make estimates of density consistent with average figures for whole enumeration areas.

Another innovation in the study of population was offered in 1954 by Lester K. Klimm of the University of Pennsylvania (Klimm, 1954). He pointed to the existence of large empty areas within such a long-settled region as the northeastern states. He described empty areas as follows:

> These empty areas are not used for farming, no one lives in them, they contain virtually no recreational or commercial structures. Most of the surface is in woods or brush, ranging from ''barrens,'' burnt-over land, or bog to large tracts of managed commercial forest in Maine, New Hampshire, and New York and extensive areas of state and national forest. Forestry and recreation are the principal present uses. Where they have resulted in structures, the areas occupied have been classified as not being empty [Klimm, 1954:325].

His map on a scale of 1/2,000,000 shows large continuous empty areas, especially in northern Maine and New Hampshire, in the Adirondacks, the Allegheny Plateau, the Catskills and Poconos, and in the New Jersey Pine Barrens. He also shows a patchwork of smaller empty areas scattered in many parts of the region. It is clear from an examination of this study that the first step in preparing a map of population (by whatever method population is to be shown) must be to mark off the empty areas. Of course, the population data by census districts entirely obscures this kind of information.

These are only a few examples of the many experiments with method that were tried and then discussed in seminars and at meetings of the Association.

Definition of the Field

The Commission on the Social Studies in the Schools appointed by the American Historical Association invited Bowman to represent geography and subsequently to write a book on the relation of that branch of science to the social studies. Bowman wrote *Geography in Relation to the Social Sciences* (1934). The content of the book includes: "By Way of Definition," "Measurement in Geography," "Population and Land Studies," "Technique in Geographical Analysis," "Regional Geography," "Economic and Political Bearings," "Conclusions." Bowman had for some years been predisposed toward undertaking such a book. He had been frequently asked to define geography by schoolteachers, fellow geographers and by university administrators. There was a void in the literature concerning the scope and nature of the field. Points of view conflicted and there was little philosophical cohesion to geography. Bowman revealed his thought and feelings on the scope of geography:

This world is made up of regions and each region has its own personality, its own set of significant conditions. A Tibetan yak driver, an Egyptian fellah, an Uros fisherman, an Argentine hacendado, a Kansas farmer, a Peace River pioneer—each lives in a world whose conditions and outlook are almost completely unlike the others. To apprehend those earth qualities, conditions, outlines, measured components, and interactions that enable us to look understandingly at man in relation to the pervasive elements of his complex regional environment—these are the most distinctive as they are the culminating purposes of geographical research . . . [1934:4]

Bowman did not wish to impose an orthodoxy upon workers in the field, but he did suppose that a consensus was desirable. The book represented his own geographical point of view. Five years later another book was published that synthesized many geographers' viewpoints concerning the nature of geography. Occasionally a book is published that stands as a landmark in the history of geographic thought. Such a one is *The Nature of Geography* (Hartshorne, 1939). Hartshorne, then at the University of Minnesota, had been a graduate student at Chicago and a participant in the spring conferences. His published studies during the 1920s and 1930s ranged widely over the field, including studies of agricultural regions, transportation and urban development, climate, and studies of the factors in the location of manufacturing industries. He also published a paper on racial distributions in the United States and on some fundamental concepts in political geography (Hartshorne, 1927, 1932, 1938, 1950). Field studies on the boundary problems of the upper Silesian industrial district (Hartshorne, 1934) excited his curiosity about boundary questions in general. When he was granted a sabbatical leave in 1938–39 together with financial aid from the Social Science Research Fund of the University of Minnesota, he planned to make a field survey of European boundary problems. But 1938 was no time for an American geographer to be examining European boundaries with notebooks, maps, and camera—for the events leading up to World War II had started. Before leaving for Europe he had submitted a paper to the *Annals* regarding certain methodological questions. In Vienna, hoping that he might yet be able to carry on field studies, he received several letters from Derwent Whittlesey, then editor of the *Annals,* suggesting additional materials that could be added to his paper. He made use of the library at the University of Vienna to consult new sources of information. But as time went on and conditions grew worse rather than better, he focused his attention on the many documentary materials available in European libraries and also carried on interviews with leading geographers. The result was a book of nearly 500 pages (Hartshorne, 1979).

Hartshorne describes his purpose as follows:

The detailed examination of the nature of geography which this paper endeavors to present is not based on any assumption that geography is or ought to be a science—or that it ought to be anything other than it is. Assuming only that geography is some kind of knowledge concerned with the earth, we will endeavor to discover exactly what kind of knowledge it is. Whether science or an art, or in what particular sense a science or an art, or both, are questions which we must face free of any value concepts of titles. . . .

The writer's concern . . . is to present geography as other geographers see it—or have seen it in the past. If we wish to keep on the track—or return to the proper track . . . —we must first look back of us to see in what direction that track has led. Our first task will be to learn what geography has been in its historical development [Hartshorne, 1939:205, 207].

Hartshorne's book was widely acclaimed and is generally accepted as an authoritative account of the points of view of the major builders of geographical ideas. It is a product of careful scholarship. But in the course of time colleagues and graduate students in seminars all across the country attempted to identify the positive conclusions regarding the nature of the field, and this proved to be very difficult. Hartshorne had either quoted or paraphrased some 300 methodological writings, some of which departed from what he identified as the mainstream of geographic scholarship. The continued discussion of these fundamental questions raised certain doubts and challenges that required clear answers (Schaefer, 1953; Hartshorne, 1955, 1958). Hartshorne undertook to provide a restatement of the positive conclusions to be drawn about the nature of geography. His *Perspective on the Nature of Geography* (1959) is organized around ten topics, each of which is the subject of a chapter:

1. What is meant by "geography as the study of areal differentiation"?
2. What is meant by the earth's surface?
3. Is the integration of heterogeneous phenomena a peculiarity of geography?
4. What is the measure of significance in geography?
5. Must we distinguish between human and natural factors?
6. The division of geography by topical fields—the dualism of physical and human geography.
7. Time and genesis in geography.
8. Is geography divided between systematic and regional geography?
9. Does geography seek to formulate scientific laws or to describe individual cases?
10. The place of geography in a classification of the sciences.[16]

The conclusions that Hartshorne reached (1959) are summarized in the following statements (note that the earlier ones are modified somewhat by the later ones):

[16]Hartshorne sidesteps the question: Is geography an art or a science? "Whether such a field is to be called 'science' is a semantic question, depending on what particular definition is given to a word on which there is much disagreement" (1959:11). He also insists that writers on methodology should read carefully the writings of others of whom they are critical.

Geography is concerned to provide accurate, orderly, and rational description and interpretation of the variable character of the earth surface [p. 21].

The earth surface is the outer shell of the earth where lithosphere, hydrosphere, atmosphere, biosphere, and anthroposphere are intermingled. This is the geographer's universe [pp. 22–25].

The goal of geography, the comprehension of the earth surface, involves therefore the analysis and synthesis of integrations composed of interrelated phenomena of the greatest degree of heterogeneity of perhaps any field of science [p. 35].

Any phenomenon, whether of nature or of man, is significant in geography to the extent and degree to which its interrelations with other phenomena in the same place or its interconnections with phenomena in other places determines the areal variations of those phenomena, and hence the totality of areal variation, measured in respect to significance to man [p. 46].

In describing and analyzing individual features and elements, we are free to utilize whatever categories of classification are empirically significant to the study of their interrelationships, without concern for the abstract distinction between those of human origin and those of natural origin [p. 64].

The traditional organization of geography by topics into two halves, "physical" and "human," and the division of each half into sectors based on similarity of the dominant phenomena in each, is of relatively recent origin and has proven detrimental to the purpose of geography—the comprehension of the integrations of phenomena of diverse character which fill areas in varying ways over the earth [p. 79].

. . . historical studies of changing integrations are essentially geography rather than history as long as the focus of attention is maintained on the character of areas, changing in consequence of certain processes, in contrast to the historical interest in the processes themselves [p. 107].

Geographic studies do not fall into two groups (topical and regional) but are distributed along a gradual continuum from topical studies of the most elementary integration at one end to regional studies of a most complete integration at the other [p. 144].

[Regarding the question of the nomothetic or the idiographic approach:] We start with "observation"—sensory description, often assumed to be the sole meaning of "description." We proceed to "analysis"—the description of the several parts of what has been observed as they appear to be related to each other. Next we state a hypothesis of relationships among the elements and processes. If sound, we have arrived at a higher level of knowledge—"cognitive description" of the elements and interrelationships among them [p. 171].

[Regarding Hettner's concept of geography as a chorological science:] Acceptance of the concept is in no way essential to geographic work. But students who cannot accept the particular characteristics empirically demonstrated as essential to geography, because they cannot understand that necessity, repeatedly attempt to change the subject to fit their view of what a science should be. The long history of such attempts demonstrates that their only effects are the personal frustration and professional unhappiness of those who try to fit a square peg into a round hole [p. 181].

To comprehend these areal variations fully we must dip back into past relationships of the factors involved, and those whose interest so directs them may reach as far back into history as the availability of data may permit. Release from the necessity of focusing our attention on the relations between two particular groups of features, human and nonhuman, permits a wider expansion of interest and at the same time a more effective coherence of the entire field. The opportunity to develop generic studies leading to scientific principles is present in the many forms of topical geography. Likewise, the unlimited number of unique places in the world, each of which is important and intellectually significant at least to those who live there, provides an inexhaustible field for those most interested in this type of research [p. 183].

American Geography: Inventory and Prospect

As the fiftieth anniversary of the founding of the Association of American Geographers approached, some geographers felt that this was a time for stocktaking. World War II had created an unprecedented demand for trained geographers, and those engaged in any of the numerous branches of war work had to devise new methods and make use of new, and often unfamiliar, materials. In 1949 at a meeting in Evanston, Illinois, attended by the chairman of several committees that had been appointed by the National Research Council to discuss different aspects of geography, it was decided to undertake a series of symposia for the discussion of geographical questions and eventually to publish a book on the results. Preston E. James and Clarence F. Jones were appointed to direct the project. With funds provided by the Social Research Council and the National Research Council a number of conferences were set up, each one to consider the controversial problems of the various parts of the field. The whole program called for wide discussion throughout the profession. The original drafts of chapters were read critically by members of the committees and were also presented to sessions of the Association and to university seminars throughout the country. The resulting book, therefore, represents the combined thoughts of between 1 and 200 geographers. In fact, a major accomplishment of the project was stimulation of widespread discussion throughout the profession of the objectives, methods, and concepts of geography (James and Jones, 1954). The following is a list of the chapters and the principal author of each:

1. The Field of Geography, Preston E. James
2. The Regional Concept and the Regional Method, Derwent Whittlesey
3. Historical Geography, Andrew H. Clark
4. The Geographic Study of Population, Preston E. James
5. Settlement Geography, Clyde F. Kohn
6. Urban Geography, Harold H. Mayer
7. Political Geography, Richard Hartshorne
8. The Geography of Resources, J. Russell Whitaker
9. The Fields of Economic Geography, Raymond E. Murphy

      Marketing Geography, William Applebaum
      Recreational Geography, K. C. McMurry
10. Agricultural Geography, Harold H. McCarty
11. The Geography of Mineral Production, Raymond E. Murphy
12. The Geography of Manufacturing, Chauncy D. Harris
13. Transportation Geography, Edward L. Ullman
14. Climatology, John Leighly
15. Geomorphology, Louis C. Peltier
16. The Geographic Study of Soils, Carleton P. Barnes
17. The Geographic Study of Water on the Land, Peveril Meigs, III
18. The Geographic Study of the Oceans, C. J. Burke and Francis E. Elliott
19. Plant Geography, A. W. Küchler
20. Animal Geography, L. C. Stuart
21. Medical Geography, Jacques M. May
22. Physiological Climatology, D. H. K. Lee
23. Military Geography, Joseph A. Russell
24. Field Techniques, Charles M. Davis
25. The Interpretation of Air Photographs, Hibberd V. B. Kline, Jr.
26. Geographic Cartography, Arthur H. Robinson

It is important that no one answer was to be expected regarding controversial issues. The purpose was to identify differences of opinion and, so far as possible, to eliminate differences among the definitions of words. The various chapters did not necessarily record an accepted version of geography, but they did express the variety of interests and points of view that were actually held by members of the profession at midcentury. Not the least important was the extensive bibliography of geographical writings included in each chapter (Wooldridge, 1956; Wooldridge and East, 1958).

    Other articles analyzing or summarizing the geographical work of the period between World War I and the decade of the 1950s include: J. Brunhes (1925), C. Sauer (1927), A. E. Parkins (1934), R. Hartshorne (1935), C. Colby (1936), G. Pfeifer (1938), K. T. Whittemore (1972), W. B. Fairchild (1979). Five books that reveal much American geography during this period include Taylor's collection of essays on *Geography in the Twentieth Century,* which was discussed earlier (Taylor, 1951); J. K. Wright's story of the activities of the American Geographical Society from 1851 to 1951 (Wright, 1952); collections of the more significant papers of C. O. Sauer (Leighly, 1963), and J. K. Wright (Wright, 1966); and the history of the Association of American Geographers (James and Martin, 1979).

CHAPTER 15

Applied Geography

Interest in planning for the classification and use of the land and other natural resources is not new to our science; in fact, it is one of the most persistent interests in American geography.

The period from World War I to the decade of the 1950s also witnessed a notable increase in the application of geographical knowledge and skills to the study of practical problems of government or business. Of course there has never been a time when the search for knowledge about the earth as the home of man has not been undertaken for practical purposes as well as for the satisfaction of intellectual curiosity. There are few fields of learning in which the relevance of concepts and specialized methods to practical needs can be more clearly demonstrated. There are many examples of such applications: as when Strabo wrote his geography for the use of Roman administrators, or when Varenius wrote his special studies of Japan and Siam for the merchants of Amsterdam, or when Maury made use of his wind and current charts to formulate improved sailing directions. One of the distinctive characteristics of American geography has been the tradition of the resource inventory—studies of this sort were made long before Thomas Jefferson gave Lewis and Clark specific instructions about the kinds of information needed; such practical

The quotation above is from Charles C. Colby in his presidential address to the Association of American Geographers (Colby, 1936:30).

342

purposes motivated the Great Surveys of the American West and the numerous surveys of resources along proposed rail lines. But the men who led these surveys were not trained as geographers—they had to work out their own objectives and methods as the work proceeded.

As the number of professionally trained geographers gradually increased, a certain proportion of the younger generation—then as now—expressed impatience with theoretical studies and demanded that geographic investigations be clearly relevant to the practical problems involving public or private policy. In the 1920s and 1930s there were some geographers who were disenchanted with the experiments with field methods applied to small areas; and there was a demand for some tangible connection with the ''overriding'' economic, social, or political problems of the day. There was no clear answer to the question about how one might demonstrate a tangible connection, so each scholar who decided to use geographical studies for practical ends had to formulate his own justification. The result was the appearance of a wide variety of studies in what might be called ''applied geography,'' in the sense that the purpose was to provide the basis for planning remedial action.

The first large breakthrough in the use of professionally trained geographers to study practical problems came during World War I and its aftermath. During the 1920s and 1930s not only were certain wartime projects continued and completed but new kinds of applied research were undertaken. A large number of geographers were called into both military and civilian service during World War II, and since then opportunities for employment in various branches of government and also in private business firms have grown rapidly. Many studies involving the geographical analysis of location or of areal spread are still made by nongeographers, but in the contemporary period there is a growing appreciation of the value of the professional training that geographers have.

IN WORLD WAR I

With United States participation in World War I and the Peace Conference that followed, fifty-one association members contributed their abilities.[1] Five members (W. W. Atwood, H. A. Gleason, H. E. Gregory, A. E. Parkins, and R. H. Whitbeck) offered instruction to groups of soldiers and the Students Army Training Corps (SATC) program; five members (C. F. Brooks, A. J. Henry, A. McAdie, J. Warren Smith, and R. DeC. Ward) contributed research findings on aerography, meteorology and climatology; ten members (N. A. Bengtson, A. H. Brooks, R. M. Brown, G. E. Condra, N. H. Darton, G. R. Mansfield, F. E. Matthes, O. E. Meinzer, P. S. Smith, and T. W. Vaughan) pursued research concerning the location of

[1] ''War Services of Members of the Association of American Geographers,'' *Annals AAG,* 9 (1919):53–70.

minerals; four members (S. W. Cushing, J. W. Goldthwait, E. Huntington, and L. Martin) were attached to intelligence; H. Bingham advised on aviation, W. Bowie advised on map projections, E. E. Free advised on gases, W. Churchill and D. W. Johnson wrote propaganda tracts, and G. E. Nichols functioned as botanical advisor on sphagnum moss for the American Red Cross (much used as a substitute for cotton in absorbent surgical dressing). It was at this time that the international significance of commodities began to make itself felt.

Commodity Studies

One of the first uses of the knowledge and skills of trained geographers was in connection with certain commodity studies. In the fall of 1917 the Division of Planning and Statistics of the U.S. Shipping Board was asked to make a survey of imports. The problem was that there were not enough ships to carry the supplies of war in addition to normal trade. The board was asked to classify all imports into one of three categories: (1) those commodities so necessary to the war effort that all available supplies must be imported; (2) those commodities essential to the war effort, but not in such short supply that all available production had to be imported; and (3) those commodities not required for military or civilian needs. As usual in such wartime agencies, the answer was expected the next morning. The chairman of the research group was Dean Edwin F. Gay of the Harvard Business School. By chance Walter S. Tower, the Chicago geographer, was in Washington, D.C., at the time, and he was asked to help in this rush job because he had been teaching a course at Chicago in the geography of commerce. It is reported that ''by midnight Gay realized that Tower knew more about commodities and where they could be obtained than any of the others'' (Colby, 1955:13). The next day Tower was appointed head of the Commodities Section of the Shipping Board and was asked to bring together a staff of experts in this field.[2] Among others he selected Vernor C. Finch of Wisconsin, coauthor with Oliver E. Baker of *Geography of the World's Agriculture* (Finch and Baker, 1917); William H. Haas of Northwestern University, a former student at Chicago; and George B. Roorbach, an assistant professor of economic geography at the University of Pennsylvania. Later Charles C. Colby of

[2]Walter S. Tower received the B.A. degree from Harvard in 1903 and the M.A. in 1904. In 1906 he completed the Ph.D. at the University of Pennsylvania with a dissertation, *A Regional and Economic Geography of Pennsylvania*. He taught economic geography at the University of Pennsylvania from 1906 to 1911, at which time he became a member of the Department of Geography at the University of Chicago. He was promoted to professor in 1916, but in 1917 he was granted leave of absence to work for the War Shipping Board in Washington. He never returned to a university post. From 1919 to 1921 he was trade advisor to the Consolidated Steel Company. From 1921 to 1924 he was the U.S. Commerical Attaché in London. From 1924 to 1933 he was with the Bethlehem Steel Company. From 1933 to 1940 he was executive secretary of the American Iron and Steel Institute and from 1940 to 1952 he was president of the institute. He retired in 1952 and lived in Carmel, California, until his death at the age of eighty-eight in 1969.

Chicago also joined the staff. These university people found what so many have had to rediscover in later years—that intelligence reports must be summarized in simple, unambiguous language on page one. Colby turned out to be especially adept at extracting the "meat" from a report and writing useful summaries. Those who remember his presence at many later meetings of geographers will recall his amazing ability to summarize long and involved discussions in terse sentences so clear that discussants wondered what they had been arguing about. Colby became a very effective writer of proposals to be submitted for financing to government agencies, an ability that he no doubt possessed before 1917, but it was certainly improved by his wartime experience.

The discovery that geographers had useful skills for commodity studies led to the establishment of another Division of Planning and Statistics, this time in the War Trade Board, also under Dean Gay. The geographer selected to head the research unit was Harlan H. Barrows of Chicago. For his staff Barrows recruited J. Russell Smith of the University of Pennsylvania, Ray H. Whitbeck of Wisconsin, and Nels A. Bengtson of Nebraska. Other work, too, was undertaken in conjunction with the war effort.[3]

The Inquiry

In September, 1917, President Wilson directed his close friend and advisor, Col. E. M. House, to set up an organization that would gather the most complete collection of information possible and prepare it for use at the coming Paris Peace Conference. Under the direction of President S. E. Mezes of the College of the City of New York, some 150 persons were recruited, including historians, economists, geographers, journalists, and others with special knowledge of particular areas. The group, which became known as The Inquiry, carried on its research in the building of the American Geographical Society in New York City. The work of The Inquiry was made possible by its access to the library and map collections of the society.[4]

The subjects studed by The Inquiry included the political and diplomatic history of Europe; international law, including the geographic interpretation of problems of territorial waters and of interconnections across frontier zones; economics and economic geography; physiography in relation to strategic boundaries; and

[3]W. M. Davis prepared *A Handbook of Northern France* (1918) describing the terrain features of the war zones, illustrated with his incomparable pen and ink sketches. The book was printed in pocket size. Some 4000 copies were supplied to infantry officers at the front, while an additional 5000 copies were supplied to YMCA libraries for the use of troops.

Ellsworth Huntington and H. E. Gregory wrote and edited *The Geography of Europe* (New Haven: Yale University Press, 1918), to which seventeen other geographers made contributions and which was intended to serve as an up-to-date textbook for courses in the SATC in the universities. The work was sponsored by the National Research Council.

[4]"The American Geographical Society's Contribution to the Peace Conference," *Geographical Review*, 7 (1919):1-10. See also Lawrence E. Gelfand (1963) and Arthur Walworth (1976).

many other more detailed investigations of major problem areas where plebiscites were to be carried out. A major part of the work was in the field of cartography in which a map-making program of unprecedented size and detail was undertaken. First a set of new base maps was made, showing prewar political boundaries, the complete drainage system, the roads and railroads, and the cities and towns. Some of the maps were on scales of 1/1,000,000 or 1/3,000,000. These were the general maps of Europe as a whole or the somewhat more detailed maps of the Balkans. But there were also a great number of very large-scale maps, such as the map of Alsace-Lorraine on a scale of 1/250,000. For several places of critical importance, block diagrams were drawn showing the geological structure on the sides of the block and the terrain on its surface. All these maps were available at the Paris Peace Conference to be used for the study of various boundary proposals. On these maps information was plotted to show population density, ethnic composition, agriculture, industrial centers, mineral resources, and many other things needed by the Paris Peace Conference (Wright, 1952:200). The maps were also made available to universities in the United States, where courses in war aims were being offered.

Isaiah Bowman, who had been the director of the American Geographical Society since 1915, supervised the geographical studies. He brought together a number of geographers to work with him. Mark Jefferson, his former teacher at Ypsilanti, became Chief Cartographer.[5] Most of the geographers were assigned to study various topics and regions of Europe. Fortunately, the society had already published the results of the relation between language and the division of that continent into separate states, which had been done by Leon Dominian (1917). But much more detailed information would be needed when new political boundaries were to be drawn. A few members of The Inquiry worked on the collection of material concerning other parts of the world: Bailey Willis reported on the problem areas of Latin America, especially on the background of the Tacna-Arica dispute between Peru and Chile; H. L. Shantz looked for information about the plant resources of Africa; C. F. Marbut was assigned the task of compiling a soil map of Africa; and J. Warren Smith (of the U.S. Weather Bureau) prepared maps of climatic elements.

The Paris Peace Conference

On December 4, 1918 The Inquiry specialists, their assistants (together with the materials they had gathered) and numerous officials, including President Woodrow Wilson, sailed for France on U.S.S. *George Washington*. At Paris Bowman

[5]For biographical data on Bowman, see pp. 298-301; on Jefferson, see pp. 296-298. Other geographers who worked on The Inquiry were O. E. Baker, N. M. Fenneman, W. L. G. Joerg, C. F. Marbut, E. C. Semple, H. L. Shantz, and B. Willis. Army officers assigned to The Inquiry were Major D. W. Johnson and Major Lawrence Martin as well as two nongeographers, Captain W. C. Farabee and Captain S. K. Hornbeck.

was given the title of Chief Territorial Specialist of the American Commission to Negotiate Peace; Mark Jefferson was appointed Chief Cartographer (C. Stratton and A. K. Lobeck were his assistants). At Paris the map became everything and it is said that there was coined the aphorism, "One map is worth ten thousand words." The map became the international language of the conference and the Americans were best prepared to make these maps. Copies of the American-made maps concerning European matters were reduced in size and entered in the Black Book, and maps of colonial matters were entered into the Red Book of the American delegation. These books were constantly being consulted by leaders and diplomats of many of the delegations at the conference (Martin, 1966).

With the map and the abstract principle of Wilsonian justice, some 3000 miles of new boundaries were created about the former states of central Europe (Rhoads, 1954).

Wartime Projects Published Later

The information gathered and digested during the war and its aftermath was not entirely buried in government documents. Isaiah Bowman wrote a book concerning the problem areas of the world that remained for many years the most authoritative study in political geography—*The New World* (1921). In this book Bowman did not attempt to build a theory of political geography, but rather to analyze in informative detail the particular problems of particular regions, with adequate description of the local setting and the historical background so that the reader understood what was going on in the postwar world. It served its purpose admirably.[6] Also H. L. Shantz and C. F. Marbut published a monograph on the vegetation and soils of Africa, which for many years remained a definitive statement on that subject (Shantz and Marbut, 1923). D. W. Johnson's report on the relation of military strategy and tactics to the terrain features was published in 1921. This was the first substantial contribution by an American to military geography (Johnson, 1921).

For many years after the war Bowman and the American Geographical Society were involved in boundary studies and a program of mapping. While the Paris Peace Conference settled the affairs of Europe, the Latin American countries suffered poorly marked boundaries. Guatemala and Honduras asked the United States to settle their dispute over their common boundary; and Robert Lansing, the Secretary of State, turned to the man with whom he had dealt on such matters in Paris. He asked Bowman to arrange a study of the Guatemala-Honduras boundary and suggest a solution. The American Geographical Society organized a research team under the direction of Major Percy H. Ashmead to make a map of the area, showing not only the details of the terrain but also the distribution of people and their ways of using

[6]The Department of State placed a copy of the book in each of the U.S. consular offices around the world, and the Carnegie Endowment for International Peace distributed copies to the leading centers of teaching and research in international affairs (Wright, 1952:255).

the land. The survey was made in 1919, and a suggested solution was submitted to Mr. Lansing. The negotiations took fourteen years, but the settlement (which was accepted in 1933) was based on the maps and recommendations of the society's report.

The Millionth Map of Hispanic America

Bowman's work in Peru as well as his experience with the Guatemala-Honduras problem brought to light a real lack of geographical information on Latin America as a whole. There was no reliable map of the region. With the methods of surveying then in use it would have taken many decades and vast sums of money to produce a useful map. But Bowman knew that there were a large number of original surveys in manuscript form that had been made by private companies for a variety of purposes. He proposed that the American Geographical Society should undertake a major research program leading to the compilation of a map of Hispanic America on a scale of 1/1,000,000, conforming to the standards and format of the International Map of the World that was originally proposed by Albrecht Penck. Raye R. Platt, reviewing the completion of the Millionth Map in 1946, of which he had been director for twenty-three years, quoted the annual report of the Council of the American Geographical Society for 1920 as follows:

> The first step in the development of this program aims at the review and classification of all available scientific data of a geographical nature that pertains to Hispanic America.... The work will involve the compilation of maps—topographic and distributional—on various scales, but always including sheets on the scale of 1/1,000,000 which will conform to the scheme of the International Map.... The undertaking is an ambitious one, but the Society is happy to say that assurances of cooperation have been given by the whole group of Hispanic American countries in a cordial spirit that augurs well not only for the immediate scientific results but also for the fostering of mutual understanding and sympathetic relations toward which the field of geography offers a peculiarly fortunate approach [Platt, 1946:2].

As part of the map of Hispanic America project, the society supported the publication of a series of research studies. Some were based on field surveys, such as the reports on European colonies in Chile, Argentina, and Brazil by Mark Jefferson (1921, 1924, 1926); the studies of land-settlement problems in Mexico and Chile by George M. McBride (1923, 1936); or the additional studies of Peru carried out by O. M. Miller (1929). The study of the central Andes by A. G. Ogilvie, the Scottish geographer, was compiled from the drawings of the Millionth Map and from Bowman's copious notes (Ogilvie, 1922).[7] The map was used to help adjudicate disputes between Chile and Peru in 1925, Bolivia and Paraguay in 1929,

---

[7]Also based on field study was Bowman's *Desert Trails of Atacama* (Bowman, 1924).

Colombia and Peru in 1932, and Colombia and Venezuela in 1933. And the Hispanic America Map, completed in 1946 in 107 sheets (over 300 square feet in extent) was a contribution to the Millionth Map of the World, then in progress.

LAND CLASSIFICATION STUDIES

Another quite different application of geography to the solution of practical problems has to do with studies of land quality and land use. It had long been recognized that detailed information concerning land resources was needed if plans for better resource use were to be properly guided. The destruction of the land through improper use, which the public began to talk about in the early 1970s, had been reported by geographers a century before—notably by George Perkins Marsh and Nathaniel Southgate Shaler. Ratzel used the expressive German term, *Raubbau*, or robber economy, to describe a form of land use that destroys the land base. As we learned in Chapter 7, efforts to classify land in terms of its potential use began in the early days of the independent United States and were carried to new levels of utility in the Great Surveys of the West and in the work of the U.S. Geological Survey in the latter part of the nineteenth century. Major steps were taken, however, to enlarge the scope of land classification studies and to improve methods during the 1920s and 1930s. The discussions of method in the annual field conferences led directly to practical applications, especially in the program of resource inventory carried out in the state of Michigan. So important was the work of the Michigan Land Economic Survey that the record needs to be presented in some detail.

The Michigan Land Economic Survey

To understand the problems of public policy that made the Land Economic Survey important we must review the conditions of land and land use in Michigan at the end of World War I. From the point of view of its natural features, the state of Michigan can be divided into two quite different parts. To the south of a line drawn roughly from Saginaw Bay on the east to Muskegon on the west (Fig. 30), Michigan is a part of the productive agricultural plains of the Middle West. The soils are mostly loams, only in a few small spots is the surface too steep for cultivation, and the growing season is long enough to permit the ripening of grain crops. But north of this line the physical character of Michigan is very different. Here the land is made up of a deep accumulation of glacial deposits—moraines, till plains, and sandy outwash plains. The Upper Peninsula is similar to the northern part of the Lower Peninsula as far west as Marquette. West of this city the knobby crystalline rock hills of the Canadian Shield form the surface. Whereas the southern part of Michigan was once covered by a broadleaf forest similar to that of Ohio and eastern Indiana, the north country was covered by one of the finest stands of white pine to be found in America. The white pine, intermingled with broadleaf species, formed a

Figure 30. Areal coverage of land classification maps in Michigan, July, 1939.

1. Bay	8. Crawford	15. Montmorency	22. Menominee
2. Roscommon	9. Oscoda	16. Alpena	23. Alger
3. Ogemaw	10. Benzie	17. Emmet	24. Delta
4. Mason	11. Grand Traverse	18. Cheboygan	25. Schoolcraft
5. Lake	12. Leelanau	19. Presque Isle	26. Luce
6. Alligan	13. Antrim	20. Iron	27. Chippewa
7. Kalkaska	14. Otsego	21. Dickinson	

dense and almost unbroken forest cover. When settlers came into southern Michigan they cleared the broadleaf forests and established farms, creating what became a part of the Hay and Dairy Belt. But the first penetration of the country north of Saginaw Bay came after the Civil War and was based on lumbering rather than farming. From the stands of white pine came the wood that was used to build most of the houses and other structures of the prairie states.

In those days no questions were raised about the way private interests made use of natural resources. The lumberman was there to cut down trees and transport the logs to the sawmills. No one thought of requiring the replanting of cutover lands; and the lumberman removed all the trees, leaving none to provide for reseeding. The slashings left after the tree trunks were cleared of branches were left lying on the ground, and the wood-burning locomotives were not equipped with screens to catch sparks. Forest fires were frequent and destructive, especially when there were high winds after a long dry period. On October 8, 1871, on the same day that the famous Chicago fire started, a forest fire started near Petoskey on Lake Michigan. The same conditions that made the Chicago fire so destructive—high winds after a long period of drought—made the forests almost ready to explode. During the following days the fire swept on a broad front all the way across the Lower Peninsula to the shores of Lake Huron. It has been estimated that more trees were destroyed by fire than were cut by the lumberman's ax.

When forest fires and lumbering had completed their work, no trees were left on the completely denuded land. A scrubby second growth of brush appeared in some places; in others there was only a growth of low plants that barely covered the charred remains of the forest. The large lumber companies tried to sell their land to farmers, and it was actually offered for sale at $2 an acre. There were some farmers who thought this was too good a bargain to miss. The result was a thin scattering of isolated farms and a few small towns. But large parts of the lumber properties could not be sold and were simply abandoned. In Michigan, when an owner fails to pay his taxes for seven years, the land reverts to state ownership. By 1910 not only had the lumber companies moved away but even some of the farmers had found nothing but disillusionment in the poor sandy soils and the short growing season. Only in widely scattered localities were there small groups of rural people struggling to hold on. Already there were some people who believed that the greatest values to be found in the north country were in the wild game animals—fish, game birds, and deer. The problem of idle land was already critical because the people of the southern part of the state were forced to carry the tax burden of supporting the scattered settlements of the north.

Two men provided the leadership that resulted in effective action. One was Carl O. Sauer, a member of the staff of the Department of Geology and Geography at the University of Michigan. Sauer's interest in the field survey of land quality and use dated back to his days as a graduate student at Chicago and his association there with Wellington D. Jones, recently returned from a survey of northern Patagonia with Bailey Willis. When Sauer joined the Michigan faculty in 1915 he found the state facing a serious practical problem but lacking the kind of specific information needed to formulate a remedial policy. The agricultural experts wanted only to find ways to make farming pay; the forestry experts wanted only to plant trees; the hunting and fishing clubs wanted to keep everyone else out. And people who were familiar with the character of the north country knew that it contained a great diversity of physical conditions that made the adoption of any one general policy impossible. There was need for just the kind of attention to the mapping of significant differences from place to place that geographers in that period had been talking about. Sauer's plan for a land classification survey represented the application of ideas generated in professional discussion to real practical problems (Sauer, 1919, 1921).

At Ann Arbor Sauer met a forester and naturalist named Parish S. Lovejoy, who was also deeply concerned about the problem of the cutover lands. Lovejoy had the knowledge and the commitment to become a kind of gadfly to stir various groups into action—the meetings of scientists, the hunting clubs, the members of the state legislature. Speaking before such groups, Lovejoy pointed out eloquently that a third of the state of Michigan was bankrupt and that the situation was spreading. Some kind of public policy, he insisted, must be adopted, and quickly. But no policy will be worth anything, he continued, unless it is based on an accurate and detailed knowledge of the relevant facts. The gist of what Lovejoy had to say is contained in the annual report of the Michigan Academy of Sciences for 1921.[8] At a special session of the Academy on "Michigan's Idle Lands," held in 1920, the nature of the critical situation in the north country was spelled out, and a program of action was presented. A resolution was adopted and sent to the State Department of Conservation recommending immediate action on setting up a land survey.

The newly formed Department of Conservation took the recommended action. Securing the cooperation of the U.S. Department of Agriculture, the University of Michigan, and the Michigan Agricultural College (now Michigan State University), funds were made available for an experimental field study of Charlevoix County to be carried out in the summer of 1922. The results proved the value of the information to be obtained, and the Michigan Land Economic Survey was created. The survey was directed to make a detailed inventory of the cutover lands, together with reports on the current economic situation by counties. Because the inventory of land and land use was to be used not only as a basis for developing some kind of policy but to guide programs of land management, the mapping had to be done in great detail; yet, since there was no previous experience to indicate the best method to use, much of the work had to be improvised to meet clear and specific needs. Here is what one of the field surveyors had to say about the procedures:

[8]P. S. Lovejoy. "The Need for a Policy for the Cut-Over Lands of Michigan," 22nd *Annual Report of the Michigan Academy of Sciences* (1921):5-7.

Field operations in the first year were experimental, and many changes [in categories to be mapped] were later made. . . . Field crews consisted of a cover and base mapper together with a soil and slope mapper. Most of the field time was taken up with boundary delineation and the determination of soil types. One section [one square mile] per day was considered a good field accomplishment. The instruments were a compass together with a soil augur; the only basic maps were the General Land Office plats, more than half a century old, showing the section corners of which little evidence remained, and drainage features where these crossed the section lines. . . . The original field scale was eight inches to the mile, subsequently changed to four inches [quoting Horace Clark in Davis, 1969:18-19].

After the first few field sessions a more-or-less established procedure was followed (Barnes, 1929). The work was done by soil specialists, foresters, and graduate students in geography from the University of Michigan. After a summer in the field with compass and notebook, most of the survey teams were able to pace a straight line through brush and swamps and could usually find the weather-beaten stakes that had been set out to mark the section corners by the General Land Office surveyors fifty years before. Students of geography who experienced this kind of practical training had no trouble with the identification of areal associations or the concept of the unit area.

In fact, the concepts of the unit area and of the land type were the major professional contributions of the survey. Remember that in the 1920s the experimental studies of small areas, such as the study of Montfort by V. C. Finch, had not yet been made. The fractional code system, which implies the existence of a unit area—uniform with respect to the physical land and the land use or cover—had not been devised. The field operations of the survey resulted in a series of maps of individual elements: lay of the land (showing five categories of slope); soil types (based on the standard definitions of the U.S. Soil Survey); drainage features; cover (including wild vegetation, crops, or planted pastures and also abandoned farms); population; political organization; assessed valuation; tax delinquency; land ownership and, where taxes were still being paid, the intention of the owner in maintaining possession of the land; and trade areas.

Each of these items was plotted on a separate map. But, when the maps were compared in the office, it became clear that certain conditions were found repeatedly to form the same associations. Not only could certain natural land types be identified (repeated associations of slope, soil, drainage, and wild cover) but also certain economic conditions were found to have a high correlation with certain land types. Wade DeVries, a land economist, may have been the first to call attention to these associations (DeVries, 1927, 1928). But it remained for J. O. Veatch, a soil specialist, to define the types and to prepare a statewide scheme of the regions (Veatch, 1930, 1933, 1953).

Meanwhile Lee Roy Schoenmann provided an example of how the information gathered by the survey could be used to establish local zoning rules for rural areas.

He took the maps for Alger County and presented them at a series of meetings with local business people and farmers. As a result the community set up its own restrictions on land use, designating certain areas for farming and pasture, other areas for fish and wildlife preserves, still other areas for reforestation (Schoenmann, 1931).

Only about half of the counties of the original cutover area had been mapped when the survey was brought to an end in 1933 during the Great Depression. Enough work had been done in different parts of the north country to demonstrate that much of the area could not be managed by private owners. The completed maps (Fig. 30) provided enough data so that land management planning could be extended to neighboring counties. Furthermore, by 1933 the method of field-mapping by pacing with a compass was outmoded through the use of vertical air photographs (Chapter 17). Also there were certain pressure groups in the state that wanted to see the survey cease publishing this kind of information. For example, real estate interests were attempting to sell land on the shores of the numerous inland lakes, but not all the lake shores were sandy and suitable for summer homes. The survey classified the lake shores, showing where they were sandy and where boggy. Pressures similar to those used against Powell in his surveys of the West reappeared in Michigan in the 1930s. Enough had been accomplished, however, to make it clear that the geographers, with their methods of identifying areal associations and with their experience in the analysis of the interplay of diverse processes (which we would now describe as spatial systems), were in a position to make a distinctive contribution to the expanding field of land classification and land-use planning. (McMurry, 1936).

Further Development of the Land Classification Idea

The land classification idea was picked up by many individuals and many agencies of the federal and state governments. To reduce costs, mapping was done on air photographs and there was continued experimentation with different scales and categories. Several previously unmapped Michigan counties were examined by these new methods. By 1940 the Subcommittee on Land Classification (Charles C. Colby, chairman) of the Land Committee of the National Resources Planning Board reported the existence of seventy-two separate land classification projects then being carried on by forty-six agencies of the federal government and by twenty-eight state agencies (Colby, 1941).

Also in the planning field was Harlan H. Barrows on the Water Resources Committee of the National Resources Planning Board. Barrows made three major contributions to water-planning studies. First, he insisted on the use of clear, simple English in the writing of reports for publication. "He slashed the Federalese, the sodden engineering description, and the politically ambiguous prose that came within his reach" (Colby and White, 1961:398). Second, this insistence on clear writing meant also clear thinking: he played an important role in the formulation of policy regarding multiple-purpose river-development projects that were developed during the 1930s. Third, he designed the procedure for integrated regional studies as an essential basis for policy planning. Between 1935 and 1938 he drew up the plan that was accepted by the states of Colorado, New Mexico, and Texas for the allocation of the water of the upper Rio Grande. He was successful in working out similar cooperative solutions for the water of the Pecos River, the Red River of the North, and the Columbia Basin. The impact of his efforts in these river-basin projects is summarized as follows:

> In subsequent years, the report of those investigations shaped technical analysis of water projects over many other areas. Moreover, the list of questions posed for investigation still is an incisive classification of resource-use problems in an irrigated area. His outline for the investigations became the springboard for studies and policy discussions cutting across all the relevant disciplines and levels of jurisdiction. He continued the same type of analysis as a consultant to the Department of the Interior concerned with resource development problems in Alaska and in the Central Valley of California [Colby and White, 1961:398–399].

In connection with the planning for economic development in the Tennessee Valley by the Tennessee Valley Authority, geographers were employed in the Department of Regional Planning Studies under the direction of G. Donald Hudson. The whole river basin was surveyed by the unit area method to provide an inventory of the nature and extent of land resources. In addition, an inventory of scenic resources was made; and a special group was assigned to study the proposed reservoir sites to determine the extent of lands to be purchased by the Authority— the so-called taking line studies. The idea was to avoid taking so much of the land of an individual farmer that he could not make a living with what was left. The taking line was to leave workable farm units beyond the immediate shore of the reservoir, which was to be used for recreation. Hudson described the method for defining and mapping unit areas in 1936 (Hudson, 1936).

In 1941 the Land Committee was searching for less costly methods of survey-ing the land and land use of an area as a basis for improving the economy of depressed areas. Charles C. Colby and Victor Roterus were appointed as consultants to prepare an inexpensive method for gathering the necessary information. Their proposal was published in 1943, along with a Livelihood Area map of the United States (Colby and Roterus, 1943). The so-called area analysis method was based on an outline to be filled in by field observation and was summarized under four main headings: (1) the employment pattern, (2) the conditions affecting employment and income (natural resources, economic activities, and institutions), (3) directions of desirable readjustment, and (4) the proposed program of remedial action. Between 1941 and 1943 the method was applied to many small areas in different parts of the

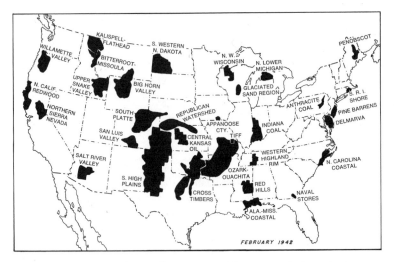

Figure 31. Areas of the United States reported by area analysis method.

United States and proved useful in the guidance of efforts to rebuild depressed economies (Fig. 31).

Since World War II the whole approach to studies of this kind has been revolutionized through the use of computer programs, with data supplied from new remote sensing devices. These changes are discussed in Chapter 17.

Land Classification Studies in Latin America

Land classification studies are of special importance in countries with developing economies, such as those of Latin America. But before the 1950s there were very few Latin American geographers trained in the methods of the field survey. The first major attempt to apply land classification methods to a Latin American country was directed by geographers from the United States or trained in the United States. This was in Puerto Rico.

The Puerto Rico Rural Land Classification Program was carried out between June, 1949 and August, 1951. Here was a small island, some 3435 square miles in area, which in 1950 had a density of population of 642 people per square mile. With a purely agricultural economy such a density could not be adequately supported; furthermore, the greater part of the land of high potential productivity was used to grow sugar cane, while basic foods had to be imported. But the island government adopted a policy of providing for the rapid improvement of the economy by making better use of the land to produce crops and by investing heavily in new manufacturing industries to be scattered throughout the island and connected to ports by new all-weather roads.[9] Rafael Picó (who holds the Ph.D. in geography from Clark University) was chairman of the Puerto Rico Planning Board. He understood better than most Latin Americans at that time that economic planning had to be based on detailed knowledge of the resource base. Picó asked G. Donald Hudson, then chairman of the Department of Geography at Northwestern Univeristy, to help in planning such a survey. In March, 1949 Hudson and his Latin American specialist at Northwestern, Clarence F. Jones, went to Puerto Rico to work out plans. It was decided that the island should be covered by maps on the very large scale of 1/10,000. Northwestern University would supply each year some advanced graduate students, either from their own university or from other graduate departments around the United States. Each of these graduate students would be joined by a Puerto Rican student to form a field team, and each team was assigned a section of the island to survey. Hudson, Jones, and Picó made an experimental traverse across the island to test out the categories of land and land use. Mapping was done on vertical air photographs and unit areas were identified by the fractional code method. The first field team started work in July, 1949.[10]

The Puerto Rico survey has demonstrated the value of this kind of inventory. Land redistribution in some other parts of Latin America (where it was done with no mapped information) has brought disastrous results. In Puerto Rico the information gathered by the survey was used as the basis for the replanning of land use; crops were matched to favorable land types, or they were removed from unsuited lands, such as steep slopes, where they brought destructive erosion. The survey information was used to plan the routes of new roads and to locate the numerous small manufacturing plants in relation to population and accessiblity. The success of Operation Bootstrap was in no small measure based on the existence of reliable, detailed knowledge about the land quality and existing land use.

Since that time many somewhat similar surveys have been undertaken in Latin America, some by the Organization of American States, some by the Agency for International Development, and some by Latin American government agencies, as in Brazil. Some surveys, as in Chile, were done by private agencies in the United States on contract.

[9]Governor Muñoz Marin was Puerto Rico's first elected governor in 1948. Puerto Rico in 1952 became the Commonwealth of Puerto Rico, freely associated with the United States. Governor Muñoz Marin undertook to develop the island's economy in what has been called Operation Bootstrap.

[10]The mapping was done on a scale large enough so that every plot of land, used or unused, could be shown. Eight categories of land use were identified: (1) cropped land, (2) pasture and harvested forage, (3) forest and brush, (4) nonproductive land, (5) rural public community service land, (6) land used for quarrying or mining, (7) urban and manufacturing, (8) miscellaneous, such as canals, water storage tanks, roads, railroads, farm buildings, and so on. The physical characteristics of the land, which appeared as the denominator of the fraction, included soil types (as defined by the U.S. Soil Survey), degree of slope, conditions of drainage, rate of erosion, and amount of stoniness and rock exposure. See *Rural Land Classification Program of Puerto Rico* (Evanston, Ill.: Northwestern University Studies in Geography, 1952). See also Jones and Picó, 1955, and Jones and Berrios, 1956.

THE PIONEER BELT STUDIES

Intended also to provide knowledge on which to formulate policy were the studies of pioneer settlement initiated by Isaiah Bowman. In 1925 he turned his attention to problems relating to the thinly populated areas on the margins of settlement. He submitted a proposal to the National Research Council (NRC) to obtain funds for studies of pioneer areas. After two years of consideration by a special committee and by the Division of Geology and Geography of the NRC, the project was recommended to the Social Science Research Council. In 1931 both councils endorsed Bowman's program, and the Council of the American Geographical Society also gave its support.

Bowman outlined the nature of the problem (Bowman, 1932). Pioneering in the 1930s, he wrote, is not at all like the pioneering of the past century when new settlers depended almost entirely on their own muscles. Nowadays pioneers want the latest machinery, the best medical services, and well-developed facilities to connect them with markets. Yet pioneer zones are always experimental. When prices for farm products are low, men may seek new lands where the price per acre is lower than in settled communities; but these same pioneers may be driven back from low-priced land by droughts. Pioneer belt studies are not solely concerned with the possibilities of new settlement: they may also be concerned with the need to withdraw from less favorable places. In Malthus's time more food could be produced by moving farmers onto new lands and creating new agricultural communities. But by the 1930s to increase the supply of food without increasing the prices was done by reducing the number of farmers and withdrawing them from the marginal lands. Productivity in the modern world is increased by concentration in the more accessible and better suited areas and withdrawal from the remote and marginal areas. But all such changes must be applied in particular places. Bowman proposed to study pioneer movements all around the world and to identify certain general conditions: not only the kinds of physical conditions considered favorable but the attitudes and objectives that led people to become pioneers, and the economic, social, and political institutions that could best support pioneers. But Bowman also proposed to investigate the particular and unique conditions in specific pioneer areas, knowledge of which would be essential to the formulation of policy. His proposal ranged widely over all the fields of the social sciences and was essentially interdisciplinary in concept.

Several studies of pioneer belts in general (and also of specific ones) were published during the 1930s. Bowman's book *The Pioneer Fringe* (Bowman, 1931) stated the nature of the problem and offered examples from the western United States, Canada, Australia, Southern Africa, Siberia, Mongolia, Manchuria, and South America. A volume containing twenty-seven cooperative studies of particular pioneer regions followed the next year (Joerg, 1932). Finally Bowman, aided substantially by Karl Pelzer, summarized the results of the whole undertaking in a report on the world's potential pioneer areas (Bowman, 1937). Meanwhile, pioneer studies were vigorously pursued in Canada under the direction of the Canadian Pioneer Problems Committee headed by W. A. Mackintosh of Queens University. Under the general title, *Canadian Frontiers of Settlement*, edited by Mackintosh and Joerg), eight separate volumes were published, starting in 1934 (Innis, 1935).

Bowman gave a convocation address at the University of Western Ontario in 1937 in which he summarized his own point of view toward geography in the 1930s:

> Within limits that have varied widely in time, geography has set itself the task of understanding man's relation to the earth, and I shall presently attempt to explain that phrase with some precision. Always there must be food and clothing, tolerable if not optimum temperature ranges for both man and the things he requires, transport needs and desires, and, unhappily, wars and famines, for a time at least, as well as great conquests and conditional conquests of at least the local and immediate in the environment. Out of this play of forces—by no means either infinite or hopelessly complex— man is progressively creating and experimenting, and the chief experiment is himself. He is changing himself as well as the world as he goes along [Bowman, 1938:2].

For a more complete statement of Bowman's point of view, see his *Geography in Relation to the Social Sciences* (Bowman, 1934).

Geography in World War II

The demand for the services of geographers in World War II far exceeded the supply of experienced and properly trained professionals. Geographers were needed in all the kinds of work performed during World War I and also in many research studies. Geographers worked as commissioned officers or as noncommissioned draftees assigned to intelligence agencies. Geographers in large numbers came as civilians either for full-time positions in war agencies or for short-term specific studies.

By 1943 there were over 300 geographers working in Washington, D.C. These included 75 geographers in the Research and Analysis Branch of the Office of Strategic Services (initially known as the Coordinator of Information), 46 in the War Department, 23 in the Intelligence Division (G2), and an additional 23 in the Army Map Service. The office of the Geographer, Department of State, had 13 geographers; 15 were employed in the Board on Geographic Names; 12 in the Office of Economic Warfare; 12 within the Department of Agriculture. In addition, 8 geographers were employed by the Geological Survey and 6 by the Coast and Geodetic Survey. There were 5 geographers in the Weather Bureau, 4 in the Map Division of the Library of Congress, and 18 others were scattered among a variety of agencies (these figures do not include draftsmen or others engaged in producing maps and charts). Approximately 25 other geographers were employed in posts overseas.

Some of these geographers helped to prepare the compilations of information about countries or parts of countries, either as a basis for planning military operations or as a guide to military government after the war. The Joint Army-Navy Intelligence Studies (JANIS) brought together many kinds of data with which geographers had no previous experience. But a very important part of the JANIS program consisted of the compilation and publication of many detailed maps of special features. Large numbers of geographers in the Office of Strategic Services were assigned to the cartographic work, while others worked on the various countries where JANIS handbooks were needed. Many geographers, also, worked on special problems and prepared background reports for the guidance of those who were responsible for decisions.

A few examples of the kinds of work geographers did can be offered. One had to do with the kinds of uniforms and equipment needed in different environments. In 1940 the Quartermaster General of the army had three sets of uniforms for military use: temperate, torrid, and frigid! When troops occupied the Aleutian Islands they were equipped with temperate zone uniforms, but, when these proved to be quite inadequate, it was clear that Aristotle's climatic zones were no longer useful. The Quartermaster General established a research laboratory at Natick in Massachusetts to test different kinds of equipment under a great variety of artificially produced climatic conditions. The problem was to identify the important differences of climate and other environmental conditions, then not only to find the kinds of equipment best suited to them but also to find out in detail where such environments would be encountered all over the earth. The result was the so-called clothing atlas. This atlas specifies in detail by means of a complicated key the variety of equipment necessary to carry on field operations in the world's many kinds of environment. This work was continued and expanded after the war.

Before the landings in Normandy many geographers, including those in the intelligence branch of the army, were busy making detailed studies of the beaches and the terrain behind beaches. Johnson's *Battlefields of the World War* had become an historical document, for the changed technology of warfare rendered his interpretation of the significance of terrain obsolete. When warfare became mechanized, the pattern of paved roads became more significant than the arrangement of hills and valleys: small villages where paved roads came together became more important than cuestas. The basic point was that an army operating on foot can move as rapidly off the road as on it; but a mechanized army moves with great speed on a road, regardless of slope, and very much more slowly off the road. This is another example of the general principle suggested earlier that the significance of the physical and biotic features of the earth changes with changes in the attitudes, objectives, and technical skills of man himself.

In the Pacific theater there was a serious lack of any reliable information about the character of beaches or the terrain. Geographers were set to work combing the literature to find descriptions or old photographs. Missionaries and tourists had provided some information, but it was scattered and hidden in much irrelevant detail. Yet maps were actually compiled and published showing in amazing detail the arrangement of coral reefs, cliffs, roads, caves, and other features of military importance.

One group of geographers received special training in the study of transportation facilities. What were the essential items of equipment in a port that would determine its capacity to handle traffic? A trained port engineer had to explain such matters. Geographers learned that in some of the ports of western Europe the tidal range was so great that gates had to be provided to keep the water from draining out at low tide. The condition of the gates was critical. Then, what about the conditions of roads and railroads? This group of geographers, together with expert photographers from the motion picture studios of California, undertook to provide descriptions, photographs, and maps showing the condition of transportation facilities immediately after the armies started their advance eastward. The materials gathered proved to have great practical utility for those commanders in charge of logistics.

A very important function that could be performed only by an experienced regional specialist was the interpretation of capabilities and intentions of foreign countries. Unfortunately, the number of geographers who had specialized in the study of foreign areas before World War II was quite inadequate for the demand. Those geographers who had called themselves regional specialists had focused on parts of the United States or Latin America. The number of geographers who had specialized in European, Asian, or African countries was very small. As a result the work of foreign area interpretation was done by language specialists, historians, or others who happened to have a familiarity with areas in question. The further result was that many persons—geographers and others—who were assigned to such positions proved inept and unreliable. E. A. Ackerman, pointing a critical finger at what he described as inadequate professional training in the systematic aspects of geography, summarized the role of geographers in the war effort:

In the three years from 1941 to 1944 American geographers dealt almost constantly with a series of difficult professional problems. The profession as a whole may take pride in the manner in which these situations were met. Both the well-known and the previously obscure showed skill, imagination, energy, and unselfishness as they perspired over wartime tasks. Our techniques advanced, and the prestige of the profession increased notably during those years. Scholars and administrators who had scarcely heard of geography before Pearl Harbor are now familiar with its methods and its results. Geography unquestionably has wider recognition than ever before in this country.

However, an assessment would hardly be honest if one were to stop with praise of our recent performance. Wartime experience has high-lighted a number of flaws in theoretical approach and in the past methods of training men for the profession. It is no exaggeration to say that geography's wartime achievements are based more on individual ingenuity than on thorough, foresighted training. The geographer perfectly or even adequately trained for the specialty into which he was thrown has generally been an exception. The unfamiliarity of most young American geographers with foreign

geographic literature; their almost universal ignorance of foreign languages; their bib-
liographic ineptness; and their general lack of systematic specialties are but a few points
which may be cited in proof. All these were just as regular a source of difficulty as the
strangeness of the problems and the pressure under which we worked [Ackerman,
1945:121-122].

Ackerman arrived at the important conclusion that a major source of difficulty
in the preparation of geographers before World War II was a widespread belief in
the essential duality of the subject. In many places it was felt that a geographer
might become either a regional specialist without any training in a systematic field
or that a geographer might become a specialist in any one of a number of systematic
fields. This is the duality that the German geographers had resolved and that
Hartshorne had attacked in 1939. It was the duality that participants in the annual
field conferences deplored. But wartime experience with the employment of many
poorly trained or partly trained people proved that the conceptual structure of geography was still not widely understood.

Geography in the White House

President Franklin D. Roosevelt had a keen interest in geography. In 1921 he
had been elected to the Council of the American Geographical Society. Since that
time he had retained a fascination with the subject and had developed a substantial
knowledge of atlases. Pursuant to the German annexation of Austria in March, 1938
and increased anti-Semitic activity by the Nazis, Roosevelt began to think about
resettlement of European Jewry—and other refugees—on a large scale. He held
private meetings with Isaiah Bowman, exploring possibilities as to where several
million such people might be relocated. As a result of these meetings, Bowman
arranged for a team of workers (including geographers) to make feasibility studies
of refugee settlement in different parts of the world. In addition, President
Roosevelt initiated the M Project (M for anonymity), which, under Bowman's
direction, and through the person of Henry Field, provided some 666 studies in
20,000 pages and an *Atlas of Population and Migration Trends* (Martin, 1980).

Bowman worked in the Department of State three days a week and was frequently called on to advise Sumner Welles, Cordell Hull, and the President. He was
made a member of the Stettinius Mission to London (1944), the Dumbarton Oaks
Conference (1944), and the San Francisco Conference (1945). At all stages leading
to the creation of the United Nations Charter, the geographical point of view was
found to be of value (Martin, 1980).

OTHER APPLICATIONS OF GEOGRAPHY

The applications of geography to practical problems took place in many other
sectors before the decade of the 1950s. One of these was in marketing research for
private business firms. In 1931 William Applebaum was working on a thesis on

secondary commercial centers of Cincinnati. In Cincinnati Applebaum began to focus his attention on the location factor in the development of outlying retail market centers. He also learned that the Kroger Company was looking for the best places to locate planned supermarkets. Applebaum turned his attention to the selection of sites for Kroger, and his results proved so useful that the company became interested in his methods. He went to work for Kroger to apply his method to the selection of other supermarket locations. Since 1931, and especially since World War II, most business firms engaged in selling to the public have added market research departments; and the demand for persons with geographic training to work in the field of market research has increased rapidly.

What does a geographer do when he works on a problem of retail store location? He makes maps of the distribution of potential customers. But the population maps that are made by counting the number of people in census districts lack the relevant detail needed for such studies. It is necessary to plot on a map the arrangement of people along specific streets and to know the patterns of their daily trips to work or to retail stores. Often a store location proves much better on one side of the street than on the other, depending on the customers' routes to work. It is also necessary to map the areas from which other competing stores draw their customers. Because geographers are usually familiar with the making and use of detailed maps and with the map analysis of location problems, their contribution to the study of market areas has become widely recognized and appreciated (Applebaum, 1952).

Since World War II the use of geographers in marketing research problems has been greatly extended. In 1961 a whole issue of *Economic Geography* was devoted to a series of papers detailing examples of this kind of applied research.[11] The latest mathematical procedures have been applied to this kind of investigation (Applebaum and Cohen, 1961). Applebaum continued to publish on matters relating to marketing research until 1974. Meanwhile, other imaginative applications of geographic methods have been made to the study of the operations of large corporations (McNee, 1961). Meanwhile, economic geographers have been making studies in the location of economic activity ranging from iron and steel plants to flour milling.

Geographers such as Gilbert White have made studies of natural hazards, including major river floods (Kates, 1962; White 1973). Studies of military matters from the geographical point of view have been made by Joseph A. Russell and others. Yet other geographers have worked with agencies of the United Nations, have made environmental studies in the wake of ever-increasing industrial pollution (F. W. McBride formed his own agency), and have studied the effects of

[11]*Economic Geography* 37 (1961): Saul B. Cohen, "Location Research Programming for Voluntary Food Chains," 1–11; Bart J. Epstein, "Evaluation of an Established Planned Shopping Center," 12–21; Howard L. Green, "Planning a National Retail Growth Program," 22–32; Harold R. Imus, "Projecting Sales Potentials for Department Stores in Regional Shopping Centers," 33–41; Jack C. Ransome, "The Organization of Location Research in a Large Supermarket Chain," 42–47; William Applebaum, "Teaching Marketing Geography by the Case Method," 48–60.

weather and climate upon humans (physiological climatology). Especially noteworthy in the latter case is the work of E. Huntington and D. H. K. Lee (Martin 1974). C. W. Thornthwaite applied his remarkable knowledge of climatology and other aspects of the physical environment to the dairy industry in New Jersey with remarkable success (1931, 1933). L. D. Stamp demonstrated the value of land utilization study (1931, 1952). H. H. Bennett studied soil erosion in relation to the productivity of the land (1928). E. L. Ullman was a member of the board of directors of Amtrack. M. I. Glassner has advised the government of Nepal in negotiating a transit treaty with India and has functioned as consultant to the UN Development Programme for the land-locked countries of Asia. Professional geographers have been employed by the Bureau of the Census since the 1920s.

There are almost endless applications of the geographical point of view to the problems of the real world.

PART THREE

CONTEMPORARY

World War II was a shattering experience for those who
were directly involved; and the repercussions of that
experience have changed the quality of life all around the
globe, even in the most remote places. The technological
innovations of the war and postwar periods are the most
noticeable changes; but there have also been revolutionary
changes of attitudes and objectives. The list of new ways of
looking at life is a long one, including such innovations as
the abandonment of colonialism, the active promotion of the
ideals of racial equality, the spread of the emotions of hatred
and frustration as larger and larger numbers of individuals
find their living space reduced below the threshold of easy
adjustment. Suddenly large numbers of people have become
aware of the population explosion and that mankind seems
to be rushing toward its own destruction. The new
technology of instant and worldwide communication itself
guarantees that the affluent and genteel ways of living can
no longer be supported in the presence of large
concentrations of poor and ignorant people.

The world of scholarship could not escape the waves of
innovation. And change has not been restricted to the

sciences and medicine but has swept over the humanities. In
the first place, the new electronic devices have made
possible complex mathematical computations never before
attempted. But there has also been a fundamental revision of
the process of education. Children in the 1960s were
indoctrinated with a new concern for science and
mathematics, and this indoctrination was accompanied by a
decrease of emphasis on skills in the art of writing and in the
interpretation of history. Few young people today learn
Latin, and few are exposed to the classics of literature in the
original. A new generation of students moved into the
universities in the 1960s ready to believe the findings of
scientists and influenced much less than formerly by
thoughtful and artistic writing. The new generation of
students accepted none of the traditional standards of
thought and action without critical reappraisal, even to the
point of threatening the continued existence of the
university as a community of scholars elevated above the
threat of outside interference with the freedom of thought
and speech. But then, in the early 1970s, came the
inevitable reaction to what seemed like too much emphasis
on science and mathematics. A new movement toward the
search for insights through art, music, literature,
philosophy, and history marked a turn of the tide.

Geography, as a field of learning, has been caught up
in this worldwide ebb and flow. The period of the 1960s
witnessed an enthusiastic swing by the younger generation
toward the use of the language of mathematics because it
could be made to express ideas more precisely than the
language of literature. There was an attack—which often
went too far—on "mere empirical description" without
adequate efforts to formulate abstract models. But the
formulation of theory and the deduction of hypotheses
requires rigorous thinking: not everyone was able to reach
the goals prescribed by the few; and there were many who
felt that the claims for the scientific approach had not been
fulfilled. Not everyone realized that the followers of the
literary tradition, like those who followed the mathematical
tradition, would have to accept a rigorous intellectual
discipline. The observer might say that it would, indeed, be
unfortunate if the turn of the tide should reduce the very real
gains in precision resulting from the innovations of the

contemporary period. But he would also say that it would be unfortunate if the focus of attention on scientific and quantitative methods should leave no place for those who still want to grapple with interpretation of the region.

The last three chapters of this book constitute an effort to put contemporary thought into perspective. Chapter 16 reviews some of the basic geographic concepts that still seem to be relevant to the world of the 1980s. Chapter 17 describes some of the new technologies of observation and analysis. Chapter 18 attempts to portray as objectively as possible the interplay of tradition and innovation in contemporary geographic scholarship.

CHAPTER 16

The Concept of Occupied Space

. . . In the long run, relativity may prove a more important factor in language than in physics. Its impact has caused thoughtful men everywhere to look at their words, to question the validity of their concepts. In the domain of physics, chemistry, biology, relativity has been responsible for an unprecedented crop of young geniuses, due to the sudden expansion of understanding which its concepts promote. To see the outside world primarily in terms of relations rather than in terms of absolute substances and properties seems to develop an intellectual keenness hitherto unknown.

When primitive man asked what it was like on the other side of the range of hills that formed his horizon he was formulating the question that has motivated the study of geography all through the ages. He was calling for the same basic elements of geographical study that still concern the more sophisticated scholars in the second half of the twentieth century (Lowenthal and Bowden, 1976.) We can identify geographical questions, no matter who asks them, because they deal with the significance of relative location on the face of the earth. They deal with such measures as distance, direction, extent, or density; and, because the features observed anywhere must be seen as the momentary expression of ongoing processes, geographic questions must also deal with the location of sources of innovation, the

The quotation above is from Stuart Chase, *The Tyranny of Words* (New York: Harcourt, Brace, 1938), p. 117.

diffusion of innovation, the measures of accessibility, the succession of things that occupy space, and other derivatives of relative location.

Curiosity about such questions has distinguished what we call geographical inquiry; and occupied space on the face of the earth has provided the subject matter with which these geographical questions are concerned (Ley, 1977). Meanwhile, the methods of finding answers to these questions and the kinds of answers that are professionally acceptable have changed many times as new paradigms of scholarship replace old ones. We may now review and enlarge upon some of the concepts regarding occupied space that were presented in Chapter 1 and that have been exemplified in succeeding chapters.

THE FACE OF THE EARTH

As Hartshorne points out in his chapter, "What Is Meant by the Earth Surface?" (Hartshorne, 1959:22–25), the restriction of geography to the outer shell of the earth is an idea of relatively recent origin. Until the end of the eighteenth century there was no good reason to distinguish different fields of study. Humboldt's *Kosmos* included not only a study of earth space but also celestial space. Since Richthofen's statement concerning the field of geography, however, there has been a general agreement on this restriction. Geographers examine the face of the earth as the home of man. It is understood that the face of the earth is not to be taken literally. In general, the subject matter of geography is derived from what Sten De Geer described as the zone of overlap between lithosphere, atmosphere, hydrosphere, biosphere, and anthroposphere (De Geer, 1923).

The face of the earth, so defined, is not the exclusive domain of geography. Observations made on the face of the earth furnish the empirical data used in all the scholarly works of mankind. But the distinguishing trait of geography is to be found in the kinds of questions scholars ask about man's universe—questions about the differentiation of the face of the earth resulting from differences in occupied space (De Jong, 1962:4, 191, 193).

What Is Space?

Philosophers have speculated about the nature of space since that prehistoric time when language was developed to symbolize abstract ideas. Is space an objective reality apart from space-filling objects? Is space a receptacle for things or an attribute of them? Aristotle believed that space is the logical condition for the existence of things. Newton thought of space as an objective reality, but intrinsically void. Berkeley saw space as a mental construct based on the coordination of sight and sound. Kant described space as an a priori form into which sensuous experience necessarily falls, providing, therefore, for the physical classification of knowledge. All spatial concepts, said Einstein, are derived from sense experience dealing with material bodies. Traditionally, space has been described as boundless, extending in all directions, and of indefinite divisibility. Geometry, which is the pure science of space, was—until recently—based on experience in the measurement of the face of the earth and so was built around the three dimensions of length, breadth, and thickness. But Einstein was able to demonstrate that the presence of matter actually warps space, requiring a modification of some aspects of Euclidean geometry.

The kind of space that geographers are concerned about is earth space (Downs, 1970). Earth space is boundless, but it does not extend in all directions. Rather it is spherical in shape and is closed—perhaps a very small replica of the curved celestial space that Einstein revealed mathematically. Earth space is occupied by all the material and nonmaterial things that exist together at the face of the earth. And the complex associations of material and nonmaterial things in earth space are the momentary reflections of continuing processes of change, or sequences of events. There are physical processes that are described by the so-called laws of physics and chemistry; there are biotic processes that are described by the concepts of biology; and there are cultural processes that are much less perfectly described by the models formulated by the several social sciences. But each kind of process is also modified by the presence of other things and events of unlike origin that exist together in mutual interaction in earth space. The interconnections among things and events of unlike origin on the face of the earth form systems of functionally related parts. Guyot referred to the earth as a "great individual organization," meaning a "man-environment system." Any system in which one or more functionally important variables is spatial is described as a "spatial system" (Wilbanks and Symanski, 1968:83).

Earth space can be indefinitely subdivided into segments of various sizes. When such a segment of earth space is set off by boundaries it is known as an *area*. Here it is necessary to distinguish between areas that are arbitrarily separated segments of earth space and a special kind of an area known as a *region*. A region is identified by specified criteria, and its boundaries are determined by these criteria. These are the areas that Ritter said were to be studied in terms of the particular characteristics resulting from the associated phenomena, interrelated with each other, which fill the areas. These are the special kinds of areas with which geographers are concerned.

THE REGIONAL CONCEPT

The *regional concept* is the term we use to refer to the mental image of an earth's surface differentiated by an exceedingly complex fabric of interwoven strands and produced by diverse but interrelated processes. This is not the relatively unsophisticated concept of the earth's surface as made up of a "mosaic of spaces," each forming a unit of area (Gibson, 1978; Paterson, 1974). The mosaic of spaces idea suggests that a geographer's job is to identify and define these unit areas and make an inventory of their contents. This relatively simple concept of the nature of the face of the earth has not dominated geographical thinking for at least fifty years—yet there are some who still use this concept of a regionally divided earth as a basis for teaching (Minshull, 1967:26–37). Similarly the word, chorology, which refers to geographical study approached from the regional concept, should not, as it often is, be confused with the mosaic of spaces idea. These distinctions are so important that we will examine them more closely (De Jong, 1962).

The Region

In spite of the confusion caused by the use of the word, region, in English there does not seem to be a good substitute for it. In common English usage a region is an uninterrupted area possessing some kind of homogeneity in its core, but lacking clearly defined limits. Furthermore, the word is usually applied to an area large enough to form a major subdivision of a continent. But in the technical language here suggested, the word, region, is applied to an area of any size throughout which there is some kind of homogeneity as specified by the criteria adopted to define it.

During the five years or so when geographers in the United States were discussing the ideas to be included in *American Geography: Inventory and Prospect* (James and Jones, 1954), a committee, with Derwent Whittlesey as chairman, was appointed to consider the history and underlying philosophy of regional study. The committee reported as follows:

> The committee came to see the region as a device for selecting and studying areal groupings of the complex phenomena found on the earth. Any segment or portion of the earth surface is a region if it is homogeneous in terms of such an areal grouping. Its homogeneity is determined by criteria formulated for the purpose of sorting from the whole range of earth phenomena the items required to express or illuminate a particular grouping, areally cohesive. So defined, a region is not an object, either self-determined or nature-given. It is an intellectual concept, an entity for the purpose of thought, created by the selection of certain features that are relevant to an areal interest or problem and by the disregard of all features that are considered to be irrelevant [Whittlesey in James and Jones, 1954:30].

The committee on regional geography discussed the use of the word, region, in this special and technical sense. Was the regional method so much discredited by past practice that the name could not be recaptured? The term, regional geography, had been used to distinguish a new teaching method from the traditional political geography in which an encylopedic collection of data, organized by political units, was presented to be memorized. Regions, in this sense, were just another receptacle. John K. Wright described it as the "trashcan approach" because it consisted of describing the contents of an arbitrarily defined container. George H. T. Kimble described regional geographers as "trying to put boundaries that do not exist around areas that do not matter" (Kimble, 1951:159). E. A. Wrigley, defining regional studied as describing the close connections between an agricultural community and the land it occupies, decided that regional geography could not deal with modern locational analysis (Wrigley, 1965). The attempt to endow an old term with new meaning could not be other than confusing (Minshull, 1967). Yet the committee could agree on no alternative.[1]

This confusion over the word, region, and its adjective, regional, has served to strengthen the belief that topical and regional studies are two different approaches that can never be combined. Topical studies are those in which a scholar pursues the examination of one element or one group of related elements over the whole world. Regional studies are those in which a variety of elements are examined in their mutual interdependence in some specific segment of earth space. This is the distinction Varenius made between general and special geography; but Varenius pointed out that these two views of geography were mutually interdependent. Here is what Whittlesey had to say about the relations between topical and regional studies:

> The study of a topical field in geography involves the identification of areas of homogeneity, which is the regional approach: the study of regions that are homogeneous in terms of specified criteria makes use of the topical approach, because the defining criteria are topical [Whittlesey in James and Jones, 1954-31. See also Hartshorne, 1959:108-145].

Regional Analysis

In any attempt to describe and make sense (in terms of the symbols of the profession) out of the complex interwoven fabric of the face of the earth, there are certain basic elements and a number of derivatives that must be considered (Nystuen, 1963; Ackerman, 1965; Taaffe, 1970; Isard, 1956, 1975). These basic elements include location, distance, direction, spread or extent, and succession (which adds the dynamic dimension to momentarily static phenomena). These are the same elements that the Greek geographers of the classical period learned to observe and measure. Derived from the basic elements are such concepts as pattern (in the geographical sense of spatial arrangement), circulation (as in flows of water, air, goods, people, or messages), diffusions (as when an innovation spreads from a source), successions (as when there are successive periods of sequent occupance), accessibility (which is a measure of contiguity or contact), and other attributes that command the attention of geographers.

The basic elements and their derivatives must be analyzed through the manipulation of the symbols that geographers use. When things that exist as part of the

[1]It is perhaps unfortunate that Whittlesey's term, *compage*, which he used to embrace the sum total of man-environment relations, could not have been applied to the concept of the region as then defined.

complex face of the earth are transferred as symbols to maps—which is a traditional geographical method of study—the resulting reduction in scale changes them from volumes to two-dimensional geometric forms. On maps these symbols form patterns of points, lines, or areas. For example, houses, which are actually volumes, appear as points on most maps, often with out-of-scale symbols (because, on most maps used for geographical purposes, the scale is too small to permit the plotting of the area occupied by the house). Rivers or roads, which actually have length, breadth, and depth, appear on most maps as lines with out-of-scale breadth and no depth. Political boundaries are geometric lines with length and no breadth. Such things as categories of surface features, climate, vegetation, land use, or political organization appear on maps as areas with length and breadth but no thickness. Most map analyses, therefore, deal with a two-dimensional world.

Geographers distinguish three different kinds of areal distributions on maps, each of which must be identified and analyzed with somewhat different procedures (James, 1952). There are distributions which extend without interruption over the surface of the earth and which differ from place to place in intensity or degree. These are called *continuities,* or continuous distributions. To bring out the pattern of variation within a continuity, selected values are plotted on the map as points; and then lines are drawn to connect points of equal value. For example, contours are drawn through points of equal elevation; and from the arrangement and spacing of these contours it is possible to "see" the shape of the surface and the degree of its slope. Or lines may be drawn through points of equal air temperature, thereby revealing the pattern of temperature differences (Fig. 19, p. 123). Such lines are called *isarithms* (lines of equal arithmetic value).[2]

There are also distributions that are not continuous but occupy discrete areas, individually distinct. These are called *discontinuities,* or discontinuous distributions. The lines on maps showing discontinuities show the extent of the spread of a category of phenomenon, and they separate the area occupied by one category from the area occupied by another. Examples of such discrete units are land and water, forest and grassland, land used for wheat and land used for maize, soil types, or political areas. To reveal such a pattern of distribution on a map, a particular category of phenomenon (such as a kind of land use or a soil type) must be defined by specific criteria on nominal or ordinal scales. A line drawn around an area that meets the criteria is a boundary between different kinds of things.

The third kind of distribution makes use of ratio scales. For example, the individuals in a population occupy discrete areas. But to analyze the density of their distribution, the count of individuals (as by the census) is summed up in enumeration areas. The result is a ratio of people to area—population per square mile. But the figure of density thus derived depends on the shape and size of the enumeration area. Such a distribution is known as a *contingent distribution.* To reveal such a

[2]For example, isohypse, isotherm, isobar, isohyet, isobath.

pattern on a map, lines are drawn through the enumeration areas to show equal ratios (see Fig. 35, p. 400). Such lines are called *isopleths* (literally, lines of equal fullness).

These three kinds of lines, which reveal the patterns of distribution, must be interpreted quite differently. The lines on a map of discrete distributions, each of which is measured on a nominal scale, separate areas occupied by different kinds of things. To be sure, within each category, there is a certain permissible range of difference. Usually there is a core in which the characteristics of a category are most completely observable, but toward the margins of the discrete area the characteristics merge with those of neighboring categories. The boundary line is drawn where the limits are set by the specified criteria, as on any regional map (Fig. 29, p. 334). The phenomena to be found on the two sides of such lines differ in kind from each other. But the isarithms used to reveal the pattern of variation within a continuity are in no sense boundaries between different kinds of things—for the same kind of thing is found on both sides of a line. The isarithms only show the direction of greatest difference in intensity or degree, which is always at right angles to the line. For example, the difference in color on a hypsometric map (showing altitudes above sea level) between green and brown is not to be interpreted as a change from a nearly level plain to a hilly or mountainous country. Yet untrained secondary-school teachers and even professional geographers who are careless in their methods sometimes use contours to separate different categories of surface features (James, 1937).

The isarithms that bring out the pattern of variation of intensity or degree in a continuity can be drawn very precisely. To be sure, on an ordinary contour map certain points are carefully located and measured for elevation on an interval scale, and the contours are then interpolated between these measured points. But modern remote sensing equipment (Chapter 17) makes the placing of contours possible with great precision. The isopleths, based on the ratios within a contingent distribution, cannot be located so precisely. The isopleths pass through areas of equal ratio, but the exact position of the line within the area cannot be identified. Isopleths, like isarithms, reveal the direction of greatest change—which is always at right angles to the line. But isopleths are in no sense boundaries between different kinds of things.[3]

Kinds of Regions

Whittlesey's committee identified a variety of different kinds of regions (Whittlesey in James and Jones, 1954:32–47). The uniform region is a discrete distribution that is defined in terms of specified criteria and homogeneous throughout in terms of these criteria. Measurements on which such maps are based make use of nominal or ordinal scales, and the boundaries that outline the extent of the

[3]For discussions of the methods of pattern interpretation see Jones, 1930; Wright, 1937; Mackay, 1951; Taaffe, 1970:37–49. See also discussions of the method of transforming a contingent distribution into a dot map in Taaffe, 1970:39.

area within which the criteria are met are lines separating different kinds of things. Examples of such regions are the unit areas on Finch's map of Montfort (Fig. 29) or the Cotton Belt as defined in 1927 by O. E. Baker (Baker, 1927). Uniform regions are defined in terms of single features or associations of several features. Whittlesey defined the *compage* as a uniform region in terms of all the features, natural or man-made, that are related to the human occupance of area (Whittlesey, 1956).

The nodal region, or functional region, is an area that is tied functionally to a node or to several nodes. The definition of such a region involves the measurement of movements, or spatial interaction—for example, the area functionally connected with a central place; the service area of a market; the area reached by a newspaper; or the territory that is tied to a center of government by the lines of political authority. The nodal region is bounded by lines that mark the disappearance or weakening of the tie to its own focus in favor of some other focus. The lines that mark the extension of a nodal region are similar to isopleths in that they do not separate different kinds of things. Furthermore, the lines of connection to the node usually run at right angles to the boundary of the region.[4]

Resolution Levels

It is important to understand that any kind of region represents a generalization of the actual complexity of the face of the earth. Since no two microscopic points in earth space are identical, the only way the earth could be exactly reproduced would be to draw a pattern, like a dress pattern, on a scale of 1/1. This would be isomorphic mapping. When a geographer makes a map he is reducing the measurements of distance by some kind of ratio, which is homomorphic mapping. The basic fact is that the face of the earth is very much larger than man the observer; and, to gain any useful knowledge about occupied space, it is necessary to define categories through the selection of things to map, omitting reference to many other things that occur together in the same area.

For example, suppose that a group of geographers in the field were mapping land use. Before them is a square, fenced-in area planted with maize. The group maps this as a homogeneous, uniform region, based on the criterion of the growing crop. But a close examination of a part of this field brings to light two interesting facts: first, there is more bare ground between the rows of maize than the total proportion of the field in which maize is actually growing; second, since the field is not well cultivated, there are more ragweed plants than maize plants by actual count. Yet no one would suggest mapping it as a ragweed field. Why not? Because, if we are studying land use, we select for mapping those features that are relevant to our purpose. If we were studying the geography of hay fever, we might prefer to

[4]For additional discussion of the idea of the region and its importance in geographic study, see Robinson, 1953; Gilbert, 1960; Grigg, 1965; Haggett, 1966:241–276; Minshull, 1967, 1970.

map it as a ragweed field. In a sense, then, all regions are hypotheses. They are drawn for particular purposes, and they are judged good if they fulfill the purposes. The regional method involves the selection of criteria to define categories of regions, the testing of these criteria, and the identification of the criteria that most effectively select from the complex totality of interrelated elements just those things that are relevant to a particular problem. There is no such thing as a ''true region.'' The region exists only as an intellectual concept, useful for a specific purpose and to be judged suitable or unsuitable on the basis of the light it throws on the problem being investigated.[5]

The regions that are defined at one scale tend to disappear when the scale of mapping is increased or decreased, as when the field of maize is examined closely enough to include the ragweed plants. In studying land use, for example, the usual forty-acre field of the American Corn Belt occupies a square quarter inch on maps with a scale of an inch to the mile (1/63,360) (Davis in James and Jones, 1954:504 505). But to show the variations of soil or other features within a forty-acre field would require mapping on a much larger scale—such as the maps on 1/10,000 of the Puerto Rico survey (James, 1952:206-215).

These are not the smallest units of area that could be mapped. If a geographer were the size of an ant, he would recognize regions within an area of a few square feet. But even these ant-sized regions could be further subdivided if they were examined still more closely. There is no such thing as a ''unit area'' in the sense of an area so uniform that it cannot be further subdivided. The identification of a homogeneous area of whatever size is a generalization made by selecting relevant items. Geographers do not make ant-sized maps because geographers are not the size of ants. Both the markings on the face of the earth resulting from human action and the size of the regions designed to reveal the patterns of these markings are related to the physical stature of man himself.

If one is examining the broader patterns of land use or land quality over a wider area—say the American Midwest—maps of much smaller scale are required if any overview of the patterns of distribution is to be made. At map scales of 1/1,000,000 or smaller, the specific fields of maize or wheat cannot be shown. Instead, the specific crops are generalized into ''types of farming areas'' or even general agricultural regions, such as the Corn Belt. The Corn Belt can be defined, as it was by O. E. Baker, as including all those counties that produce 3000 bushels of corn per

[5]The Soviet geographers do not accept the idea of a region as an intellectual concept. A region is a combination of interrelated elements occupying a discrete territory. It exists objectively in the natural environment. The division of the Soviet Union into twenty-one regions by GOSPLAN for purposes of economic development has created regions that are ''real'' in the administrative sense, just as the territory included under the jurisdiction of one government is a ''real'' political region. The field of maize exists objectively, not merely as an intellectual concept. But the point of view developed by the author insists that if the region is viewed in greater detail, its homogeneity breaks up. In this sense the region is an intellectual concept, depending for its existence on the criteria that define it.

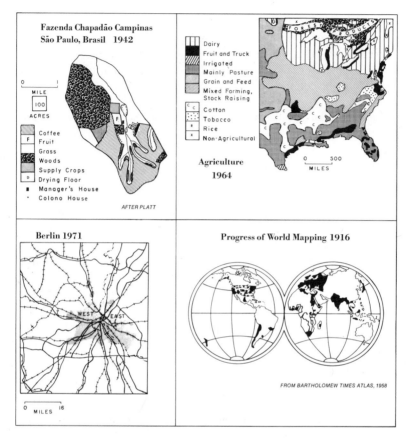

Figure 32. Resolution levels.

square mile. Similarly the soil types, as defined by the soil survey, occupy areas much too small to appear on maps of 1/1,000,000. But the soil geographers generalize soil types into soil associations defined as groups of soil types that are repeatedly found associated in the same areas. Soil associations can be shown on these maps of smaller scale.[6]

In other words, geographical studies may be carried on at different degrees of generalization (Fig. 32). This is what David Harvey calls the *resolution level* (Harvey, 1969:452). The resolution level suggests that the face of the earth is being observed by satellite imagery. At a very low level of resolution, only the broadest divisions of the face of the earth can be observed. By raising the resolution level— that is, by focusing more sharply on smaller areas—regional divisions that are less

[6]See the chapters by Carleton P. Barnes in James and Jones, 1954:391-393, and by Charles M. Davis, ibid.: 504-505.

highly generalized can be identified. It is possible to focus so sharply that a golf ball can be identified on an area the size of a golf green—which would represent an ant-sized observation. The term resolution level would seem to be synonymous with the "degree of generalization" (James, 1952:205–215).[7]

THE CONCEPTUAL STRUCTURE OF GEOGRAPHY

In formulating and finding the answers to questions concerning occupied space, geographers have developed *conceptual structures*—that is, frameworks of ideas about the character of the earth's surface and the methods of deriving geographical understandings from the study of it. The methods of formulating conceptual structures in geography do not differ from the methods followed by scholars in other fields of learning. Furthermore, geography is not the only field that is handicapped by the confusion of word meanings, and helpful perspective is gained by examining some of the discussions of theory and method produced in other disciplines. As a result of this confusion each scholar who deals with the philosophy of his field must make clear in one way or another how he uses his word symbols, and, when he introduces a new word symbol, he must make some effort to inform the reader how the new symbol differs in meaning from symbols already in use (Cole and King, 1968:522; Amedeo and Golledge, 1975).[8]

Percepts and Concepts

In this book we started with percepts and concepts. A *percept* is an empirical observation—an observation made through the senses and based on experience. An

---

[7]David Harvey concludes his book, *Explanation in Geography,* with the following: "I suspect, thus, that geography is moving into a phase in which it ranges less over a whole spectrum of resolution levels mainly because other disciplines have encroached. International trade and inter-cultural differences are firmly in the hands of economists and anthropologists, social interaction on housing estates firmly in the province of the sociologist. It is perhaps easier now to identify the typical resolution level of the geographer than it was some thirty years ago. But here I am speculating. Nevertheless I am prepared to suggest that another basic tenet of geographical thought is that its domain is defined in terms of a regional resolution level. Any phenomenon that exhibits significant variation at that resolution level is likely to be the subject of investigation by the geographer" (Harvey, 1969:484–485).

Note that Harvey uses the word, regional, in the popular sense. Note also that what he calls the "regional resolution level" would seem to be the equivalent of what the author has described as the chorographic scale. The author cannot agree that geographical studies are no longer appropriate at either the topographic or the global scales. Geographical studies are undertaken at different resolution levels depending on the purpose to be served. Certainly questions concerning the significance of relative location can be—and are being—investigated along the whole continuum of resolution levels from topographic (where the individual man-made structures can be seen) to the global (where major regional divisions of the earth can be examined). In 1952 the author recommended more attention to studies at the chorographic scale (James, 1952:215).

[8]See Braithwaite, 1953; Rapoport, 1959; Nagel, 1961; Brown, 1963; Nystuen, 1963; Bridgman, 1964; Meehan, 1965; Manners and Kaplan, 1968; Shafer, 1969; Wallace, 1969; Minshull, 1970.

empirical observation is sometimes called a factual statement. But as we pointed out in Chapter 1, the features on the face of the earth that are actually perceived are closely influenced if not determined by concepts. A *concept,* as we use the word, is a mental image of a thing or event. As such, the word concept is a general term that stands for a whole hierarchy of ideas ranging from simple generalizations to abstract theory. The words used to describe this hierarchy are for the most part poorly defined, overlapping, and are often used by the same author in different senses. The hierarchy suggested here is somewhat modified from the diagram (Fig. 33) presented by Walter L. Wallace in *Sociological Theory* (1969:ix). At the base are observations of the kind we call percepts. The circular form of the diagram indicates the important part played by hypotheses in determining observations. But proceeding in the opposite direction, the observations—which must always be of unique things—are generalized in what Wallace calls empirical generalizations. This requires scaling and measurement applied to the basic observations. Then, through inductive logic and no inconsiderable amount of intuition and imagination, the empirical generalizations can be further abstracted into theory. Theory, however, cannot be tested by direct observation. A theory is judged "good" if it permits the formulation of numerous hypotheses by deduction. The hypotheses can then be tested by observation and hopefully will yield new information from hitherto unexpected observations.

Geographers (and scholars in other fields) are accustomed to use other terms in various parts of this simple circular diagram. For example, David Harvey speaks of empirical laws (Harvey, 1969:31). What exactly is a law, and what can be done with it? Braithwaite defines a law as "a generalization of unrestricted range in space and time"—in other words, a generalization of universal applicability (Braithwaite, 1960:12). In this sense a law would be introduced on the theoretical side of empirical generalizations. The distinction between a mere generalization and a law would rest on two distinctions: an empirical generalization applies in a particular place or at a particular time, whereas a law is universal; and a law is embraced as one aspect of a still more abstract theory.[9]

There are two major difficulties with the use of the word, law. In the first place, whether we use it in a sophisticated sense or not, the fact remains that some people, even some scholars, use the word, law, in the anthropomorphic sense: a law governs events and to break a law is wrong and must result in punishment. Of course in moments of clarity it is recognized that a scientific law does not govern events—it is governed by events, of which it is a generalization. The word has, and still can, result in obscurity of thought. But even more important is the fact that a

[9]Cole and King recognize six different meanings attached to the word, law (Cole and King, 1968:522). Golledge and Amedeo also identify six kinds of law, none of which is exactly equivalent to any of Cole and King's laws (Golledge and Amedeo, 1968). Examining other writings on these matters, one finds a great emotional respect for a concept defined as a law but almost no common ground for defining the concept (Minshull, 1970:119–129).

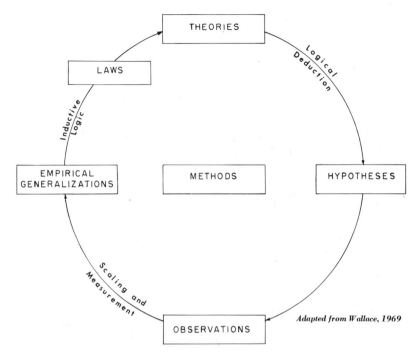

Figure 33. Diagram of concepts (Adapted from Wallace, 1969.)

law in the strict sense as defined by Braithwaite could scarcely be formulated from geographical evidence. If geography deals with things and events on the face of the earth, then its generalizations cannot be universally true or at least cannot be verified as universals from geographic observations. The only laws that could be defined as universals would be those of physics and chemistry; yet even in physical science there is a certain indeterminacy that makes necessary the use of probability concepts—these are *stochastic* laws. But even when they cannot predict every situation, they are universally probable. In the social and behavioral sciences, however, the only way a law can be identified as universally applicable is to define man's universe as the things and events located on the surface of one medium-sized planet in a very unusual relationship to a medium-sized sun.

Nevertheless, the word, law, is so widely used by geographers and others that to abandon it would not seem possible. As Braithwaite points out, very good results can be derived from treating some more abstract empirical generalizations as if they were laws. In such a case a law is more widely applicable than a generalization. Nevertheless, it is important to keep these distinctions in mind and not to claim more for geographical study than can be delivered.

Pattern Concepts and Process Concepts

The mental images geographers develop regarding the arrangement of things on the face of the earth are *pattern concepts*. They are empirical generalizations. It is recognized that at whatever scale or resolution level one uses to look at the earth's surface, certain kinds of associations of features of diverse origin are found repeatedly in similar situations. Soil associations, for example, are groups of soil types that are found again and again in the same locations relative to each other. Bowman's regional diagrams, which he used to give a generalized picture of the arrangement of surface features in the Andes of southern Peru, offer another example of pattern concepts (Figs. 27 and 28). At a global scale the generalized continent devised by W. Köppen to show the orderly arrangement of climates with reference to latitude and land and water distribution is another example of a pattern concept (Fig. 22, p. 183). Köppen's generalized continent is shown occupying a hemisphere, with the greatest east-west expanse of land about 70° north and with the smoothed outline of a land mass tapering to the southern extremity at latitude 55° south. On this continent (with differences of altitude removed) Köppen showed the ideal, or generalized, arrangement of his climatic types. This same procedure was used to show the generalized arrangement of habitats, with reference to latitude and to position on the western side, interior, or eastern side of a continental land mass (Fig. 34).[10]

Process concepts are those that provide a generalized picture of sequences of events. Examples of such concepts include the model of the cycle of erosion formulated by W. M. Davis. Chorley and Haggett have edited a book in which a variety of kinds of models are illustrated: models of physical systems; models of socioeconomic systems; and models of mixed systems (Chorley and Haggett, 1967). They distinguish two varieties of models: those that are deterministic in that if specified conditions exist at a particular moment of time, it is possible to specify what conditions existed at an earlier time or will exist at a later time; and stochastic models, which specify sequences of events within a certain range of probability.[11] In any case they provide an ideal picture of a process against which the actual

[10]W. Köppen, "Versuch einer Klassifikation der Klimate, vorzugsweise nach ihren Beziehungen zur Pflanzenwelt," *Geographische Zeitschrift*, 6 (1900):593–611. The eight habitats shown in Fig. 34 are defined in terms of associations of climate, water, vegetation, and soil (the mountain habitats are omitted because they do not fit on this generalized continent). The habitats are: I. Dry lands; II. Tropical forest lands; III. Tropical woodlands and savannas; IV. Mediterranean lands; V. Mid-latitude mixed forests; VI. Mid-latitude grasslands; VII. Boreal forests and woodlands; VIII. Polar lands. Each of these habitats occupies a particular location relative to latitude and the land-water factor.

[11]The term, model, is very difficult to define because it is used in so many different senses. Harvey quotes one opinion that "a model can be a theory, or a law, or a relationship, or a hypothesis, or an equation, or a rule" (Harvey, 1969:145). This seems to be a particular way of formulating a process concept. Models have proved to be of great utility in geographical studies (Chorley and Haggett, 1967; Cole and King, 1968:463–520).

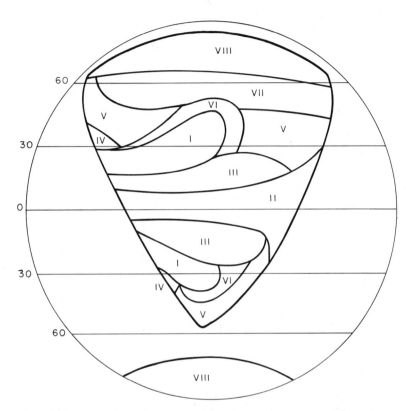

Figure 34. The generalized pattern of habitats. (From James, 1966. Reprinted by permission of the publisher.)

departures from a norm can be measured. It was concerning such process models that Wellington D. Jones is credited with the remark: "models tell us what would happen if things do not happen the way they do."

Geography: Nomothetic or Idiographic?

The idea that any field of learning could be restricted either to a study of unique things or to the formulation of general concepts, but not both, is a remarkable example of the "tyranny of words." The terms, idiographic and nomothetic, were first used by Wilhelm Windelbrand in 1894. Once the word symbols were provided, the existence of a dichotomy could be proclaimed (Sidall, 1961). The dichotomy was turned loose in the literature of geography in 1953 by Fred K. Schaefer. Sauer pointed out in 1925 that although geography was formerly devoted to descriptions of unique places as such, the geographers had for a long time been seeking to formulate illuminating generalizations about the earth and man's place on it (Sauer, 1925:27). But in the 1950s, after both Hettner and Hartshorne had been incorrectly quoted as saying that geography is essentially idiographic and is not concerned with general concepts, this characterization gained an amazingly wide acceptance (Harvey, 1969:50-51).[12]

Such a distinction would never have occurred to anyone experienced in the out-of-door study of the face of the earth. Of course, every observation has to do with things that are unique in time and place. But it is not even possible to identify any one feature as unique until there is some kind of empirical generalization with which to compare it. Geographers have always been concerned with the question of what things to observe. How does one select specific features to observe out of the complex fabric of interwoven strands that makes up the face of the earth? How does one identify things that are significant or relevant? Only in relation to concepts: hypotheses, empirical generalizations, and, hopefully, some kind of generalization that could be treated as if it were a law (Burton, 1963:156). A very fundamental part of the scientific method consists in learning how to distinguish the relevant from the irrelevant, and this cannot be done without a framework of ideas. Geographers have always observed unique things; but they have also sought to formulate those illuminating concepts that make sense out of the apparent disorder of indirectly related parts.

Description and Explanation

There is confusion, also, over the meaning of the words, description and explanation. Davis used the term, explanatory description, to refer to the use of his model of an ideal cycle of erosion as a frame of reference for the description of landforms. To describe means to transfer observations of things and events on the face of the earth into symbols—either word symbols, or cartographic symbols, or mathematical symbols. Robert Brown defines the word, explanation, as an effort to remove impediments of understanding, an effort "to deprive puzzles, mysteries, and blockages of their force, hence of their existence" (Brown, 1963:40-44). Percy Bridgman wrote that, "explanation consists of analyzing our complicated systems into simpler systems in such a way that we recognize in the complicated systems the interplay of elements already so familiar to us that we accept them as needing no

[12]Peter Haggett quotes Hartshorne's *Nature of Geography* as follows: " . . . no universals need be evolved, other than the general law of geography that all its areas are unique" (p. 468) (Haggett, 1966:2-3). In the preceding paragraph, which is not quoted, Hartshorne writes: "For the interpretation of its findings it [regional geography] depends upon generic concepts and principles developed in systematic geography. Furthermore, by comparing different units of area that are in part similar, it can test and correct the universals developed in systematic geography" (p. 467). Both Hettner and Hartshorne insisted that geography is both idiographic and nomothetic, as indeed almost all other fields of learning must be (Hartshorne, 1959:146-172).

explanation'' (Bridgman, 1964:63). David Harvey puts it more simply: "The purpose of an explanation may be regarded as making an unexpected outcome an expected outcome, of making a curious event seem natural or normal'' (Harvey, 1969:13).

There are of course many different ways to make a curious event seem normal or to change the confused complexity of things on the face of the earth into some kind of order. There was a time when a satisfactory explanation for observed features could be provided by reference to God's divine plan; but the more that is known about the complexity of man's universe, the more rigid become the requirements for acceptable explanations. Robert Brown lists seven quite different kinds of explanation for human behavior (Brown, 1963). David Harvey, in a major contribution to the clarification of geographic ideas, analyzes the explanatory procedures available to geographers (Harvey, 1969). It seems, however, that it might be possible to include all these procedures under two categories: temporal and conceptual.

In the temporal approach, explanation is given by adding the time dimension to descriptions of observed patterns and locations. In some cases this is adequately done by showing the origin of what is to be explained—the genetic form of explanation. In other cases it is more satisfactory to trace the changes through time—the evolutionary or developmental form of explanation. In some cases both of these are combined, as when one goes back to origins and traces developments. The difficulty is that actual identification of an origin is seldom possible; for there are always antecedents that must help to explain so-called origins. Basically, the temporal approach shows the changes through time of which the observed features at any moment of time are like snapshots or like the frames in a moving picture. It has been accepted practice to explain the geographical arrangements at any one time by describing the sequences of events of which they are the momentary results. This is the method of historical geography (A. H. Clark in James and Jones, 1954:70–105). This type of explanation provides an important means of testing hypotheses and checking on generalizations based on observation, and as such it is an important and indeed an essential part of the whole program of geographic scholarship.[13]

In the 1950s and 1960s a strong reaction against the temporal method of seeking explanations in favor of the conceptual approach set in. The movement seems to have started at certain universities in the United States (Washington, Northwestern, and the Department of Regional Science at Pennsylvania) and then spread to Cambridge and Bristol in England, with strong support from Lund in Sweden. The temporal method was relegated to a less sophisticated stage of scholarship both by Harvey (Harvey, 1969) and by Haggett (Haggett, 1966). The difference between the temporal and conceptual approaches is illustrated by the story of Newton and the apple. Newton, the story goes, was faced with alternatives when the apple fell on his head:

[13]Examples of this kind of explanation in the contemporary period may be found in Meinig, 1962, 1965, 1968, 1969; Ward, 1964; Kniffen, 1965; Jordan, 1967, 1969.

Had he asked himself the obvious question: why did that particular apple choose that
unrepeatable instant to fall on that unique head, he might have written the history of an
apple. Instead of which he asked himself why apples fell and produced the theory of
gravitation. The decision was not the apple's but Newton's [quoted in Haggett,
1966:2].[14]

The story is clever and, for the uninitiated, clearly convincing. Of course one
should not bother with unique events but only seek general theory. Yet the fact is
that some individual scholars are deeply and productively concerned with the search
for valid explanations in geography through studies in historical perspective, most
of which are not properly labeled as trivial. Others are as deeply concerned with the
search for theory. Each scholar is inclined to think that the road he is following
toward the horizon is the only straight road. Yet we must acknowledge that for
every Newton or Darwin there are thousands of "lesser" scholars searching for
sequences of events by which the validity of theory may be tested. From among the
many scholars so engaged have come most of the hypotheses the implementing of
which has brought progress (Haggett, 1966:277). The continued development of
geography as a field of learning would not be served by the elimination or the
denigration of either the temporal or the conceptual methods.

Explanation through the conceptual approach was at one time focused on the
search for repeated cause and effect relations. A prior cause had to be related to a
later effect. In the eighteenth century David Hume argued that it was not possible to
demonstrate such a relationship except through experience with repeated examples
of the same sequence of events (Ducasse, 1969).[15] Furthermore, the philosophers
and metaphysicians became enmeshed in hopeless discussions of the problem of
free will as opposed to determinism.[16] The principle of causation is by no means
discarded; but it is now generally recognized that simple cause and effect connections do not exist and that, in fact, the interconnections among things and events on
the earth are much more complex than was formerly appreciated.

A dangerous but sometimes very useful kind of conceptual explanation is
found in the use of analogy (Chorley, 1964). This is what Herbert Spencer did when
he developed the concept of Social Darwinism, or what Ratzel did when he described the state as a pseudoorganism. Explanations based on analogy in these cases
had the effect of obscuring rather than clarifying things. On the other hand Faraday's description of the electromagnetic field in terms of the motions of fluids was a
highly successful clarification. In contemporary times geographers who make use of

---

[14]The story, which was included in a monograph on geomorphology by R. J. Chorley, was written
by M. Postan, "The Revulsion from Thought," *Cambridge Journal,* 1 (1948):395–408.

[15]David Hume, *Enquiry Concerning Human Understanding* (London, 1748).

[16]There is a long list of papers that discuss the problem of determinism in geography, especially
environmental determinism. See Platt: 1948; Spate, 1952, 1958; Hofstadter, 1955; Montefiore and
Williams, 1955; Sprout and Sprout, 1956; Herbst, 1961; Lewthwaite, 1966.

gravity models, using mathematical formulas derived from Newton's law of gravitation to predict the volume of interconnections between two cities, offer an apparently successful use of the analogy method. Another stimulating use of analogy is suggested by William Bunge in his search for a single rule to predict shifts of highways, rivers, or shopping centers (Bunge, 1966:27–33). The high land values along a main artery of transportation can be likened to the natural levees along a graded river. Imaginative suggestions of this kind, whether they prove valid or not, are urgently needed in geographical studies.

In the contemporary period the search for explanation through the conceptual approach has been focused on functional relations as opposed to cause and effect relations (Manners and Kaplan, 1968:212–216). The existence of systems of functionally interconnected parts has been known for thousands of years, but until the age of the electronic computer they could not be handled operationally. A system can be defined logically as a set of elements so closely interconnected that a change in any one of them results in changes in all the others. When one element of a system can be identified as spatial—that is, related to location and extension on the face of the earth—it becomes a spatial system. It is now widely accepted that a major concern of geography is with spatial systems (to be discussed in Chapter 18). The area occupied by a functionally interconnected system is, by definition, one kind of region, as Robert S. Platt pointed out in 1928. There have been some stimulating studies along these lines and at various resolution levels (Philbrick, 1957).

Fundamental to the study of systems in geography is the use of probability theory. As long as a system is self-perpetuating because at least one element remains in a steady state, the various components of the system can be treated in terms of functions. Stochastic models of such systems now open new ranges of understanding.

THE SECOND HALF OF THE TWENTIETH CENTURY

Fundamental innovations swept through all the fields of scholarship in the second half of the twentieth century. Geographic study felt the impact of change like all other fields. For many generations the basic paradigm of scientific procedure was to take complex groupings of phenomena apart to study each part separately, in isolation from the complicating impact of the total environments of particular places. Great advances in understanding resulted from this procedure; and the several academic disciplines were identified in terms of the processes they examined. Yet already in the 1920s and 1930s unexpected progress was being made by scholars from different disciplines who collaborated in the border zones between disciplines. Such new fields as biochemistry and social psychology began to appear. By midcentury the tendency was for the various disciplines to take on a problem orientation and often disregard the arbitrary divisions of knowledge.

Geographers were not quite ready for these changes. During the 1920s and 1930s their major concern had been to gain status as an independent field. In a few cases individual geographers collaborated across fields—especially on the borders between geography and anthropology. Area studies, or interdisciplinary investigations of the problems and conditions of particular culture regions, were actively promoted by some geographers, especially in the Latin American area. But so few geographers became active in area studies that many such programs were organized with no geographers included on their staffs. Not until the 1950s did geographers in large numbers feel the impact of innovation and start moving with the tide. (Bennett and Chorley, 1978; Chapman, 1977)

By midcentury the changes were proceeding rapidly. There were new techniques of observation (including the use of space satellites), new techniques of analyzing data (including the use of the electronic computer), and new techniques of preparing maps. These innovations revolutionized geographic procedures and required the rapid construction of new training programs and new specializations. The use of mathematics spread rapidly and threatened to separate the geographic profession (like many others) into two separate groups, each talking a different language and each intolerant of the other. But this is progress in geography. Intellectual growth has been evolutionary in nature. And time is needed to recall the best of older hard-won traditions and to blend this with the extract of what is most valuable from the ''revolution'' of the last twenty years or so. Traditional regional syntheses and description remain of inestimable worth in different parts of the world. The new advances in the field can aid in further development of this work. Especially is this of value in developing countries, where the man-land hypothesis has not yet been rendered obsolete by man's cosseted, mechanistic existence. And even in the developed world the large urbanized populations remain dependent on the physical environment for sustenance and the breath of life.

Notwithstanding the epistemological debate provided by the (new, newer) newest geography, the concept of occupied terrestrial space remains central to the locus of our inquiry (Glacken, 1956). Understanding of the interplay of innovation and tradition is vital if the two roles are to emerge as complementary and not adversarial in mentality. Discussion of these developments forms the subject matter of the last two chapters.

CHAPTER 17

New Methods of Observation and Analysis

The idea that thought is the measure of all things, that there is such a thing as utter logical rigor, that conclusions can be drawn endowed with inescapable necessity, that mathematics has an absolute validity and controls experience—these are not the ideas of a modest animal. Not only do our theories betray the somewhat bumptious traits of self-appreciation, but especially obvious through them all is the thread of incorrigible optimism so characteristic of human beings.

. . . When will we learn that logic, mathematics, physical theory, are all only inventions for formulating in compact and manageable form what we already know, and like all inventions do not achieve complete success in accomplishing what they were designed to do, much less complete success in fields beyond the scope of the original design, and that our only justification for hoping to penetrate at all into the unknown with these inventions is our past experience that sometimes we have been fortunate enough to be able to push on a short distance by acquired momentum?

The industrial revolution has finally caught up with the study of geography. For thousands of years the techniques of observation and the methods of analysis remained essentially the same. Men on foot, men on horseback, men in horse-drawn carriages, and men in canoes and sailing ships traveled through different parts of the world. They recorded the things they saw by direct observation, analyzing their observations into component parts to clarify their descriptions, and testing their

The quotation above is from Percy W. Bridgman, *The Nature of Physical Theory* (New York: John Wiley & Sons, 1964), pp. 135–136.

389

hypotheses regarding processes by additional direct observation. A geographer was one who gained more than common satisfaction in visiting unfamiliar and out-of-the-way places and returning to interpret the circumstances observed in the strange new worlds beyond the far horizons. There never was anyone who did this kind of work better than Alexander von Humboldt, the last great universal scholar. But now, and especially since World War II, there has been a revolution in the technology of observation and analysis.

There are certain dates that mark these technological advances. In 1950 the U.S. Bureau of the Census installed its first electronic computer, known as UNIVAC. In 1957 the Soviet Union sent up its first successful satellite, *Sputnik I*. In 1962 the first geodetic satellite was placed in orbit, and thereafter the age-old problem regarding the shape of the earth was resolved with a degree of accuracy never possible before. In 1964 the first weather satellite *Nimbus I* was launched and *Nimbus II* in 1966. The result was information concerning the state of the earth's atmosphere that brought synoptic meterology from an art to a science.

The computer came just in time.[1] With this electronic device mathematical computations can be carried out in seconds that would have required something like 4000 years for an individual with pencil and paper. Areas of the earth's surface defined as homogeneous by specified and measured criteria can be outlined in minutes from satellite data fed into computers. Furthermore, for every piece of information about the face of the earth that Humboldt could command, the student of geography today has tens of thousands of items of information to overwhelm the traditional methods of analysis. Data banks can store this information, which can be recalled in moments. The incredible transformation of the technology is, in fact, so incredible that a large number of scholars raised in the older traditions seem unprepared to move forward into the new world. When, in 1969, a man explored the face of the moon in full sight of television viewers it seemed just like another bit of science fiction.

THE TECHNOLOGY OF OBSERVATION

Although the ancient Greeks understood the method of calculating the size and shape of the earth, they lacked the instruments to make measurements with sufficient accuracy. Not until the seventeenth century were the methods of a national mapping program worked out by the Cassinis in France. Newton's and Huygen's deduction that the earth must be flattened at the poles was challenged by the Cassinis on the basis of their measurement of the Paris meridian across France. This

[1]The first electronic computer was made by Eckert and Mauchly at the University of Pennsylvania in 1946. It was designed for the Army Ordnance Corps to calculate ballistic trajectories that involved the numerical solution of very difficult differential equations. It was called ENIAC.

led to the expeditions of La Condamine and Maupertuis and the confirmation of the polar flattening.

The advance of the modern technology of measurement would not have been possible without several international agreements and several technical innovations made in the seventeenth and eighteenth centuries. The pendulum clock was one of these and another was the development of a chronometer that could be used at sea. And there had to be some kind of international agreement concerning units of measurement—a type of agreement that the Greeks and Romans were never able to reach. It was not until 1791 that the French adopted the standard meter as 1/10,000,000 of the meridian from the equator to the pole passing through Paris. A rod consisting of an alloy of metals that minimizes changes of length owing to temperature changes is kept in an air-conditioned vault in Paris, and this rod has been accepted throughout the world as the standard of linear measurement. Even the length of the foot, still the popular unit of measurement in the United States, is determined by comparison with the French meter.

As a result of these agreements and the technical improvements of the contemporary period, the measurement of the shape of the earth has now been done in great detail. The observations of satellite motions every hour or so reveals minute differences in the shape of the earth, and the observation of these minute differences in motion has been made possible by a new kind of satellite tracking camera devised and constructed by the Smithsonian Astrophysical Observatory (King-Hele, 1967). The earth is now described as slightly pear-shaped with a bulge south of the equator. A map of departures from a geoid-shaped sea level shows bulges in western Europe, to the north of New Guinea, and between Africa and Antarctica. The average equatorial diameter of the earth is 12,756.38 km (7926.42 miles), and the average polar diameter is 12,713.56 km (7899.83 miles). The equatorial circumference is now measured as 40,075,51 km (24,902.45 miles).

Mapping the Earth

The method of making a large-scale map of a country that was developed by the Cassinis remained the standard procedure until the 1930s. G. R. Crone describes the steps that had to be taken in making a national survey as follows:

1. Determination of mean sea level, at one point at least, to which all altitudes are referred;
2. A preliminary plane table reconnaissance to select suitable points for the triangulation, and the erection of beacons over them;
3. Determination of initial latitude, longitude, and azimuth (for direction) which will "tie" the map to the earth surface;
4. Careful measurement of the base or bases with tape or wire of a special alloy;
5. Triangulation, the theodolite being used to observe horizontal angles from the base and beaconed points, and to measure altitudes by readings of the vertical angles;

6. Calculation of the triangulation and heights, and the transference of the trig points to the sheets issued to plane tablers;

7. The filling-in on the sheets by plane tablers of the required topographical detail—contour lines, rivers, woods, settlements, routes, and names. [Crone, 1950:152.]

By the end of the nineteenth century almost all the countries of Europe had been covered by topographic (large-scale) maps; but each country used its own scale and its own projection, and each included different categories of features. There were some fifteen different prime meridians in use. Since the maps were not comparable, no overall topographic coverage for Europe was available. Outside of Europe small parts of North America had been mapped, there was a fairly complete survey of India, but elsewhere large-scale maps were few (Robinson, 1956).

Two important steps were taken in the late nineteenth century. In 1884 a conference was convened in Washington, D.C., to discuss the adoption of a single prime meridian. As a result, twenty-five nations agreed to use the meridian of Greenwich in England (the astronomical observatory on the Thames, now in the suburbs of London) as the 0° meridian and to measure longitude east and west of Greenwich. During the nationalistic days of the 1930s many countries gained stature (in their own eyes) by adopting a prime meridian through their own capital city; but by 1950 almost all the world's maps made use of the Greenwich meridian. And second, at the Fifth International Geographical Congress held in Vienna in 1891 a proposal was made by Albrecht Penck that an international map of the world should be made on a scale of 1/1,000,000, using uniform symbols and conforming to agreed standards. The "millionth map" is still far from complete, but it is being extended more rapidly as the need for it becomes clearer.

Meanwhile, the production of large-scale maps (on scales of 1/100,000 or larger) continues slowly. Arthur H. Robinson summarized the world mapping situation in 1956:

> The need for such maps is continually increasing, for, as the world's population increases and life becomes more complex, more and more planning becomes necessary to insure adequate food supplies and transportation facilities. This means the extension of soil surveys, land-use surveys, water-supply surveys, erosion surveys, population surveys, and a host of others. None of these can be carried on adequately without the basic topographic map as a point of departure. The use of the topographic map as a base is by no means limited to those activities in which detailed maps are generally conceded to be indispensable; it also serves an important function as a source of information for a wide range of interests, from "marketing" maps to "treasure" maps [Robinson, 1956:296].

Vertical Air Photography

Vertical air photography came into use during World War I. Even in the Civil War, photographs had been taken from balloons for military purposes; but the first vertical photographs for intelligence purposes were used during the trench warfare on the western front. Skilled photo-interpreters learned how to identify objects in the pictures when they were viewed from directly overhead and how to penetrate the screen of camouflage intended to hide military objects from overhead viewers.

The applications of vertical air photography to geographical studies were appreciated even during the war. In 1917 the first vertical photograph (of a part of Paris) was published in the *Geographical Review* to demonstrate the utility of such pictures for geographical studies (Woodhouse, 1917:337). After the war the airplane was used to experiment with air photography for geographical purposes, and several pioneer studies were published during the 1920s. In 1920 Willis T. Lee, a geomorphologist on the staff of the U.S. Geological Survey, published a paper demonstrating how air pictures could be used to interpret the landforms and settlement patterns of the Coastal Plain in the eastern United States (Lee, 1920; also Wright, 1952:330-334). In 1921 Jules Blache, a professor of geography at Grenoble, made use of photographs taken from airplanes to interpret the conditions of life in Morocco, even beyond the territory then controlled by France (Blache, 1921). In 1922 the American Geographical Society published a book by Willis T. Lee showing a variety of landforms and settlement types as seen from the air (Lee, 1922).

A pioneer in the development of new techniques for using photographs to make topographic-scale maps was the Scottish cartographer, O. M. Miller. An artillery officer during World War I, Miller appreciated the potential uses of air photography not only for spotting artillery targets but also for making maps for peaceful purposes. In 1923 Miller joined the staff of the American Geographical Society to offer training in this work in the School of Surveying under the direction of Alexander Hamilton Rice (Wright, 1952:320-322). Miller had completed such a course of study at the Royal Geographical Society in London. The purpose was to train explorers in the techniques of field-mapping. In 1926 when Miller was in Zurich he witnessed a demonstration in the laboratories of the Wild Instrument Company of a device for plotting topographic maps from stereoscopic pairs of air photographs. When adjacent pictures were overlapped by about 60 percent, the eye, viewing the pictures through a specially designed pair of lenses, could see the terrain and settlement features in relief. The stereoscopic plotting instrument permitted the map-maker to draw contour lines by following the levels viewed in the photographs. Furthermore, it was possible to use oblique air photographs after applying the corrections devised by Miller (Miller, 1931). Thereafter both vertical and oblique air photographs were used to make topographic maps much more rapidly and at less cost per square mile than had ever been possible before.[2]

---

[2]After these pioneering experiments the field of photogrammetry, a branch of engineering, developed more and more precise methods of producing maps from air photographs. It improved both cameras and methods of plotting the maps to overcome errors of perspective, including eventually even the effect of differences in altitude of the earth's surface.

The Use of Air Photographs for Plotting Geographical Data

In addition to plotting points and lines for the purpose of making a large scale base map, geographers were also intrigued by the possibility of plotting the areas occupied by the phenomena they wanted to study. At the Ann Arbor meeting of the Association of American Geographers in 1922, W. L. G. Joerg of the American Geographical Society showed how vertical photographs could be used to make maps of American cities in much greater detail than was provided by the usual topographic map.[3]

Nevertheless, the field-mapping of the Michigan Land Economic Survey during the 1920s and 1930s was done in the traditional manner by field surveyors equipped with plane tables, compasses, and alidades. The plane table was set up on a tripod and oriented by the compass (with care to make sure that the compass was not deflected by a pocket knife or a wire fence). The alidade was used to sight at distant objects and to draw the line of sight on the map. Distances were measured by pacing. Although the geographers were aware of the new techniques for field-mapping, there were not enough airplanes properly equipped with cameras to apply these improved methods.

The use of vertical air photographs was a major breakthrough in field-mapping techniques. The first experiment in the use of vertical air photography to map the vegetation and land use of a small area was done by K. C. McMurry, then chairman of the Department of Geography at the University of Michigan.[4] The experimental area was Isle Royale in Lake Superior. The entire island was photographed—after a delay of more than a month because forest fire haze made photography impossible. McMurry mounted the photographs in the form of a mosaic, and from them he identified certain areas he thought were representative of differing vegetation and land use categories. The type areas made up about 25 percent of the total area. He then went into the field to make maps of these areas by traditional methods. The comparison of the photographs with the ground maps made it possible to extend the mapping to the whole island by photo-interpretation (Russell, Foster, and McMurry, 1943).

By this time the geographers who met at the annual spring field conferences were ready to try mapping geographical data directly on the photographs. Using the photographs in the field made it possible to identify slope, soil, drainage, vegetation, land use, and settlement features and to plot these features directly on the pictures in relation to the visible objects picked up by the camera. The first major land and land use inventory to use the photographs in this way was the survey of the area administered by the Tennessee Valley Authority carried out under the direction of G. Donald Hudson. With the photographs to provide exact outlines of fields and other features, it was possible to plot information about land use, physical land

[3]Reported in the *Annals AAG,* 13 (1923):211.
[4]Reported in the *Annals AAG,* 22 (1932):69.

conditions, crop yields, land value, market areas of towns, and other relevant information much more rapidly and with much greater accuracy than could be done by the traditional methods used in the Michigan survey. After that time, mapping geographical information on vertical air photographs became the standard procedure.

Radar and Infrared Imagery

World War II produced another and even more far-reaching breakthrough in the use of new devices for the remote sensing of the face of the earth. And in many ways the history of the application of the new devices to geographical research was repeated. In the 1920s the Michigan Land Economic Survey had to make use of already outmoded methods of field-mapping; but even in the 1970s the new devices for mapping all kinds of important features are still not available to the great majority of geographers. Yet it is clear that the profession is on the threshold of a new era of low-cost field observation, with a new command of detail and a variety of resolution levels. We shall describe only two of the new devices: the side-looking airborne radar (SLAR) and infrared color film. Both are dependent on the new high-altitude reconnaissance planes.

The airborne radar is capable of scanning large areas rapidly and, if necessary, repeatedly. The plane carries its own source of energy, which permits the emission of thousands of pulses of electromagnetic energy per second. Some of the energy is reflected back to the airplane from the earth's surface. A specially devised camera with a continuous film strip records the radar imagery from a cathode-ray tube. Robert B. Simpson has summarized the capabilities of SLAR as follows:

> Assuming the availability of a properly equipped aircraft, almost any task within the capability of SLAR can be accomplished more cheaply by it. A recent study for the U.S. Agency for International Department indicated that the most pressing elements of a topographic map production program for a typical underdeveloped country could be accomplished by SLAR in one-quarter to one-tenth the time, and at one-quarter to one-tenth the cost of a conventional mapping program.

Among the types of material that such a survey could provide an underdeveloped country are the following:

1. An area mosaic. . . .
2. Natural regions overlays, including those showing geomorphology, and other aspects of geology, vegetation, soil, land utilization, and a variety of other critical economic and scientific parameters, such as the size and shape of drainage basins.
3. Information necessary to determine the areas suitable or practices which will upgrade the economy, such as areas suitable for irrigated floodplain agriculture.
4. Suggestions as to location of mineral resources. . . .
5. A basis for the selection of routes and sites, such as for roads, terminals, industrial plants and railways, in detail appropriate to the scale.

6. Data from which to determine areas requiring later large-scale Class-A map coverage. . . .

Meanwhile, SLAR surveys can provide the data for the production of small-scale, Class-A planimetric sheets, as well as that for the interim production of medium-scale sheets [Simpson, 1966:96].

Another exciting new source of geographical data is the use of infrared imagery of the 4.5 to 5.5 micron wavelength band. The cameras are carried in high-altitude airplanes that permit the rapid survey of large areas of the land surface. Already such photography has been applied to forest resource inventories; not only do the pictures permit estimates of the available lumber in a forest but the extent of forest damage by disease, for the diseased trees appear in a distinctive shade of red. It is possible also to make studies of agricultural land use, including types and conditions of crops and differences of cultivation practices (Olson, 1967). With adjustments of camera and film it is possible to focus on soil types, differences of soil productivity, or variations in the availability of water.

These new instruments make possible a census of the whole of the world's population, covering any particular populations at frequent intervals. The patterns of metropolitan areas can be exactly delimited and the movements of goods and people can be plotted either between urban centers or within urban areas.

Satellite Imagery

The year 1957 witnessed the beginning of another major advance in man's long and continuous efforts to increase his knowledge of the physical character of the earth. Starting on July 1 and continuing for eighteen months, seventy nations cooperated in the International Geophysical Year (IGY). Simultaneous observations carried on with standardized procedures were made all around the world for the purpose of finding answers to a variety of geophysical questions. Why do magnetic storms disrupt communications? Can the positions of the continents and ocean basins be mapped so accurately that minute movements can be measured? Are glaciers receding and ice caps melting? The observations provided new information for meteorology, geodetic surveys, ionospheric physics, glaciology, oceanography, seismology, and other branches of geophysics as well as for the study of geomagnetism, gravity measurements, aurora and air glow, solar activity, and cosmic rays. The results provided a spectacular increase in knowledge of physical processes. Then on October 4, 1957 the Soviet engineers put *Sputnik I* in orbit and on January 31, 1958 the United States sent up *Explorer I*. Thereafter a large number of satellites have been placed in orbit, each to perform certain specific functions. Satellite imagery now provides a wealth of new data concerning the face of the earth.

The satellites can do many things of interest to geographers. Satellites of the *Tiros* and *Nimbus* series are providing new views of the earth's atmosphere. Every day—or every hour, if necessary—the receiving stations on the earth get TV pictures of the earth's cloud cover, thus making possible not only a general view of atmospheric circulation but also a day-to-day synoptic view of the storm patterns. Weather forecasting takes on a new dimension. From these same satellites temperatures can be recorded all over the earth by the use of infrared energy detectors. The earth, as a great system of complex parts kept going by radiation from the sun, can now be observed as a whole and the inputs of energy exactly measured. The cloud patterns, from which atmospheric circulation can be read, show no signs of Maury's wind zones (pp. 150-154), but they do clearly reveal the more or less permanent oceanic whirls and the cold fronts that push into them on the poleward sides (Barrett, 1970).

NEW ANALYTIC PROCEDURES

All this new technology of observation has been matched in the contemporary period by new analytic procedures. The so-called quantitative revolution means that many of the younger generation of geographers have discovered the value of mathematics and especially of mathematical statistics. But an essential part of the revolution consists in the use of the electronic computer as an analytic device.

The Use of Mathematical Concepts and Statistical Procedures

Following the notable success of econometrics in the 1930s in making the description of economic processes more precise and providing for the objective testing of economic hypotheses, the use of mathematics began to spread into other social and behavioral sciences. Geographers, however, lagged behind the others, in part because the statistical procedures developed in other fields were not directly applicable to the analysis of spatial factors. It remained for mathematically minded geographers and geographically minded economists to develop statistical procedures specifically for the study of geographical problems.

The use of mathematics in geography is not really new. Of course even in the days of Thales and Eratosthenes there was a branch of geography known as *mathematical geography,* but this had to do chiefly with studies of the form of the earth and the relation of the earth to the celestial bodies. In the modern period *mathematical geography* is again a branch of the field, but it refers to the use of mathematical concepts and statistical procedures for the study of occupied space. The two usages of these words should not be confused. Even in the latter sense, however, statistical procedures have been used in almost every generation during the modern period. Ellsworth Huntington made use of statistical analyses to give plausible support for his otherwise verbal hypotheses regarding the effects of climate. In 1937 John K. Wright used mathematics to provide a quantitative measurement of the variations in the intensity of phenomena over the earth and of degrees of correspondence between two or more phenomena (Wright, 1937). In 1939 M. G. Kendall published a paper analyzing the covariance among ten crops in forty-eight counties of England in terms of productivity (Kendall, 1939). After World War II papers pointing to the importance of using quantitative procedures appeared more frequently (Weaver, 1954, 1956). Of special influence was the paper by John Q. Stewart, published in 1947, in which he demonstrated the application of mathematics to the study of urban hierarchy and to other problems of population distribution. Here is what Stewart had to say about the utility of mathematics:

> The way of progress is obstructed by the opinion, common among authorities on economics, politics, and sociology, that human relationships never will be described in mathematical terms. There may be some truth in this as regards the doings of individual persons. Even the physicist has given up the idea that the behavior of individual particles can be precisely described thus and necessarily contents himself with the discussion of averages. But the time to emphasize individual deviations is after the general averages have been established, not before [Stewart, 1947:461].

But in spite of these and other individual efforts in the use of statistical procedures, there were no large numbers of followers. It may be said that the quantitative revolution had its beginning when William L. Garrison offered the first seminar for the training of graduate students in geography in the use of mathematical statistics. This was at the University of Washington in 1955.[5]

There are numerous benefits to be derived from the use of mathematical concepts and statistical procedures in geographical studies. Mathematics provides a clear way to avoid the old problem of tracing cause and effect relations. The differential equations in calculus may be used to describe the way variables change through time, but they do not specify an antecedent cause and subsequent effect. Given the condition of a phenomenon at any one time, it is possible to describe its condition either at an earlier or later time. The theory of functions may prescribe that where A exists, B also exists, but it does not have anything to say about one being the cause of the other.

A very important benefit to clarity of thought resulting from the use of mathematical concepts is that deterministic models can be replaced by stochastic models (Lewis, 1965). Since 1927, when Werner K. Heisenberg, a German physi-

[5]The seminar during the first few years included Brian J. L. Berry, William Bunge, Michael F. Dacey, Arthur Getis, Duane F. Marble, Robert W. Morrill, John D. Nystuen, and Waldo Tobler. Torsten Hägerstrand was a visiting scholar for one semester. From the University of Washington and the University of Lund, training in quantitative procedures spread to numerous universities. In Britain the chief center for the new approach was at Bristol University where Haggett was invited to become a second professor of geography. The new procedures spread rapidly in the Soviet Union and other parts of the world where geography was developed on the Soviet model. The economist Walter Isard established the Department of Regional Science at the University of Pennsylvania, replacing the Department of Geography. The Regional Science Association now has branches in many countries.

cist, formulated the principle of indeterminacy in physics, the basic concept has spread to other fields of study. In physics this principle demonstrates that it is impossible to measure with full accuracy at the same time both the position and the velocity of an electron. On the other hand, when large numbers of electrons are measured, the probability that they will occupy certain positions and be moving with certain velocities becomes more and more predictable. The principle has obvious applications to geography.

Probability theory provides the mathematical foundation on which statistical analysis is built (King, 1969:32). Models based on probability are stochastic models. Statistics are also designed to permit drawing interfaces from a set of observations by consideration of a sample. The inferences are then used to develop empirical generalizations, models, and hypotheses.

As an example of the statistical approach to an old problem, consider the procedure geographers use in defining the degree of correspondence in the extent of area occupied by two phenomena. This was formerly done by the inspection of superimposed maps from which it was possible to identify in qualitative terms the degree of correspondence. It was also possible to identify the problem areas where the two phenomena did not correspond. The results, however, were not precise (McCarty and Salisbury, 1961). In 1957 Arthur H. Robinson and Reid Bryson demonstrated what might be done to compare the density of rural population and the average annual rainfall in a specific area—Nebraska (Robinson and Bryson, 1957). From an inspection of the two maps (Fig. 35a and 35b), it is clear that the density of rural population is lower where the average rainfall is less. But there are some exceptions. The two phenomena do not vary exactly in the same way, therefore, it seems that some other factors beside rainfall are important in explaining the density of population. The statistical comparison of two unlike quantities (population density and rainfall) requires a special procedure to make the figures comparable. After plotting a pattern of random points over the state, readings of the population density and of the rainfall are recorded for each point. These data are then plotted on a scatter diagram where one scale measures population density and the other measures rainfall (Fig. 35d). From the equation that best fits the distribution of dots on the diagram it is possible to discover the density that should be expected for each amount of rainfall. A map showing the departures from the expected densities identifies the problem areas (Fig. 35c). The technique of multiple regression analysis makes possible the measurement of such areal associations (Robinson, 1962).

In this study of Nebraska, two contingent distributions were compared—one contingent on the area of enumeration, the other contingent on the period of observation. But geographers often want to compare the correspondence of two discrete or discontinuous distributions—for example, land use and soil type. The area to be analyzed can be divided into unit areas, similar to those revealed by the fractional code system used in Finch's study of the Montfort area (Fig. 29). The existence of a

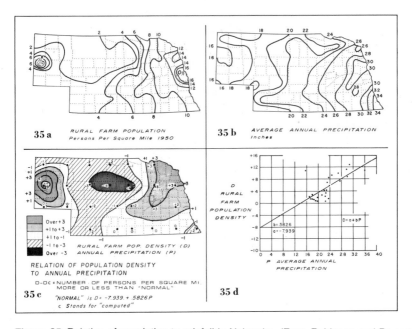

Figure 35. Relation of population to rainfall in Nebraska. (From Robinson and Bryson, 1957.)

concentration of certain kinds of land use on certain kinds of soil can be identified by counting the unit areas. By the use of a random sampling of land use and soil at selected points (the chi-square method), the degree to which a hypothetical correlation agrees with the facts can be determined. This method gives a number that increases with the absolute amount of difference between a hypothetical association and an actual, observed distribution. By this method the areal spread of a region can be determined quantitatively.

William Bunge, in his book *Theoretical Geography,* provides a demonstration of the benefits of using sampling procedures by making a partial restudy of Finch's Montfort area (pp. 333-335). A table of random numbers was translated into coordinates to provide a map of random points. At each point the land use was recorded. By noting the occurrence or nonoccurrence of one form of land use (grassland), Bunge arrived at the estimate that this kind of cover was to be found on 21 percent of the area. Finch had calculated that grassland covered 24.1 percent of the area. Which figure is more nearly correct? The use of sampling from random points permits the geographer to calculate his probable error. Furthermore, by using random points plotted on vertical air photographs and then checking the necessary information at each point in the field, it is not necessary to make a complete map of the whole area. Finch spent 120 days in the field, and the preparation of his land use table by planimetering the field map took a year. Bunge estimated that by using the random sampling method the required information could be acquired in three days, and with much greater accuracy.

> With random sampling techniques, it is entirely feasible to obtain good estimates of continental or global land use percentages on a modest budget (Bunge, 1966:104–107).

Geographers have been accustomed to pointing out that the nature of their field of study does not permit the use of controlled experiments. This is no longer true. Statistical procedures offer the equivalent of a geographical laboratory. For example, Waldo Tobler shows how the distorting effects of transportation, terrain, and other conditions can be eliminated on maps to permit effective testing of the concept of central places (Tobler, 1959). When the interaction of several features associated in a spatial system is to be analyzed, a covariance analysis permits the student to keep one element constant while other elements vary in relation to it. It is possible to measure the variations in relation to constant factors so that geographers can now measure functional relations where the distributing effects of "other things" that are not equal are eliminated.[6]

The Electronic Computer

Mathematical procedures offer a way of thinking about geographical problems. But to carry out all the computations that are required as well as to store the vast amount of new information where it can be recalled when needed would not have been possible without the electronic computer, which first became available in the 1950s. Computer programming has now became a new tool of research, indispensable for anyone working with quantitative techniques (Kao, 1963). On the other hand, it is important to underline what the computer cannot do. Unlike the human brain, the computer can never detect a mistake and rectify it. It can only make use of the categories of information fed into it and carry out computations as commanded by its programmer. There is still the basic requirement that scholars must think logically and must pose their questions within the framework of accepted geographic concepts. If silly questions are asked, silly answers will come out.

In 1967 Duane F. Marble, one of those who studied with Garrison in the Washington seminar, and in 1970 a professor at Northwestern University, published a compendium of twenty-eight computer programs for the guidance of geographers (Marble, 1967). There are eight chief kinds of programs, which were described by Forest R. Pitts as follows:

General Mathematics
  Regular Markov chains
  Simple absorbing Markov chain

---

[6]For recent books describing these statistical techniques, see Duncan, Cuzzort, and Duncan, 1961; Haggett, 1966; Cole and King, 1968; King, 1969; and Taaffe, 1970.

General Statistics

Chi-square interactions

Multivariate groupings of observations using a defined distance criterion

Central movements, skewness, and kurtosis of distributions

Individual terms of the negative binomial probability law

Individual terms of the Poisson probability law, and dispersed probability law

Maximum likelihood parameters of a negative binomial distribution, given the observed
frequency distribution (chi-square values are also computed for these data)

Spatial Statistics

Number of points within rectangular cells of a rectangular region

Tests randomness of a two-category nominal scaling of contiguity measures

Extends previous program to the K-color case

Various computations on any rectangular array of data

Fits a first, second, and third-degree polynomial trend surface to a set of three-
dimensional values

Describes a point set mapped onto a torus

Contiguity measures for an evaluation of randomness in

the arrangement of values of a property

General Geography

Great circle distances and azimuths between any two points on the surface of the earth or
the surface of the moon

Counting on a ring and sector basis of a series of points given in Cartesian coordinates

Minimum path networks in a transportation system of large size

Node accessibility indices of Shimbel and Katz for a moderate-sized transportation network

Spatial Simulation

Synthetic data of Hägerstrand models 1 and 2

Summarizes means, variances, and standard deviations for cell locations for the first
program

Graphic plotting of summary measures

Cartographic Routines

Constructs a contour map from data points transferred to a grid [see also computer
programs for some thirty map projections written by Waldo R. Tobler and Clyde P.
Patton, not included in Marble's book].

Utility Routines

Reforming data decks

Subroutines

Matrix inversion

Generator of normal random variables

Generator of uniform random numbers

Determining whether a specified point lies within or on one of the boundaries of a defined
polygon

[*Geographical Review*, 1968, 58:509–510]

These various new procedures for asking geographical questions concerning
occupied space on the face of the earth, for formulating useful empirical generaliza-

tions concerning the things observed, and for formulating and testing hypotheses and hypothetical models are all part of the latest "new geography" of the contemporary period. As in all previous periods when new methods of scholarly procedure are proposed, those who accept the changes (and especially those who make use of the new jargon) think of themselves as a new breed of more progressive and more sophisticated scholars who have achieved, or are about to achieve, the status of "scientists." In the past there have been numerous examples of the failure of one generation to study the problems and the answers of earlier generations. The result, on occasion, has been the persistence of avoidable error. The question we must now face is this: How much of what geographers are concerned about in the 1970s represents innovation in thinking? How much is a continued improvement in the finding of answers to traditional questions? This we will consider in the final chapter.

CHAPTER 18

Innovation and Tradition

The advantages of mathematical models—unambiguity, possibility of strict deduction, verifiability by observed data—are well known. This does not mean that models formulated in ordinary language are to be despised or refused. A verbal model is better than no model at all, or a model which, because it can be formulated mathematically, is forcibly imposed upon and falsifies reality.

The appearance of a "new geography" has been loudly and persistently proclaimed by almost every generation of geographers since ancient times. Usually the adjective only indicates that there is some new information at hand; but occasionally there are genuine innovations of technique or method or in the concepts that provide an enlarged comprehension of some kind of order in earth space. Sometimes a "new geography" means that a whole new world has been brought to light. As long as there is hope for progress in scholarship, the number of all possible worlds awaiting discovery is infinite. But it is always wise to distinguish what is new from what is not new and by examining the record to avoid, as much as possible, the persistence of old error.

In the 1970s, when the existence of another "new geography" was loudly proclaimed, there is more need than ever before to raise the question: "What is

The quotation above is from Ludwig von Bertalanffy, *General System Theory, Foundations, Development, Applications* (New York: George Braziller, 1968), p. 24.

New?'' (Dickenson and Clarke, 1972). In the preceding chapter we have followed some of the unprecedented changes in the technology of observation and analysis that not only provide geographers with more information than has ever been available before but also provide a means of storage and recall and a way to carry out complex analyses. What kinds of questions do geographers ask of these new data? And, have the basic purposes of geographic study been given a new direction? The geography of the 1970s and early 1980s turns out to be a mixture of innovation and tradition that makes the study of the history of geographical ideas highly important (Chorley and Haggett, 1965; Chorley, 1973; Taylor 1976).

A review of what geographers have done in the past reveals the persistence of certain kinds of avoidable error (James, 1967). One major source of repeated error seems to be the common failure of too many geographers to read what other geographers, past and present, have written. Strangely, this appears to be a characteristic of all fields of scholarship, and is not restricted to the contemporary period. One reason why Strabo's books on geography were found almost intact was that his contemporaries did not read what he had so laboriously written. Again and again we come across examples of general concepts formulated by scholars of one generation that have outlived their usefulness in illuminating the arrangement of things on the earth. Notable is the persistence of Aristotle's ideas concerning the difficulties of life in the so-called torrid zone even after the attack on these ideas began in the thirteenth century. Another example is the continued use of Maury's wind zones. Peter Haggett suggests that progress is marked ''by the sound of plummeting hypotheses'' (Haggett, 1966:277). The difficulty is that some hypotheses do not plummet soon enough but remain as obstacles to confuse later generations.

Geographers, like scholars in other fields of learning, have been caught in certain semantic traps. Because man, alone among the animals, possesses language in which abstract ideas can be represented by symbols, he very easily confuses the symbol with the ''reality'' for which it stands (Bertalanffy, 1965). Yet the nature of word symbols, which are not always precisely defined and which carry heavy burdens of connotation, makes possible the development of professional controversy that is largely based on differing interpretations of word meanings. For example, what is the ''order'' that we seek in our universe? ''Order'' and ''chaos'' exist as concepts in the human mind, conjured up by the use of these word symbols. Perhaps what we call chaos is really a kind of natural order not yet comprehended. What do we mean by cause and effect? Actually the meaning of causality has never been resolved by the philosophers. Only recently has the search for functional relationships replaced simplistic cause and effect, and strict determinism has given way to the search for probabilities (Haggett, 1966:23-27).

The acceptance of many ''dichotomies'' is another example of a semantic trap. A dichotomy exists when two opposites are defined as mutually contradictory, such as good and evil, or reason and faith. But a dichotomy does not exist when one of the alleged opposites forms a subordinate part of the other or when one is derived from the other. Furthermore, a dichotomy may exist for some people and not for others, depending on certain basic attitudes of the culture.

A dichotomy that is embedded in our culture is the dualism of man and nature, which has long been accepted in geographic thought. From the Judeo-Christian teaching comes the directive that man should establish his conquest over nature. The teleologists had no doubts that the all-wise creator had built the natural world for the special benefit of man. This separation of the natural world and the human world was the cornerstone of the conceptual structure developed by the Social Darwinists. Yet for a majority of the world's inhabitants such a dichotomy cannot exist. Among the Buddhists and Hindus, for example, man is a part of nature, not separate from it. The individual hopes to be absorbed into the universe after overcoming his ignorance, his lusts, and his angry reactions to frustration.

Among the dichotomies that exist because of the meaning given to word symbols and that have been harmful to the clarity of geographical thought, we may underline five: (1) that geography must be either idiographic or nomothetic, but not both; (2) that physical and human geography are separate branches of study with different conceptual structures; (3) that geography must be either topical or regional; (4) that geography must be either deductive or inductive; and (5) that geography as a field of study must be classified either as a science or as an art. The fact that geographic writings may be placed in all these categories destroys the validity of the dichotomies.

A REVIEW OF THE TRADITIONS

As we have seen, methodological discussions involving all these dichotomies and others started in the 1870s at the time when faculties were being established in universities to offer advanced training in geography. Since the first new appointments were people who had never been trained in a graduate school of geography, each had to define the field to his own satisfaction. In the United States most of the presidents of the Association of American Geographers presented their ideas about the scope and method of geography in their presidential addresses. Before turning to some of the current conceptions, we will review earlier statements, taking guidance from William Pattison, who has suggested that American definitions in general have reflected the history of research work and that this work has exhibited an essential unity attributable to a small number of distinct but affiliated traditions (Pattison, 1964).

Pattison proposes four traditions: (1) an earth science tradition, (2) a man-land tradition, (3) an area studies tradition, and (4) a spatial tradition. He maintains that although all four have found expression throughout the past century of American geography—as continuing parts of a general legacy of Western thought—each has tended to enjoy a time of preference. The earth science tradition, for example, was particularly prominent in the thinking of researchers shortly before the founding of the profession:

As Mackinder of Oxford has recently expressed it, geography is the study of the present in the light of the past. When thus conceived it forms a fitting complement to geology, which, as defined by the same author, is the study of the past in the light of the present [Davis, 1888].

The discernment of the meaning of surface features gives soul and sense to that too often soulless and senseless study, geography, for there is significance in every cape and every estuary, in every cataract and every delta. . . . In this phase the new geology is the new geography [Chamberlin, 1892].

The man-land tradition next rose to dominance. Interpretations of it changed during the time of ascendancy as these two definitions show:

Any statement is of geographical quality if it contains . . . some relation between an element of inorganic control and one of organic response [Davis, 1906].

Geographers . . . define their subject as dealing solely with the mutual relations between man and his natural environment. . . . Thus defined geography is the science of human ecology [Barrows, 1923].

The area studies tradition, especially favored in geographic work of the mid-twentieth-century years, is strongly represented in these assertions:

Our fundamental definition of geography [is] the study of the areal differentiation of the world [Hartshorne, 1939: 242].

Geography is . . . the field of study that deals with the associations of phenomena that give character to particular places, and with likenesses and differences among places [James and Jones, 1954: 6].

Finally, the spatial tradition, vigorously pursued in ensuing years through studies of geometry and movement, is selected for emphasis in these statements:

The main contribution of the geographer is his concern with space and spatial interaction [Ullman, 1953: 56].

The contemporary stress is on geography as the study of spatial organization, expressed as patterns and processes [Taaffe, 1970: 5-6].

Through the same sequence of decades, many of the important pronouncements on geography drew attention more to the complementarity of the traditions than to their distinctiveness, as these examples demonstrate:

Geography treats the man-environment system primarily from the point of view of space in time. It seeks to explain how subsystems of the physical environment are organized on the earth's surface, and how man distributes himself over the earth in his space relation to physical features and to other men [Ackerman, 1965: 1].

> Modern geography has continued to give attention to the same questions as did primitive folk, the significance of position, the togetherness of things, the areal distribution of entities and aggregations, the utility of the environment [Sauer, 1966: 60].

Acceptance of any of these positions, as David Harvey points out, depends in part on one's own philosophy of geography (Harvey, 1969: 3–8). Most of them are both innovative and traditional; each expresses a piece of the whole concept of geography as a field of learning [cartography and behavioral geography have been suggested as fifth and sixth traditions (Blaut, 1979)]. As fashions in words change, graduate students are presented with new programs of scholarly procedure and invited to abandon old ones (Berry, 1973). In the meantime frustration with the effort to provide widely accepted logical definitions leads to an increasing attention to operational definitions, which gives added strength to the old quip that, "Geography is what geographers do." It would seem to be a dissipation of professional energy to find so many printed pages devoted to such questions as whether geography studies areal differentiation or spatial interaction, or whether geography looks for similarities or differences among places, or whether geography is merely descriptive or seeks to explain things. Geography, like all other branches of learning, seeks answers to its questions by all these routes, but none of them exclusively. An individual geographer may claim to devote his life to describing the unique character of places, yet it is logically impossible to identify a characteristic as unique without some measure of the general. Out of this maze of word traps, however, it is possible to distinguish innovations and to record progress toward the goals set up in all these statements. It is important to seek what is new and place it in balance with what is traditional.

SYSTEMS

The newest "new geography" has to do with spatial systems. Edward A. Ackerman was one of the first geographers to point to the rise of systems research throughout the scientific world after World War II. All science, said Ackerman, is concerned with four "overriding" problems:

1. The particular structure of energy and matter (physics);
2. The structure and content of the cosmos (astronomy, astrophysics, geophysics);
3. The origin and physical unity of life forms (biological sciences);
4. The functioning of systems of multivariables, such as life systems and social systems (worked cooperatively by all the sciences).

[Ackerman, 1963:434.]

Geographers, Ackerman continued, must find the concept of a system of many different but interdependent variables ready-made for the meaningful study of "all humanity and its natural environment." Is this what Ritter was feeling for when he wrote about "coherent relationships," or Guyot when he wrote of "the great life system," or Hartshorne when he wrote about "integrations," or Platt when he referred to "'process-patterns of dynamic social relations'"? How did it happen that only after midcentury did the concept of the system as a unit of study receive overwhelming acceptance?

General System Theory

Anatol Rapoport defines a system as:

> a whole (a person, a state, a culture, a business firm) which functions as a whole because of the interdependence of its parts [Rapoport in Buckley, 1966:xvii].

The words are new, but the mental image of such structures of interconnected parts goes back at least to the Greek philosophers. But before World War II, organized complexity could only be contemplated qualitatively or described as a "balance of nature."

The scholar who is credited with presenting the first outlines of a general theory of systems is Ludwig von Bertalanffy (Bertalanffy, 1951a, 1951b, 1956, 1962, 1968; see also Boulding, 1956). His account of how he developed these ideas is instructive. When he started his professional career as a biologist in the 1920s he found his colleagues seeking more knowledge about the nature of organisms by dissecting them into smaller and smaller parts. It struck him that until the organism was examined as a structure of interdependent parts no real understanding of the "laws" governing organic life could be gained. From thinking about biological organisms, he broadened his views by recognizing the existence of other kinds of systems to which the concept of systems behavior could also be applied. When he presented these ideas at a seminar in philosophy at the University of Chicago in 1937, the scholarly world was not yet ready for the broad approach. In the 1930s the tendency was to make more and more minute analyses and to be skeptical of general theory. Almost the only field devoted to the development of general theory was physics. Most scientists were seeking simple, one-way cause and effect sequences.

This trend toward the isolation of minute problems for investigation was challenged during World War II and has been definitely reversed in the contemporary period. During the war, scholars were called on to work on strategic problems and to formulate policy recommendations involving complex issues. They found that they could not find answers to the questions with which they were confronted as long as they remained within the confines of single disciplines. After the war the value of the interdisciplinary approach to a variety of nonmilitary problems was at last fully recognized. The scholarly world was ready for Bertalanffy's ideas.[1]

---

[1]In 1948 Norbert Wiener published his book *Cybernetics* (Cambridge; M.I.T. Press). Cybernetics (literally, steersmanship) is a science of communications that seeks to move information and directives through administrative channels more efficiently. From this new approach have come the application of automation to industry and the computerized analysis of complex administrative problems in government. Since the first formulation by Wiener, the objectives of this field of study have been broadened. Cybernetics is a special application of general system theory.

General system theory seeks to identify the characteristics that are common to many different kinds of systems. There are three fundamental aspects of all systems: structure, functioning, and evolution (being, acting, and becoming). Where systems are isolated in laboratories or symbolically isolated by statistical procedures, they become closed and irreversible; but on the face of the earth, systems are open and reversible as they receive inputs of energy or information and send forth outputs. As more and more scholars began to study different kinds of systems, it was discovered that all systems, however defined, behave in certain predictable ways. For example, it was noted that the growth curve of organisms (the S-curve) is mathematically very similar to growth curves for the spread of innovations, for economic development, or for populations.[2] General system theory seeks for the abstract properties that can be applied to all systems. Such *isomorphisms* form the basic structure of general system theory and can be used to predict the working of drainage systems (Chorley, 1962), ecosystems (DeLaubenfels, 1970:112–120), political systems (Cohen and Rosenthal, 1971), economic systems, and many others.

Consider the isomorphism between the behavior of heat in a thermodynamic system and the movement of information in an economic system. The second law of thermodynamics states that in any heat system that undergoes irreversible change (without inputs or outputs of energy) there must be a loss of the energy available to do work. The mathematical factor that measures the unavailable energy in a thermodynamic system is called *entropy*. In a closed thermodynamic system entropy increases. In an economic system in which information concerning markets or new technology or other matters is being communicated, there is a loss of information because communication is not perfect. The increase of entropy in a thermodynamic system and the loss of information in an economic system may be described by the same mathematical formulas.

Spatial Systems

Geographers are especially concerned with any systems that involve, as functionally important variables, such spatial elements as location, distance, direction, extent, density, succession, or derivatives of these. Any system of which one or more functionally important variables are spatial is a spatial system (Wilbanks and Symanski, 1968). Such systems are not the same as regions, although a spatial system may provide the criteria by which a region is defined and identified.

Systems, like regions, require redefinition when the student moves from one resolution level to another. Harvey points out that the elements of a system that can be defined at one resolution level may become subsystems themselves when the resolution level is raised. Whole new systems come into focus. Or, by lowering the resolution level, systems may be defined at the global scale. By combining the geographer's concern with specifically defined segments of earth space with the

---

[2]Note the use of this concept in the studies of population growth and decline by Stanley D. Dodge in the 1930s (p. 327).

system analyst's concern with the operation of functionally interconnected sets of elements, new insights into the nature of order on the face of the earth may be expected. There should also be an important feedback into general system theory as a result of focusing attention on the spatial aspects of systems. General system theory offers an exciting new prospect for the study of those complex integrations of things and events that have always stimulated the curiosity of geographically minded students. Now, for the first time in the history of geographical inquiry, the available technology seems to be matched to the potential power of method. It is important at this time to take stock of the spatial theory available in the contemporary period.

THE CUTTING EDGES OF GEOGRAPHIC RESEARCH

In 1963 the National Academy of Sciences/National Research Council appointed an ad hoc committee within the Division of Earth Sciences to consider the potential contribution of geographic research to the general progress of science.[3] The committee identified four "problem areas and clusters of research interest"— physical geography, cultural geography, political geography, and a fourth field in which traditional economic, transportation, and urban geography are grouped under "location theory."

Physical geographers study the physiographic-biotic system as man's habitat, or environment (Ackerman, 1965:14–22). The environment, however, can be examined from several different points of view: in terms of what the inhabitants perceive it to be; in terms of the identification of desirable changes to be brought about by human action; or in terms of a static setting in which man momentarily finds himself. The judicious selection of significant parameters of the natural environment is the task of the physical geographer. Particular stress is laid on the system relations among such elements as surface features, air, water, soil, and biota (Kalesnik, 1964).[4]

[3]The National Academy of Sciences was established by the federal government in 1863 to advise the government (on request) on matters of science and technology. The National Research Council was set up in 1916 to promote especially critical scientific research. After World War II the Academy and the Council were combined. The ad hoc Committee on Geography, appointed in 1963, was made up of E. A. Ackerman (Carnegie Institution of Washington), chairman; Brian J. L. Berry (University of Chicago); Reid A. Bryson (University of Wisconsin); Saul B. Cohen (then at Boston University, later at Clark University); Edward J. Taaffe (Ohio State University); William L. Thomas, Jr. (California State University at Hayward); and M. Gordon Wolman (Johns Hopkins University).

[4]*The Science of Geography* (Washington, D.C.: Ackerman, 1965) includes eleven pages of references to current writings on these four problem areas. Since World War II, especially since 1960, the number of geographers and the volume of notable contributions to geography have both increased until it is no longer possible in a book of this size to refer even to a representative sample. The student should consult the *Bibliographie géographique (Paris: Armand Colin,), Current Geographical Publications* (New York American Geographical Society), and *A Geographical Bibliography for American College Libraries* (Association of American Geographers, Commission on College Geography, Publication No. 9).

Cultural geographers seek an understanding of the interactions between human societies and those features of the human habitat that have been produced or modified by human action (Ackerman, 1965:23–31). Attention is focused on the differences from place to place in the ways of life of human communities. Two different methods of study are commonly in use: developmental, focusing attention on the origin and diffusion of cultures and on cultural growth and retrogression; and functional, focusing on the short-term processes of cultural interaction, spatial organization, and flow or movement. Many of these studies are in the literary tradition and use the method of historical geography (Meinig, 1962, 1968, 1969).

Political geography is the study of the interaction between political processes and geographical areas (Ackerman, 1965:31–44). A major theme is the effect of the spatial arrangement of elements relevant to the operation of political processes. The territorial phenomena of political systems are studied over a wide range of resolution levels—from supranational political organizations to the nation state, to the political subdivisions of the state, to metropolitan urban communities, and to local or regional special-purpose administrations (Kasperson and Minghi, 1969; McColl, 1969; Cohen and Rosenthal, 1971; Soja, 1971).

The *Science of Geography* describes location theory as follows:

> Recent development in all three traditional subjects [economic, urban, and transportation geography] has involved extensive application of mathematical methods to facilitate refinement of theory, and a higher level of generalization than existed before has emerged. As a result, it would appear that the three traditional fields have been joined in a problem area which we entitle . . . location theory studies. It is of special interest in our discussion because these studies include: (a) research on the qualities of space in a theoretical framework; (b) application of formal systems methods to space relations study; and (c) integration of at least three of the spatial subsystems of culture. . . . It is also of interest for the extent to which its methods and concepts have found practical applications in a revitalized "applied geography" directed publicly to problems of urban and regional planning, and privately to marketing analysis for plant and store location [Ackerman, 1965:44].

Still another survey of the role of geography as a behavioral and social science was carried out by the Panel on Geography of the Behavioral and Social Science Survey and published in 1970 (Taaffe, 1970).[5] This report provides illustrative

[5]The Behavioral and Social Science Survey was carried out between 1967 and 1969 under the auspices of the Committee on Science and Public Policy of the National Academy of Sciences and the Problems and Policy Committee on the Social Science Research Council. A general volume, *The Behavioral and Social Sciences: Outlook and Needs* (Englewood Cliffs, N.J.: Prentice-Hall, 1969), discusses the relations among the disciplines, broad questions of utilization of the social sciences by society and makes specific recommendations for public and university policy. The Panel on Geography was made up of Edward J. Taaffe (Ohio State University), chairman; Ian Burton (Toronto); Norton Ginsburg (University of Chicago); Peter R. Gould (Pennsylvania State University); Fred Lukermann (University of Minnesota); and Phillip L. Wagner (Simon Fraser University)

studies in six different geographical fields (spatial distributions and interrelation-ships, circulation, regionalization, central-place systems, diffusion, and environ-mental perception). It also discusses research methods and shows how they are applied to studies of locational analysis, cultural geography, urban studies, and environmental and spatial behavior. It discusses geography and public policy and goes into some detail on the status and trends of the profession in manpower, research, and research training.

Both of these reports recognize that it is in the general range of locational analysis that geographers in the contemporary period have made the largest ad-vances toward the formulation of general concepts. Whether these concepts are to be identified as empirical generalizations, laws, or theories may be left to the judgment of professional geographers. To illustrate the kinds of innovation that have been made, we will discuss four chief examples: (1) the gravity model, (2) central-place concepts, (3) diffusion concepts, and (4) regional identification.

The Gravity Model

As early as 1929 W. J. Reilly, studying retail marketing problems, postulated that the movement of persons between two urban centers would be proportional to the product of their populations and inversely proportional to the square of the distance between them.[6] In 1949 this empirical generalization was refined by the economist, G. K. Zipf, who formulated the principle of least effort in human behavior (Zipf, 1949/1965). But it was the astrophysicist, John Q. Stewart, who is often credited with being the first to point out the isomorphic relationship of these concepts with Newton's law of gravitation (Stewart, 1947). Thereafter this concept became known as the gravity model.[7] William Warntz, working with Stewart, also borrowed analogy models from physics in his studies of population potential (Warntz, 1959b, 1964). He suggested that the mathematics of population potential is the same as that which describes a gravitational field, a magnetic potential field, and an electrostatic potential field.

The question is sometimes raised whether the use of analogy models is accept-able. There have, of course, been examples of the misuse of isomorphisms; on the other hand, there have been many very successful uses of them. William Bunge offers the following observation:

---

[6]W. J. Reilly, *Methods for the Study of Retail Relationships* (Austin: University of Texas Press, 1929).

[7]It is always hazardous to identify any one scholar as being the first to formulate a concept. It seems that very much the same notion was expressed by Bishop George Berkeley in 1713; and both the notion of the gravity model and the population potential were specifically stated by Henry C. Carey (1793–1879) in 1868. Carey wrote: "Gravitation is here, as everywhere else in the material world, in the direct ratio of the mass, as in the inverse one of the distance" (McKinney, 1968:103).

It is an observed fact that once theory is produced it often can be applied to a variety of subjects. In this sense, there is unity of knowledge. To give this assertion substance some examples appropriate to geography are offered.

Consider Enke's paper [Stephen Enke in *Econometrica,* 1951] "Equilibrium among Spatially Separated Markets: Solution by Electric Analogue." Can electricity be expected to behave like a spatial economic system, as he insists? Yes, because it has been found that the underlying mathematics can be translated into certain carefully selected aspects of both subjects. A second illustration of the borrowing of theories is available from Beckmann's "A Continuous Model of Transportation" [in *Econometrica,* 1952]. It is suggested by hydrodynamics. Can water be expected to behave like a spatial economic system? Again, it is the mathematics that can be made to fit features of both sets of phenomena. If social scientists are somewhat defensive because they have been borrowing heavily from mathematics and from theory first used in other fields, they can draw some comfort from the knowledge that there is reciprocity. Programming, first applied in social science, is now being used in designing electric networks [Bunge, 1966:4].

The gravity model, however, has to be refined for maximum applicability to studies of location and spatial interchange. Models describing the volume of exchange between two populations are more expressive when the populations are weighted by some such factor as income per capita, which measures the degree of economic activity. Distance between places is not a matter of simple linear distance but of rather more sophisticated measures of distance in terms of transport routes and facilities; frequency of movement by land, sea, or air; and the costs of transport (Haggett, 1966:35-40).

Central-Place Studies

Efforts to formulate some testable generalizations regarding the spacing and functions of central places have been given a large amount of attention by geographers in the contemporary period (Berry and Garrison, 1958a, 1958b; Berry and Pred, 1961; Harvey, 1969:118-119, 138; Taaffe, 1970:24-27; Johnson, 1971; Preston, 1971). The original formulation of what is sometimes called central-place theory was done in 1933 by the German geographer Walter Christaller (1933/1966) in a study of the spatial arrangement of tertiary economic functions in southern Germany. Christaller thought of his work as complementary to von Thünen's model of agricultural land use (Von Thünen, 1826/1966) and Alfred Weber's model of industrial locations (Weber, 1909/1966). He noted that "the crystallization of mass about a nucleus is part of the elementary order of things" and human settlement obeys this principle as well as physical elements. The foci or nodes around which settlement tends to cluster are what Christaller called central places, each surrounded by a complementary area with which the central place is functionally related. Based on evidence from Germany, Christaller described a "nested hierarchy" of central places ranging through seven orders. Places of the lower orders provide goods and services that are needed frequently with a minimum of travel. Places of higher orders not only provide these same goods and services but also other, more specialized goods and services that are needed less frequently and for which people are willing to travel greater distances. The higher order places have worldwide connections.

Christaller also attempted to explain the spacing and pattern of arrangement of central places. He postulated a uniform plain evenly settled by an agricultural population as the base of the hierarchy. If access to a market is the chief factor in the development of a settlement pattern, then the minimum average travel distance from the complementary area is gained if the area has the shape of a hexagon. But Christaller recognized that optimum conditions would also be found where as many central places as possible are located along a main traffic route between higher order places. Furthermore, where government administration or other purposes are to be served, a further modification of the simple market-oriented type of pattern would appear. The actual pattern to be observed in any area depends on the interplay of these three principles: marketing, traffic, and administration (Berry and Pred, 1961:15–18).

German geographers in the 1930s paid little attention to Christaller's work. It was tested by Otto Schlier in 1937 on the basis of census data (Schlier, 1937); and it was applied to the pattern of settlement in Estonia by Edgar Kant. In 1941 central-place theory was introduced to American geographers in a paper by Edward L. Ullman (1941). In 1940 a German economist named August Lösch had investigated Christaller's ideas in a book on the spatial pattern of the economy; in 1944 Lösch published a revised and much enlarged edition, which was translated into English in 1954 (Lösch, 1940, 1944/1954). Lösch found evidence to support Christaller's concepts, both with regard to the nested hierarchy of central places and also the hexagonal patterns of complementary areas. In 1953 John E. Brush published a study of the hierarchy concept as applied to the southwestern part of Wisconsin (Brush, 1953). He concluded that Christaller's work provided a norm against which to measure observed differences in various parts of the world. He saw no possibility of finding precisely the same hierarchies in different culture regions. More recently, Harvey concludes, on the basis of numerous attempts to apply the concept in various places, that "the tests have shown that actual spatial patterns do not conform to theoretical expectation" (Harvey, 1969:138). The economic theory regarding the range of a commodity—that is, the distance a customer will travel in order to make a purchase—is "inherently untestable." Nevertheless, the study of settlement patterns and hierarchies was greatly stimulated by Christaller's hypotheses, which led to many studies of specific areas and to the formulation of a number of alternative hypotheses.[8] George K. Zipf proposes a rank-order hypothesis that predicts the

[8]Brian Berry and Allan Pred published a review of the theory and a bibliography of published materials (Berry and Pred, 1961). The flow of studies has continued since 1961. Among the many, see Curry, 1964; Woldenberg, 1968; Parr and Denike, 1970.

population of urban places on the basis of the rank of any one city among all the cities of a country (Zipf, 1949/1965:374–386).

Studies of Diffusion

One of the traditional concerns of geographers (at least since Ratzel) has been the interpretation of the spread or retrogression of the things that occupy space on the face of the earth. Ratzel, in the second volume of his *Anthropogeographie*, described the patterns of population and culture that had resulted from the process of diffusion from centers of origin. Nor should we forget the contribution that Ellen C. Semple made to the explanation of culture patterns. She discussed the various ways culture change could be brought about—by conquest, infiltration, influence, and many other ways. Semple proceeded to offer two important empirical generalizations in verbal form:

> In general, however, any piecemeal or marginal location of a people justifies the question as to whether it results from encroachment, dismemberment, and consequently national or racial decline. This inference as a rule strikes the truth. The abundance of such ethnic islands and reefs—some scarcely distinguishable above the flood of the surrounding population—is due to the fact that when the area of a distribution of any life form, whether racial or merely animal, is for any cause reduced, it does not merely contract but breaks up into detached fragments. . . . Ethnic or political islands of decline can be distinguished from islands of expansion by various marks. When survivals of an inferior people, they are generally characterized by inaccessible or unfavorable geographic location. . . . The scattered islands of an intrusive people, bent upon conquest or colonization, are distinguished by a choice of sites favorable to growth and consolidation, and by the rapid extension of their boundaries until that consolidation is achieved; while the people themselves give signs of the rapid differentiation incident to adaptation to a new environment [1911:164–165].

Geographers have traditionally sought explanations of diffusion through the deciphering of historical processes. By plotting patterns resulting from the diffusion process on maps, it has been possible to identify centers of origin and directions of spread. Biogeographers and cultural anthropologists have used the cartographic method to throw light on the subjects they were investigating. Carl Sauer's thought-provoking hypotheses regarding agricultural origins and dispersals use the cartographic method to illuminate prehistoric problems (Sauer, 1952). Fred B. Kniffen uses maps of house types to reveal directions of migration, just as Kurath did with word usage and pronunciation in the late 1940s (Kurath, 1949; Kniffen, 1965). The list of substantial contributions to this kind of geographical study is long and growing rapidly.[9]

The use of mathematical models to describe and predict the diffusion of inno-

[9]The rural sociologist, Everett M. Rogers (*Diffusion of Innovations,* New York: The Free Press,

vations was started by the Swedish geographer, Torsten Hägerstrand. He developed two kinds of models: first an inductive model (empirical generalization) to describe the characteristics of innovation waves; and second a stochastic model in which he made use of a Monte Carlo simulation to predict the probability of innovation spread. Individuals, he postulated, are more likely to be informed about an innovation the closer they are to the source of the innovation. Distance, of course, is not mere linear distance, but also a measure of contiguity and contact. In 1965 R. S. Yuill made use of simulation models to show the probable patterns of spread around several kinds of barriers (Yuill, 1966; see also Bunge, 1966:112-132; Haggett, 1966:59-60; Morrill, 1970).[10] Other applications of mathematical procedures to the study of innovation diffusions include Edward Soja's study of the economic and political modernization of Kenya (Soja, 1968) and Lawrence Brown's general discussion of diffusion processes (Brown, 1968; Brown and Moore, 1969).

These studies of the diffusion process have been done at different resolution levels. Hägerstrand was working at a high resolution level that permitted him to focus on specific individuals and made possible certain general concepts regarding individual human behavior. Yuill and Soja were working at a somewhat lower resolution level that focused on groups rather than individuals. It is also possible to study diffusion processes at a very low resolution level—at a global scale. Interestingly, the concepts and models found useful at one resolution level are not necessarily useful at other levels. Diffusion theory can be enriched by the examination of a variety of levels.

Studies of Regions

Mathematical geographers have made important and imaginative applications of spatial theory to the identification and definition of regions. Many of the basic concepts concerning regions have been presented verbally (Whittlesey in James and Jones, 1954:19-68). But these verbalizations can be given greater precision by translating them into mathematical terms. Peter Haggett devotes a chapter to the application of mathematical methods to a variety of regional problems, such as the identification of regional cores, regional limits, the definition of nodal regions, and the question of whether a particular piece of territory should be assigned to one region or to a neighboring one. Haggett discusses the identification of regions at different resolution levels and the use of sampling procedures to identify regional characteristics (Haggett, 1966:241-276).[11]

1962), makes no mention of contributions by geographers to diffusion studies. He mentions Walter H. Kollmorgen as a rural sociologist (p. 32), and refers to Hägerstrand (pp. 154, 298) for his use of game theory.

[10]Robert E. Nunley has developed an electronic device that can simulate a diffusion to show the flow patterns around barriers (Nunley, 1971).

[11]Zobler, 1958; Berry, 1961; Nystuen and Dacey, 1961; Siebert, 1967; Lankford, 1969; King, 1969;194-215; Johnston, 1970; Taaffe, 1970:18-24.

In the contemporary period there is an increasing awareness of the need for an interdisciplinary approach to regional problems (Thompson, 1966; Parkes, 1980). The volume of information and the complexity of the questions asked have made it almost impossible for any one scholar to master all the relevant knowledge about a region, even a small one. The alternative is to bring together groups of scholars with diverse backgrounds and skills to focus on specific sets of regional problems.

An example of this kind of interdisciplinary approach is the appearance of a new professional field—regional science. In 1954, under the leadership of the American economist Walter Isard, a group of economists, geographers, and other social scientists and engineers formed the Regional Science Association, which is described as

> ... an international association devoted to the free exchange of ideas and viewpoints with the objective of fostering the development of theory and method in regional analysis and related spatial and areal studies [Isard, 1956, 1960].[12]

The regional science movement has gained support all around the world (Olsson, 1965). It proves to be especially attractive in countries in which geographical studies are applied to practical problems. The countries of the underdeveloped world find this approach of great value in the guidance of development programs. At the International Geographical Congress in New Delhi in 1968 the Commission on Quantitative Methods was constituted to meet the following needs:

> a. the continuing need to review the state of the art in terms readily comprehensible to a world-wide and often nontechnical audience.
> b. the need for a comprehensive "users manual" of quantitative techniques, to provide the basis for sound training of the next generation of geographers and the retooling of present professionals so inclined.
> c. the need to continue stimulating papers exploring the unsolved technical questions of spatial analysis. [*Economic Geography*, 46 (1970):212.]

Studies of Practical Problems

The concepts and methods of geography are clearly "relevant" to the practical problems faced by government administrators or business executives. In both the United States and the Soviet Union geographers have long been accumstomed to

---

[12]At the University of Pennslyvania the Department of Geography was renamed the Department of Regional Science with Walter Isard as chairman. Its new orientation is described in the announcement of the graduate program for 1971–72: "The Regional Science Department . . . has two basic orientations: one is toward the underlying theory of location and spatial interaction of human activities in their economic, social, and political contexts; the other is toward the development of techniques of analysis for regional systems to provide guidelines for public policy and private decisions. Both the research and study in this field rely heavily on mathematical models and quantitative methods. Much of the underlying theory in regional science is based on existing and newly developing social science theory."

working on research that was expected to contribute to the solution of such problems. In the 1920s and 1930s it was sometimes asserted in the United States that geographers were very useful in supplying information or giving advice but that, when the time of decision arrived, the geographers were seldom consulted. Pierre George in France has recently suggested that geographers are beyond their range of professional competence when called on to decide between alternative policies, but he has been successful in arranging for the establishment of a chair of applied geography at the Sorbonne to remedy this deficiency (Meynier, 1969:186, 188).

All around the world geographers are being employed in various kinds of planning agencies. Norton Ginsburg points out that in Japan interdisciplinary teams of scholars work on urban planning problems and that geographers work effectively on these teams. Pedro Geiger writes that the geographers in the Instituto Brasileiro de Geografia have been collaborating with scholars in other fields in the preparation of national plans for the more effective use of resources. The General Secretariat of the Organization of American States published a monograph demonstrating the utility of investigations of physical resources as a basis for planning economic development.[13] In the United Nations Secretariat there is a cartographic unit where geographers are employed on various research undertakings as members of interdisciplinary teams.

The list of positions held by geographers at all levels of government in the United States as well as in many business firms is too long even to summarize. They are to be found in almost every agency of the federal government and on the staffs of most urban planning commissions.

One example of the contribution of geographers to the economic improvement of areas of poverty in the United States is to be found in the studies of Appalachia.[14] This is a region of high unemployment, especially in the coal fields that were largely mechanized during the 1940s and 1950s. Furthermore, it is a region dissected by many small streams that provide numerous "hollows," or headwater valleys, where small settlements are almost isolated from the outside world. The Appalachian Regional Development Commission decided that one major source of poverty in the region was the lack of urban centers capable of providing the services and the potential labor force to support economic growth. The problem was to select a limited number of urban centers with "growth potential" and by carefully planned investment to build a hierarchy of central places. Bruce Ryan reports the situation in 1970 as follows:

> For many geographers, the unnerving fascination of this approach to regional planning lies in its brazen intention to rejuvenate a central place system that serves over 18 million people. Using real cities, and to all intents "experimentally," it proposes to

[13] *Physical Resource Investigations for Economic Development, A Casebook of OAS Field Experience in Latin America, ed.* Kirk B. Rodgers (Washington, D.C., 1969).

[14] In 1965 a program of regional development was set up to improve opportunities for employment and to provide better facilities for transportation, education, and the maintenance of health.

manipulate what has been a theoretical dream among geographers for forty years [Ryan, 1970:118].

A paper by Donald A. Blome describes a model of the relationship between the hierarchy of central places in Appalachia and the stream pattern. The model predicts the size, number, and location of the central places (Blome, 1970).

Geographers in universities have also contributed studies in applied geography. There is a growing literature on environmental perception as a guide to behavior—for example, the perception of the flood hazard in certain river floodplains (Lowenthal, 1967; Harrison, 1969). Environmental destruction owing to soil erosion or air pollution is the subject of numerous studies (Leighton, 1966). Special mention should also be made of the work of John R. Borchert in urban and regional planning in the state of Minnesota.

In the Soviet Union, where the tendency is to bend all scholarship toward practical purposes, the economic geographers have found the new mathematical and statistical procedures of the utmost importance in economic planning. Robert G. Jensen and Gerald J. Karaska, reviewing the new Soviet interest in regional science, report as follows:

> One of the more noteworthy results of the mathematical thrust in Soviet economic geography has been a vastly increased interest in similar developments in the West. The neutrality and generality of the mathematical approach appears to reduce significantly the restrictive overtones of ideology and creates in its place a bond of mutual interest which transcends national boundaries. The mathematical approach, therefore, appears as an especially valuable means of facilitating interaction among Soviet and American geographers. Indeed, this has already been made evident by Soviet participation in the international meetings of the Regional Science Association and in the quantitative sessions of the International Geographical Union [Jensen and Karaska, 1969:141].

The Population Problem

In 1970 Wilbur Zelinsky, a specialist in population geography at Pennsylvania State University, pointed to the critical importance of examining the results of the continued growth of the world's population and the continued expansion of industrial production. As this "growth syndrome" looms ahead as perhaps the major problem of the twentieth and twenty-first centuries, scholars in a variety of fields have contributed studies of different aspects of the problem (Trewartha, 1969). There have been studies by sociologists, demographers, economists, political scientists, ecologists, agronomists—and by geographers, Zelinsky, Kosiński, and Prothero, 1970).

> The geographic approach is one of several needed to understand and treat those troubles everyone now recognizes as stemming from rapid population growth and uneven de-

velopment in the less advanced regions of the world. Precisely the same is true of the afflictions growing out of the unceasing accumulation of human beings and things in the affluent nations, problems glimpsed only dimly as yet or in terms of isolated facets. Geographic analysis is hardly the single magic nostrum for these ills, but it is hard to imagine any workable therapy that excludes it [Zelinsky, 1970:498].

Zelinsky's paper questions the traditional Western belief in the benevolence of continued growth. The problems that are being faced in the economically underdeveloped countries cannot be solved, he argues, until they have been faced and solved in the more advanced countries—first of all in the United States. Geographers can offer diagnoses describing the nature of the illness with which mankind is afflicted; or they can become prophets, predicting the likely results of various remedial policies; or they can join with others as the architects of the utopia that must be built if utter disaster is to be avoided. He concludes as follows:

> A thorough review of the present status of human geography, and of population geography in particular, would reveal how woefully deficient we are in terms of practitioners, in terms of both quantity and quality, how we are still lacking in relevant techniques, but most of all that we are totally at sea in terms of ideology, theory, and proper institutional arrangements. Even if we were to be showered with unlimited funds tomorrow morning with which to initiate research on key aspects of the geographic diagnosis of the Growth Syndrome, it would almost certainly be impossible to put enough people with the right skills and attitudes into the proper settings. Given the hope, a not totally unrealistic one, that this picture will alter radically before the end of this decade, I have stated a number of difficult, but ultimately operational, research themes, practical daydreams that could inflame the imagination [Zelinsky, 1970:529].

REAPPRAISAL AND PROSPECT

After World War II, and especially since 1960, there was a worldwide emphasis on science and mathematics at the expense of history, language, and literature. The traditional teaching of Latin in the schools was either dropped or greatly reduced (Warntz, 1959a; Meynier, 1969:118). In the United States the National Defense Education Act of 1958 provided federal funds to the schools to improve the teaching of mathematics, foreign languages, and sciences that provide a background for engineering. In 1964 the act was amended to include six additional fields, one of which was geography.

This widespread emphasis on science and mathematics had a major impact on geography. All over the world there was an increasing use of quantitative techniques and a renewed search for useful theoretical models. Peter Gould summarizes the methodological developments of the 1960s under six headings:

    I. New ways of tackling old problems
    II. Developments in area sampling and data gathering

III. Multivariate analysis and the adaptation of inferential and descriptive statistical methods

IV. Developments in formal spatial models at theoretical and applied levels

  1. Geometric and graph theoretic models of transportation networks

  2. Geographical applications of linear programming models

  3. Simulation models and spatial diffusion

  4. Spectral methods and geographic research

  5. Geographical approaches to general systems modelling

  V. Behavioral geography: research on environmental perception and spatial behavior

VI. Methodological and mathematical developments peculiarly geographical

[Gould, 1969:2]

Gould described the decade of the 1960s as one of the greatest periods of intellectual ferment in the whole history of geography. Others went further, proclaiming that a new paradigm for geographical study had replaced the traditional verbal and descriptive studies (Chojnicki, 1970:213).

It should be remembered that the decade of the 1960s was one of social, economic, and political turmoil. The new geography of the late 1960s and the 1970s began to reveal expressions of disillusionment. Something of a retreat from the spatial analysis theme and the quantitative approach began to emerge. Reappraisal of the geography of the 1960s had led to this. Findings in urban geography had led some of the leading theoreticians to assume professional posts in the realm of planning. But planning for whom, for what? Whose values should be adopted? (Harvey, 1973, 1974). Geographers were now brought face to face with the substance of power. And it was realized perhaps more clearly than ever before that geography is value laden (Smith, 1977). Dissent was expressed with the founding of *Antipode: A Radical Journal of Geography* and the founding of Socially and Ecologically Responsible Geographers (SERGE). This part of the radical geography movement sought to revise the geography of capitalism using the traditional though ever-changing instruments of the discipline (Peet, 1977, 1978). Another sector of the radical geography movement, believing that a different foundation of values was a necessary prerequisite for the erection of the alternative geography founded the Union of Socialist Geographers. These geographers were led to the assumptions of Marxist theory (Folke, 1972).

A more conformal reaction to scientism in the wake of sputnik and to the enthusiasm for hard science that resulted was a return to a thorough-going humanism within human geography. Historical geography, studies in perception, inquiry into geosophy [a study of people's nonscientific geographical beliefs (Lowenthal and Bowden, 1976)], a reappraisal of geography as human ecology, and a new-found appreciation of the regional concept have also characterized geography in the United States since the middle 1960s. Geographers began to study the recent history of their discipline to see if the path of intellectual evolution could reveal order in the eclectic and plural complexity of the geography now current.

This excursus into the history of geographical thought may do much to bring order to geographers thought and perspectives (Freeman 1961; Gregory 1978). To facilitate such a study of the recent past a "commission on the history of geographical thought" has been created by the International Geographical Union. From this commission emerged the publication, *Geographers: Biobiliographical Studies*. The Archive and Association History Committee was founded (1971), the seventy-fifth anniversary of the Association was recorded (James and Martin, 1979), a special issue of the *Annals* of the association—"Seventy-Five Years of American Geography"—was published, and more papers concerning the history of geography are now published or being read at professional meetings. Maynard Weston Dow has taken films of geographers being interviewed (and has also filmed geographic occasions of note, e.g., seminars and banquet addresses) and amassed a treasury of the visual and oral lore of the field in a unique medium (Dow, 1974). G. Martin wrote life and thought studies of M. Jefferson (1968), E. Huntington (1973), and I. Bowman (1980) that when integrated with the W. M. Davis volume by Chorley, Beckinsale, and Dunn (1973), help reveal much of the little-written story of American geography in the first half of the twentieth century. Yet, while these thrusts are helpful in revealing something of our intellectual journey, unity of the field remains illusory (Stoddart 1967; Bird 1975, 1977, 1978).

Swings of the academic pendulum can be observed in the interplay between the two basic traditions—mathematical and literary. It would be quite erroneous to equate the use of mathematics with a nomothetic objective and the use of verbal language with an idiographic objective. Actually mathematics in many cases provides a notably more precise descriptive method. Studies in literary form may provide exciting innovative approaches to the formulation of concepts. Verbalized and nonquantitative studies in historical geography have nevertheless led the way in the approaches to environmental perception (Meinig, 1962:207; Brookfield, 1969). In the early 1970s the professional periodicals were carrying contributions in both the mathematical and the literary tradition—neither seemed about to be abandoned. Whether the field of geography could ever accept a single paradigm was as open to doubt in the 1970s as it had been in the 1920s and 1930s. In view of the experiences of workers in other professional fields of science, this was an advantage; for there is the danger, when only one set of symbols is used, of becoming entrapped in built-in ambiguities and self-references (Bronowski, 1966:7). But the use of both mathematical and word symbols could help to guard against such obscurities of thought.

Meanwhile, the field of geography remains too little known to the general public or to workers in other scholarly fields (Brewer, 1978). Although trained geographers are being sought in increasing numbers,[15] the great majority of Ameri-

[15]The Taaffe committee reported that a continuing shortage of trained geographers poses a serious professional problem in the face of increased numbers of jobs in universities, government, and private business (Taaffe, 1970:104–130). In 1971, when there was a decline in the number of new openings in universities, the demand for geographers continued in government and business. The Association of American Geographers listed many opportunities in its monthly list of "Jobs in Geography."

cans still have only the vaguest idea about what geographers do. Frequently the newspapers run special articles on the geographical illiteracy of American young people, and there is a flurry of excitement in the schools to see that pupils memorize more place-names. This kind of teaching was outmoded in Ritter's time. Although geographers are necessarily concerned with place-names (unless their work is purely theoretical), place-names by themselves cannot be called geography. In the public mind, also, geography is another name for "popular description and travel." Marvin W. Mikesell has reported on how geography is viewed by other social scientists:

> . . . it is probably fair to say that most of the geographic works known to scholars in other fields are not regarded by geographers themselves as indicative of their current interests. Among anthropologists and historians, the voice of American geography is undoubtedly Ellsworth Huntington, whose theories of climatic influence have been obsolete for more than thirty years. Among political scientists, geography is most commonly identified with the various schools of "geopolitics" that flourished in the 1940s and are no longer taken seriously by political geographers. It would be difficult to prove that geographic thought is more clearly perceived by economists, although central-place theory and other formulations of location theory have some currency among economists concerned with urban and regional planning. Sociologists are perhaps unique among social scientists in having a more accurate perception of modern trends in geographic research, although this awareness is confined very largely to the work of urban geographers [Mikesell, 1969:240–241].

Looking at how geographers themselves view their own field, we find that diversity of approach has always characterized geographic study. There was plenty of diversity in ancient Greece, where most of the traditional currents of geographic thought originated. Diversity was emphasized in the summary of geographic thought provided by Humboldt and Ritter. And for the past century, when the nature of geography as a field of learning and its relation to other fields of learning have been debated at length, differences in the way geography is verbally described have become even more diverse. This state of affairs has bothered some scholars and from time to time there have been efforts to provide narrow definitions of the field that exclude considerable numbers of active workers, past and present, from membership in the profession. These efforts have not been successful. In 1956 Carl O. Sauer wrote:

> We continue properly to be, as I have said that we have been always, a diverse assemblage of individuals, hardly to be described in terms of dominance of any one kind of aptitude or temperament, mental faculty, or emotional drive, and yet we know that we are drawn together by elective affinity. It is about as difficult to describe a geographer as it is to define geography, and in both cases I am content and hopeful. With all shortcomings as to what we have accomplished, there is satisfaction in knowing that we have not really prescribed limitations of inquiry, method, or thought upon our associates. From time to time there are attempts to the contrary, but we shake them off after a while and go about doing what we most want to do. . . .

It seems appropriate therefore to underscore the unspecialized quality of geography. The individual worker must try to gain whatever he can of special insights and skills in whatever most absorbs his attention. Our overall interests, however, do not prescribe the individual direction. We have a privileged status which we must not abandon. Alone or in groups we try to explore the differentiation and interrelation of the aspects of the earth. We welcome whatever work is competent from whatever source, and claim no proprietary rights. In the history of life the less specialized forms have tended to survive and flourish, whereas the functional self-limiting types have become fossils. Perhaps there is meaning in this analogy for ourselves, that many different kinds of minds and bents do find congenial and rewarding association, and develop individual skills and knowledge. We thrive on cross-fertilization and diversity [Sauer, 1956:292–293].

The decision regarding what approaches to geography will survive and what ones will disappear can only be made by the coming generations of scholars. As Stephen E. Toulmin points out, progress and change in any scientific field do not take place because the great scholars of an older generation change their minds, rather they are the result of the younger generation's breaking with the traditions of their teachers (Toulmin, 1967). If a particular approach—whether mathematical or literary, genetic or conceptual—is to flourish it must be attractive to younger scholars who are in training. If large numbers of new students are attracted to a particular approach to the study of geography, that aspect of the field will be nourished and will progress rapidly; those aspects of geography that do not attract young people must eventually disappear.

What makes some branches of a field like geography attractive? To be attractive it must do one of two things. First it must make a clear contribution, widely recognized, to the overriding problems with which mankind is faced in the 1980s. It must help to find solutions to problems of poverty, hunger, injustice, violence, and warfare. If it fails to have any relevance to these problems, it cannot long attract the attention of young scholars. If the scholars who are devoted to the formulation of abstract concepts turn away from the real world and contemplate only self-images, the continued growth of a much-needed conceptual framework will be disastrously affected. Julian Jaynes puts it this way:

... as science folds back on itself and comes to be scientifically studied, it is being caricatured into a conformity which is nonsense, into a neglect of its variety which is psychotic, into a nagging and insistent attention to its cross-discipline similarities which are of trivial importance [1966:94].

Or if geographers are ever satisfied with writing mere descriptions of particular places without reference to broadly unifying concepts, the field will be doomed. A powerful way to make geographic study attractive to the younger generation is to demonstrate clearly the nature of its contribution to the solution of major problems. To fail to come to grips with the need for making practical application of abstract concepts is to face the danger of permitting geographic work to become trivial.

Another way to attract young people is to encourage their disciplined curiosity and to show that geographic study can satisfy curiosity by providing verifiable answers to geographic questions. As Sauer points out, curiosity must not be placed within prescribed limits, nor must the methods of satisfying it be restricted to any one procedure. Geography thrives by giving young minds the freedom to identify enigmatic questions and to seek answers. To ensure that curiosity is disciplined, the paradigms that have proved useful are taught to students, but always with the encouragement to challenge and to innovate.

We have a long and dignified heritage of geographic study—of effort to identify order in occupied space on the face of the earth in terms of the symbols we adopt to guide our thinking. We need to move forward without repeating the unnecessary errors of the past, yet always with the courage to formulate new hypotheses and to see hypotheses we formulate challenged and perhaps destroyed. There can never be an end to this, for the kind of order we conceive changes with the change of symbols or with the kinds of questions we ask. Always there is the challenge to see whether there is another new world to describe and explain lying just beyond the horizon.

References

The following references are arranged alphabetically by chapter. References in the text are by author and date of publication. As the publication of scholarly works in geography continues to increase in volume so greatly, the references contained herein can include only selected items that are representative of the whole body of writings on geography. For more complete coverage of the literature, the reader is referred to such standard international bibliographies as the *Bibliographie géographique internationale*, published annually (since 1923) by Armand Colin in Paris. In the United States the American Geographical Society (Broadway at 156th St., New York, N.Y. 10032) has published, since 1938, its *Current Geographical Publications: Additions to the Research Catalogue of the American Geographical Society*, which is issued monthly except for July and August.

Annals AAG refers to the *Annals of the Association of American Geographers*.

CHAPTER 1

Bertalanffy, L. von. 1968. *General System Theory: Foundations, Development, Applications*. New York: George Braziller.

Brown, R., 1963. *Explanation in Social Science*. Chicago: Aldine.

Ducasse, C. J. 1969. *Causation and the Types of Necessity*. New York: Dover.

Glacken, C. J. 1967. *Traces on the Rhodian Shore, Nature and Culture in Western Thought from Ancient Times to the End of the Eighteenth Century.* Berkeley and Los Angeles: University of California Press.

Krauss, R. M. 1968. "Language as a Symbolic Process in Communication." *American Scientist,* 56:265-278.

CHAPTER 2

Aristotle. *Metaphysica.* Trans. D. E. Gershenson and D. A. Greenberg, 1963. Vol. 2, *The Natural Philosopher.* Pp. 5-55. New York: Blaisdell.

Berger, H. 1903. *Geschichte der Wissenschaftlichen Erdkunde der Griechen.* Leipzig: Veit.

Boyce, R. R. 1977. *The Trade of Tyre: Anomaly of the Ancient World.* Seattle: Seattle Pacific College.

Búnbury, E. H. 1883. *A History of Ancient Geography Among the Greeks and Romans from the Earliest Ages till the Fall of the Roman Empire.* 2 vols. London: John Murray.

Casson, L. 1959. *The Ancient Mariners. . . .* New York: Macmillan.

Glacken, C. J. 1956. "Changing Ideas of the Habitable World." In W. L. Thomas, ed., *Man's Role in Changing the Face of the Earth.* Pp. 70-92. Chicago: University of Chicago Press.

―――. 1967. *Traces on the Rhodian Shore, Nature and Culture in Western Thought from Ancient Times to the End of the Eighteenth Century.* Berkeley and Los Angeles: University of California Press.

Heidel, W. A. 1937. *The Frame of Ancient Greek Maps.* New York: American Geographical Society.

Honigmann, E. 1939. *Die sieben Klimata.* Heidelberg: Winter.

Ninck, M. 1945. *Die Entdeckung von Europa durch die Griechen.* Basel: Benno Schwabe.

Popper, K. R. 1945/1962. *The Open Society and Its Enemies.* New York: Harper & Row

Sarton, G. 1952. *A History of Science, Ancient Science Through the Golden Age of Greece.* Cambridge, Mass.: Harvard University Press (reprinted New York: John Wiley & Sons, 1964).

―――. 1959. *A History of Science, Hellenistic Science and Culture in the Last Three Centuries* B.C. Cambridge, Mass.: Harvard University Press (reprinted New York: John Wiley & Sons, 1965).

Schamp, H. 1955-56. "Die Turm der Winde in Athen und die Luftkörperklimatologie." *Die Erde,* 7-8:119-128.

Strabo. *The Geography of Strabo.* Trans. H. L. Jones, 1917. New York: G. P. Putnam's Sons.

Thomson, J. O. 1965. *History of Ancient Geography.* New York: Biblo & Tannen.

Tozer, H. F. 1897. *A History of Ancient Geography.* Cambridge: At the University Press (reprinted New York: Biblo & Tannen, 1964).

CHAPTER 3

Ahmad, N. 1947. *Muslim Contributions to Geography.* Lahore: Muhammad Ashraf.

Bagrow, L., and Skelton, R. A. 1964. *History of Cartography.* Cambridge, Mass.: Harvard University Press. The original book by Leo Bagrow, *Geschichte der Kartographie*

(Berlin: Safari-Verlag, 1951), was translated into English by D. L. Paisey in 1960. The present book was revised and enlarged by R. A. Skelton.

Beazley, C. R. 1949. *The Dawn of Modern Geography.* 3 vols. New York: Peter Smith (original publication, London: John Murray 1897-1906).

Cassidy, V. H. 1968. *The Sea Around Them: The Atlantic Ocean,* A.D. *1250.* Baton Rouge: Louisiana State University Press.

Glacken, C. J. 1956. ''The Changing Ideas of the Habitable World.'' In W. L. Thomas, ed., *Man's Role in Changing the Face of the Earth.* Pp. 70-92. Chicago: University of Chicago Press.

_____. 1967. *Traces on the Rhodian Shore, Nature and Culture in Western Thought from Ancient Times to the End of the Eighteenth Century.* Berkeley and Los Angeles: University of California Press.

Goldstein, T. 1965. ''Geography in Fifteenth Century Florence.'' In John Parker, ed., *Merchants and Scholars: Essays in the History of Exploration and Trade.* Pp. 9-32. Minneapolis: University of Minnesota Press.

Hsieh, Chiao-min. 1968. ''The Chinese Exploration of the Ocean—A Study in Historical Geography.'' *Chinese Culture* (Taiwan), 9:123-131.

Ibn-Batuta. *The Travels of Ibn-Battuta,* A.D. *1325-1354.* Trans. C. Defrémery and B. R. Sanguinetti, 1958. Cambridge: At the University Press.

Ibn-Khaldun. *The Muqaddimah.* Trans. Franz Rosenthal, 1958. New York: Pantheon Books.

Kimble, G. H. T. 1938. *Geography in the Middle Ages.* London: Methuen.

Mirsky, J., ed. 1964. *The Great Chinese Travelers.* New York: Pantheon Books.

Morison, S. E., trans. and ed. 1963. *Journals and Other Documents on the Life and Voyages of Christopher Columbus.* New York: Heritage Press.

_____. 1971. *The European Discovery of America, the Northern Voyages.* New York: Oxford University Press.

_____. 1974. *The Southern Voyages.* New York: Oxford University Press.

Needham, J. 1963. ''Poverties and Triumphs of the Chinese Scientific Tradition.'' In A. C. Combie, ed., *Scientific Change.* Pp. 117-153. New York: Basic Books.

Needham, J., and Ling, W. 1959. *Science and Civilization in China.* Vol. 3, *Mathematics and the Sciences of the Heavens and the Earth.* Cambridge: At the University Press.

Nunn, G. E. 1924. *The Geographical Conceptions of Columbus.* New York: American Geographical Society, Research Series No. 14.

Polo, M. *The Travels of Marco Polo* (revised from Marsden's translation, edited and with an introduction by Manuel Komroff). New York: Liveright, 1930.

Sauer, C. O. 1968. *Northern Mists.* Berkeley and Los Angeles: University of California Press.

Sykes, P. 1961. *A History of Exploration from the Earliest Times to the Present Day.* New York: Harper Bros.

Taylor, E. G. R. 1957. *The Haven-Finding Art: A History of Navigation from Odysseus to Captain Cook.* New York: Abelard-Schuman.

Thomson, J. O. 1965. *History of Ancient Geography.* New York: Biblo & Tannen.

Tillman, J. P. 1971. *An Appraisal of the Geographical Works of Albertus Magnus and His Contributions to Geographical Thought.* Ann Arbor: University of Michigan, Department of Geography, Publication No. 4.

Wright, J. K. 1925. *The Geographical Lore at the Time of the Crusades*. . . . New York: American Geographical Society, Research Series No. 15.

CHAPTER 4

Babcock, W. H. 1922. *Legendary Islands of the Atlantic, a Study in Medieval Geography*. New York: American Geographical Society, Research Series No. 8.

Bagrow, L., and Skelton, R. A. 1964. *History of Cartography*. Cambridge, Mass. Harvard University Press (see the reference in Chapter 3).

Beazley, C. R. 1895. *Prince Henry the Navigator*. New York: G. P. Putnam's Sons.

Brown, L. A. 1960. *Map Making: The Art that Became a Science*. Boston: Little, Brown.

Crone, G. R. 1950. *Maps and Their Makers, an Introduction to the History of Cartography*. New York: Capricorn Books.

Davies, A. 1967. "Columbus Divides the World." *Geographical Journal*, 133:337-344.

Debenham, F. 1960. *Discovery and Exploration—An Atlas—History of Man's Wanderings*. New York: Doubleday & Co.

Friis, H. R., ed. 1967. *The Pacific Basin, a History of its Geographical Exploration*. New York: American Geographical Society, Special Publication No. 38.

Hakluyt, R. *Hakluyt's Voyages*. Ed. I. R. Blacker, 1965. New York: Viking Press.

Hale, J. R. 1966. *Age of Exploration*. New York: Time Inc.

Hanson, E. P., ed. 1967. *South from the Spanish Main*. New York: Delacorte Press.

Morison, S. E. 1942. *Admiral of the Ocean Sea, A Life of Christopher Columbus*. New York: Little, Brown.

————, trans. and ed. 1963. *Journals and Other Documents on the Life and Voyages of Christopher Columbus*. New York: Heritage Press.

Nunn, G. E. 1924. *The Geographical Conceptions of Columbus, a Critical Consideration of Four Problems*. New York: American Geographical Society, Research Series No. 14.

Oliveira Martins, J. P. *The Golden Age of Prince Henry the Navigator*. Trans. J. J. Abraham and W. E. Reynolds, 1914. London: Chapman & Hall (Portuguese title: *Os Filhos de D. João I*, Lisbon, 1901).

Parker, J., ed. 1965. *Merchants and Scholars, Essays in the History of Exploration and Trade*. Minneapolis: University of Minnesota Press.

Parks, G. B. 1928. *Richard Hakluyt and the English Voyages*. New York: American Geographical Society, Special Publication No. 10.

Penrose, B. 1952. *Travel and Discovery in the Rennaissance, 1420-1620*. Cambridge, Mass.: Harvard University Press.

Quill, H. 1966. *John Harrison, the Man Who Found Longitude*. London: Pall Mall.

Rogers, F. M. 1962. *The Quest for the Eastern Christians: Travels and Rumor in the Age of Discovery*. Minneapolis: University of Minnesota Press.

Sauer, C. O. 1966. *The Early Spanish Main*. Berkeley and Los Angeles: University of California Press.

Skelton, R. A. 1969. Captain James Cook, After Two Hundred Years. A commemorative address before the Hakluyt Society. London: The British Museum.

Stokes, E. 1970. "European Discovery of New Zealand Before 1642, a Review of the Evidence." *The New Zealand Journal of History*, 4:3-29.

Sykes, P. 1961. *A History of Exploration from the Earliest Times to the Present Day*. New York: Harper Bros.

Taylor, E. G. R. 1957. *The Haven-Finding Art: A History of Navigation from Odysseus to Captain Cook*. New York: Abelard-Schuman.

Tooley, R. V. 1949. *Maps and Map-Makers*. New York: Crown.

CHAPTER 5

Adickes, E. 1924-25. *Kant als Naturforscher*. 2 vols. Berlin: W. de Gruyter.

Baker, J. N. L. 1955a. "Geography and Its History." *Advancement of Science*, 12:188-198.

———. 1955b. "The Geography of Bernhard Varenius." *Transactions and Papers, Institute of British Geographers*, 21:51-60.

———. 1963. "Major James Rennel, 1742-1830, and His Place in the History of Geography." In *The History of Geography*. Pp. 130-157. New York: Barnes & Noble

Berget, A. 1913. "La répartition des terres et des mers et la position du pole continental de la terre." *Revue de géographie*, 7:1-36.

Beythien, H. 1898. *Eine neue Bestimmung des Pols der Landhalbkugel*. Kiel: Lipsius & Tischer.

Chorley, R. J., Dunn, A. J., and Beckinsale, R. P. 1964. *A History of the Study of Landforms, or the Development of Geomorphology*. Vol. 1, *Geomorphology Before Davis*. London: Methuen.

Dainville, F. de. 1970. "From the Depths to the Heights: Concerning the Marine Origins of the Cartographic Expression of Terrestrial Relief by Numbers and Contour Lines." *Surveying and Mapping*, 30:389-403. (Translated from the French by A. H. Robinson and M. Carlier.)

Dickinson, R. E., and Howarth, O. J. R. 1933. *The Making of Geography*. Oxford: The Clarendon Press.

Glacken, C. J. 1960. "Count Buffon on Cultural Changes of the Physical Environment." *Annals AAG*, 50:1-21.

———. 1967. *Traces on the Rhodian Shore, Nature and Culture in Western Thought from Ancient Times to the End of the Eighteenth Century*. Berkeley and Los Angeles: University of California Press.

Hartshorne, R. 1939. *The Nature of Geography, a Critical Survey of Current Thought in the Light of the Past*. Lancaster, Pa.: Association of American Geographers.

Jefferson, T. 1787. *Notes on the State of Virginia*. London: John Stockdale.

Kimble, G. H. T. 1938. *Geography in the Middle Ages*. London: Methuen.

Kriesel, K. M. 1968. "Montesquieu: Possibilistic Political Geographer." *Annals AAG*, 58:557-574.

May, J. A. 1970. *Kant's Concept of Geography and Its Relation to Recent Geographical Thought*. Toronto: University of Toronto, Department of Geography, Research Paper No. 4.

Parks, G. B. 1928. *Richard Hakluyt and the English Voyages*. New York: American Geographical Society, Special Publication No. 10.

Partsch, J. 1891. "Philipp Clüver, der Begründer der historischen Länderkunde, ein Beitrag zur Geschichte der geographischen Wissenschaft." *Geographische Abhandlung*, 5(2). (47 pages.)

Peschel, O. 1865. *Geschichte der Erdkunde bis auf A.v. Humboldt und Carl Ritter.* Munich: J. G. Cotta.

Playfair, J. 1802. *Illustrations of the Huttonian Theory of the Earth* (reprinted New York: Dover, 1956).

Pollard, A. W., ed.1964. *The Travels of Sir John Mandeville.* New York: Dover.

Sauer, C. O. 1974. "Foreword to Historical Geography." *Annals AAG,* 31:1-24.

Taylor, E. G. R. 1948. "The English Worldmakers of the Seventeenth Century and Their Influence on the Earth Sciences." *Geographical Review,* 38:109-112.

_____. 1950. "The Origin of Continents and Oceans, A Seventeenth Century Controversy." *Geographical Journal,* 116:193-198.

Thrower, N. J. W. 1969. "Edmund Halley as a Thematic Geo-Cartographer." *Annals AAG,* 59:652-676.

Tooley, R. V. 1949. *Maps and Map Makers.* New York: Crown.

Tuan, Yi-fu. 1968. *The Hydrologic Cycle and the Wisdom of God: A Theme in Geoteleology.* Toronto: University of Toronto, Department of Geography, Research Publications.

Wagner, H. 1920-1922. *Lehrbuch der Geographie,* 10th ed. Hanover: Hahnsche Buch handlung. (Part 1, 1920; Part 2, 1921; Part 3, 1922.)

Warntz, W. 1964. *Geography Now and Then, Some Notes on the History of Academic Geography in the United States.* New York: American Geographical Society, Research Series No. 25.

CHAPTER 6

Beck, H. 1959-61. *Alexander von Humboldt.* Vol. 1 (1959), *Von der Bildungsreise zur Forschungsreise, 1769-1804.* Vol. 2 (1961), *Vom Reisewerk zum "Kosmos," 1804-1859.* Wiesbaden: Franz Steiner.

Bögekamp, H. 1863. "An Account of Prof. Ritter's Geographical Labors." In W. L. Gage, trans., *Geographical Studies by the Late Professor Carl Ritter of Berlin.* Pp. 33-51. Boston: Gould & Lincoln.

Dickinson, R. E. 1969. *The Makers of Modern Geography.* London: Routledge & Kegan Paul.

Fröbel, J. 1831. "Einige Blicke auf den jetsigen formellen Zustand der Erdkunde." *Annalen der Erd-, Völker-, und Staatenkunde,* 4:493-506.

Gage, W. L., trans. 1863. *Geographical Studies by the Late Professor Carl Ritter of Berlin.* Boston: Gould & Lincoln.

Guyot, A. H. 1860. "Carl Ritter." *Journal of the American Geographical and Statistical Society,* 2:25-63.

Hartshorne, R. 1939. *The Nature of Geography, a Critical Survey of Current Thought in the Light of the Past.* Lancaster, Pa.: Association of American Geographers.

_____. 1958. "The Concept of Geography as a Science of Space, from Kant and Humboldt to Hettner." *Annals AAG,* 48:97-108.

Humboldt, A. von. 1793. *Florae fribergensis subterraneas exhibens.* Berlin: H. A. Rottman.

_____. 1805-1834. *Voyage aux régions équinoxiales du Nouveau Continent.* Paris. (See footnote on pp. 120-121 for titles of the thirty volumes.)

_____. 1808. *Ansichten der Natur, mit wissenschaftlichten Erläuterung.* 2nd ed., 1849. Stuttgart: Cotta.

_____. 1814-25. *Relation historique du voyage au régions équinoxiales du Nouveau Continent* (Vols. 28-30, 1805-1834). English translation by H. M. Williams, 1825. *Personal Narrative of Travels in the Equinoxial Regions of the New Continent During the Years 1799-1804.* 5 vols. Paris, German translation by H. Hauff, 1859-60. *Alexander von Humboldt's Reise in die Aequinoctial Gegenden des neuen Continents.* 4 vols. Stuttgart.

_____. 1845-62. *Kosmos: Entwurf einer physischen Weltbeschreibung.* 5 vols. Stuttgart: Cotta. (Vol. 1, 1845; Vol. 2, 1847; Vol. 3, 1850; Vol. 4, 1858; Vol. 5, 1862.) English translation by E. C. Otté, London: H. G. Bohn, 1849-58.

Kellner, L. 1963. *Alexander von Humboldt.* London: Oxford University Press.

Kramer, F., 1959. "A Note on Carl Ritter, 1779-1859." *Geographical Review,* 49:406-409.

Meyer-Abich, A. 1967. *Alexander von Humboldt in Selbstzeugnissen und Bilddokumenten.* Rowohlt: Kurt Kisenberg.

Ritter, C. 1822-59. *Die Erdkunde, im Verhältniss zur Natur und zur Geschichte des Menschen, oder allgemeine vergleichende Geographie als sichere Grundlage des Studiums und Unterrichts in physikalischen und historischen Wissenschaften.* 19 vols. Berlin: G. Reimer.

_____. 1852. *Einleitung zur allgemeinen vergleichenden Geographie, und Abhandlungen zur Begründung einer mehr wissenschaftlichen Behandlung der Erdkunde.* Berlin: G. Reimer.

_____. 1862. *Allgemeine Erdkunde.* Berlin: G. Reimer.

Schultz, J. H., ed. 1959. *Alexander von Humboldt: Studien zu seiner universalen Geisteshaltung.* Berlin: W. de Gruyter.

Terra, H. de. 1955. *The Life and Times of Alexander von Humboldt, 1769-1859.* New York: Alfred A. Knopf.

CHAPTER 7

Bartlett, R. A. 1962. *Great Surveys of the American West.* Norman, Okla.: University of Oklahoma Press.

Beck, H. 1956. "Heinrich Berghaus und Alexander von Humboldt." *Petermanns Geographische Mitteilungen,* 100:4-16.

Brown, R. H. 1951. "A Letter to the Reverend Jedidiah Morse, Author of *The American Universal Geography.*" *Annals AAG,* 41:188-198.

Colby, C. C. 1936. "Changing Currents of Geographic Thought in America." *Annals AAG,* 26:1-37.

Commission on History of Geographical Thought, International Geographical Union. 1972. *Geography Through a Century of International Congresses.*

Coues, E., ed. 1893. *History of the Expedition Under the Command of Lewis and Clark.* New York (republished in 3 vols., New York: Dover, 1965).

Curti, M. 1943. *The Growth of American Thought.* New York: Harper Bros.

Darrah, W. C. 1951. *Powell of the Colorado.* Princeton, N.J.: Princeton University Press.

Darwin, C. R. 1842. *The Structure and Distribution of Coral Reefs,* 2nd ed. 1874. London: John Murray (3rd ed., New York: D. Appleton, 1889).

_____. 1859. *On the Origin of Species by Means of Natural Selection, or The Preservation of Favoured Races in the Struggle for Life.* London: John Murray.

Davis, W. M. 1924. "The Progress of Geography in the United States." *Annals AAG,*
14:159–215.
_____. 1928. *The Coral Reef Problem.* New York: American Geographical Society, Spe-
cial Publication No. 9.
Dillon, R. 1965. *Meriwether Lewis.* New York: Coward-McCann.
Dunbar, G. S. 1978. *Elisée Reclus: Historian of Nature.* Hamden; Conn.: Shoe String Press.
Gilbert, G. K. 1878. *Report on the Geology of the Henry Mountains.* Washington, D.C.:
Department of the Interior.
_____. 1890. *Lake Bonneville.* Washington, D.C.: U.S. Geological Survey, Monograph
No. 1.
Ginsburg. N. 1972. "The Mission of a Scholarly Society." *Professional Geographer.*
24:1–6.
Glick, T. F., ed. 1974. *The Comparative Reception of Darwinism.* Austin: University of
Texas Press.
Goode, J. P. 1927. "The Map as a Record of Progress in Geography." *Annals AAG,*
17:1–14.
Guyot, A. 1849. *The Earth and Man: Lectures on Comparative Physical Geography in Its
Relation to the History of Mankind.* Boston: Gould & Lincoln.
_____. 1873. *Physical Geography.* New York: Scribner, Armstrong & Co.
Hartshorne, R. 1939. *The Nature of Geography, a Critical Survey of Current Thought in the
Light of the Past.* Lancaster, Pa.: Association of American Geographers.
James, P. E. 1964. "A New Concept of Atmospheric Circulation." *Journal of Geography,*
63:245–250.
_____. 1969. "The Significance of Geography in American Education." *Journal of Geog-
raphy,* 68:473–483.
_____. 1979. "John Wesley Powell: 1834–1902." In *Geographers: Biobibliographical
Studies.* Vol. 3, pp. 117–124. London: Mansell.
James, P. E., and Jones, C. F., eds. 1954. *American Geography, Inventory and Prospect.*
Syracuse, N.Y.: Syracuse University Press.
Leighly, J. 1938. "Methodological Controversy in Nineteenth Century German Geography." *Annals AAG,* 28:238–258.
_____. 1949. "Climatology Since the Year 1800." *Transactions of the American Geophys-
ical Union,* 30:658–672.
_____. 1977. "Matthew Fontaine Maury: 1806–1873." In *Geographers: Biobibliographi-
cal Studies.* Vol. 1, pp. 59–63. London: Mansell.
Libby, W., Jr. 1884. "The Life and Scientific Work of Arnold Guyot." *Bulletin of the
American Geographical Society,* 16:194–221.
Lorenz, E. N. 1966. "The Circulation of the Atmosphere." *American Scientist,* 54:402–
420.
Lowenthal, D. 1958. *George Perkins Marsh, Versatile Vermonter.* New York: Columbia
University Press.
Lurie, E. 1960. *Louis Agassiz: A Life in Science.* Chicago: University of Chicago Press.
Marsh, G. P. 1864. *Man and Nature, or Physical Geography as Modified by Human Action.*
New York: Charles Scribner (republished, David Lowenthal, ed.; Cambridge, Mass.:
Harvard University Press, 1965).

————. 1874. *The Earth as Modified by Human Action*, 2nd ed. 1885. New York: Charles Scribner.

Maury, M. F. 1850. "On the General Circulation of the Atmosphere." *Proceedings of the American Association for the Advancement of Science*, 3:126–147.

————. 1851. *Explanations and Sailing Directions to Accompany the Wind and Current Charts*. Washington, D.C.: C. Alexander.

————. 1855. *The Physical Geography of the Sea*. New York: Harper Bros.

Meinig, D. W. 1955. "Isaac Stevens: Practical Geographer and Historian." *Geographical Review*, 45:542–558.

Powell, J. W. 1878. *Report on the Lands of the Arid Region of the United States with a More Detailed Account of the Lands of Utah. Washington, D.C.*: 45th Congress, 2d Session.

————. 1885. "The Organization and Plan of the United States Geological Survey." *American Journal of Science*, 29:93–102.

Reclus, E. 1867–68. *La terre, description des phénomènes de la vie du globe*. 2 vols. Paris: Hachette.

————. 1869. *Histoire d'un ruisseau*. Paris: J. Hetzel.

————. 1876–94. *Nouvelle géographie universelle, la terre et les hommes*. 19 vols. Paris: Hachette. English translation by E. G. Ravenstein and A. H. Keane. *The Earth and Its Inhabitants*. London: 1878–94.

————. 1880. *Histoire d'une montagne*. Paris: J. Hetzel.

————. 1905–1908. *L'homme et la terre*. Paris: Librairie Universelle.

Shafer, R. J., ed. 1969. *A Guide to Historical Method*. Homewood, Ill.: Dorsey Press.

Shaler, N. S. 1905. "Earth and Man: An Economic Forecast." *International Quarterly*, 10:227–239.

————. 1912. *Man and the Earth*. New York: Duffield & Co.

Sinnhuber, K. A. 1959. "Carl Ritter, 1779–1859." *Scottish Geographical Magazine*, 75:152–63.

Stegner, W. 1954. *Beyond the Hundredth Meridian: John Wesley Powell and the Second Opening of the West*. Boston: Houghton Mifflin.

Stoddart, D. R. 1966. "Darwin's Impact on Geography." *Annals AAG*, 56:683–698.

Thomas, W. L., ed. 1956. *Man's Role in Changing the Face of the Earth*. Chicago: University of Chicago Press.

Voeikov, A. I. 1901. "De l'influence de l'homme sur la terre." *Annales de géographie*, 10:97–114, 193–215.

Wagner, H. 1880. "Bericht über die Entwicklung der Methodik der Erdkunde." *Geographisches Jahrbuch*, 8:523–598.

Wheeler, G. M. 1885. *Report upon the Third Geographical Congress and Exhibition at Venice, Italy, 1881. Accompanied by Data Concerning the Principal Land and Marine Surveys of the World*. Washington, D.C.: House Executive Document 270, 48th Congress, 2d Session.

Williams, F. L. 1963. *Matthew Fontaine Maury, Scientist of the Sea*. New Brunswick, N.J.: Rutgers University Press.

Wright, J. K. 1951. "The Field of the Geographical Society." In *Geography in the Twentieth Century*. Pp. 543–565. G. Taylor, ed., New York: Philosophical Library.

————. 1952. *Geography in the Making: The American Geographical Society, 1851–1951*. New York: The American Geographical Society.

_____. 1953. "The Open Polar Sea." *Geographical Review*, 43:338-365.

_____. 1961. "Daniel Coit Gilman: Geographer and Historian." *Geographical Review*, 51:381-399.

CHAPTER 8

Bartels, D., and Peucker, T. 1969. "German Social Geography Again." *AAG*, 59:596-98.

Beck, H. 1957. "Geographie und Reisen im 19. Jahrhundert: Prolegomena zu einer allgemeinen Geschichte der Reisen." *Petermanns Geographische Mitteilungen*, 101:1-14.

Bobek, H. 1948. "Stellung and Bedeutung der Sozialgeographie." *Erdkunde*, 2:118-125.

Bobek, H., and Schmithüsen, J. 1949. "Die Landschaftsbegriff im logischen System der Geographie." *Erdkunde*, 3:112-120.

Büttner, M. 1978. "Bartholomaus Keckermann: 1572-1609." In *Geographers: Biobibliographical Studies*. Vol. 2, pp. 73-79. London, Mansell.

_____. 1979. "Philipp Melanchthon: 1497-1560." In *Geographers: Biobibliographical Studies*. Vol. 3, pp. 93-97. London: Mansell.

Büttner, M., and Burmeister, K. H. 1979. "Sebastian Munster: 1488-1552." In *Geographers: Biobibliographical Studies*. Vol. 3, pp. 99-106. London, Mansell.

Christaller, W. 1933. *Die zentralen Orte in Süddeutschland*. Jena: Gustav Fischer. Trans. C. W. Baskin, *Central Places in Southern Germany*. Englewood Cliffs, N.J.: Prentice-Hall, 1966.

Dickinson, R. E. 1969. *The Makers of Modern Geography*. London: Routledge & Kegan Paul.

Fischer, E., Campbell, R. D., and Miller, E. S. 1967. *A Question of Place: The Development of Geographic Thought*. Arlington, Va.: Beatty.

Gradmann, R. 1931a. "Das länderkundliche Schema." *Geographische Zeitschrift*, 37:540-548.

_____. 1931b. *Süd-Deutschland*. 2 vols. Stuttgart: J. Engelhorn.

Hahn, E. 1892. "Die Wirtschaftsformen der Erde." *Petermanns Geographische Mitteilungen*, 38:8-12.

_____. 1896. *Die Haustiere und ihre Beziehungen zur Wirtschaft des Menschen*. Leipzig: Duncker & Humblot.

_____. 1919. *Von der Hacke zum Pflug, Garten, und Feld: Bauern und Hirten in unserer Wirtschaft und Geschichte*. Leipzig: Quelle & Meyer.

Hajdu, J. G. 1968. "Toward a Definition of Post-War German Social Geography." *Annals AAG*, 58:397-410.

Hard, Gerhard. 1969. "Die Diffusion der 'Idee der Landschaft': Praliminarien zu einer Geschichte der Landschaftsgeographie." *Erdkunde*, 23:249-364.

_____. 1970. "Was ist eine Landschaft? Etymologie als Denkform in der geographischen Literatur." In D. Bartels, ed., *Wirtschafts- und Sozialgeographie*. Pp. 66-84. Berlin/Cologne: Kiepenheuer and Witsch.

Hartke, W., ed. 1960. *Denkschrift zur Lage der Geographie*. Wiesbaden: Franz Steiner.

Hartshorne, R. 1939. *The Nature of Geography, a Critical Survey of Current Thought in the Light of the Past*. Lancaster, Pa.: Association of American Geographers.

————. 1958. "The Concept of Geography as a Science of Space, from Kant and Humboldt to Hettner." *Annals AAG*, 48:97-108.

————. 1959. *Perspective on the Nature of Geography.* Chicago: Rand McNally.

Harvey, D. 1969. *Explanation in Geography.* London: Edward Arnold.

Hassert, K. 1905. "Friedrich Ratzel, Sein Leben und Wirken." *Geographische Zeitschrift,* 11:305-325, 361-380.

Hettner, A. 1895. "Geographische Forschung und Bildung." *Geographische Zeitschrift,* 1:1-19.

————. 1905. "Das Wesen und die Methoden der Geographie." *Geographische Zeitschrift,* 11:549-553.

————. 1907. *Grundzüge der Länderkunde.* Vol. 1, *Europa.* Rev. eds., 1923, 1932. Vol. 2, *Die Aussereuropäische Erdteile.* Rev. eds., 1923, 1926. Leipzig: Teubner.

————. 1927. *Die Geographie—ihre Geschichte, ihr Wesen, und ihre Methoden.* Breslau: Ferdinand Hirt.

Jager, H. 1972. "Historical Geography in Germany, Austria and Switzerland." In A. R. H. Baker, ed., *Progress in Historical Geography.* Pp. 45-62. New York: John Wiley & Sons.

James, P. E. 1934. "The Terminology of Regional Description." *Annals AAG,* 24:78-92.

————. 1936. "The Geography of the Oceans: A Review of the Work of Gerhard Schott." *Geographical Review,* 26:664-669.

Joerg, W. L. G. 1922. "Recent Geographical Work in Europe." *Geographical Review,* 12:431-484.

Köppen, W. 1923. *Die Klimate der Erde, Grundriss der Klimatologie* (revised and enlarged, 1931). Berlin: W. de Gruyter.

————. 1936. "Das geographische System der Klimate." In W. Köppen and R. Geiger, eds., *Handbuch der Klimatologie.* Vol. 1, Part C. Berlin: Gebrüder Borntraeger.

Kramer, F. L. 1967. "Eduard Hahn and the End of the 'Three Stages of Man.' " *Geographical Review,* 57:73-89.

Krebs, N. 1923. "Natur- und Kulturlandschaft." *Zeitschrift der Gesellschaft für Erdkunde zu Berlin.* Pp. 81-94.

Kuhn, T. S. 1963. "The Function of Dogma in Scientific Research." In A. C. Crombie, ed., *Scientific Change.* Pp. 347-369. New York: Basic Books.

Lautensach, H. 1952. "Otto Schlüter's Bedeutung für die Methodische Entwicklung der Geographie." *Petermanns Geographische Mitteilungen,* 96:219-231.

————. 1964. *Iberische Halbinsel.* Munich: Keysersche Verlagsbuchhandlung.

Martin, G. J. 1974. "A Fragment on the Pencks(s)—Davis Conflict." *Special Libraries Association. Geography and Map Division.* Bulletin 98, pp. 11-27.

Passarge, S. 1919-20. *Die Grundlagen der Landschaftskunde.* 3 vols. Hamburg: L. Friederichsen.

————. 1923. *Die Landschaftsgürtel der Erde.* Breslau: Ferdinand Hirt.

————. 1930. "Sesen und Grenzen der Landschaftskunde." In *Herman Wagner Gedenkschrift, Ergebnisse und Aufgaben der geographischen Forschung* (Ergänzungsheft). *Petermanns Geographische Mitteilungen,* 209:29-44.

Peschel, O. 1865. *Geschichte der Erdkunde bis auf A.v.Humboldt und Carl Ritter.* Munich: J. G. Cotta.

_____. 1870. *Neue Probleme der vergleichenden Erdkunde als versuch einer Morphologie der Erdoberfläche*. Leipzig: Duncker & Humblot.

_____. 1879. *Physische Erdkunde*. Ed. Gustav Leipoldt. Leipzig: Duncker & Humblot.

Pfeifer, G. 1965. "Geographie Heute?" In *Festschrift Leopold G. Scheidl zum 60 Geburstag*. Pp. 78–90. Vienna: Ferdinand Berger & Söhne.

Philippson, A. 1904. *Das Mittelmeergebiet* 4th ed. 1922. Leipzig: Teubner.

Ratzel, F. 1882–91. *Anthropogeographie*. Vol. 1, *Grundzüge der Anwendung der Erdkunde auf die Geschichte*, 2nd ed., 1889; 3rd ed., 1909. Vol. 2, 1891, *Die geographische Verbreitung des Menschen*, 2nd ed., 1912. Stuttgart: J. Engelhorn.

_____. 1897. *Politische Geographie, oder die Geographie der Staaten, des Verkehrs, und der Krieges*, 2nd ed., 1903; 3rd ed., 1923. Munich and Berlin: R. Oldenbourg.

_____. 1898. *Deutschland, Einführung in die Heimatkunde*. Leipzig: Grunow.

Richthofen, F. von. 1877–1912. *China: Ergebnisse eigener Reisen und darauf gegründte Studien*. 5 vols. Berlin: Dietrich Reimer.

_____. 1883. *Aufgaben und Methoden der heutigen Geographie* (Akademische Antrittsrede). Leipzig: Veit.

_____. 1886. *Führer für Forschungsreisende*. Berlin: Robert Oppenheim.

Sauer, C. O. 1971. "The Formative Years of Ratzel in the United States." *Annals AAG*, 61:245–254.

Schaefer, F. K. 1953. "Exceptionalism in Geography: A Methodological Examination." *Annals AAG*, 43:226–249.

Schlüter, O. 1906. *Die Ziele der Geographie des Menschen* (Antrittsrede). Munich: R. Oldenbourg.

_____. 1920. "Die Erdkunde in ihrem Verhältnis zu den Natur- und Geisteswissenschaften." *Geographische Anzeiger*, 21:145–152, 213–218.

_____. 1928. "Die analytische Geographie der Kulturlandschaft erläutert am Beispiel der Brücken." *Zeitschrift der Gesellschaft für Erdkunde zu Berlin*, Sonderband. Pp. 388–411.

Schmieder, O. 1966. "Die deutsche Geographie in der Welt von Heute." *Geographische Zeitschrift*, 54:207–222.

Schmithüsen, J. 1959. "Das System der geographischen Wissenschaft." In *Festschrift Theodor Kraus. . . .* Pp. 1–14. Bad Godesberg.

_____. 1963. *Was ist eine Landschaft?* Erdkundliches Wissen, No. 9. Wiesbaden: Franz Steiner.

Schott, G. 1912. *Geographie des Atlantischen Ozean*, 4th ed. 1942. Hamburg: C. Boysen.

_____. 1935. *Geographie des Indischen und Stillen Ozeans*. Hamburg: C. Boysen.

Semple, E. C. 1911. *Influences of Geographic Environment, on the Basis of Ratzel's System of Anthropo-Geography*. New York: Henry Holt.

Smith, T. R., and Black, L. D. 1946. "German Geography: War Work and Present Status." *Geographical Review*, 36:398–408.

Spencer, H. 1864. *Principles of Biology*. 2 vols. New York: D. Appleton.

_____. 1876–96. *The Principles of Sociology*. 3 vols. New York: D. Appleton.

Speth, W. W. 1977. "Carl Ortwin Sauer on Destructive Exploitation." *Biological Conservation*, 11:145–160.

Spethmann, H. 1931. *Das länderkundliche Schema in des deutschen Geographie*. Berlin: Reimar Hoffing.

Steinmetzler, J. 1956. *Die Anthropogeographie Friedrich Ratzels und ihre ideenges-chichtlichen Wurzein.* Bonn: Geographische Abhandlung.

Storkenbaum, W., ed. 1967. *Zum Gegenstand und zur Methode der Geographie.* Darmstad: Wissenschaftliche Buchgesellschaft.

———, ed. 1969. *Sozialgeographie.* Darmstad: Wissenschaftliche Buchgesellschaft.

Troll, C. 1947. "Die geographische Wissenschaft in Deutschland in dem Jahren 1933 bis 1945: Eine Kritik und Rechtfertigung." *Erdkunde,* 1:3-48.

———. 1949. "Geographical Science in Germany during the Period 1933-1945: A Critique and Justification." Trans. and ed. by E. Fischer. *Annals AAG,* 39:100-137.

———. 1950. "Die Geographische Landschaft und (ihre) Erforschung." *Stadium Generale,* 3:163-181.

———. 1966. "Hermann Lautensach." *Erdkunde,* 20:243-252.

Van Valkenburg, S. 1951. "The German School of Geography." In G. Taylor, ed., *Geography in the Twentieth Century.* Pp. 91-115. New York: Philosophical Library.

Wagner, H. 1880. "Bericht über die Entwicklung der Methodik der Erdkunde." *Geographisches Jahrbuch,* 8:523-598.

———. 1920. "Geschichte der Methodik der Geographie als Wissenschaft." *Lehrbuch der Geographie,* 1:17-25.

Waibel, L. 1933. "Was verstehen wir unter Landschaftskunde?" *Geographische Anzeiger,* 34:197-207.

Wanklyn, H. 1961. *Friedrich Ratzel, a Biographical Memoir and Bibliography.* Cambridge: At the University Press.

CHAPTER 9

Beaujeu-Garnier, J. 1951. *Le Morvan et sa bordure.* Paris: Armand Colin.

———. 1956-1958. 2 vol. *Géographie de la population.* Paris: Librairie de Medici.

———. 1976. *Methods and Perspectives in Geography.* London, New York: Longmans (translated by J. Bray).

Blanchard, R. 1906. *La Flandre: étude géographique de la Plaine Flamande in France, Belgique, et Pays-Bas.* Paris: Armand Colin.

Brunhes, J. 1910. *La Géographie humaine.* Paris: Armand Colin.

Buttimer, A. 1971. *Society and Milieu in the French Geographic Tradition.* Chicago: Rand McNally.

Capot-Rey, R. 1946. *Géographie de la circulation sur les continents.* Paris: Gallimard.

Claval, P. 1964. *Essai sur l'évolution de la géographie humaine.* Cahiers de géographie de Besançon, No. 12. Paris: Les Belles Lettres.

———. 1972. *La pensée géographique. Introduction à son histoire.* Paris: Sedes.

———. 1975. "Contemporary Human Geography in France." *Progress in Geography,* 7:253-292.

Comité National de Géographie. 1972. *Recherches géographiques en France.* Montréal and Paris.

Demangeon, A. 1905. *La Picardie et les régions voisines, Artois, Cambrésis, Beauvaises.* Paris: Armand Colin.

de Martonne, E. 1902. *La Valachie, essai de monographie géographique.* Paris: Armand Colin.

———. 1909. *Traité de géographie physique.* Revised and enlarged, 1913, 1920; 3 vols. 1925-27. Paris: Armand Colin.

———. 1917. "The Carpathians: Physiographic Features Controlling Human Geography." *Geographical Review,* 3:417-437.

———. 1927. "Regions of Interior-Basin Drainage." *Geographical Review,* 17:397-414.

Denaix, A. 1827. *Essais de géographie méthodique et comparative.* Paris

———. 1841. *Géographie prototype de la France. Paris: Piquet.*

Dickinson, R. E. 1969. *The Makers of Modern Geography.* London: Routledge & Kegan Paul.

Dunbar, G. S. 1978. *Elisée Reclus: Historian of Nature.* Hamden, Conn.: Shoe String Press.

Fischer, E., Campbell, R. D., and Miller, E. S. 1967. *A Question of Place, the Development of Geographic Thought.* Arlington, Va.: Beatty.

Gallais, J. 1967. *Le delta intérieur du Niger. Étude de géographie régionale.* Dakar: IFAN.

Gallois, L. 1908. *Régions naturelles et noms de pays: étude sur la région parisienne.* Paris: Armand Colin.

Gottmann, J. 1946. "French Geography in Wartime." *Geographical Review,* 36:80-91.

Harrison-Church, R. J. 1951. "The French School of Geography." In G. Taylor, ed., *Geography in the Twentieth Century.* Pp. 70-90. New York: Philosophical Library.

Joerg, W. L. G. 1922. "Recent Geographical Work in Europe." *Geographical Review,* 12:431-484.

Levainville, J. 1909. *Le Morvan.* Paris: Armand Colin.

L'Information géographique. 1957. *La Géographie française au millieu du XXe siècle.* Paris: Bailliere & Fils.

Martin, G. J. 1964. "The Region in French Geographic Thought, c. 1900-1930." In *Papers of the Michigan Academy of Science, Arts & Letters,* 49:325-332.

McDonald, J. R. 1964. "Current Controversy in French Geography." *Professional Geographer,* 16:20-23.

———. 1965. "Publication Trends in a Major French Geographical Journal." *Annals AAG.* 55:125-139.

———. 1975. "Current Trends in French Geography." *Professional Geographer.* 17:15-18.

McKay, D. V. 1943. "Colonialism in the French Geographical Movement, 1871-1881." *Geographical Review.* 33:214-232.

Meynier, A. 1952. "Cinquante ans de géographie française." In *Volume jubilaire du laboratoire de géographie de Rennes.* Pp. 47-52.

———. 1969. *Histoire de la pensée géographique en France.* Paris: Presses universitaires de France.

———. 1972. *La Pensée géographique francaise contemporaine.* Presses universitaires de Bretagne.

Monbeig, P. 1952. *Pioneers et planteurs de São Paulo.* Paris: Armand Colin.

Sautter, G. 1966. *De l'Atlantique au fleuve Congo: une géographie du souspeuplement.* 2 vols. Paris: Imprimerie Nationale.

Sion, J. 1908. *Les paysans de la Normandie orientale.* Paris: Armand Colin.

Sorre, M. 1913. *Les Pyrénées méditerranéennes.* Paris: Armand Colin.

———. 1948. *Les fondements de la géographie humaine.* Paris: Armand Colin.

Vallaux, C. 1906. *La basse Bretagne.* Paris: Armand Colin.

———. 1925. *Les sciences géographiques,* 2d ed. 1929. Paris: Armand Colin.

Vidal de la Blache, P. 1899. "Leçon d'ouverture du cours de géographie." *Annales de géographie*, 8:97–109.

_____. 1903. *Tableau de la géographie de la France*. Vol. 1, E. Lavisse, ed. *Histoire de France*. Paris: Hachette. Published separately as *La France: tableau géographique*. 1908. Paris: Hachette.

_____. 1913. "Des caractères distinctifs de la géographie." *Annales de géographie*, 22:289–299.

_____. 1917. *La France de l'Est: Lorraine-Alsace*. Paris: Armand Colin.

_____. 1922. *Principes de géographie humaine*. Ed. E. de Martonne. Paris: Armand Colin. Trans. M. T. Bingham, *Principles of Human Geography*. 1926. New York: Henry Holt.

Vidal de la Blache, P., and Gallois, L., eds. 1927–48. *Géographie universelle. 15 vols.* Paris: Armand Colin.

CHAPTER 10

Baker, A. R. H. 1972. "Historical Geography in Britain." In A. R. H. Baker, ed., *Progress in Historical Geography*. Pp. 90–110. Devon: David & Charles.

Baker, J. N. L. 1963. *The History of Geography*. New York: Barnes & Noble.

Beresford, M. W. 1954. *The Lost Villages of England*. New York: Philosophical Library.

Chisholm, G. G. 1889. *Handbook of Commercial Geography*, 18th ed. 1966. L. D. Stamp and S. C. Gilmour, eds. London: Longmans, Green.

Chisholm, M. 1962: *Rural Settlement and Land Use*. London: Hutchinson University Library.

_____. 1975: *Human Geography: Evolution or Revolution?* Harmondsworth: Penguin Books.

Chorley, R. J., ed. 1973a. *Directions in Geography*. London: Methuen.

_____. 1973b. "Geography as Human Ecology." In R. J. Chorley ed., *Directions in Geography*. Pp. 155–170. London: Methuen.

Chorley, R. J., and Haggett, P., eds. 1965. *Frontiers in Geographical Teaching*. London: Methuen.

Chorley, R. J., Beckinsale, R. P., and Dunn, A. J. 1973. *The History of The Study of Landforms, or the Development of Geomorphology. Volume 2 The life and work of William Morris Davis*. London: Methuen.

Chorley, R. J., Dunn, A. J., and Beckinsale, R. P. 1964. *The History of the Study of Landforms, or the Development of Geomorphology. Vol. 1, Geomorphology Before Davis*. London: Methuen.

Clark, A. H. 1954. "Historical Geography." In P. E. James and C. F. Jones, eds., *American Geography: Inventory and Prospect*. Pp. 70–105. Syracuse, N.Y.: Syracuse University Press.

Clayton, K. M., ed. 1964. *A Bibliography of British Geomorphology*. London: George Philip & Son.

Cole, M. M. 1960. *South Africa*. London: Methuen.

Coleman, A. 1961. "The Second Land Use Survey: Progress and Prospects." *Geographical Journal*, 127:168–186.

Crone, G. R. 1964. "British Geography in the Twentieth Century." *Geographical Journal*, 130:197–220.

Darby, H. C., ed. 1936. *A Historical Geography of England Before A.D. 1800.* Cambridge: At the University Press.

———. 1940b. *The Medieval Fenland.* Cambridge: At the University Press.

———. 1940a. *The Draining of the Fens.* Cambridge: At the University Press.

———. 1951. "The Changing English Landscape." *Geographical Journal,* 117:377–398.

———. 1952. *The Domesday Geography of Eastern England.* Cambridge: At the University Press.

———. 1953: "On the Relations of Geography and History." *Transactions and Papers, Institute of British Geographers,* 19:1–11.

———. ed. 1973. *A New Historical Geography of England.* London: Cambridge University Press.

———. 1977. *Domesday England.* London: Cambridge University Press.

Davies, W. K. D. 1972. "Geography and the Methods of Modern Science." In W. K. D. Davies, ed., *The Conceptual Revolution in Geography.* Pp. 131–139. London: University of London Press.

Dickinson, R. E. 1969. *The Makers of Modern Geography.* London: Routledge & Kegan Paul.

———. 1976. *Regional Concept: The Anglo-American Leaders.* London: Routledge & Kegan Paul.

Dickinson, R. E., and Howarth, O. J. R. 1933. *The Making of Geography.* Oxford: The Clarendon Press.

Dryer, C. R. 1920. "Mackinder's 'World Island' and Its American 'Satellite'." *Geographical Review,* 9:205–207.

East, W. G. 1935. *An Historical Geography of Europe.* London: Methuen.

———. 1951. "Historical Geography." In S. W. Wooldridge and W. G. East, eds., *The Spirit and Purpose of Geography.* Pp. 80–102. London: Hutchinson University Library.

Fawcett, C. B. 1919. *The Provinces of England,* rev. ed. 1960. London: Hutchinson University Library.

———. 1932. "Distribution of Population in Great Britain." *Geographical Journal,* 79:100–116.

Fischer, E., Campbell, R. D., and Miller, E. S. 1967. *A Question of Place, the Development of Geographic Thought.* Arlington, Va.: Beatty.

Fleure, H. J. 1917. "Régions humaines." *Annales de géographie,* 26:161–174.

———. 1919. "Human Regions." *Scottish Geographical Magazine,* 35:94–105.

Fox, C. 1932. *The Personality of Britain: Its Influence on Inhabitant and Invader in Prehistoric and Early Historic Times.* Cardiff, Wales: National Museum.

Freeman, T. W. 1950. *Ireland: A General and Regional Geography.* London: Methuen.

———. 1961. *A Hundred Years of Geography.* Chicago: Aldine.

———. 1967. *The Geographer's Craft.* New York: Barnes & Noble.

———. 1974. *The British School of Geography.* Unpublished manuscript. 12 pages.

———. 1977. "Hugh Robert Mill 1861–1950." In *Geographers: Biobibliographical Studies.* Vol. 1, pp. 73–78. London: Mansell.

———. 1980. *A History of Modern British Geography.* New York: Longmans.

———. Forthcoming. "On RGS History." In *Geography, Yesterday and Tomorrow,* E. H. Brown, ed. London: Oxford University Press.

Galton, F. 1855. "Notes on Modern Geography." In *Cambridge Essays*. Pp. 79–109. London: Parker.

Geikie, A. 1865. *The Scenery of Scotland Viewed in Connection with Its Physical Geology*, 2nd ed. 1887. London: Macmillan.

Gilbert, E. W. 1933. *The Exploration of Western America, 1800–1850, An Historical Geography*. Cambridge: At the University Press.

———. 1960. "The Idea of the Region." *Geography*, 45-157–175.

———. 1972. *British Pioneers in Geography*. Newton Abbot: David and Charles.

Grigg, D. B. 1977. "Ernst Georg Ravenstein: 1834-1913." In *Geographers: Biobibliographical Studies*. Vol. 1, pp. 79-82. London, Mansell.

Haggett, P. 1966. *Locational Analysis in Human Geography*. New York: St. Martin's Press.

Haggett, P., and Chorley, R. J. 1967. Models, Paradigms, and the New Geography." In *Models in Geography*, Pp. 19-42. London: Methuen.

Harrison-Church, R. J., et al. 1964. *Africa and the Islands*. New York: John Wiley & Sons.

Herbertson, A. J. 1905. "The Major Natural Regions: An Essay in Systematic Geography." *Geographical Journal*, 25:300-312.

Hogarth, D. G. 1902. *The Nearer East*. New York: D. Appleton.

Hoskins, W. G. 1955. *The Making of the English Landscape*. London: Hodder & Stoughton.

Jay, L. J. 1979. "Andrew John Herbertson: 1865-1915." In *Geographers: Biobibliographical Studies*. Vol. 3, pp. 85-92. London: Mansell.

Joerg, W. L. G. 1922. "Recent Geographical Work in Europe." *Geographical Review*, 12:431-484.

Johnston, R. J. 1976. "Anarchy, Conspiracy and Apathy: The Three 'Conditions' of Geography." *Area*, 8:1-3.

———. 1978. "Paradigms and Revolutions or Evolution: Observations in Human Geography Since the Second World War." *Progress in Human Geography*, 2:189-206.

———. 1979. *Geography and Geographers: Anglo-American Human Geography Since 1945*. New York: John Wiley & Sons.

Keltie, J. S. 1921. *The Position of Geography in British Universities*. New York: American Geographical Society.

Kimble, G. H. T. 1951. "The Inadequacy of the Regional Concept." In L. D. Stamp and S. W. Wooldridge, eds., *London Essays in Geography*. Pp. 151-174. Cambridge: Harvard University Press.

Kirwan, L. P. 1964. "The R.G.S. and British Exploration, a Review of Recent Trends." *Geographical Journal*, 130:221-225.

Longrigg, S. H. 1963. *The Middle East, a Social Geography*. Chicago: Aldine.

Lyde, L. W. 1913. *The Continent of Europe*. Londong: Macmillan.

Mackinder, H. J. 1887. "On the Scope and Methods of Geography." *Proceedings of the Royal Geographical Society*, 9:141-174.

———. 1902. *Britain and the British Seas*. New York: D. Appleton.

———. 1904. "The Geographical Pivot of History." *Geographical Journal*, 23:421-437.

———. 1919. *Democratic Ideals and Reality*. New York: Henry Holt (republished 1942).

Middleton, D. 1977. "George Adam Smith: 1856-1942." In *Geographers Biobibliographical Studies*. Vol. 1, pp. 105-106. London: Mansell.

Mill, H. R. 1891. *The Realm of Nature*. London: John Murray.

_____. 1896. "A Proposed Geographical Description of the British Isles." *Geographical Journal*, 7:345-365.

_____. 1900. "A Fragment of the Geography of England—Southwest Sussex." *Geographical Journal*, 15:205-227, 353-378.

_____. 1951. *An Autobiography*. London: Longmans, Green.

Minshull, R. 1967. *Regional Geography, Theory and Practice*. London: Hutchinson University Library.

Monkhouse, F. J. 1959. *A Regional Geography of Western Europe*. London: Longmans, Green.

Newbigin, M. I. 1926. *Canada: The Great River, the Lands, and the Men*. New York: Harcourt, Brace.

Ogilvie, A. G., ed. 1928. *Great Britain: Essays in Regional Geography*. Cambridge: At the University Press.

Oughton, M. 1978. "Mary Somerville: 1780-1872." In *Geographers: Biobibliographical Studies*. Vol. 2, pp. 109-111. London: Mansell.

Priestley, R., Adie, R. J., and Robin, G. deQ. 1964. *Antarctic Research, a Review of British Scientific Achievement in Antarctica*. London: Butterworth.

Prothero, R. M., ed. 1969. *A Geography of Africa: Regional Essays on Fundamental Characteristics, Issues and Problems*. New York: Praeger.

Robson, B. T., and Cooke, R. U. 1976. "Geography in the United Kingdom, 1972-1976." *Geographical Journal*, 142:3-72.

Roxby, P. M. 1916. "Wu Han, the Heart of China." *Scottish Geographical Magazine*, 32:266-278.

_____. 1925. "The Distribution of Population in China." *Geographical Review*, 15:1-24.

_____. 1926. "The Theory of Natural Regions." *Geographical Teacher*, 13:376-382.

_____. 1930. "The Scope and Aims of Human Geography." *Scottish Geographical Magazine*, 46:276-299.

_____. 1938. "The Terrain of Early Chinese Civilization." *Geography*, 23:225-236.

Stamp, L. D. 1947. *The Land of Britain, Its Use and Misuse*, 2nd ed. 1950. London: Longmans, Green.

Stamp, L. D., and Wooldridge, S. W., eds. 1951. *London Essays in Geography, Rodwell Jones Memorial Volume*. Cambridge: Harvard University Press.

Stoddart, D. R. 1965. "Geography and the Ecological Approach: The Eco-system as a Geographic Principle and Method." *Geography*, 50:242-51.

Sykes, P. 1934. *A History of Exploration from the Earliest Times to the Present Day*, 2nd ed. 1935; 3rd ed. 1950. London: Routledge & Kegan Paul (reprinted, New York: Harper Bros., 1961).

Taylor, E. G. R. 1930. *Tudor Geography, 1485-1593*. London: Methuen.

_____. 1934. *Late Tudor and Early Stuart Geography*. London: Methuen.

Teggart, F. J. 1919. "Geography as an Aid to Statecraft, an Appreciation of Mackinder's 'Democratic Ideals and Reality'." *Geographical Review*, 8:227-242.

Thompson, J. M. 1929. *Historical Geography of Europe, 800-1789*. London: The Clarendon Press.

Unstead, J. F., 1916. "A Synthetic Method of Determining Geographical Regions." *Geographical Journal*. 48:230-249.

_____, et al. 1937. "Classifications of Regions of the World." *Geography*, 22:253-282.

Watson, J. W., and Sissons, J. B., eds. 1964. *The British Isles, a Systematic Geography.* London: Thomas Nelson.

Wise, M. J. 1977. "On Progress and Geography." *Progress in Human Geography,* 1:1–11.

Wooldridge, S. W. 1956. *The Geographer as Scientist, Essays on the Scope and Nature of Geography.* London: Thomas Nelson.

Wooldridge, S. W., and East, W. G. 1951. *The Spirit and Purpose of Geography.* London: Hutchinson University Library.

Wrigley, E. A. 1965. "Changes in the Philosophy of Geography." In R. J. Chorley and P. Haggett, eds., *Frontiers in Geographical Teaching.* Pp. 3–20. London: Methuen.

Wrigley, G. M. 1950. "Hugh Robert Mill: An Appreciation." *Geographical Review,* 40:657–660.

CHAPTER 11

Bagrow, L., and Skelton, R. A. 1964. *History of Cartography.* Cambridge: Harvard University Press (see reference in Chapter 3).

Chappell, J. E., Jr. 1965. "Marxism and Geography." *Problems of Communism,* 14:12–22.

Esakov, V. A. 1978. "Dmitry Nikolaevich Anuchin, 1843–1923." In *Geographers: Biobibliographical Studies.* Vol. 2 pp. 1–5. London: Mansell.

Fedosseyev, I. A. 1978. "Alexander Ivanovitch Voyeikov, 1842–1916." In *Geographers: Biobibliographical Studies.* Vol. 2 pp. 135–141. London: Mansell.

French, R. A. 1968. "Historical Geography in the USSR." *Soviet Geography: Review and Translation,* 9:551–553.

———. 1969. "Lifting the Iron Curtain." In R. U. Cooke and J. H. Johnson, eds., *Trends in Geography: An Introductory Survey.* Pp. 268–274. Oxford: Pergamon Press.

Fuchs, R. J. 1964. "Soviet Urban Geography—An Appraisal of Postwar Research." *Annals AAG,* 54:276–289.

Fuchs, R. J., and Demko, G. J., eds. 1977. *Theoretical Problems of Geography* by V. A. Anuchin (introduction by D. J. M. Hooson) Columbus: Ohio State University Press.

Gerasimov, I. P. 1966. "The Past and Future of Geography." *Soviet Geography,* 7; September, 3–14.

———. 1968a. "Constructive Geography: Aims, Methods and Results." *Soviet Geography,* 9:739–755.

———. 1968b. "Fifty Years of Development of Soviet Geographic Thought." *Soviet Geography,* 9:238–252.

Gerasimov, I. P., et al., eds. 1962. *Soviet Geography, Accomplishments and Tasks.* (A symposium of fifty chapters by fifty-six leading Soviet geographers, edited by a committee of the Geographical Society, I. P. Gerasimov, chairman. English translation by L. Ecker, edited by C. D. Harris.) New York: American Geographical Society.

Glinka, K. D. 1914. *Die Typen der Bodenbildung, ihre Klassifikation und geographische Verbreitung.* Berlin: Trans. C. F. Marbut, 1927. *The Great Soil Groups of the World and Their Development.* Ann Arbor, Mich.: Edwards Bros.

Harris, C. D. 1970. *Cities of the Soviet Union, Studies in Their Functions, Size, Density, and Growth.* Chicago: Rand McNally.

Hooson, D. J. M. 1959. "Some Recent Developments in the Content and Theory of Soviet Geography." *Annals AAG, 49:73–82.*

_____. 1962. "Methodological Clashes in Moscow." Annals AAG, 52:469-475.

_____. 1968. "The Development of Geography in Pre-Soviet Russia." Annals AAG, 58:250-272.

_____. In press. "A Review of V. A. Anuchin, Osnovy Prirodopolzovaniya: Teoreticheskii aspekt." Soviet Geography.

Ilyina, T. D. 1977. "Vladimir Leontyevitch Komarov: 1869-1945." In Geographers: Biobibliographical Studies. Vol. 1, pp. 55-58. London: Mansell.

Isachenko, A. G. 1968. "Fifty Years of Soviet Landscape Science." Soviet Geography, 9:402-407.

Kalesnik, S. V. 1958. "La géographie physique comme science et les lois géographiques générales de la terre." Annales de géographie, 67:385-403.

_____. 1968. "The Development of General Earth Science in the U.S.S.R. During the Soviet Period." Soviet Geography, 9:393-402.

Konstantinov, O. A. 1968. "Economic Geography in the U.S.S.R. on the 50th Anniversary of Soviet Power." Soviet Geography, 9:417-424.

Markov, K. K. 1968. "Methodological Principles of the Curriculum of a Geography Faculty." Soviet Geography, 9:358-367.

Matley, I. M. 1966. "The Marxist Approach to the Geographical Environment." Annals AAG, 56:97-111.

Nikitin, N. P. 1966. "A History of Economic Geography in Prerevolutionary Russia." Soviet Geography, 7:3-37.

Pokshishevskiy, V. V. 1966. "Relationships and Contacts Between Prerevolutionary Russian and Soviet Geography and Foreign Geography." Soviet Geography, 7:56-76.

Ryabchikov, A. M. 1968. "Geography at Moscow University over the Last 50 Years (1917-1967)." Soviet Geography, 9:343-357.

Saushkin, Y. G. 1962. "Economic Geography in the U.S.S.R." Economic Geography, 38:28-37.

_____. 1966. "A History of Soviet Economic Geography." Soviet Geography, 7:3-104.

Semenov Tyan-Shanski, P. P. 1900. La Russie Extra-Européene et Polaire. Paris: P. Dupont.

Semenov Tyan-Shanski, V. P. 1928. "Russia: Territory and Population, a Perspective on the 1926 Census." Geographical Review, 18:616-640.

Szava-Kovats, E. 1966. "The Present State of Landscape Theory and Its Main Philosophical Problems." Soviet Geography, 7:28-40.

Voeikov, A. I. 1901. "De l'influence de l'homme sur la terre." Annales de géographie, 10:97-114, 193-215.

Volskiy, V. V. 1963. "On Some Problems of Theory and Practice in Economic Geography." Soviet Geography, 4:14-25.

Zvonkova, T. V., and Saushkin, Y. G. 1968. "Problems of Long-Term Geographic Prediction." Soviet Geography, 9:755-765.

CHAPTER 12

Almagià, R. 1929. "The Repopulation of the Roman Campagna." Geographical Review, 19:529-555.

_____. 1959. L'Italia. Turin: Editrice Torinese.

Anrick, C. J. 1923. "A Popular Geographic Club of Sweden, The Swedish Touring Club and Its Activities." *Geographical Review*, 13:608-612.

Awad, M. 1954. "The Assimilation of Nomads in Egypt." *Geographical Review*, 44:240-252.

Birukawa, S. 1950. "Agricultural Regions of Japan Based on a New System." *Memoirs of the Otsuka Geographical Society*, 6:237-244 (in Japanese).

Chatterjee, S. P. 1964/1968. *Fifty Years of Science in India: Progress of Geography.* Calcutta: Indian Science Congress Association.

————. 1968. *Progress of Geography in India.* Calcutta: 21st International Geographical Congress.

Chojnicki, Z. 1970. "Prediction in Economic Geography." *Economic Geography*, 46:213-222.

Chu, C. C. 1926. "Climatic Pulsations During Historic Time in China." *Geographical Review*, 16:274-282.

Clark, A. H. 1949. *The Invasion of New Zealand by People, Plants and Animals: The South Island.* New Brunswick, N.J.: Rutgers University Press.

Cressey, G. B. 1934. *China's Geographic Foundations.* New York: McGraw-Hill.

————. 1955. *Land of the 500 Million, A Geography of China.* New York: McGraw-Hill.

Cvijić, J. 1918a. "Hydrographie souterraine et évolution morphologique du Karst." *Recueil des Travaux de l'Institute de Géographie Alpine*, 6:375-426.

————. 1918b. *La Péninsule Balkanique.* Paris: Armand Colin.

Dainelli, G. 1929. "The Italian Colonies." *Geographical Review*, 19:404-419.

Dalla Vedova, G. 1881. "Il concetto populare e il concetto scientifico della geografía." *Bulletino della Società Geografica Italiana*, 18:5-27.

Deffontaines, P. 1938. "The Origin and Growth of the Brazilian Network of Towns." *Geographical Review*, 28:379-399.

De Geer, S. 1908. "Befolkningens fördelning på Gottland." *Ymer*, 28:240-253.

————. 1919. *Karta över befolkningens fördelning i Sverige.* Stockholm, Wahlström & Widstrand.

————. 1922a. "A Map of the Distribution of Population in Sweden: Method of Preparation and General Results." *Geographical Review*, 12:72-83.

————. 1922b. "Storstaden Stockholm ur geografisk synpunkt." *Svenska Turistföreningen Arsskrift*, 155-168.

————. 1923a. "On the Definition, Method, and Classification of Geography." *Geografiska Annaler*, 5:1-37.

————. 1923b. "Greater Stockholm, A Geographical Interpretation." *Geographical Review*, 13:497-506.

————. 1927. "The American Manufacturing Belt." *Geografiska Annaler*, 9:233-259.

————. 1928a. "Das geologische Fennoskandia und das geographische Baltoskandia." *Geografiska Annaler*, 10:119-139.

————. 1928b. "The Subtropical Belt of Old Empires." *Geografiska Annaler*, 10:205-244.

De Jong, G. 1962. *Chorological Differentiation as the Fundamental Principle of Geography, an Inquiry into the Chorological Conception of Geography.* Groningen, Netherlands: J. B. Wolters.

Denis, P. 1909. *Le Brésil au XXe Siècle.* Paris: Armand Colin.

————. 1920. *La république Argentine, La Mise en valeur du pays.* Paris: Armand Colin.

Fischer, E., Campbell, R. D., and Miller, E. S. 1967. *A Question of Place, the Development of Geographic Thought.* Arlington, Va.: Beatty.

Fraser, J. K. 1967. "Requiem or Rennaissance?" *Geographical Bulletin,* 9:i–iii.

Freeman, T. W. 1967. *The Geographer's Craft.* New York: Barnes & Noble.

Fukui, E. 1933. "Climatic Divisions of Japan." *Geographical Review of Japan,* 9:1–19, 109–127, 195–219, 271–300 (in Japanese).

Granö, J. G. 1929. "Reine Geographie: eine methodologische Studie beleuchtet mit Beispielen aus Finnland und Estland." *Acta Geographica* (Helsinki), 2(2).

Hägerstrand, T. 1953. *Innovationsförloppet ur korologisk synpunkt.* Lund: C. W. K. Gleerup. Trans. A. Pred, 1967, *Innovation Diffusion as a Spatial Process.* Chicago: University of Chicago Press.

Hall, R. B. 1937. "Tokaido: Road and Region." *Geographical Review,* 27:353–377.

Hall, R. B., and Noh, T. 1970. *Japanese Geography: A Guide to Japanese Reference and Research Materials.* Ann Arbor: University of Michigan Press.

Hamelin, L.-E. 1962. "Petite histoire de la géographie dans le Québec et à l'université Laval." *Cahiers de géographie de Québec,* 13:137–152.

Hsieh, Chiao-min. 1959. "The Status of Geography in Communist China." *Geographical Review,* 49:535–551.

Huzayyin, S. 1956. "Changes in Climate, Vegetation, and Human Adjustment in the Saharo-Arabian Belt with Special Reference to Africa." In W. L. Thomas, ed., *Man's Role in Changing the Face of the Earth.* Pp. 304–323. Chicago: University of Chicago Press.

Joerg, W. L. G. 1922. "Recent Geographical Work in Europe." *Geographical Review,* 12:431–484.

Kant, E. 1953. "Migrationernas klassifikation och problematik," In *Svensk Geografisk Arsbok.* Pp. 180–209. (Reprinted and translated as "Classification and Problems of Migrations." In P. L. Wagner and M. W. Mikesell, eds., 1962, *Readings in Cultural Geography.* Pp. 341–354. Chicago: University of Chicago Press.)

Kazakova, O. N. 1966. "The Development of Theory of Geography and Landscape Science in East and West Germany." *Soviet Geography,* 7:40–47.

Kikolski, B. 1964. "Contemporary Research in Physical Geography in the Chinese People's Republic." *Annals AAG,* 54:181–189.

Leszczycki, S. 1963. "The Development of Geography in the People's Republic of China." *Geography,* 48:139–154.

Lewthwaite, G. R. 1966. "Environmentalism and Determinism: A Search for a Clarification." *Annals AAG,* 56:1–23.

Mabogunje, A. L. 1968. *Urbanization in Nigeria.* London: University of London Press.

Marinelli, O. 1919. "The Regions of Mixed Populations in Northern Italy." *Geographical Review,* 7:129–148.

———. 1922. *Atlante dei tipi geografici desunti dai rilievi al 25,000 e al 50,000 dell' Instituto Geografico Militare.* Florence: Instituto Geografico Militare.

Marinelli, O., and Dainelli, G. 1912. *Risultati scientifici di un viaggio nella Colonia Eritrea.* Florence: Instituto Geografico Militare.

Michotte, P. 1921. "L'Orientation nouvelle en géographie." *Bulletin de la Sociétè Royale Belge de Géographie,* 45:5–43.

Milone, F. 1955. *L'Italia nell' economia delle sue regioni.* Turin: Edizioni Scientifichi Einaudi.

Neef, E., ed. 1956. *Das Gesicht der Erde,* 2nd ed. 1962. Leipzig: Brockhaus.

―――. 1967. *Die theoretischen Grundlagen der Landschaftslehre.* Gotha: Hermann Haack.

Nordenskjöld, O. 1920. Geografisk Forskning og geografiske Opdagelser i det nittende Aarhundrede. Copenhagen: Nordsik Forlag.

Ogasawara, Y. 1950. *Land Use of Japan.* Tokyo: Bulletin of the Geographical Survey Institute.

Robinson, J. L. 1967. "Growth and Trends in Geography in Canadian Universities." *Canadian Geographer,* 11:216-229.

Sanders, E. M. 1921. "The Cycle of Erosion in a Karst Region (after Cvijić)." *Geographical Review,* 11:593-604.

Sandru, I., and Cucu, V. 1966. "The Development of Geographical Studies in Rumania." *Geographical Journal,* 132:43-48.

Schultze, J. H. 1955. *Die Naturbedingten Landschaften der Deutschen Demokratischen Republik.* (Ergänzungsheft). *Petermanns Geographische Mitteilungen,* 257.

Seki, T. 1930. *The Outline Soil Map of Japan, 1/5,000,000.* Tokyo: Tokyo Agricultural Experiment Station.

Shimomura, H. 1926-27. "Physiographic Provinces of Japan." *Geographical Review of Japan,* 2:1027-1039; 3:327-335, 863-873 (in Japanese).

Spate, O. H. K. 1958. "The End of an Old Song? The Determinism-Possibilism Problem." *Geographical Review,* 48:280-282.

Sporck, J. A., ed. 1967. *Mélanges de géographie physique, humaine, économique, appliquée offerts à M. Omer Tulippe.* Gembloux, Belgium: J. Duculot.

Tanaka, Keiji. 1927. "Geographical Units of Japan." *Geographical Review of Japan,* 3:1-2 (in Japanese).

Tatham, G. 1951. "Environmentalism and Possibilism." In G. Taylor, ed., *Geography in the Twentieth Century.* Pp. 128-162. New York: Philosophical Library.

Taylor, G. 1926. "The Frontiers of Settlement in Australia." *Geographical Review,* 16:1-25.

―――. 1937. *Environment, Race and Migration: Fundamentals of Human Distribution, with Special Sections on Racial Classification and Settlement in Canada and Australia.* Toronto: University of Toronto Press.

―――. 1941. *Australia, a Study of Warm Environments and Their Effect on British Settlement.* London: Methuen.

―――. 1942. "Environment, Village and City, a Genetic Approach to Urban Geography, with Some References to Possibilism." *Annals AAG,* 32:1-67.

―――. 1946. "Future Population in Canada—A Study in Technique." *Economic Geography,* 22:67-74.

―――. 1947. *Canada, A Study of Cool Continental Environments and Their Effects on British and French Settlement.* London: Methuen.

―――., ed. 1951. *Geography in the Twentieth Century.* New York: Philosophical Library.

―――. 1958. *Journeyman Taylor: The Education of a Scientist.* London: Robert Hale.

Waibel, L. 1948. "Vegetation and Land Use in the Planalto Central of Brazil." *Geographical Review*, 38:529-554.

———. 1950. "European Colonization in Southern Brazil." *Geographical Review*, 40:529-547.

Watanabe, A. 1970. "Regional Divisions of Japan." *Ochanomizu University Studies in Arts and Culture*, 23:87-129.

Wiens, H. J. 1961. "Development of Geographical Science, 1949-1960." In S. H. Gould, ed., *Sciences in Communist China*. Washington, D.C.: American Association for the Advancement of Science, Publication No. 68.

William-Olsson, W. 1940. "Stockholm: Its Structure and Development." *Geographical Review*, 30:420-438.

Wissler, C. 1920. "Arctic Geography and Eskimo Culture: A Review of Steensby's Work." *Geographical Review*, 9:125-138.

Yoshikawa, T. 1953. "Recent Trends of Geographical Research in Japan." *Geographical Review of Japan*, 26:620-674 (in Japanese).

CHAPTER 13

Aay, H. Forthcoming. "Textbook Chronicles: Disciplinary History and The Growth of Geographic Knowledge." In B. Blouet, Ed., *The Evolution of Academic Geography in the United States*. Hamden, Conn.: Shoe String Press.

Barrows, H. H. 1962. *Lectures on the Historical Geography of the United States, as Given in 1933*. Ed. W. A. Koelsch. Chicago: University of Chicago, Department of Geography.

Baulig, H. 1950. "William Morris Davis: Master of Method." *Annals AAG*, 40:188-195.

Beckinsale, R. P. Forthcoming. "W. M. Davis and American Geography: 1880-1930." In B. Blouet, Ed., *The Evolution of Academic Geography in the United States*. Hamden, Conn.: Shoe String Press.

Bowman, I. 1911. *Forest Physiography: Physiography of the United States and Principles of Soils in Relation to Forestry*. New York: John Wiley & Sons.

———. 1916. *The Andes of Southern Peru: Geographical Reconnaissance Along the Seventy-third Meridian*. New York: Henry Holt.

———. 1924. *Desert Trails of Atacama*. New York: American Geographical Society.

Brigham, A. P. 1903. *Geographic Influences in American History*. Boston: Ginn & Co.

———. 1915. "Problems of Geographic Influence." *Annals AAG*, 5:3-25.

———. 1924. "The Association of American Geographers." *Annals AAG*, 14:109-116.

Bryan, K. 1935. "William Morris Davis—Leader in Geomorphology and Geography." *Annals AAG*, 25:23-31.

Butzer, K. W. 1964. *Environment and Archeology*. Chicago: Aldine.

Carter, G. F. 1950. "Isaiah Bowman, 1878-1950." *Annals AAG*, 40:335-350.

Chamberlin, R. T. 1931. "Memorial to Rollin D Salisbury." *Bulletin of the American Geographical Society*, 42:126-138.

Chamberlin, T. C. 1897. "The Method of the Multiple Working Hypotheses." *Journal of Geology*, 5:837-848.

Chappell, J. E. Jr. 1970. "Climatic Change Reconsidered: Another Look at 'The Pulse of Asia.'" *Geographical Review*, 60:347-373.

Chorley, R. J. 1965. "A Re-evaluation of the Geomorphic System of W. M. Davis." In R. J. Chorley and P. Haggett, eds., *Frontiers in Geographic Teaching*. Pp. 21–38. London: Methuen.

Chorley, R. J., Dunn, A. J., and Beckinsale, R. P. 1964. *The History of the Study of Landforms, or the Development of Geomorphology*. Vol. 1, *Geomorphology Before Davis*. New York: John Wiley & Sons.

Colby, C. C. 1933. "Ellen Churchill Semple." *Annals AAG*, 23:229–240.

————. 1955. "Narrative of Five Decades." In *A Half Century of Geography—What Next?* (Papers presented at the alumni reunion, June 5, 1954.) Chicago: University of Chicago, Department of Geography.

Davis, W. M. 1899a. "The Geographical Cycle." *Geographical Journal*, 14:481–504.

————. 1899b. "The United States of America." In H. R. Mill, ed., *The International Geography*. New York: D. Appleton.

————. 1905. "The Opportunity for the Association of American Geographers." *Bulletin of the American Geographical Society*, 37:84–86.

————. 1906. "An Inductive Study of the Content of Geography." *Bulletin of the American Geographical Society*, 38:67–84 (reprinted in Davis, 1909).

————. 1909. *Geographical Essays*. Ed. D. W. Johnson. Boston: Ginn & Co.

————. 1910. "Experiments in Geographical Description." *Bulletin of the American Geographical Society*, 42:401–435.

————. 1911. "The Colorado Front Range, A Study in Physiographic Presentation." *Annals AAG*, 1:21–84.

————. 1912. *Die erklärende Beschreibung der Landformen*. Trans. Rühl. Leipzig: Teubner.

————. 1915. "The Principles of Geographic Description." *Annals AAG*, 5:61–105.

————. 1919. "Passarge's Principles of Landscape Description." *Geographical Review*, 8:266–273.

————. 1922. "Peneplains and the Geographical Cycle." *Bulletin of the Geological Society of America*, 33:587–598.

————. 1924. "The Progress of Geography in the United States." *Annals AAG*, 14:159–215.

————. 1928. *The Coral Reef Problem*. New York: American Geographical Society.

————. 1930a. "The Origin of Limestone Caverns." *Bulletin of the Geological Society of America*, 41:475–628.

————. 1930b. "Rock Floors in Arid and Humid Climates." *Journal of Geology*, 38:1–27, 136–158.

————. 1932. "A Retrospect of Geography." *Annals AAG*, 22:211–230.

Davis, W. M., and Daly, R. A. 1930. "Geology and Geography, 1858–1928." In S. E. Morison, ed., *The Development of Harvard University, 1869–1929*. Pp. 307–328. Cambridge: Harvard University Press.

de Martonne, E. 1909. *Traité de géographie physique*. Paris: Armand Colin.

Dryer, C. R. 1915. "Natural Economic Regions." *Annals AAG*, 5:121–125.

————. 1924. "A Century of Geographic Education in the United States." *Annals AAG*, 14:117–149.

Dunbar, G. S. 1978. "George Davidson: 1825–1911." In *Geographers: Biobibliographical Studies*. Vol. 2, Pp. 33–37. London: Mansell.

Emerson, F. V. 1908-1909. "A Geographic Interpretation of New York City." *Bulletin of the American Geographical Society*, 40:587-612, 726-738; 41:3-20.

Fenneman, N. M. 1914. "Physiographic Boundaries Within the United States." *Annals AAG*, 4:84-134.

———. 1916. "Physiographic Divisions of the United States." *Annals AAG*, 6:19-98 (with folded map on a scale of 1/7,000,000).

———. 1919. "The Circumference of Geography." *Geographical Review*, 7:168-175.

Friis, H. R. Forthcoming. "The Role of Geographers and Geography in the Federal Government: A Brief History, 1774-1905." In B. Blouet, Ed., *The Evolution of Academic Geography in the United States*. Hamden, Conn.: Shoe String press.

Gilbert, G. K. 1878. *Report on the Geology of the Henry Mountains*. Washington, D.C.: Department of the Interior.

———. 1886. "The Inculcation of the Scientific Method by Example." *American Journal of Science, 3rd ser., 31-284-299.*

Hann, J. 1903. *Handbook of Climatology*, Part I. Trans. R. DeC. Ward. New York, Macmillan.

Hartshorne, R. 1939. *The Nature of Geography, A Critical Survey of Current Thought in the Light of the Past*. Lancaster, Pa.: Association of American Geographers.

Huntington, E. 1907. *The Pulse of Asia*. Boston: Houghton Mifflin.

———. 1915. *Civilization and Climate*. New Haven, Conn.: Yale University Press.

———. 1924. *The Character of Races as Influenced by Physical Environment, Natural Selection, and Historical Development*. New York: Charles Scribner.

———. 1945. *Mainsprings of Civilization*. New York: John Wiley & Sons.

Huntington, E., and Cushing, S. W. 1920. *Principles of Human Geography*. New York: John Wiley & Sons.

James, P. E. 1978. "Albert Perry Brigham: 1855-1932." In *Geographers: Biobibliographical Studies*. Vol. 2, pp. 13-19. London: Mansell.

James, P. E., and Martin, G. J. 1979. "On AAG History." *The Professional Geographer*, 31:353-357.

Jefferson, M. 1909. "The Anthropography of Some Great Cities; A Study in Distribution of Population." *Bulletin of the American Geographical Society*, 41:537-566.

———. 1915. "How American Cities Grow." *Bulletin of the American Geographical Society*, 47:19-37.

———. 1928. "The Civilizing Rails." *Economic Geography*, 4:217-231.

———. 1939. "The Law of the Primate City." *Geographical Review*, 29:226-232.

Joerg, W. L. G. 1914. "The Subdivision of North America into Natural Regions: A Preliminary Inquiry." *Annals AAG*, 4:55-83.

Jones, W. D., and Sauer, C. O. 1915. "Outline for Field Work in Geography." *Bulletin of the American Geographical Society*, 47:520-525.

Koelsch, W. A. 1976. *Lectures on the Historical Geography of the United States as Given in 1933* [by Harlan H. Barrows]. New York: Oxford University Press.

———. 1979a. "Nathaniel Southgate Shaler: 1841-1906." In *Geographers: Biobibliographical Studies*. Vol. 3, Pp. 133-139. London: Mansell.

———. 1979b. "Wallace Walter Atwood: 1872-1949." In *Geographers: Biobibliographical Studies*. Vol. 3, Pp. 13-18. London: Mansell.

———. Forthcoming. "The New England Meteorological Society: 1884-1896. A Study in

Professionalization.'' In B. Blouet, ed., *The Evolution of Academic Geography in the United States*. Hamden, Conn.: Shoe String Press.

Krug-Genthe, M. 1903. "Die Geographie in die Vereinigten Staaten." *Geographische Zeitschrift*, 9:626-637, 666—685.

Lee, D. H. K. 1954. "Physiological Climatology." In P. E. James and C. F. Jones, eds., *American Geography, Inventory and Prospect*. Pp. 470-483. Syracuse, N.Y.: Syracuse University Press.

Lewis, G. M. Forthcoming. "Amerindian Antecedents of American Academic Geography.'' In B. Blouet, Ed. *The Evolution of Academic Geography in the United States*. Hamden, Conn.: Shoe String Press.

Martin, G. J. 1968. *Mark Jefferson, Geographer*. Ypsilanti: Eastern Michigan University Press.

————. 1973. *Ellsworth Huntington: His Life and Thought*. Hamden, Conn.: Shoe String Press.

————. 1977. "Isaiah Bowman: 1878-1950." In *Geographers: Biobibliographical Studies*. Vol. 1, pp. 9-18. London: Mansell.

————. 1980. *The Life and Thought of Isaiah Bowman*. Hamden, Conn.: Shoe String Press.

Mayo, W. L. 1965. *The Development and Status of Secondary School Geography in the United States and Canada*. Ann Arbor, Mich.: University Publishers.

Pattison, W. D. Forthcoming. "The Salisbury Commitment." In B. Blouet, Ed., *The Evolution of Academic Geography in the United States*. Hamden, Conn.: Shoe String Press.

Powell, J. W. 1896. "Physiographic Regions of the United States." In *Physiography of the United States*. Washington, D.C.: National Geographic Society, Monograph No. 1.

Roorbach, G. B. 1914. "Trend of Modern Geography—A Symposium." *Bulletin of the American Geographical Society*, 46:801-816.

Rowley, V. M. 1964. *J. Russell Smith: Geographer, Educator, and Conservationist*. Philadelphia: University of Pennsylvania Press.

Salisbury, R. D. 1907. *Physiography*. New York: Henry Holt.

Salisbury, R. D., and Alden, W. C. 1899. *The Geography of Chicago and Its Environs*. Chicago: University of Chicago Press.

Salisbury, R. D., and Atwood, W. W. 1908. *The Interpretation of Topographic Maps*. Washington, D.C.: U.S. Geological Survey, Professional Paper 60.

Salisbury, R. D., Barrows, H. H., and Tower, W. S. 1912. *The Elements of Geography*. New York: Henry Holt.

Semple, E. C. 1897. "The Influence of the Appalachian Barrier upon Colonial History." *Journal of School Geography*, 1:33-41.

————. 1901. "The Anglo-Saxons of the Kentucky Mountains." *Geographical Journal*, 17:588-623; reprinted in the *Bulletin of the American Geographical Society*, 42(1910):561-594.

————. 1903. *American History and Its Geographic Conditions*. Boston: Houghton Mifflin. Revised by the author with C. F. Jones, 1933.

————. 1911. *Influences of Geographic Environment*. New York: Henry Holt.

————. 1915. "The Barrier Boundary of the Mediterranean Basin and Its Northern Breaches as Factors in History." *Annals AAG*, 5:27-59.

————. 1927. Templed Promontories of the Ancient Mediterranean." *Geographical Review*, 17:353-386.

_____. 1931. *The Geography of the Mediterranean Region, Its Relation to Ancient History*. New York: Henry Holt.

Sherwood, M. 1977. "Alfred Hulse Brooks: 1871-1924." In *Geographers: Biobibliographical Studies*. Vol. 1, Pp. 19-23. London: Mansell.

Smith, J. R. 1913. *Industrial and Commercial Geography*. New York: Henry Holt.

Speth, W. W. 1978. "The Anthropogeographic Theory of Franz Boas." *Anthropos*, 73:1-31.

Strahler, A. N. 1950. "Davis' Concepts of Slope Development Viewed in the Light of Recent Quantitative Investigations." *Annals AAG*, 40:209-213.

Tower, W. S. 1910. "Scientific Geography: The Relation of Its Contents to Its Subdivisions." *Bulletin of the American Geographical Society*, 42:801-825.

Visher, S. S. 1948. "Memoir to Ellsworth Huntington, 1876-1947." *Annals AAG*, 38:38-50.

Ward, R. DeC. 1908. *Climate, Considered Especially in Relation to Man*. New York: G. P. Putnam's Sons.

_____. 1925. *Climates of the United States*. Boston: Ginn & Co.

Ward, R. DeC., and Brooks, C. F. 1936. *The Climates of North America*. In W. Köppen and R. Geiger, eds., *Handbuch der Klimatologie*, Vol. 2, Part J. Berlin: Borntraeger.

Warntz, W. Forthcoming. "*Geographia Generalis* and the Early Development of Academic Geography in the United States." In B. Blouet Ed., *The Evolution of Academic Geography in the United States*. Hamden, Conn.: Shoe String Press.

Whittlesey, D. S. 1935. "Dissertations in Geography Accepted by Universities in the United States for the Degree of Ph.D. as of May, 1935." *Annals AAG*, 25:211-237.

Wright, J. K. 1952. *Geography in the Making, the American Geographical Society 1851-1951*. New York: American Geographical Society.

Wrigley, G. M. 1951. "Isaiah Bowman." *Geographical Review*, 14:7-65.

CHAPTER 14

Atwood, W. W. 1935. "The Increasing Significance of Geographic Conditions in the Growth of Nation-States." *Annals AAG*, 25:1-16.

Atwood, W. W., and Mather, K. F. 1932. *Physiography and Quaternary Geology of the San Juan Mountains, Colorado*. Washington, D.C.: U.S. Geological Survey, Professional Paper 166.

Aurousseau, M. 1921. "The Distribution of Population: A Constructive Problem." *Geographical Review*, 11:563-592.

Baker, O. E. 1921. "The Increasing Importance of the Physical Conditions Determining the Utilization of Land for Agricultural and Forest Production in the United States." *Annals AAG*, 11:17-46.

_____. 1923. "Land Utilization in the United States: Geographic Aspects of the Problem." *Geographical Review*, 13:1-26.

Barnes, H. E. 1925. *History and Prospects of the Social Sciences*. New York: Alfred A. Knopf.

Barrows, H. H. 1923. "Geography as Human Ecology." *Annals AAG*, 13:1-14.

_____. 1962. *Lectures on the Historical Geography of the United States, as Given in 1933*. Ed. W. A. Koelsch. Chicago. University of Chicago, Department of Geography.

Bowman, I. 1934. *Geography in Relation to the Social Sciences*. New York: Charles Scribner.

Broek, J. O. M. 1932. *The Santa Clara Valley, California: A Study in Landscape Changes*. Utrecht: The University of Utrecht.

_____. 1938. "The Concept of Landscape in Human Geography." *Comptes rendus de la congrés internationale de géographie*. Vol. 2, Sec. 3a: 103-109. Amsterdam.

Brown, R. H. 1943. *Mirror for Americans, Likeness of the Eastern Seaboard, 1790-1810*. New York: American Geographical Society.

_____. 1948. *Historical Geography of the United States*. New York: Harcourt, Brace.

Browning, C. E. 1970. *A Bibliography of Dissertations in Geography, 1901 to 1969*. Chapel Hill: University of North Carolina, Department of Geography, Studies in Geography No. 1.

Brunhes, J. 1925. "Human Geography." In H. E. Barnes, ed., *History and Prospects of the Social Sciences*. New York:

Bushong, A. Forthcoming. "Geographers and Their Mentors: A Genealogical View of American Academic Geography." In B. Blouet, ed., *The Evolution of Academic Geography in the United States*. Hamden, Conn.: Shoe String Press.

Carter, G. F. 1945. *Plant Geography and Culture History in the American Southwest*. New York: Viking Fund Publications in Anthropology, No. 5.

Clark, A. H. 1949. *The Invasion of New Zealand by People, Plants, and Animals: The South Island*. New Brunswick, N.J.: Rutgers University Press.

_____. 1954. "Historical Geography." In P. E. James and C. F. Jones, eds., *American Geography, Inventory and Prospect*. Pp. 70-105. Syracuse, N.Y.: Syracuse University Press.

Colby, C. C. 1924. "The California Raisin Industry." *Annals AAG*, 14:49-108.

_____. 1936. "Changing Currents of Geographic Thought in America." *Annals AAG*, 26:1-37.

Dodge, S. D. 1932. "The Vermont Valley: A Chorographical Study." *Papers of the Michigan Academy of Science, Arts and Letters*. 17:241-274.

_____. 1933. "A Study of Population in Vermont and New Hampshire." *Papers of the Michigan Academy of Science, Arts and Letters*, 18:131-136.

_____. 1935. "A Study of Population Regions in New England on a New Basis." *Annals AAG*, 25:197-210.

Dryer, C. R. 1920. "Genetic Geography: The Development of the Geographic Sense and Concept." *Annals AAG*, 10:3-16.

Fairchild, W. B. 1979. " 'The Geographical Review' and the American Geographical Society." *Annals AAG*, 69:33-38.

Fenneman, N. M. 1919. "The Circumference of Geography." *Annals AAG*, 9:3-11.

Finch, V. C. 1933. *Montfort: A Study in Landscape Types in Southwestern Wisconsin*. Chicago: Geographical Society of Chicago, Bulletin 9.

_____. 1939. "Geographical Science and Social Philosophy." *Annals AAG*, 29:1-28.

Goldthwait, J. W. 1927. "A Town That Has Gone Downhill." *Geographical Review*, 17:527-552.

Hall, R. B. 1934. "The Cities of Japan: Notes on Distribution and Inherited Forms." *Annals AAG*, 24:175-200.

Harris, C. D. 1979. "Geography at Chicago in the 1930s and 1940s." *Annals AAG*, 69:21-32.

Hartshorne, R. 1927. "Location as a Factor in Geography." *Annals AAG*, 17:92–99.

———. 1932. "The Twin City District: A Unique Form of Urban Landscape." *Geographical Review*, 22:431–442.

———. 1934. "Upper Silesian Industrial District." *Geographical Review*, 24:423–438.

———. 1935. "Recent Developments in Political Geography." *American Political Science Review*. 29:785–804, 943–966.

———. 1938. "Racial Maps of the United States." *Geographical Review*, 28:276–288.

———. 1939. *The Nature of Geography, a Critical Survey of Current Thought in the Light of the Past*. Lancaster, Pa.: Association of American Geographers.

———. 1948. "On the Mores of Methodological Discussion." *Annals AAG*, 38:113–125.

———. 1950. "Functional Approach to Political Geography." *Annals AAG*, 40:95–130.

———. 1955. " 'Exceptionalism in Geography' Re-examined." *Annals AAG*, 45:205–244.

———. 1958. "The Concept of Geography as a Science of Space, from Kant and Humboldt to Hettner." *Annals AAG*, 48:97–108.

———. 1959. *Perspective on the Nature of Geography*. Chicago: Rand McNally.

———. 1979. "Notes Toward a Bibliobiography of 'the Nature of Geography.' " *Annals AAG*, 69:63–76.

Hayes, E. C. 1908. "Sociology and Psychology; Sociology and Geography." *American Journal of Sociology*, 14:371–407.

Hewes, L. 1946. "Dissertations in Geography Accepted by Universities in the United States and Canada for the Degree of Ph.D., June, 1935, to June, 1946, and Those Currently in Progress." *Annals AAG*, 36:215–247.

———. 1950. "Some Features of Early Woodland and Prairie Settlement in a Central Iowa County." *Annals AAG*, 40:40–57.

Huntington, E. 1924. "Geography and Natural Selection." *Annals AAG*, 14:1–16.

James, P. E. 1927. "A Geographic Reconnaissance of Trinidad." *Economic Geography*, 3:87–109.

———. 1929. "The Blackstone Valley, a Study in Chorography in Southern New England." *Annals AAG*, 19:67–109.

———. 1931. "Vicksburg, a Study in Urban Geography." *Geographical Review*, 21:234–243.

———. 1967. "On the Origin and Persistence of Error in Geography." *Annals AAG*, 57:1–24.

James, P. E., and Jones, C. F. 1954. *American Geography, Inventory and Prospect*. Syracuse, N.Y.: Syracuse University Press.

James, P. E., Jones, W. D., and Finch, V. C. 1934. "Conventionalizing Geographic Investigation and Presentation." *Annals AAG*, 24:77–122.

James, P. E., and Martin, G. J. 1979. *The Association of American Geographers: The First Seventy-Five Years, 1904–1979*. Washington, D.C.: Association of American Geographers.

James, P. E., and Mather, E. C. 1977. "The Role of Periodic Field Conferences in the Development of Geographical Ideas in the United States." *The Geographical Review*, 67:446–461.

Jastrow, J., ed. 1936. *The Story of Human Error*. New York: Appleton-Century.

Jefferson, M. 1917. "Some Considerations on the Geographical Provinces of the United States." *Annals AAG*, 7:3–15.

———. 1939. "The Law of the Primate City." *Geographical Review*, 29:226–232.

Joerg, W. L. G. 1914. "The Subdivisions of North America into Natural Regions: A Preliminary Inquiry." *Annals AAG*, 4:55-83.

_____. 1936. "The Geography of North America: A History of Its Regional Exposition." *Geographical Review*, 26:640-663.

Johnson, D. 1929. "The Geographic Prospect," *Annals AAG*, 19:167-231.

Jones, W. D. 1930. "Ratios and Isopleth Maps in Regional Investigation of Agricultural Land Occupance." *Annals AAG*, 20:177-195.

Jones, W. D., and Finch, V. C. 1925. "Detailed Field Mapping in the Study of the Economic Geography of an Agricultural Area." *Annals AAG*, 15:148-157.

Klimm, L. E. 1954. "The Empty Areas of the Northeastern United States." *Geographical Review*, 33:325-345.

Kniffen, F. B. 1931. "Lower California Studies III: The Primitive Cultural Landscape of the Colorado Delta." *University of California Publications in Geography*, 5:43-66.

_____. 1932. "Lower California Studies IV: The Natural Landscape of the Colorado Delta." *University of California Publications in Geography*, 5:149-244.

Koelsch, W. A. 1969. "The Historical Geography of Harlan H. Barrows." *Annals AAG*, 59:632-651.

Kuhn, T. S. 1962. *The Structure of Scientific Revolutions*. Chicago: University of Chicago Press.

Leighly, J., ed. 1963. *Land and Life, a Selection from the Writings of Carl Ortwin Sauer*. Berkeley and Los Angeles: University of California Press.

_____. 1976. "Carl Ortwin Sauer, 1889-1975." *Annals AAG*, 66:337-348.

Lewthwaite, G. R. 1966. "Environmentalism and Determinism: A Search for Clarification." *Annals AAG* 56:1-23.

McCarty, H. H. 1940. *The Geographic Basis of American Economic Life*. New York: Harper Bros.

_____. 1942. "A Functional Analysis of Population Distribution." *Geographical Review*, 32:282-293.

Martin, A. F. 1951. "The Necessity for Determinism." *Transactions and Papers, Institute of British Geographers*. 17:1-12.

Meigs, P. 1935. "The Dominican Mission Frontier of Lower California." *University of California Publications in Geography*, 7:1-192.

Parkins, A. E. 1918. *The Historical Geography of Detroit*. Lansing: Michigan Historical Commission.

_____. 1934. "The Geography of American Geographers." *Journal of Geography*, 33:221-230.

Parsons, J. J. 1949. *Antioqueño Colonization in Western Colombia*. Berkeley and Los Angeles: University of California Press, Ibero-Americana 32.

Peattie, R. 1929. "Andorra: A Study on Mountain Geography." *Geographical Review*, 19:218-233.

_____. 1940. *Geography in Human Destiny*. New York: George W. Stewart.

Pfeifer, G. "Regional Geography in the United States Since the War; a Review of Trends in Theory and Method." Trans. J. B. Leighly, 1938. American Geographical Society, Mimeographed pub. No. 2.

Platt, R. S. 1928. "A Detail of Regional Geography: Ellison Bay Community as an Industrial Organism." *Annals AAG*, 18:81-126.

_____. 1931. "An Urban Field Study: Marquette, Michigan." *Annals AAG*, 21:52-73.

_____. 1933. "Magdalena Atlipac: A Study in Terrene Occupancy in Mexico." *Bulletin of the Geographical Society of Chicago*, 9:45–75.

_____. 1935. "Field Approach to Regions." *Annals AAG*, 25:153–174.

_____. 1946. "Problems of Our Times." *Annals AAG*, 36:1–43.

_____. 1948. "Environmentalism Versus Geography." *American Journal of Sociology*, 53:351–358.

_____. 1959. *Field Study in American Geography, The Development of Theory and Method Exemplified by Selections*. Chicago: University of Chicago, Department of Geography, Research Paper No. 61.

Popper, K. R. 1959. *The Logic of Scientific Discovery*. London: Hutchinson.

Sauer, C. O. 1924. "The Survey Method in Geography and Its Objectives." *Annals AAG*, 14:17–33.

_____. 1925. "The Morphology of Landscape." *University of California Publications in Geography*, 2:19–53.

_____. 1927. "Recent Developments in Cultural Geography." In E. C. Hayes, ed., *Recent Developments in the Social Sciences*. Pp. 154–212. Philadelphia: J. B. Lippincott.

_____. 1931. "Cultural Geography." In *Encyclopedia of the Social Sciences*. Vol. 6, pp. 621–623. New York: Macmillan.

_____. 1932. *The Road to Cibola*. Berkeley and Los Angeles: University of California Press, Ibero-Americana 3.

_____. 1941. "Foreword to Historical Geography." *Annals AAG*, 31:1–24.

_____. 1952. *Agricultural Origins and Dispersals*. New York: American Geographical Society, Bowman Memorial Lectures, Ser. 2.

_____. 1956. "The Agency of Man on the Earth." In W. L. Thomas, ed., *Man's Role in Changing the Face of the Earth*. Pp. 46–69. Chicago: University of Chicago Press.

_____. 1966a. "On the Background of Geography in the United States." *Heidelberger Studien zur Kulturgeographie, Festgabe für Gottfried Pfeiffer*, 15:59–71.

_____. 1966b. *The Early Spanish Main*. Berkeley and Los Angeles: University of California Press.

Sauer, C. O., and Brand, D. D. 1932. *Aztatlán, Prehistoric Mexican Frontier on the Pacific Coast*. Berkeley and Los Angeles: University of California Press, Ibero-Americana 1.

Schaefer, F. K. 1953. "Exceptionalism in Geography: A Methodological Examination." *Annals AAG*, 43:226–249.

Spencer, J. E. 1939. "Changing Chungking: The Rebuilding of an Old Chinese City." *Geographical Review*, 29:46–60.

Speth, W. W. Forthcoming. "Berkeley Geography, 1923–1933." In B. W. Blouet, ed., *The Origins of Academic Geography in the United States*. Hamden, Conn.: Shoestring Press.

Stanislawski, D. 1946. "The Origin and Spread of the Grid-Pattern Town." *Geographical Review*, 36:105–120.

_____. 1947. "Early Spanish Town-Planning in the New World." *Geographical Review*, 37:94–105.

_____. 1975. "Carl Ortwin Sauer, 1889–1975." *The Journal of Geography*, 74:548–554.

Taylor, G., ed. 1951. *Geography in the Twentieth Century*. New York: Philosophical Library.

Thoman, R. S. 1979. "Robert Swanton Platt, 1891–1964." In *Geographers: Biobiblio graphical Studies*. Vol. 3, pp. 107–116. London: Mansell.

Trewartha, G. T. 1979. "Geography at Wisconsin." *Annals AAG,* 69:16-21.

West, R. C. 1952. *Colonial Placer Mining in Colombia.* Baton Rouge: Louisiana State University Press.

Whitaker, J. R. 1954. "The Way Lies Open." *Annals AAG,* 44:231-244.

Whitbeck, R. H. 1926. "Adjustments to Environment in South America: An Inter-play of Influences." *Annals AAG,* 16:1ff.

Whitbeck, R. H., and Thomas, O. J. 1932. *The Geographic Factor.* New York: Century.

Whittemore, K. T. 1972. "Celebrating Seventy-Five Years of the 'Journal of Geography' 1897-1972." *Journal of Geography,* 71:7-18.

Whittlesey, D. S. 1925. "Field Maps for the Geography of an Agricultural Area." *Annals AAG,* 15:187-191.

_____. 1927. "Devices for Accumulating Geographic Data in the Field." *Annals AAG,* 17:72-78.

_____. 1929. "Sequent Occupance." *Annals AAG,* 19:162-165.

_____. 1935. "Dissertations in Geography Accepted by Universities in the United States for the Degree of Ph.D. as of May, 1935." *Annals AAG,* 25:211-237.

_____. 1939. *The Earth and the State.* New York: Henry Holt.

Wooldridge, S. W. 1956. *The Geographer as Scientist.* London: Thomas Nelson.

Wooldridge, S. W., and East, W. G. 1958. *The Spirit and Purpose of Geography.* London: Hutchinson.

Wright, J. K. 1936. "A Method of Mapping Densities of Population, with Cape Cod as an Example." *Geographical Review,* 26:103-110.

_____. 1952. *Geography in the Making: The American Geographical Society, 1851-1951.* New York: American Geographical Society.

_____. 1966. *Human Nature in Geography.* Cambridge: Harvard University Press.

CHAPTER 15

Ackerman, E. A. 1945. "Geographic Training, Wartime Research, and Immediate Professional Objectives." *Annals AAG,* 35:121-143.

Applebaum, W. 1952. "A Technique for Constructing a Population and Urban Land Use Map." *Economic Geography,* 28:240-243.

Applebaum, W., and Cohen, S. B. 1961. "The Dynamics of Store Trading Areas and Market Equilibrium." *Annals AAG,* 51:73-101.

Barnes, C. P. 1929. "Land Resource Inventory in Michigan." *Economic Geography,* 5:22-35.

Bennett, H. H. 1928. "The Geographical Relation of Soil Erosion to Land Productivity." *Geographical Review,* 18:579-605.

Bowman, I., 1921. *The New World, Problems in Political Geography.* New York: World Book Co.

_____. 1924. *Desert Trails of Atacama.* New York: American Geographical Society, Special Publication No. 5.

_____. 1931. *The Pioneer Fringe.* New York: American Geographical Society.

_____. 1932. "Planning in Pioneer Settlement." *Annals AAG,* 22:93-107.

_____. 1934. *Geography in Relation to the Social Sciences.* New York: Charles Scribner.

_____. 1937. *The Limits of Land Settlement, a Report on Present-Day Possibilities.* New York: Council on Foreign Relations.

_____. 1938. "Geography in the Creative Experiment." *Geographical Review,* 28:1–19.

Colby, C. C. 1936. "Changing Currents of Geographic Thought in America." *Annals AAG,* 26:1–37.

_____, ed. 1941. *Land Classification in the United States.* Washington, D.C.:Report of the Land Committee to the National Resources Planning Board.

_____. 1955. "Narrative of Five Decades." In *A Half Century of Geography—What Next?* (Papers presented at the alumni reunion, June 5, 1954.) Chicago: University of Chicago, Department of Geography.

Colby, C. C., and Roterus, V. 1943. *Area Analysis—A Method of Public Works Planning.* Washington, D.C.: Technical Paper No. 6 of the Land Committee, National Resources Planning Board.

Colby, C. C., and White, G. F. 1961. "Harlan H. Barrows, 1877–1960." *Annals AAG,* 51:395–400.

Davis, C. M. 1969. "A Study of the Land Type." In *The Michigan Land Economic Survey.* Pp. 15–41. Ann Arbor: Office of Research Administration, Project No. 08055.

Davis, W. M. 1918. *A Handbook of Northern France.* Cambridge: Harvard University Press.

DeVries, W. 1927. "An Economic Survey of Chippewa County, Michigan." *Papers of the Michigan Academy of Science, Arts and Letters,* 8:255–268.

_____. 1928. "Correlation of Physical and Economic Factors as Shown by the Michigan Land Economic Survey Data." *Journal of Land and Public Utility Economics,* 4:295–300.

Dominian, L. 1917. *The Frontiers of Language and Nationality in Europe.* New York: American Geographical Society.

Finch, V. C., and Baker, O. E. 1917. *Geography of the World's Agriculture.* Washington, D.C.: U.S. Department of Agriculture.

Gelfand, L. E. 1963. *The Inquiry.* New Haven: Yale University Press.

Hudson, G. D. 1936. "The Unit Area Method of Land Classification." *Annals AAG,* 26:99–112.

Innis, H. A. 1935. "Canadian Frontiers of Settlement: A Review." *Geographical Review,* 25:92–106.

Jefferson, M. 1921. *Recent Colonization in Chile.* New York: American Geographical Society, Research Series No. 6.

_____. 1926. *Peopling the Argentine Pampa.* New York: American Geographical Society, Research Series No. 16.

Joerg, W. L. G., ed. 1932. *Pioneer Settlement, Cooperative Studies by Twenty-Six Authors.* New York: American Geographical Society.

Johnson, D. W. 1921. *Battlefields of the World War, Western and Southern Fronts: A Study in Military Geography.* New York: American Geographical Society, Research Series No. 3.

Jones, C. F. and Berrios, H. 1956. See original.

Jones, C. F., and Picó, R., eds. 1955. *Symposium on the Geography of Puerto Rico.* Rio Piedras: University of Puerto Rico Press.

Kates, R. W. 1962. *Hazard and Choice Perception in Flood Plain Management.* Chicago: University of Chicago, Department of Geography, Research Paper 78.

McBride, G. M. 1923. *The Land Systems of Mexico*. New York: American Geographical Society, Research Series No. 16.

———. 1936. *Chile: Land and Society*. New York: American Geographical Society, Research Series No. 19.

McMurry, K. C. 1936. "Geographic Contributions to Land-Use Planning." *Annals AAG*, 26:91–98.

McNee, R. B. 1962. "Centrifugal-Centripetal Forces in International Petroleum Company Regions." *Annals AAG*, 51:124–138.

Martin, G. J., ed. 1966. *Mark Jefferson: Paris Peace Conference Diary*. Ann Arbor: Michigan.

———. 174. " 'Civilization and Climate,' Revisited." *Geography and Map Division, Special Libraries Association Bulletin*, No. 96, pp. 10–17.

———. 1980. " 'The Science of Settlement' and Resettlement Schemes." In *The Life and Thought of Isaiah Bowman*. Pp. 123–139. Hamden, Conn.: Shoe String Press.

Miller, O. M. 1929, "The 1927–1928 Peruvian Expedition of the American Geographical Society." *Geographical Review*, 19:1–37.

Ogilvie, A. G. 1922. *Geography of the Central Andes*. New York: American Geographical Society, Map of Hispanic America Publication No. 1.

Platt, R. R. 1946. " The Map of Hispanic America on the Scale of 1:1,000,000.' *Geographical Review*, 36:1–28.

Rhoads, J. B. 1954. "Preliminary Inventories." *Cartographic Records of the American Commission to Negotiate Peace*, No. 68. U.S. Washington, D.C.: National Archives.

Sauer, C. O. 1919. "Mapping the Utilization of the Land." *Geographical Review*, 8:47–54.

———. 1921. "The Problem of Land Classification." *Annals AAG*, 11:3–16.

Schoenmann, L. R. 1931. "Land Inventory for Rural Planning in Alger County, Michigan." *Papers of the Michigan Academy of Science, Arts and Letters*, 16:320–361.

Schantz, H. L., and Marbut, C. F. 1923. *The Vegetation and Soils of Africa*. New York: American Geographical Society, Research Series No. 13.

Stamp, L. D. 1931. "The Land Utilization Survey of Britain." *Geographical Journal*, 78:40–53.

———. 1952. *Land for Tomorrow: The Underdeveloped World*. Bloomington: Indiana University Press.

Thornthwaite, C. W. 1931. "The Climates of North America According to a New Classification." *Geographical Review*, 21:633–655.

———. 1933. "The Climates of the Earth." *Geographical Review*, 23:433–440.

Veatch, J. O. 1930. "Natural Geographic Divisions of Land." *Papers of the Michigan Academy of Science, Arts and Letters*, 14:417–432.

———. 1933. "Classification of Land on a Geographic Basis." *Papers of the Michigan Academy of Science, Arts and Letters*, 19:359–365.

———. 1953. *Soils and Land of Michigan*. East Lansing: Michigan State University Press.

Walworth, A. 1976. *America's Moment: 1918—American Diplomacy at the End of World War I*. New York: Norton.

White, G. F. 1973. "Natural Hazards Research." In R. J. Chorley, ed., *Directions in Geography*. Pp. 193–216. London: Methuen.

Wright, J. K. 1952. *Geography in the Making, The American Geographical Society, 1851–1951*. New York: American Geographical Society.

CHAPTER 16

Ackerman, E. A. 1965. *The Science of Geography*. Washington, D.C.: National Academy of Sciences/National Research Council, Publication No. 1277.

Amedeo, D. and Golledge, R. G. 1975. *An Introduction to Scientific Reasoning in Geography*. New York: John Wiley & Sons.

Baker, O. E. 1927. "Agricultural Regions of North America." *Economic Geography*, 3:65-86.

Bennett, R. J., and Chorley, R. J. 1978. *Environmental Systems: Philosophy, Analysis and Control*. London: Methuen.

Bertalanffy, L. von. 1968. *General System Theory, Foundations, Development, Applications*. New York: George Braziller.

Braithwaite, R. B. 1960. *Scientific Explanation*. New York: Harper Bros. (Harper Torchbooks); Cambridge: At the University Press, 1953.

Bridgman, P. 1964. *The Nature of Physical Theory*. Princeton, N.J.: Princeton University Press, 1936. (Science Editions.)

Brookfield, H. 1964. "Questions on the Human Frontiers of Geography." *Economic Geography*, 40:283-303.

Brown, R. 1963. *Explanations in Social Science*. Chicago: Aldine.

Bunge, W. 1966. *Theoretical Geography*. Lund: University of Lund. (Lund Studies in Geography.)

Burton, I. 1963. "The Quantitative Revolution and Theoretical Geography." *The Canadian Geographer*, 7:151-162.

Chapman, G. P. 1977. *Human and Environmental Systems: A Geographer's Appraisal*. London and New York: Academic Press.

Chorley, R. J. 1964. "Geography and Analogue Theory." *Annals AAG*, 54:127-137.

Chorley, R. J., and Haggett, P. 1967. *Models in Geography*. London: Methuen.

Cole, J. P., and King, C. A. M. 1968. *Quantitative Geography, Techniques and Theories in Geography*. John Wiley & Sons.

De Geer, S. 1923. "On the Definition, Method, and Classification of Geography." *Geografiska Annaler*, 1923:1-37.

De Jong, G. 1962. *Chorological Differentiation as the Fundamental Principle of Geography*. Groningen, Netherlands: J. B. Wolters.

Downs, R. M. 1970. "Geographic Space Perception: Past Approaches and Future Prospects." In C. Board et al., eds., *Progress in Geography 2*. Pp. 65-108. London: Edward Arnold.

Ducasse, C. J. 1969. *Causation and the Types of Necessity*. New York: Dover.

Gibson, E. 1978. "Understanding the Subjective Meaning of Places." In D. Ley and M. S. Samuels, eds., *Man's Place: Themes in Geographic Humanism*. Pp. 138-54. Chicago: Maaroufa Press.

Gilbert, E. W. 1960. "The Idea of the Region." *Geography*, 45:157-175.

Glacken, C. J. 1956. "Changing Ideas of the Habitable World." In W. L. Thomas, eds., *Man's Role in Changing the Face of the Earth*. Pp. 70-92. Chicago: University of Chicago Press.

Golledge, R., and Amedeo, D. 1968. "On Laws in Geography." *Annals AAG*, 58:760-774.

Grigg, D. B. 1965. "The Logic of Regional Systems." *Annals AAG,* 55:465–491.

Haggett, P. 1966. *Locational Analysis in Human Geography.* New York: St. Martin's Press.

Hartshorne, R. 1959. *Perspective on the Nature of Geography.* Chicago: Rand McNally.

Harvey, D. 1969. *Explanation in Geography.* London: Edward Arnold.

Herbst, J. 1961. "Social Darwinism and the History of American Geography." *Proceedings of the American Philosophical Society,* 105:538–544.

Hofstadter, R. 1955. *Social Darwinism in American Thought.* Boston: Beacon Press.

Isard, W. 1956. "Regional Science, the Concept of Region, and Regional Structure." *Papers and Proceedings, Regional Science Association 2,* pp. 13–39.

———. 1975. *An Introduction to Regional Science.* Englewood Cliffs, N.J.: Prentice-Hall.

James, P. E. 1937. "On the Treatment of Surface Features in Regional Studies." *Annals AAG,* 27:153–176.

———. 1952. "Toward a Further Understanding of the Regional Concept." *Annals AAG,* 42:196–222.

———. 1966. *A Geography of Man,* 3rd ed. Waltham, Mass.: Ginn & Co.

James, P. E., and Jones, C. F. 1954. *American Geography, Inventory and Prospect.* Syracuse, N.Y.: Syracuse University Press.

Jones, W. D. 1930. "Ratios and Isopleth Maps in Regional Investigation of Agricultural Land Occupance." *Annals AAG,* 20:177–195.

Jordan, T. G. 1967. "The Imprint of the Upper and Lower South on Mid-nineteenth Century Texas." *Annals AAG,* 57:677–690.

———. 1969. "The Origin of Anglo-American Cattle Ranching in Texas: A Documentation of Diffusion from the Lower South." *Economic Geography,* 45:63–87.

Kimble, G. H. T. 1951. "The Inadequacy of the Regional Concept." In L. D. Stamp and S. W. Wooldridge, eds., *London Essays in Geography.* Pp. 151–174. Cambridge: Harvard University Press.

Kniffen, F. 1965. "Folk Housing: Key to Diffusion." *Annals AAG,* 55:549–577.

Lewthwaite, G. R. 1966. "Environmentalism and Determinism: A Search for a Clarification." *Annals AAG,* 56:1–23.

Ley, D. 1977. "The Personality of a Geographical Fact." *The Professional Geographer,* 29:8–13.

Lowenthal, D. 1961. "Geography, Experience, and Imagination: Towards a Geographical Epistemology." *Annals AAG,* 51:241–260.

Lowenthal, D., and Bowden, M. J., eds. 1976. *Geographies of the Mind: Essays in Historical Geosophy in Honor of John Kirtland Wright.* New York: Oxford University Press.

Mackay, J. R. 1951. "Some Problems and Techniques in Isopleth Mapping." *Economic Geography,* 27:1–9.

Manners, R. A., and Kaplan, D., eds. 1968. *Theory in Anthropology, a Sourcebook.* Chicago: Aldine.

Meehan, E. J. 1965. *The Theory and Method of Political Analysis.* Homewood, Ill.: Dorsey Press.

Meinig, D. W. 1962. "A Comparative Historical Geography of Two Railnets: Columbia Basin and South Australia." *Annals AAG,* 52:394–413.

———. 1965. "The Mormon Culture Region: Strategies and Patterns in the Geography of the American West." *Annals AAG,* 55:191–220.

_____. 1968. *The Great Columbia Plain, a Historical Geography*, 1805-1910. Seattle: University of Washington Press.

_____. 1969. *Imperial Texas, an Interpretive Essay in Cultural Geography*. Austin and London: University of Texas Press.

Minshull, R. 1967. *Regional Geography, Theory and Practice*. London: Hutchinson University Library.

_____. 1970. *The Changing Nature of Geography*. London: Hutchinson University Library.

Montefiore, A. C., and Williams, W. M. 1955. "Determinism and Possibilism." *Geographical Studies*, 2:1-11.

Nagel, E. 1961. *The Structure of Science: Problems in the Logic of Scientific Explanation*. New York: Harcourt, Brace.

Nystuen, J. D. 1963. "Identification of Some Fundamental Spatial Concepts." *Papers of the Michigan Academy of Science, Arts and Letters*, 48:373-384.

Paterson, J. H. 1974. "Writing Regional Geography." In C. Board, et al. eds. *Progress in Geography 6*. Pp. 1-26. London: Edward Arnold.

Philbrick, A. K. 1957. "Principles of Areal Functional Organization in Regional Human Geography." *Economic Geography*, 33:299-336.

Platt, R. S. 1948. "Determinism in Geography." *Annals AAG*, 38:126-132.

Prince, H. C. 1971. "Real, Imagined and Abstract Worlds of the Past." *Progress in Geography*, 3:1-86.

Rapoport, A. 1959. "Uses and Limitations of Mathematical Models in Social Science." In L. Gross, ed., *Symposium on Sociological Theory*. New York: Harper Bros.

Robinson, G. W. S. 1953. "The Geographic Region: Form and Function." *Scottish Geographical Magazine*, 69:49-58.

Sauer, C. O. 1925. "The Morphology of Landscape." *University of California Publications in Geography*, 2:19-35.

Shafer, R. J., ed. 1969. *A Guide to Historical Method*. Homewood, Ill.: Dorsey Press.

Sidall, W. R. 1961. "Two Kinds of Geography." *Economic Geography*, 37:189.

Spate, O. H. K. 1952. "Toynbee and Huntington: A Study in Determinism." *Geographical Journal*, 118:406-428.

_____. 1958. "The End of an Old Song? The Determinism-Possibilism Problem." *Geographical Review*, 48:280-282.

Sprout, H., and Sprout, M. 1956. *Man-Milieu Relationship Hypothesis in the Context of International Politics*. Princeton, N.J.: Center for International Studies.

Taaffe, E. J., ed. 1970. *Geography*. Englewood Cliffs, N.J.: Prentice-Hall.

Wallace, W. L., ed. 1969. *Sociological Theory, an Introduction*. Chicago: Aldine.

Ward, D. 1964. "A Comparative Historical Geography of Streetcar Suburbs in Boston, Massachusetts, and Leeds, England, 1850-1920." *Annals AAG*, 54:447-489.

Whittlesey, D. 1956. Southern Rhodesia—An African Compage." *Annals AAG*, 46:1-97.

Wilbanks, T. J., and Symanski, R. 1968. "What Is Systems Analysis?" *The Professional Geographer*, 20:81-85.

Wright, J. K. 1937. "Some Measures of Distributions." *Annals AAG*, 27:177-211.

Wrigley, E. A. 1965. "Changes in the Philosophy of Geography." In R. J. Chorley and P. Haggett, eds., *Frontiers in Geographical Teaching*. Pp. 3-20. London: Methuen.

CHAPTER 17

Barrett, E. C. 1970. "Rethinking Climatology: An Introduction to the Uses of Weather Satellite Photographic Data in Climatological Studies." *Progress in Geography*, 2:153-205.

Bird, J. B., and Morrison, A. 1964. "Space Photography and Its Geographic Applications." *Geographical Review*, 54:463-486.

Blache, J. 1921. "Modes of Life in the Morocco Countryside, Interpretations of Aerial Photographs." *Geographical Review*, 11:477-502.

Bunge, W. 1966. *Theoretical Geography*. Lund: University of Lund. (Lund Studies in Geography.)

Cole, J. P., and King, C. A. M. 1968. *Quantitative Geography. Techniques and Theories in Geography*. London: John Wiley & Sons.

Crone, G. R. 1950. *Maps and Their Makers, an Introduction to the History of Cartography*. New York: G. P. Putnam's Sons.

Duncan, O. D., Cuzzort, R. P., and Duncan, B. 1961. *Statistical Geography*. New York: The Free Press.

Freeman, L. C. 1965. *Elementary Applied Statistics for Students in Behavioral Science*. New York: John Wiley & Sons.

Haggett, P. 1966. *Locational Analysis in Human Geography*. New York: St. Martin's Press.

Kao, R. C. 1963. "The Use of Computers in the Processing and Analysis of Geographic Information." *Geographical Review*, 53:530-547.

Kendall, M. G. 1939. "The Geographical Distribution of Crop Productivity in England." *Journal of the Royal Statistical Society*, 102:21-62.

King, L. W. 1969. *Statistical Analysis in Geography*. Englewood Cliffs, N.J.: Prentice-Hall.

King-Hele, D. 1967. "The Shape of the Earth." *Scientific American*, 217:67-76.

Lee, W. T. 1920. "Airplanes and Geography." *Geographical Review*, 10:310-325.

———. 1922. *The Face of the Earth as Seen from the Air: A Study in the Application of Airplane Photography to Geography*. New York: American Geographical Society.

Lewis, P. M. 1965. "Three Related Problems in the Formulation of Laws in Geography." *The Professional Geographer*, 17:24-27.

McCarty, H. H., and Salisbury, N. E. 1961. *Visual Comparison of Isopleth Maps as a Means of Determining Correlations Between Spatially Distributed Phenomena*. Iowa City: State University of Iowa, Department of Geography.

Marble, D. F. 1967. *Some Computer Programs for Geographic Research*. Evanston, Ill.: Northwestern University, Department of Geography.

Miller, O. M. 1931. "Planetabling from the Air. An Approximate Method of Plotting from Oblique Aerial Photography." *Geographical Review*, 21:202-212, 660-662.

Olson, C. E. 1967. "Accuracy of Land-Use Interpretation from Infrared Imagery in the 4.5 to 5.5 Micron Band." *Annals AAG*, 57:382-388.

Robinson, A. H. 1956. "Mapping the Land." *Scientific Monthly*, 82:294-303.

———. 1962. "Mapping the Correspondence of Isarithmic Maps." *Annals AAG*, 52:414-425.

Robinson, A. H., and Bryson, R. A. 1957. "A Method for Describing Quantitatively the Correspondence of Geographical Distributions." *Annals AAG*, 47:379-391.

Russell, J. A., Foster, F. W., and McMurry, K. C. 1943. "Some Applications of Aerial Photographs to Geographic Inventory." *Papers of the Michigan Academy of Science, Arts and Letters,* 29:315-341.

Simpson, R. B. 1966. "Radar, Geographic Tool." *Annals AAG,* 56:80-96.

Stewart, J. Q. 1947. "Empirical Mathematical Rules Concerning the Distribution and Equilibrium of Population." *Geographical Review,* 37:461-485.

Stewart, J. Q., and Warntz, W. 1958. "Macrogeography and Social Science." *Geographical Review,* 48:167-184.

Taaffe, E. J., ed. 1970. *Geography.* Englewood Cliffs, N.J.: Prentice-Hall.

Tobler, W. 1959. "Automation and Cartography." *Geographical Review,* 49:526-534.

Weaver, J. C. 1954. "Crop-Combination Regions in the Middle West." *Geographical Review,* 44:175-200.

_____. 1956. "The County as a Spatial Average in Agricultural Geography." *Geographical Review,* 46:536-565.

Willow Run Laboratories. 1966. *Peaceful Uses of Earth Observation Spacecraft.* 3 vols. Ann Arbor: University of Michigan Press.

CHAPTER 18

Ackerman, E. A. 1963. "Where Is a Research Frontier?" *Annals AAG,* 53:429-440.

_____. 1965. *The Science of Geography.* Washington, D.C.: National Academy of Sciences/National Research Council, Publication No. 1277.

Barrows, H. H. 1923. "Geography as Human Ecology." *Annals AAG,* 13:1-14.

Bell, D. 1973. *The Coming of Post-Industrial Society.* New York: Basic Books.

Berry, B. J. L. 1961. "An Inductive Approach to the Regionalization of Economic Development." In N. Ginsburg, ed., *Essays on Geography and Economic Development.* Pp 78-107. Chicago: University of Chicago, Department of Geography. Research Paper No. 62.

_____. 1964. "Approaches to Regional Analysis, A Synthesis." *Annals AAG,* 54:2-11.

_____. 1973. "A Paradigm for Modern Geography." In R. J. Chorley, ed. *Directions in Geography.* Pp. 3-22. London: Methuen.

Berry, B. J. L., and Garrison, W. L. 1958a. "Alternate Explanations of Urban Rank-Size Relationships." *Annals AAG,* 48:83-91.

_____. 1958b. "A Note on Central Place Theory and the Range of a Good." *Economic Geography,* 34:304-311.

Berry, B. J. L., and Pred, A. 1961. *Central Place Studies, A Bibliography of Theory and Applications.* Philadelphia: Regional Science Research Institute.

Bertalanffy, L. von. 1951a. "General System Theory: A New Approach to the Unity of Science." *Human Biology,* 23:303-361.

_____. 1951b. "An Outline of General System Theory." *British Journal for the Philosophy of Science,* 1:134-165.

_____. 1956. "General System Theory." *General Systems,* 1:1-10.

_____. 1962. "General System Theory—A Critical Review." *General Systems,* 7:1-20.

_____. 1965. "On the Definition of the Symbol." In J. R. Royce, ed., *Psychology and the Symbol: An Interdisciplinary Symposium.* New York: Random House.

_____. 1968. *General System Theory, Foundations, Development, Applications.* New York: George Braziller.

Bird, J. H. 1975. "Methodological Implications for Geography from the Philosophy of K. R. Popper." *Scottish Geographical Magazine,* 91:153-163.

_____. 1977. "Methodology and Philosophy." *Progress in Human Geography,* 1:104-10.

_____. 1978. "Methodology and Philosophy." *Progress in Human Geography,* 2:133-140.

Blaut, J. M. 1979. "The Dissenting Tradition." *Annals AAG,* 69:157-164.

Blome, D. A. 1970. "A Spatial Model of the Urban Structure of Appalachia." *Proceedings AAG,* 2:12-16.

Boulding, K. E. 1956. "General Systems Theory—The Skeleton of a Science." *Management Science,* 2:197-208.

Bowman, I. 1934. *Geography in Relation to the Social Sciences.* New York: Charles Scribner.

Brewer. J. G. 1978. *The Literature of Geography: A Guide to Its Organization and Use.* Hamden, Conn.: Shoe String Press.

Bronowski, J. 1966. "The Logic of the Mind." *American Scientist,* 54:1-14.

Brookfield, H. C. 1969. "On the Environment as Perceived." *Progress in Geography,* 1:51-80.

Brown, L. A. 1968. *Diffusion Process and Location, A Conceptual Framework and Bibliography.* Philadelphia: Regional Science Research Institute.

Brown, L. A., and Moore, E. G. 1969. "Diffusion Research in Geography: A Perspective." *Progress in Geography,* 1:121-157.

Brush, J. E. 1953. "The Hierarchy of Central Places in Southwestern Wisconsin." *Geographical Review,* 43:380-402.

Buckley, E., ed. 1968. *Modern Systems Research for the Behavioral Scientist, A Sourcebook.* Chicago: Aldine.

Bunge, W. 1966. *Theoretical Geography.* Lund: University of Lund. (Lund Studies in Geography.)

Chamberlin, T. C. 1892. "The New Geology." *University of Chicago Weekly,* 1:7-9.

Chappell, J. E., Jr. 1970. "Climatic Change Reconsidered: Another Look at 'The Pulse of Asia.'" *Geographical Review,* 60:347-373.

Chojnicki, Z. 1970. "Prediction in Economic Geography." *Economic Geography,* 46:213-222.

Chorley, R. J. 1962. *Geomorphology and General Systems Theory.* Washington, D.C.: U.S. Geological Survey, Professional Paper 500-B.

_____, ed. 1973. *Directions in Geography.* London: Methuen.

Chorley, R. J., and Haggett, P., eds. 1965. *Frontiers in Geographical Teaching.* London: Methuen.

Christaller, W. 1933. *Die zentralen Orte in Süddeutschland.* Jena: Gustav Fischer. Trans. C. W. Baskin, 1966. *Central Places in Southern Germany.* Englewood Cliffs, N.J.: Prentice-Hall.

Cohen, S. B., and Rosenthal, L. D. 1971. "A Geographical Model for Political Systems Analysis." *Geographical Review,* 61:5-31.

Curry, L. 1964. "The Random Spatial Economy: An Exploration in Settlement Theory." *Annals AAG,* 54:138–146.

Davis, W. M. 1888. "Geographic Methods in Geologic Investigation." *National Geographic Magazine,* 1:11–26.

———. 1906. "An Inductive Study of the Content of Geography." *Bulletin of the American Geographical Society,* 38:67–84.

DeLaubenfels, D. J. 1970. *A Geography of Plants and Animals.* Dubuque, Iowa: W. C. Brown.

Dickenson, J. P. and Clarke, C. G. 1972. "Relevance and the 'Newest Geography.'" *Arena.* 4:25–27.

Dow, M. W. 1954. "The Oral History of Geography." *The Professional Geographer,* 26:430–435.

Folke, S. 1972. "Why a Radical Geographer Must be Marxist." *Antipode.* 4:13–18.

Freeman, T. W. 1961. *A Hundred Years of Geography.* London: Gerald Duckworth.

Garrison, W. L. 1959–60. "The Spatial Structure of the Economy." *Annals AAG,* 49:232–239, 471–482; 50:357–373.

Gould, P. R. 1969. "Methodological Developments since the Fifties." *Progress in Geography,* 1:1–50.

Greenberg, M. R., Carey, G. W., Zobler, L., and Hordon, R. M. 1971. "A Geographical System Analysis of the Water Supply Networks of the New York Metropolitan Region." *Geographical Review,* 61:339–354.

Gregory, D. 1978. *Ideology, Science and Human Geography.* New York: St. Martin's Press.

Guyot, A. 1872. *Physical Geography.* New York: Scribner, Armstrong & Co.

Hägerstrand, T. 1967. *Innovation Diffusion as a Spatial Process.* Trans. A. Pred. Chicago: University of Chicago Press (first published in Sweden in 1953).

Haggett, P. 1966. *Location Analysis in Human Geography.* New York: St. Martin's Press.

Harrison, J. D. 1969. *Annotated Bibliography: Environmental Perception with an Urban Emphasis.* Norman: University of Oklahoma Press.

Hartshorne, R. 1939. *The Nature of Geography, a Critical Survey of Current Thought in the Light of the Past.* Lancaster, Pa.: Association of American Geographers.

———. 1959. *Perspective on the Nature of Geography.* Chicago: Rand McNally.

Harvey, D. 1969, *Explanation in Geography.* London: Edward Arnold.

———. 1973. *Social Justice and the City.* London: Edward Arnold.

———. 1974. "What Kind of Geography for What Kind of Public Policy?" *Transactions, Institute of British Geography,* 63:18–24.

Isard, W. 1956. "Regional Science, the Concept of Region, and Regional Structure." *Papers of the Regional Science Association,* 2:13–26.

———. 1960. "The Scope and Nature of Regional Science." *Papers of the Regional Science Association,* 6:9–34.

James, P. E. 1967. "On the Origin and Persistence of Error in Geography." *Annals AAG,* 57:1–24.

James, P. E., and Jones, C. F. 1954. *American Geography, Inventory and Prospect.* Syracuse, N.Y.: Syracuse University Press.

James, P. E., and Martin, G. J. 1979. *The Association of American Geographers: The First Seventy-Five Years, 1904–1979.* Washington, D.C.: Association of American Geographers.

Jaynes, J. 1966. "The Routes of Science." *American Scientist,* 54:94-102.

Jensen, R. G., and Karaska, G. J. 1969. "The Mathematical Thrust in Soviet Economic Geography—Its Nature and Significance." *Journal of Regional Science,* 9:141-152.

Johnson, L. J. 1971. "The Spatial Uniformity of a Central Place Distribution in New England." *Economic Geography,* 47:156-170.

Johnston, R. J. 1970. "Grouping and Regionalizing: Some Methodological and Technical Observations." *Economic Geography,* 46:293-305.

Kalesnik, S. V. 1964. "General Geographic Regularities of the Earth." *Annals AAG,* 54:160-164.

Kansky, K. J. 1963. *The Structure of Transportation Networks: Relations Between Network Geometry and Regional Characteristics.* Chicago: University of Chicago, Department of Geography, Research Paper No. 84.

Kasperson, R. E., and Minghi, J. V. 1969, *The Structure of Political Geography.* Chicago: Aldine.

King, L. J. 1969. *Statistical Analysis in Geography.* Englewood Cliffs, N.J.: Prentice-Hall.

Kniffen, F. B. 1965. "Folk Housing: Key to Diffusion." *Annals AAG,* 55:549-577.

Kurath, H. 1949. *A Word Geography of the Eastern United States.* Ann Arbor: University of Michigan Press.

Lankford, P. M. 1969. "Regionalization: Theory and Alternative Algorithms." *Geographical Analysis,* 1:196-212.

Leighton, P. A. 1966. "Geographical Aspects of Air Pollution." *Geographical Review,* 56:151-174.

Linsky, A. S. 1965. "Some Generalizations Concerning Primate Cities." *Annals AAG,* 55:506-513.

Lösch, A. 1940. *Die räumliche Ordnung der Wirtschaft.* Jena: Gustav Fischer. 2nd ed., 1944, trans. W. H. Woglom, *The Economics of Location,* 1954. New Haven, Conn.: Yale University Press.

Lowenthal, D., ed. 1967. *Environmental Perception and Behavior.* Chicago: University of Chicago, Department of Geography.

Lowenthal, D., and Bowden, M. J., eds. 1976. *Geographies of the Mind: Essays in Historical Geosophy in Honor of John Kirtland Wright.* New York: Oxford University Press.

Lukermann, F. 1961. "The Role of Theory in Geographical Inquiry." *The Professional Geographer,* 13:1-6.

McColl, R. W. 1969. "The Insurgent State: Territorial Bases of Revolution." *Annals AAG,* 59:613-631.

McKinney, W. M. 1968. "Carey, Spencer, and Modern Geography." *The Professional Geographer,* 20:103-106.

MacNeish, R. S. 1964. "Ancient Mesoamerican Civilization." *Science,* 143:531-537.

Meinig, D. W. 1962. *On the Margins of the Good Earth: The South Australian Wheat Frontier, 1868-1884.* Chicago: Rand McNally.

_____. 1968. *The Great Columbia Plain, A Historical Geography, 1805-1910.* Seattle: University of Washington Press.

_____. 1969. *Imperial Texas, an Interpretive Essay in Cultural Geography.* Austin and London: University of Texas Press.

Meynier, A. 1969. *Histoire de la pensée géographique en France, 1872-1969.* Paris: Presses universitaires de France.

Mikesell, M. W. 1969. "The Borderland of Geography as a Social Science." In M. Sherif and C. W. Sherif, eds., *Interdisciplinary Relationships in the Social Sciences*. Pp. 227–248. Chicago: Aldine.

Morrill, R. L. 1970. "The Shape of Diffusion in Space and Time." *Economic Geography*, 46:259–268.

Nunley, R. E. 1971. *Living Maps of the Field Plotter, Analog Simulation of Selected Geographic Phenomena*. Washington, D.C.: Association of American Geographers, Commission on College Geography, Technical Paper No. 4.

Nystuen, J., and Dacey, M. F. 1961. "A Graph Theory Interpretation of Nodal Regions." *Papers of the Regional Science Association*, 7:29–42.

Olsson, G. 1965. *Distance and Human Interaction, A Review and Bibliography*. Philadelphia: Regional Science Research Institute.

Parkes, D. 1980. *Times, Spaces, and Places: A Chronogeographic Approach*. New York: Halsted Press.

Parr, J. B., and Denike, K. G. 1970. "Theoretical Problems in Central Place Analysis." *Economic Geography*, 46:568–586.

Pattison, W. D. 1964. "The Four Traditions of Geography." *Journal of Geography*. 63:211–216.

Peet, J. R. 1977. "The Development of Radical Geography in the United States. *Progress in Human Geography*, 1:240–63.

———. 1978. *Radical Geography*. London: Methuen.

Platt, R. S. 1957. "A Review of Regional Geography." *Annals AAG*, 47:187–190.

Preston, R. E. 1971. "The Structure of Central Place Systems." *Economic Geography*, 47:136–155.

Ryan, B. 1970. "The Criteria for Selecting Growth Centers in Appalachia." *Proceedings AAG*, 2:118–123.

Sauer, C. O. 1952. *Agricultural Origins and Dispersals*. New York: American Geographical Society.

———. 1956. "The Education of a Geographer." *Annals AAG*, 46:287–299.

———. 1966. "On the Background of Geography in the United States." In *Heidelberger Studien zur Kulturgeographie, Festgabe für Gottfried Pfeifer, Heidelberger Geographische Arbeiten*.

Schlier, O. 1937. "Die zentralen Orte des Deutschen Reichs." *Zeitschrift der Gesellschaft für Erdkunde zu Berlin*, 161–170.

Siebert, H. 1967. *Zur Theorie des regionalen Wirtschaftswachstums*. Tübingen: J. C. B. Mohr.

Soja, E. W. 1968. *The Geography of Modernization in Kenya*. Syracuse, N.Y.: Syracuse Geographical Series No. 2.

———. 1971. *The Political Organization of Space*. Washington, D.C.: Commission on College Geography, Resource Paper No. 8.

Smith, D. M. 1977. *Human Geography: A Welfare Approach*. London: Edward Arnold.

Stoddart, D. R. 1967. "Growth and Structure of Geography." *Transactions, Institute of British Geographers*, 41:1–19.

Taaffe, E. J., ed. 1970. *Geography*. Englewood Cliffs, N.J.: Prentice-Hall.

Taylor, P. J. 1976. "An Interpretation of the Quantification Debate in British Geography." *Transactions, Institute of British Geographers*, NSI. 129–142.

Thompson, J. H., ed. 1966. *Geography of New York State*. Syracuse, N.Y.: Syracuse University Press.

Toulmin, S. E. 1967. "The Evolutionary Development of Natural Science." *American Scientist*, 55:456-471.

Trewartha, G. T. 1969. *A Geography of Population: World Patterns*. New York: John Wiley & Sons.

Ullman, E. L. 1941. "A Theory of Location for Cities." *American Journal of Sociology*, 46:853-864.

———. 1953. "Human Geography and Area Research." *Annals AAG*, 43:54-66.

von Thünen, J. H. 1826. *Der isolierte Staat in Beziehung auf Landwirthschaft und Nationalökonomie*. Hamburg. Trans. Carla M. Wartenburg; ed., Peter Hall, 1966. *Von Thünen's Isolated State*. Oxford: Pergamon Press.

Warntz, W. 1959a. "Geography at Mid-Twentieth Century." *World Politics*, 11:442-54.

———. 1959b. *Toward a Geography of Price: A Study in Geo-Econometrics*. Philadelphia: University of Pennsylvania Press.

———. 1964. "A New Map of the Surface of Population Potentials for the United States, 1960." *Geographical Review*, 54:170-184.

Weber, A. 1909. *Ueber den Standort der Industrien*. Tübingen. Trans. C. J. Friedrich, 1966. *Alfred Weber's Theory of the Location of Industry*. Chicago: University of Chicago Press.

Wilbanks, T. J., and Symanski, R. 1968. "What Is Systems Analysis?" *The Professional Geographer*, 20:81-85.

Woldenberg, M. J. 1968. "Energy Flow and Spatial Order: Mixed Hexagonal Hierarchies of Central Places." *Geographical Review*, 58:552-574.

Yuill, R. S. 1966. *A Simulation Study of Barrier Effects in Spatial Diffusion Problems*. Ann Arbor: Michigan Inter-University Community of Mathematical Geographers.

Zelinsky, W. 1970. "Beyond The Exponentials: The Role of Geography in the GreatTransition *Economic Geography*, 46:498-535.

Zelinsky, W., Kosiński, L. A., and Prothero, R. M., eds. 1970. *Geography and a Crowding World*. New York: Oxford University Press.

Zipf, G. K. 1965. *Human Behavior and the Principle of Least Effort: An Introduction to Human Ecology*. New York: Hafner (first published 1949).

Zobler, L. 1958. "Decision Making in Regional Construction." *Annals AAG*, 48:140-148.

Index of Names

Aario, Leo Elino (1906–   ), Finnish geographer; head of the Institute of Geography at Helsinki, 251

Abbe, Cleveland, Sr. (1839–1916), American meteorologist who started the Weather Bureau in 1870; Editor, Monthly Weather Review, 1872–1915; charter member A.A.G., 293

Abbe, Cleveland, Jr. (1872–1934), American geologist with U.S. Geological Survey; charter member A.A.G., 293

Abbeville, Nicholas Sanson d' (1600–1667), founder of first French atlas publishing house in 17th century, 85

Ackerman, Edward A. (1911–1973), American geographer; Ph.D. Harvard, 1939; taught at Chicago, 1948–1955; since 1958 Executive Officer, Carnegie Institution, 361, 362, 408

Adams, Charles C. (1873–1955), M.S. Harvard; Ph.D. Chicago; ecologist and biogeographer; Director, New York State Museum, Albany, 1926–1943; charter member A.A.G., 293

Adams, Cyrus C. (1848–1928), American geographer; Editor, Bulletin of the American Geographical Society 1908–1915; charter member A.A.G., 293

Adickes, E., 111

Agassiz, Jean Louis Rodolphe (1807–1873), Swiss scholar; appointed Professor of Natural History at Harvard in 1848, 122, 123, 140, 147, 150, 279, 282

Ahlmann, Hans W:son (1889–1974), Swedish geographer; taught at Stockholm from 1929 to 1950, 249

Ailly, Cardinal Pierre d' (1350–1420), treatise on geography based on Ptolemy, but he discarded the idea of an enclosed Indian Ocean; the world is inhabited as far as 16° south, 47, 61, 62, 71

Al-Balkhi (10th century A.D.), in 921 published first climatic atlas, *Kitab al-Ashkal*, 50

Albertus Magnus (1193–1280), German scholar who joined the Dominican Order in 1223; lectured at Paris and Cologne; introduced Aristotelian ideas to Medieval Europe; his major geographical work was *De Natura Locorum (The Nature of Places)*, 42

Al-Biruni (972–1050), wrote geography of India, 1030, *Kitab al-Hind*, 51

Alexander the Great (356–323 B.C.), pupil of Aristotle; King of Macedonia; led Greek army eastward to the Indus River, 7, 28–29

Allenby, Lord (1860–1936), British soldier, statesman, and geographer.

Al-Idrisi, *see* Edrisi

Almagia, Roberto (1884–1962), Italian geographer; Professor at Rome, 1915–1959, 255

Al-Mamun (786–833), 49, 50

Al-Maqdisi (945–985), climatic regions based on latitude, and east–west position relative to land and water, 50

Al-Masudi (957–?), Arab scholar who described the monsoons and the relation of evaporation to rainfall, 50

Al-Rashid, Harun, 49

Amundsen, Roald (1872–1928), Norwegian explorer; discoverer of the South Pole, 1911, 160

Anaximander (610–547 B.C.), disciple of Thales in Miletus; introduced the gnomon to the Greeks; drew map of the known world to scale, 16, 18, 19

Andersson, Gunnar (1865–1928), Swedish geographer; for 25 years secretary of the Swedish Society of Anthropology and Geography, 247, 249

Anuchin, Dmitry N. (1843–1923), Russian geographer; founded Department of Geography at Moscow in 1887, 226, 227, 229, 232, 233

Anuchin, V. A. (1913–   ), Soviet geographer who argued for unity of physical and economic geography 239, 240, 243, 244

Anville, Jean Baptiste Bourguignon d' (1697–1782), French cartographer; first to

473

remove mythical decorations from maps, 85, 136

Apian, Peter (1495–1552), German mathematician and cartographer who was the teacher of Gerhard Kremer, 81

Applebaum, William (1906–1978), American geographer; specialist in marketing geography, 341, 362, 363

Arago, François (1783–1853), French physicist who developed principle of magnetism by rotation; close friend of Alexander von Humboldt, 122

Arduino, Giovanni (1713–1795), an Italian scholar who classified rocks in 1760, 101

Aristarchos of Samos (*c.* 300–*c.* 230 B.C.), believed that the earth rotates on its own axis and revolves around the sun, 91

Aristotle (384–322 B.C.), Greek philosopher who believed the world was developing toward final perfection; taught the theory of natural places for earth, air, fire, and water; founder of the Lyceum at Athens, 5, 6, 7, 24–28, 35, 37, 38, 42, 53, 61, 92, 93, 98, 139, 304, 360, 370, 405

Arnold, Thomas (1823–1900), 217, 218

Arrowsmith, John (1790–1873), British cartographer who made maps of Australia, Africa, and other continents

Arsenyev, K. I. (1789–1865), Russian geographer who pioneered regional studies, and classified cities by functions, 225

Ashmead, Percy H., 347

Atwood, Wallace W. (1872–1949), Ph.D. Chicago in geology; Professor of Physiography at Harvard, 1913–1920; President, Clark University, 1920–1946, 312, 318, 343

Augelli, John P. (1921–   ), American geographer; Ph.D. Harvard, 1949; Professor at the University of Kansas; specialist on Latin American studies, 258

Aurel Stein, Sir Mark (1863–1943), archeologist and explorer; Hungarian by birth, he acquired British citizenship

Aurousseau, Marcel (1891–   ), Australian geographer; secretary of the Committee on Geographical Names in London from 1936 to 1955, 327

Austin, Oscar Phelps (1847–1933), American economist; Chief Bureau of Statistics, Department of Commerce; charter member A.A.G., 364

Avicenna (980–1037), described formation of mountains by stream erosion, 51, 56, 100

Awad, Mohammed (1896–1967), Egyptian

geographer; completed Ph.D. at London in 1926; became Professor of Geography and head of Geography Department, Cairo University, 1936, when graduate studies were introduced there; he became the President of Alexandria University in 1953, 266

Bacon, Roger (1220–1292), "Nothing of importance can be known about things unless the places where they are, are also known," 47

Bagrow, Leo (1881–1957), historian of cartography who worked in Russia until 1918, in Berlin from 1918 to 1945, and in Stockholm 1945 to 1957; his monumental *Geschichte der Kartographie* was published in 1951 (Bagrow and Skelton, 1964), 79, 84, 85

Bain, Alexander (1818–1903), Scottish philosopher at Aberdeen; Rector of the university, 1880–1887, 323

Baker, J. N. L. (1894–1971), British specialist on the history of geography; at Oxford from 1923 to 1962, 218

Baker, Oliver E. (1883–1949), Ph.D. Wisconsin; specialist in agricultural geography, U.S. Department of Agriculture; Professor at Maryland, 1942–1949, 327, 344, 376, 377

Banks, Joseph, 138

Banse, Ewald (1883–1953), German geographer and freelance writer who demonstrated the regional unity of North Africa and Southwest Asia on basis of landscape similarity, 178

Baranskiy, N. N. (1881–1963), Soviet geographer, friend of Lenin, who helped to establish role of geography after the October Revolution of 1917, 234, 236, 237, 238, 239, 241

Barnes, Carleton P. (1903–1962), Ph.D. Clark; soil specialist in U.S. Department of Agriculture, 341

Barrett, Robert L. (1871–1969), American geographer and explorer; charter member A.A.G., 293

Barrows, Harlan H. (1877–1960), one of the leaders in the development of American geography between 1919 and 1942, when he was chairman of the Department of Geography at Chicago, 312, 313, 314, 319, 325, 345, 354

Bartholomew, John (1831–1893), Scottish cartographer; head of map publishing firm, 1856–1893, 145

dividing earth's surface into basins bordered by mountain ranges, 1752, 109, 118, 123, 128, 129, 159, 188, 189, 190

Buat, Louis Gabriel, Comte du (1734–1809), French physical geographer who worked out mathematical equations to describe equilibrium, grade, and flow of water in streams, 101

Buchanan, Keith (1919–), first Professor of Geography at Victoria University, Wellington, N.Z., 1953, 261

Buckle, H. T., British historian, 260

Buffon, Georges-Louis Leclerc, Comte de (1707–1788), French naturalist who was first to focus attention on man as an agent of change on the earth; searched inductively for generalizations in the flood of new information brought by the voyages of discovery, 105–106, 116, 149

Bunbury, Edward H. (1811–1895), British author of *History of Ancient Geography*, 1879, 37

Bunge, William W. (1928–), American geographer; outspoken advocate of the greater use of theoretical models and of the focus of attention of the social problems of the inner cities, 387, 400, 413

Burke, C. J., 341

Burnet, Thomas (*c.* 1635–1715), English clergyman who published a theory of the origin of the earth in 1681, 99, 100

Burton, Ian (1935–), Canadian geographer; Ph.D. Chicago; teaching at Toronto, 412

Büsching, Anton Friedrich (1724–1793), German Lutheran minister who had a church in St. Petersburg 1761–1766; first suggested study of landscape zones in Russia; his six-volume work on world geography (1762) contained much new information not previously available, 109–110, 111, 113, 224, 225, 228

Bushong, Allen (1931–), teaching at University of South Carolina, 306

Butland, G. J. (1910–), British geographer; since 1959 at the University of New England in Australia, 260

Buttimer, Anne (1938–), American geographer at Clark University; special emphasis on values in geography, 198.

Butzer, Karl W. (1934–), German-trained geomorphologist; D.Sc. Bonn, 1957; Professor of Geography at Chicago; special student of man-land relations in the early stages of culture development

Cabot, John, 74, 81

Cabot, Sebastian, 74

Cabral, Pedro Alvares, 74

Campbell, Marius R. (1858–1940), American geologist; studied the coal resources of the United States; charter member A.A.G., 293

Cano, Juan Sebastian, 75

Capella, Martianus (5th century A.D.), Latin translation of Plato made idea of a round earth known to European scholars, 42. *See also* Macrobius

Capot-Rey, Robert (1897–), French geographer; Professor in Faculty of Letters at Paris since 1965

Carey, Henry C. (1793–1879), 413

Carpenter, Nathanael (1589–1628), wrote first geographical work in English, 1625, 95–96, 136

Cartier, Jacques, 75

Cass, Lewis (1782–1866), American explorer of upper Great Lakes, 156

Cassini, Giovanni Domenico (1625–1712), an Italian who became a French subject in 1673; became director of the Paris Observatory in 1667; he changed his name to Jean Dominique Cassini, 87, 88, 390, 391

Cassini, Jacques (1677–1756), son of Giovanni Domenico Cassini; became director of the Paris Observatory in 1712; in 1713 measured the arc of the meridian from Dunkerque through Paris to Perpignan, 87, 390, 391

Cassini, Jacques Dominique (1748–1845), son of César François Cassini de Thury; became director of the Paris Observatory in 1784 and completed the Topographic map of France started by his father, 390, 391

Cassini de Thury, César François (1714–1784), son of Jacques Cassini; began topographic map of France, 1744; became director of the Paris observatory in 1771, 87–88

Cavendish, Henry (1731–1810), British chemist who discovered the elements in water at about the same time as Lavoisier, 128

Chabot, Georges (1890–1975), French geographer

Chaix, Émile (1855–1929), Swiss physical geographer; taught at the University of Geneva.

Chamberlin, Thomas C. (1843–1928), American geologist; President, University of Wisconsin, 1887–1892; chairman, Department of Geology, Chicago, 1892–1919, 290, 310, 313

Chang Ch'ien (2nd century B.C.), Chinese explorer who visited shores of Mediterranean in 128 B.C., 57

Ch'ang Ch'un (*b.* 1148), Chinese explorer who made contact with Genghis Khan in Samarkand, 1220

Change Heng (2nd century A.D.), Chinese cartographer who introduced grid system of latitude and longitude, 60

Chase, Stuart,

Chatterjee, Shiba Prasad (1903–   ), Indian geographer at Calcutta, 265

Cheng Ho (1371–1435), Chinese voyager who sailed his ships to the east coast of Africa and the Persian Gulf in the early 15th century, 60, 63

Chisholm, George G. (1850–1930), Scottish economic and commercial geographer at Edinburgh, 1908–1921, 204

Chisholm, Michael (1931–   ), British geographer; author of *Human Geography; Evolution or Revolution?*

Chojnicki, Zbyszko (1928–   ), Polish geographer at A. Mickiewicz University; specialist in statistical and mathematical geography, 422

Cholley, Andre (1886–1968), geographer at the University of Paris, 219

Chorley, R. J. (1927–   ), British geomorphologist; trained in geography at Oxford with R. P. Beckinsale; and in geology at Columbia with A. N. Strahler; taught at Brown University, 1954–1957, and at Cambridge University since 1958, 220, 221, 382, 423

Chou Ta-kuan (13th century A.D.), Chinese traveler who visited Cambodia in 1296, 60

Christaller, Walter (1893–1969), Ph.D. Freiburg, 1938; developed outlines of Central Place Theory in 1933, 414, 415

Chrysologue de Gy (1728–1808), a Capuchin monk whose planispheres were exhibited in Paris in 1778–1780; he was first to show land hemisphere, 1774, 110

Chu, Co-ching (1889–   ), Chinese geographer; trained at Harvard; first Director of Institute of Meteorology in the Academia Sinica, 272, 273, 274

Clark, Andrew H. (1911–1975), M.A. Toronto, Ph.D. California (Berkeley); Professor at Wisconsin 1951; historical geographer, 261, 324, 326, 340

Clark, William (1770–1838), American explorer, 342

Claval, Paul (1932–   ), French geographer; teaching at the University of Paris, 188, 198

Clayton, Henry Helm (1861–1946), American climatologist

Clements, Frederic E. (1874–1945), American plant ecologist; taught at Nebraska from 1894 to 1907; and at Minnesota from 1907 until his retirement; charter member A.A.G., 293

Cluverius (Philipp Cluver) (1580–1622), wrote historical geography of Europe and a universal geography, 1624, 95–96, 136

Cobb, Collier (1862–1934), professor of Geology at the University of North Carolina

Coffin, James H. (1806–1873), American climatologist, 155, 227

Cohen, Saul B. (1925–   ), American geographer; Ph.D. Harvard, 1954; President Queens College, C.U.N.Y.; specialist in political geography, 363, 411

Colby, Charles C. (1884–1965), Ph.D. Chicago; taught at Chicago, 1916–1949; specialist on land classification studies, 311, 314, 341, 344, 345, 354, 355

Colby, Thomas Frederick (1784–1852), lifework with the Ordnance Survey in England and Ireland

Cole, J. P. (1928–   ), British geographer; Ph.D. Nottingham, 1962; teaching at Nottingham, 258

Cole, Monica M. (1922–   ), British geographer; Ph.D. London, 1947; Professor at Bedford College, University of London; specialist in the geography of vegetation

Coleman, Alice M. (1923–   ), British geographer; reader at King's College, University of London, 220

Collie, George L., 290

Columbus, Christopher (1451–1506), born in Genoa; went on voyage for the Portuguese; with the aid of Ferdinand and Isabella of Spain, made four voyages to America, 35, 38, 46, 47, 62, 66, 70–74, 75, 80, 87

Comte, Auguste (1798–1857), 142

Condillac, Etienne Bonnot de (1715–1780), 142

Condorcet, Marquis de (1743–1794), 108

Cook, James (1728–1779), British naval captain; made three voyages to the Pacific; finally established outlines of land and water around that ocean, 39, 76, 77, 79, 86, 88, 107, 109, 110, 114, 116, 138, 199

Copernicus, Nicolaus (1473–1543), Polish astronomer; published his concept of a

sented the idea that natural and human history are obedient to the same basic laws, 110, 129, 131, 142

Herodotus (484–c. 425 B.C.), Greek historian, ethnographer, and historical geographer, 8, 11, 21, 22, 23, 24, 34, 36, 95, 268

Herz, Marcus, 113

Hettner, Alfred (1859–1941), German geographer; Ph.D. Strasbourg; 1881; taught at Leipzig, Tübingen, and Heidelberg, 1899–1928, 176, 177, 184, 186, 204, 226, 254, 271, 277, 286, 321, 384

Hewes, Leslie (1906–), American geographer; Ph.D. University of California, Berkeley, 1940; Professor Emeritus, University of Nebraska, 326

Hilgard, Eugene W. (1833–1916), occupied chair of agriculture at the University of California 1874–1904; authority on soils

Hill, Robert T. (1858–1941), American geologist; charter member A.A.G., 293

Hillary, Edmund, (1919–), explorer, mountaineer, and apiarist, 160, 207

Himly, M., 190

Hinks, Arthur Robert (1874–1945), British astronomer, cartographer, and officer of the Royal Geographical Society

Hipparchus (2nd century B.C.), first in the western world to suggest use of a grid of latitude and longitude as a basis for locating places on the face of the earth; developed, stereographic and orthographic projections; used a circle divided into 360°, 7, 33–34, 36, 71, 77

Hippocrates (5th century B.C.), Greek physician who produced the first medical geography; developed ideas of environmental influence on human behavior, 30

Hobbs, William H. (1864–1953), American geologist; established geography at Michigan, 1915, 322

Hogarth, David George (1861–1927), British specialist on Southwest Asia, 216

Holdich, Thomas H. (1842–1929), British expert on boundary problems, 220–221

Hollweg, Bethmann, 127

Holmes, William H., 157

Homer (dates unknown), the Greek poet described by Strabo as the "Father of Geography," 2, 7, 15, 16, 35, 36, 57

Hooke, Robert (1635–1703), English scientist who measured the elasticity of solid bodies; built the first effective vacuum pump, 98

Hooson, David J. M. (1926–), studied

at Oxford and London; since 1965 teaching at the University of California, Berkeley; specialist on Soviet studies, 226, 239

Hoover, Edgar Malone (1907–), American authority on the location of economic activity

Hornbeck, S. K.

House, Col. E. M., 345

Houston, Edwin J. (1847–1914), consulting electrical engineer in Philadelphia who was concerned about educational procedures in American schools, 290

Howarth, Osbert John R. (1877–1954), British geographer, 219

Hsieh, Chiao-min (1918–), Ph.D. Syracuse; taught at Dartmouth, The Catholic University of America, the University of Pittsburgh, and the Taiwan Normal University, 272

Hsüan-Tsang (A.D. 602–664), Chinese Buddhist monk who crossed Tibet into India, 57

Hudson, G. Donald (1897–), American geographer; Ph.D. Chicago, 1934; Professor Emeritus, University of Washington, 355, 357, 394

Humboldt, Alexander von (1769–1859), the great German geographer who brought the classical period of universal scholars to an end as he foreshadowed the modern period, 8, 11, 37, 107, 112–131, 119, 120, 133, 137, 138, 139, 142, 144, 146, 149, 159, 163, 173, 182, 186, 200, 225, 240, 321, 370, 390, 424

Humboldt, Wilhelm von (1767–1835), brother of Alexander; one of the founders of the University of Berlin, 9, 113, 135, 163

Hume, David (1711–1776), British philosopher and economist, 5, 110, 140

Huntington, Archer M. (1870–1955), American poet and student of the origins of Hispanic culture who contributed much to the American Geographical Society, 294

Huntington, Ellsworth (1876–1947), American geographer; taught at Yale, 1907–1915; research associate at Yale, 1920–1945; author or coauthor of 28 books; charter member A.A.G., 273, 277, 281, 295, 301–303, 318, 344, 364, 397

Hutton, James (1726–1797), Scottish geologist who was the first European fluvialist and developed ideas of uniformitarianism, 51, 101, 139

Huxley, Thomas Henry (1825–1895), British

cal Survey, 1885-1910; charter member A.A.G., 293

Meyer, Hans (1858-1929), German authority on colonial geography

Meynen, E., 185

Meynier, André (1901–), French geographer; teaching at Rennes, 198

Michelson, Albert A. (1852-1931), American physicist who measured the speed of light, 310

Michotte, Paul (1876-1940), Belgian geographer; introduced the new geography at Brussels in 1921, 254

Mikesell, M. W. (1930–), American geographer; Ph.D. Berkeley, 1959, Professor at Chicago, 424

Mill, Hugh Robert (1861-1950), British geographer who promoted survey of Great Britain; Director, British Rainfall Organization, 1901-1919, 204, 219

Mill, John Stuart (1806-1873), 142

Miller, George J. (1880–), American geographer; former editor of the Journal of Geography, 297

Miller, O. M. (1897–), specialist in surveying and mapping; trained at the Royal Geographical Society; came to the American Geographical Society in 1922 and remained until his retirement in 1968, 348, 393

Minshull, R., 214

Momsen, Richard P. (1923–), American geographer; Ph.D. from Minnesota; now at Calgary, 258

Monbeig, Pierre (1908–), French geographer; specialist on the study of Brazil; teaching at the Sorbonne, 197, 257

Monkhouse, Frank J. (1914-1975), British geographer; specialist on cartography; taught at Southampton

Montesquieu, Charles Louis de Secondat, Baron de la Brède et de (1689-1755), French political scientist, whose book *De l'esprit des lois,* was published in 1784; presented ideas about the influence of climate on politics, 260

Moodie, Arthur A. E. (1901-1970), British political geographer

Morrill, R. L. (1934–), American geographer; Ph.D. Washington, 1959; Professor at Washington, 398

Morse, Jedidiah (1761-1826), an American clergyman whose *American Universal Geography* was widely read in the schools and in the homes during the early years of the republic, 148

Münster, Sebastian (1489-1552), a Franciscan monk who was converted to the Swiss Reformed Church; wrote classical cosmography on the Strabo model, 95, 109, 113

Murphy, Raymond E. (1898–), American geographer; Ph.D. Wisconsin, 1930; Editor, *Economic Geography,* 1949-1969; specialist in urban geography, 340, 341

Musil, Alois (1868-1944), Czech explorer and Orientalist

Musset, René, 197

Nansen, Fridtjof (1861-1930), Norwegian explorer; Professor of Oceanography at Christiania (Oslo), 1908, 154, 250

Napier Shaw, Sir William (1854-1945), British meteorologist

Natoli, Salvatore J. (1929–), Educational Affairs Director, A.A.G.

Neef, Ernst (1908–), Ph.D. Heidelberg; Halle, 1951; *Technische Hochschule,* Dresden, 1966; Editor, *Petermanns Mitteilungen,* 271

Nelson, Helge M. O. (1882–), Swedish geographer; Professor at Lund from 1916 to 1947, 249

Neumann, Karl J. L. (1823-1880), Professor of geography, and ancient history at Breslau after 1865, 164

Newbigin, Marion I. (1869-1934) editor of the Scottish Geographical Magazine, 1902-1932, 207, 210, 218

Newton, Isaac (1642-1727), an English scientist who formulated laws of gravity, after first developing method of integral calculus in 1666; Professor of Mathematics at Cambridge, 1669, 87, 92, 98, 136, 138, 370, 385, 386, 387, 390

Niermeyer, Jan F. (1866-1923), Professor of geography at Utrecht, 252

Nikishov, Maxim Ivanovitch (1903-1977), Soviet expert in thematic mapping

Nikitin, N. P. 235

Nordenskjöld, Adolf Erik (1832-1901), Swedish explorer; Professor at the Swedish Military Academy in Stockholm, 1858, 247

Nordenskjöld, Erland (1877-1932), Swedish anthropologist and anthropogeographer; son of Baron Adolf Nordenskjöld, 247

Nordenskjöld, Gustav (1869-1894), Swedish anthropologist; son of Baron Adolf Nordenskjöld; provided first professional description of the Mesa Verde, 247

Nordenskjöld, Otto (1870-1928), Swedish

be done if more statistical data were available in 1662, 102

Pfeifer, Gottfried G. (1901–), German geographer; Director of the Geographical Institute at Heidelberg; Editor, *Geographische Zeitschrift,* after World War II; Professor Emeritus, Heidelberg, 341

Phei Hsiu (3d century A.D.), Chinese cartographer who was made minister of public works in A.D. 267, 60

Philbrick, Allen K. (1914–), American geographer; Ph.D. Chicago 1949; Professor, University of Western Ontario, 387

Philippson, Alfred (1864–1953), trained with Richthofen at Leipzig; Professor at Bonn, 1911–1929, 184

Picó, Rafael (1912–), Puerto Rican geographer; Ph.D. Clark, 1938; directed Puerto Rico Planning Board and was co-director of the Rural Land Classification Program, 1950–1953; Vice President, *Banco Popular,* 357

Pike, Zebulon M. (1779–1813), American explorer of the West, 156

Pinchemel, Philippe (1923–), French geographer at the Sorbonne, 198

Pitts, Forrest R. (1924–), American geographer; Ph.D. Michigan, 1955; Professor at the University of Hawaii, 401

Pius II, Pope (Aeneas Silvius Piccolomini) (1404–1464), wrote geography of Europe and Asia, doubts Ptolemy on enclosed Indian Ocean and on uninhabitability of equatorial regions, 47, 71

Plamanatz, J. P. 279

Plato (428–348 B.C.), Greek philosopher; believed the world had been created in perfection and was in process of deteriorating; stressed deductive method; accepted idea of a round earth; founder of the Academy at Athens, 5, 6, 7, 24–28, 42, 56, 105, 149

Plato of Tivoli

Platt, Elizabeth T. (1900–1943), Librarian of the American Geographical Society

Platt, Raye R. (1891–1973), American geographer; in charge of the Millionth Map of Hispanic America project at the American Geographical Society, 1923–1938, 297, 348

Platt, Robert S. (1891–1964), American geographer; Ph.D. Yale; taught at Chicago, 1919 to 1957, 239, 258, 327, 387, 408

Playfair, John (1747–1819), Scottish geologist who clarified the ideas of James Hutton, 101

Pliny the Elder (A.D. 23–79), Roman geographer; compiled a treatise on geographical writings; died during eruption of Vesuvius, 37, 42, 43

Pokshishevskiy, V. V. (1905–), Soviet geographer; Professor at the Institute of Ethnography at Moscow

Polo, Marco (1254–1323), Venetian traveler who visited China 1271–1295; his book describes the many places he visited in eastern Asia, 45–46, 49, 60, 61, 66, 71, 72, 73, 77, 92, 106, 136

Polybius, Greek historian, 34

Pomponius, Mela (1st century A.D.), first Latin writer of geography, 37, 42

Posidonius (135–50 B.C.), Greek geographer; estimated circumference of the earth to be smaller than Eratosthenes had thought; believed highest temperatures are along the tropics, not along the equator; one of the earliest philosophers to face the apparent dichotomy between factual descriptions of unique phenomena and the formulation of theoretical explanations, 34–35, 36, 46, 71, 103

Powell, John Wesley (1834–1902), American explorer and pioneer geographer; first to explore the Grand Canyon; leader of U.S. Geographical and Geological Survey of the West; second director U.S. Geological Survey, 1880–1894, 157, 158–160, 283, 303, 314, 354

Praslov, L. I. 229

Pred, Allan R. (1936–), American geographer; Ph.D. Chicago 1962; teaching at Berkeley, 415

Ptolemy (A.D. 90–168), summarized Greek knowledge of astronomy; gazeteer of places located by latitude and longitude; world map showed Indian Ocean enclosed on south by *terra incognita*; accepted smaller earth circumference from Marinus, 7, 38–39, 42, 46, 47, 51, 53, 61, 71, 72, 76, 79, 85, 86, 91, 92, 93, 95, 103, 109, 138

Pumpelly, Raphael (1837–1923), American geologist and explorer; surveyed resources along route of the Northern Pacific Railroad, 1882; exploration in Central Asia, 1903–1904, 283

Putnam, Donald F., geographer who taught at Toronto, 261

Pythagoras (6th century B.C.), Greek mathematician; believed in concept of a round earth, 6, 25, 27

Pytheas (d. 285 B.C.), explorer of Great Britain and coast of Western Europe, 29–30, 33, 36, 93

Index of Subjects